SILVER IN ORGANIC CHEMISTRY

Edited by

MICHAEL HARMATA

WILEY

A JOHN WILEY & SONS, INC., PUBLICATION

Published by John Wiley & Sons, Inc., Hoboken, New Jersey
Published simultaneously in Canada

Library of Congress Cataloging-in-Publication Data:

Silver in organic chemistry / [edited by] Michael Harmata
p. cm.
Includes bibliographical references and index.
ISBN 978-0-470-46611-7 (cloth)

10 9 8 7 6 5 4 3 2 1

This volume is dedicated to the memories of two outstanding chemists, Dr. Christopher R. Schmid and Dr. Anthony J. Shuker, both of whom succumbed to cancer at an all too early age. Their legacies live on not only in their science but also in those whom they loved, befriended, and inspired.

CONTENTS

FOREWORD

In the last two centuries, the discipline of synthesis has profoundly transformed our world, enabling access to molecules that in former times would be only scarcely or unreliably available from natural sources. Increasingly, synthesis is also being used to access new molecules, designed for function (e.g., catalysts, smart materials, self replicating materials, molecular devices, energy generation and storage systems, diagnostics, drug delivery systems, therapeutics)—many with activities superior to or often different from what nature has produced. We are no longer exclusively reliant on nature for our molecular needs. This too brings new opportunities. Whereas once the challenge in synthesis was simply to make molecules, increasingly that challenge has given away to a more demanding goal: developing strategies that provide molecules in a step economical, and green, if not ideal way. Our ability to meet this goal rests heavily not only on the refinement of existing methodology, but also on the introduction of new reactions and reagents that enable or enhance new synthetic strategies—a focus of this book.

This book explores the use of silver in organic synthesis. Silver and its salts and complexes have figured significantly in the history of chemistry, recognized for their special conductive properties, use in photography, and even biological activities. Notwithstanding the importance of these areas the broader use of silver in chemistry, and more specifically in synthesis, has lagged behind that of other coinage metals. That is changing. One now finds silver as a key component of much that is "nano," including nano-rods, spheres, sheets, clusters, prisms, membranes, plates, pillars, cubes, bowls, fibers, wires, gels, and sensors. Increasing interest is also being directed at its use and that of other coinage metals in improving synthetic procedures and in enabling new ones. This book provides an insightful overview of how silver figures in these new developments.

Professor Harmata is one of the gifted educators of our time. Through his research and books he has contributed significantly to the advancement of synthesis. For this book, he has assembled a remarkable team of thought leaders who have in their own research contributed significantly to the emerging interest in silver-based reaction science. The resultant product is a must read for those interested in synthesis. It spans impressively from the preparation and use of silver compounds to silver-catalyzed or mediated cycloadditions, rearrangements, isomerizations, group transfers, aldols, and coupling reactions to supramolecular chemistry and comparisons with other metals. It is both an educational and inspirational experience. It has impressive depth and breadth. This contribution to our community sprung in part from frustration with a rejected but clever manuscript title ("All that glitters is not gold") and the resultant motivation "to do something on behalf of silver." There is a silver lining to that cloud, as this book on silver in organic chemistry represents a brilliant contribution to the field and an educational experience that is expected to inspire new ideas and glitter for an emerging area of interest.

PAUL A. WENDER

Stanford University
April 2010

PREFACE

It was a dark and stormy night. . ..

Editors get to have some fun, don't they? This book was born out of the recognition that there existed no compilation on the power of silver in organic chemistry, particularly synthesis. I recognized this, and within less than a year, while these reviews were being written, a very nice *Chemical Reviews* issue appeared dedicated to the coinage metals and their importance to organic chemistry. That's life! Such is the pace of developments in the area of coinage metals that those reviews, and those contained herein, will need to be updated within the next few years, however. Have I just suggested that I might take on a second edition of this monograph? I must be nuts.

This book also came about because I am at times pigheaded and not the teddy bear that I am often perceived to be. Not too long ago, I tried to publish a paper that was partially entitled "All that glitters is not gold," in an effort to do some cheerleading for the silver cation. A referee thought this was an abomination, and my response was less than that of a gentleman and scholar. Fortunately, cooler heads eventually prevailed, the situation was resolved, and the paper was published: I changed the title. However, I was left with the feeling that I needed to do something on behalf of silver, and this book is the result.

My thanks go out to the authors. Through their fine efforts, a very nice monograph has been produced. If this monograph teaches and inspires, even just a little, we will have accomplished our mission.

I must thank Wiley and all the fine folks there for their help and support. My thanks go in particular to Ms. Anita Lekhwan, whose confidence in me and the project never waivered. We all need people to believe in us.

My family has been very patient with me as I put in the extra effort to bring this book to life. My deepest thanks to Judy, Gail, Diana, and Alexander.

Finally, whenever I do a project like this, I like to remind the community that I can make time for this because I have a supported research program. When I began this project, I had both NIH and NSF funding. I will retain the latter for the next few years and hopefully regain the former. A synergistic activity like this allows me to produce a teaching and learning tool and affords me the chance to interact with leading colleagues of the day. Hopefully it adds something to the community; it certainly enriches me.

MICHAEL HARMATA

University of Missouri – Columbia
April 2010

CONTRIBUTORS

Zheng-Shuai Bai, Coordination Chemistry Institute, State Key Laboratory of Coordination Chemistry, School of Chemistry and Chemical Engineering, Nanjing National Laboratory of Microstructures, Nanjing University, Nanjing 210093, China

Philippe Belmont, Equipe "Chimie Organométallique, Hétérocycles et Cibles Biologiques" (COHCB), Institut Curie et CNRS, 26 rue d'Ulm, 75005 Paris, France

Aurélien Blanc, Laboratoire de Synthèse et Réactivité Organiques, Institut de Chimie, Université de Strasbourg, 4 Rue Blaise Pascal, 67000 Strasbourg, France

David A. Capretto, Department of Chemistry, University of Chicago, 5735 South Ellis Avenue, Chicago, IL 60637, USA

Erick M. Carreira, Laboratory of Organic Chemistry, ETH Zürich, HCI H335, Wolfgang-Pauli-Strasse 10, 8093 Zürich, Switzerland

Tom G. Driver, Department of Chemistry, University of Illinois at Chicago, 845 W. Taylor St., Chicago, IL 60607, USA

Tina N. Grant, Department of Chemistry, University of Alberta, 11227 Saskatchewan Drive, T6G 2G2 Edmonton, Canada

A. Stephen K. Hashmi, Organisch-Chemisches Institut, Ruprecht-Karls-Universität, Heidelberg, Im Neuenheimer Feld 270, 69120 Heidelberg, Germany

Chuan He, Department of Chemistry, University of Chicago, 5735 South Ellis Avenue, Chicago, IL 60637, USA

Masanori Kawasaki, Medicinal Chemistry Research Laboratories, Pharmaceutical Research Division, Takeda Pharmaceutical Company Limited, 17-85, Jusohonmachi 2-chome, Yodogawa-ku, Osaka 532-8686, Japan

Zigang Li, Department of Chemistry, University of Chicago, 5735 South Ellis Avenue, Chicago, IL 60637, USA

Carl J. Lovely, Department of Chemistry and Biochemistry, The University of Texas at Arlington, 700 Planetarium Place, Arlington, TX 76019, USA

Patrick Pale, Laboratoire de Synthèse et Réactivité Organiques, Institut de Chimie, Université de Strasbourg, 4 Rue Blaise Pascal, 67000 Strasbourg, France

Rebecca H. Pouwer, School of Chemistry and Molecular Biosciences, University of Queensland, Brisbane 4072, Queensland, Australia

Wei-Yin Sun, Coordination Chemistry Institute, State Key Laboratory of Coordination Chemistry, School of Chemistry and Chemical Engineering, Nanjing National Laboratory of Microstructures, Nanjing University, Nanjing 210093, China

Alex M. Szpilman, Schulich Faculty of Chemistry, Technion—Israel Institute of Technology, 32000 Technion City, Haifa, Israel

Jean-Marc Weibel, Laboratoire de Synthèse et Réactivité Organiques, Institut de Chimie, Université de Strasbourg, 4 Rue Blaise Pascal, 67000 Strasbourg, France

Frederick G. West, Department of Chemistry, University of Alberta, 11227 Saskatchewan Drive, T6G 2G2 Edmonton, Canada

Craig M. Williams, School of Chemistry and Molecular Biosciences, University of Queensland, Brisbane, 4072 Queensland, Australia

Hisashi Yamamoto, Department of Chemistry, University of Chicago, 5735 South Ellis Avenue, Chicago, IL 60637, USA

Jin-Quan Yu, Department of Chemistry, The Scripps Research Institute, La Jolla, CA 92037, USA

1

SILVER ALKYLS, ALKENYLS, ARYLS, AND ALKYNYLS IN ORGANIC SYNTHESIS

REBECCA H. POUWER AND CRAIG M. WILLIAMS

School of Chemistry and Molecular Biosciences, University of Queensland, Brisbane, Australia

Silver in Organic Chemistry Edited by Michael Harmata

1.1 INTRODUCTION

While the coordination and inorganic chemistry of silver compounds have been prolifically documented, the use of organosilver compounds to effect useful synthetic transformations is severely underrepresented in the synthetic organic chemistry literature. This has prompted us to present a review of literature reporting synthetically useful applications of organosilver compounds in the hope of inspiring further development in this field. The majority of the literature covered in this review concentrates on silver(I) organo-species as reagents, although on some occasions silver(II) and silver "ate" complexes will be discussed, in addition to organosilver intermediates. General reviews encompassing all classes of organosilver compounds have appeared previously.[1–3]

1.2 C_{sp^3}-Ag

1.2.1 Synthesis, Stability, and Reactivity of Alkylsilver Compounds

As a result of extremely low thermal stability, alkylsilver compounds have found only a narrow range of use in organic synthesis. Procedures for the synthesis of alkylsilver compounds as anything but fleeting proposed intermediates are limited to a handful. Semerano and Riccoboni first reported the synthesis of methyl-, ethyl-, and propylsilver in 1941 (Scheme 1.1). Reaction of silver nitrate and the corresponding tetralkyllead in alcohol at $-80°C$ gave the compounds as brown precipitates that decomposed rapidly on warming to room temperature to give metallic silver and a mixture of hydrocarbons.[4] This methodology has been utilized in a limited number of investigations into the mechanism of decomposition of alkylsilver compounds.[5,6] In these cases, the presence of the alkylsilver compound, and its subsequent decomposition, is inferred from the isolation of alkyl dimers.

Two plausible mechanistic pathways have been proposed for the thermal decomposition of alkylsilver compounds: either a radically-mediated cleavage of the carbon–silver bond or a process by which the breaking of the silver–carbon bond and formation of the carbon–carbon bond are concerted. Mechanistic studies by Whitesides and coworkers in which the product ratios obtained for the thermal process

$$R_4Pb + AgNO_3 \longrightarrow RAg + R_3PbNO_3$$

Scheme 1.1

TABLE 1.1. Silver-Catalyzed Dimerization of Alkylmagesium Halides

$$\text{R–MgX} \xrightarrow{\text{BrCH}_2\text{CH}_2\text{Br, AgOTs (1 mol\%)}} \text{R–R}$$

Entry	Substrate	Product	Yield (%)
1	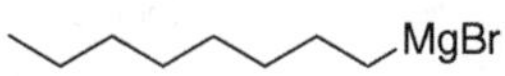	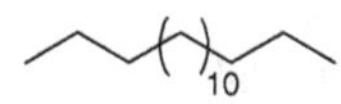	97
2	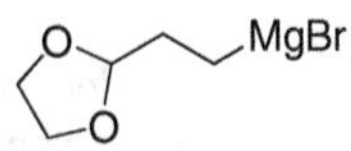	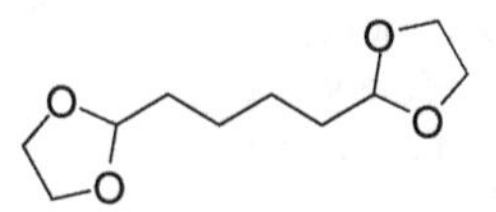	80
3	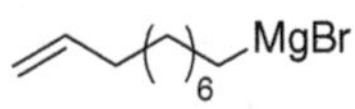	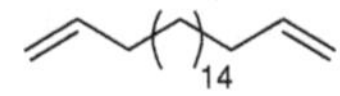	99
4	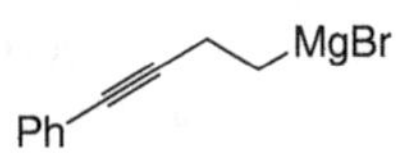	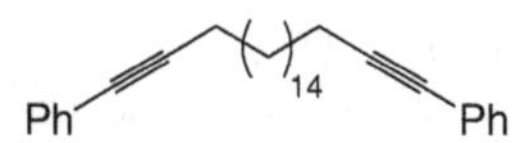	95

were compared to those for known radically-mediated reactions have suggested that a concerted process is more likely, although this has not proved to be general.[7–9]

The formation of methylsilver and dimethylargentate has been observed in the collision-induced dissociation MS3 spectrum of silver diacetate. Dimethylargentate is stable in the gas phase, and has been isolated for short periods (10 s) without significant decomposition.[10]

Alkylsilver compounds have been prepared by treatment of Grignard reagents with silver salts,[11–19] and similarly undergo oxidative homocoupling to give alkyl dimers.[11–13,19,20] Exploitation of this finding has resulted in the development of general methodology for silver-catalyzed alkyl–alkyl homocoupling of Grignard reagents (Table 1.1).[21] The catalytic cycle of this reaction is proposed to proceed via the oxidation of metallic silver with 1,2-dibromoethane to generate silver bromide (Scheme 1.2).

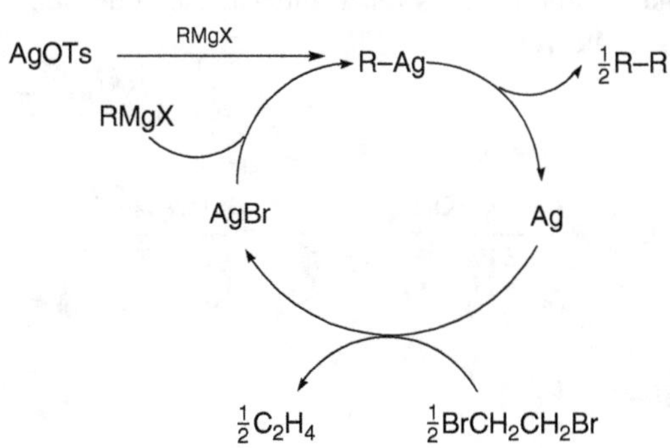

Scheme 1.2

TABLE 1.2. Silver-Mediated Ring-Closing Reaction of Bis(Alkylmagnesium Halides)

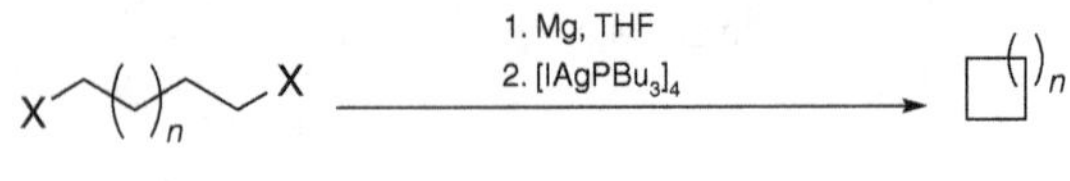

Entry	Substrate	Concentration $N \times 10^2$	Product	Yield (%)
1	1,4-Dibromobutane	5.0	Cyclobutane	83
2	1,5-Dibromopentane	2.5	Cyclopentane	83
3	1,6-Dichlorohexane	2.5	Cyclohexane	43
4	1,7-Dibromoheptane	2.5	Cycloheptane	23
5	1,8-Dibromooctane	2.5	Cyclooctane	2
6	1,9-Dichlorononane	2.5	Cyclononane	<1
7	1,10-Dibromodecane	0.77	Cyclodecane	10–15
8	1,12-Dibromododcane	0.77	Cyclodecane	10–15

Of particular note is the use of this reactivity to form small carbocycles. Whitesides and coworkers have shown that the treatment of primary bis(alkylmagnesium halides) with tributylphosphinesilver iodide produces carbocycles in a range of yields, with a strong dependence on ring size (Table 1.2). The best results were obtained for four-, five-, and six-membered rings. Although it was hoped that the aggregated nature of alkylsilver compounds would facilitate the formation of medium to large rings, compounds of this type were produced with only low yields.[8]

It has also been shown that treatment of primary bis(alkylmagnesium halides) with silver trifluoromethanesulfonate effects ring closure under mild conditions for a range of substrates, thus highlighting the generality of this reaction for producing small carbocycles (Table 1.3).

An equivalent reaction has been achieved via the treatment of hydroborated bisalkenes with alkaline silver nitrate solution (Table 1.4).[22,23] This method has been used to synthesize a number of small and medium-size carbocyclic rings in moderate to good yield. The selectivity for terminal cyclization observed for 1,6-heptadiene and 1,7-octadiene indicates that, in these cases, hydroboration of each of the alkenes occurs independently to yield acyclic boranes. It has, however, been found that both cyclic and acyclic boranes react under these conditions to yield the ring-closed products (Scheme 1.3).

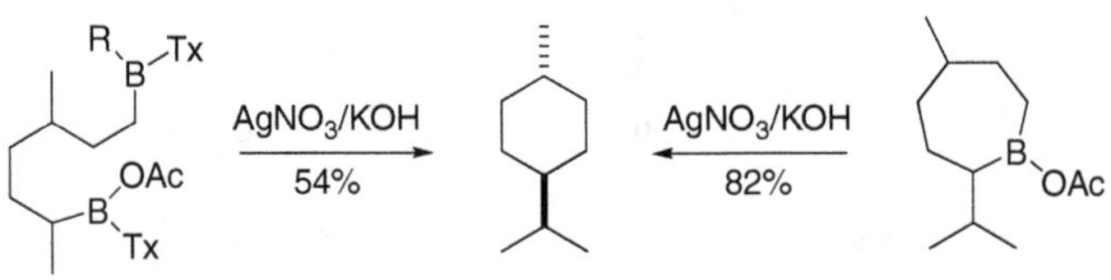

Tx = $-C(CH_3)_2CH(CH_3)_2$

Scheme 1.3

TABLE 1.3. Silver-Mediated Ring-Closing Reaction of Bis(Alkylmagnesium Halides)

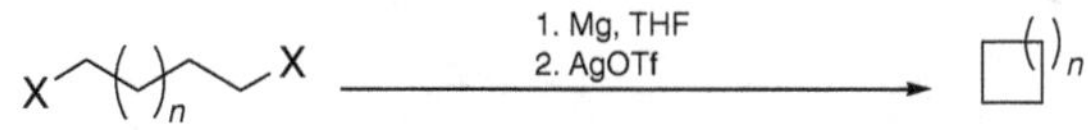

Entry	Substrate	Product	Yield (%)
1	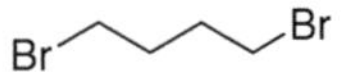		94
2	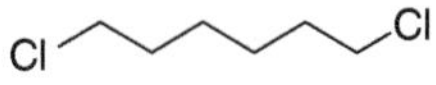	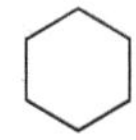	55
3	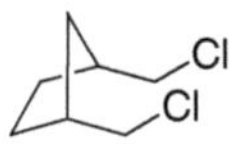		82
4	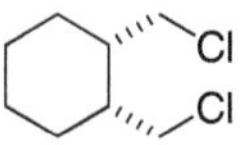	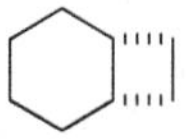	61
5	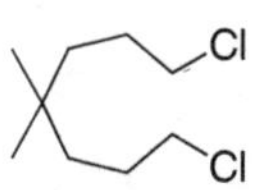	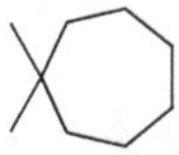	0

Intermolecular dimerization has also been effected by a comparable protocol.[24–26] Treatment of triethylborane with silver nitrate and sodium hydroxide in water at 25°C led to the rapid evolution of *n*-butane (72%), ethylene (9%), and ethane (9%). Reaction of two different alkylboranes led to statistical mixtures of dimerized and cross-coupled products. Furthermore, this strategy has been used successfully in the synthesis of olefins from dihydroborated internal acetylenes,[27] and in polymerizations of bifunctional organoboron compounds.[28]

The addition of lithium bromide significantly increases the thermal stability of alkylsilver compounds.[14] Westmijze and coworkers found that the reaction of *n*-butylmagnesium bromide, for example, with AgBr·2LiBr gave a solution of butylsilver that was stable up to −10°C, which is in stark contrast to the species obtained from the reaction with silver bromide alone, which decomposes at −60°C. This marked stabilization of the alkylsilver compounds allowed for the first meaningful use of these reagents in intermolecular reactions.

The Grignard-derived alkylsilver–lithium bromide complexes have been added to conjugated enynes and enynenitriles to give allenes via the intermediacy of allenylsilver reagents (Scheme 1.4).[14–18] The alkylsilver reagents generally reacted with the enynenitriles to give 2,3-alkadienenitriles on protolysis, while homoargentates ($R_2AgMgCl$) tended to give 2,4-alkadienenitriles. The homoargentates underwent selective *trans* addition to the enynenitriles, which is in contrast to the selective *cis* addition observed for alkylcopper reagents.[17]

TABLE 1.4. Silver-Mediated Ring Closing of Hydroborated Bisalkenes

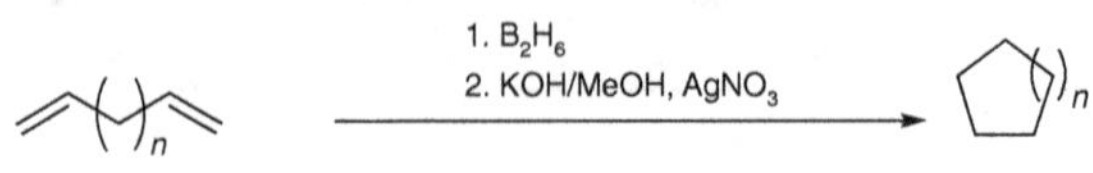

Entry	Substrate	Product	Yield (%)
1	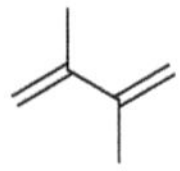		79
2	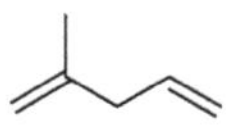	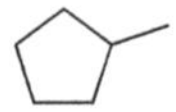	85
3	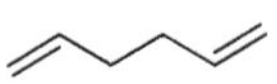		66
		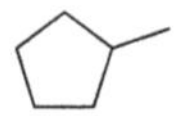	17
4			67
5	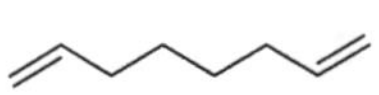	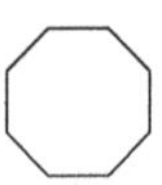	42
6	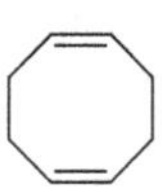	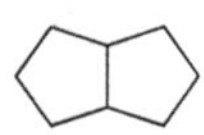	19

A number of simple alkylsilver compounds have been found to react with enynylsulfinates via a 1,5 addition to give substituted butatrienes in excellent yield (Scheme 1.5).[15] Interestingly, a different outcome was obtained for alkylcopper compounds, which underwent 1,3 addition to the enynylsulfinates.

Kauffmann and coworkers have described the reaction of alkylsilver compounds with cyclohexenone (Scheme 1.6).[29] It was found that the alkylsilver derivatives reacted preferentially, and in some cases exclusively, in the β position. A significant amount of 3-methylcyclohex-2-enol was observed for the reaction of $Me_2AgMgBr$, which has been attributed to the elimination and subsequent reaction of silver(I) hydride. Silver(I) hydride has been observed in the gas-phase fragmentation of other organosilver compounds.[30]

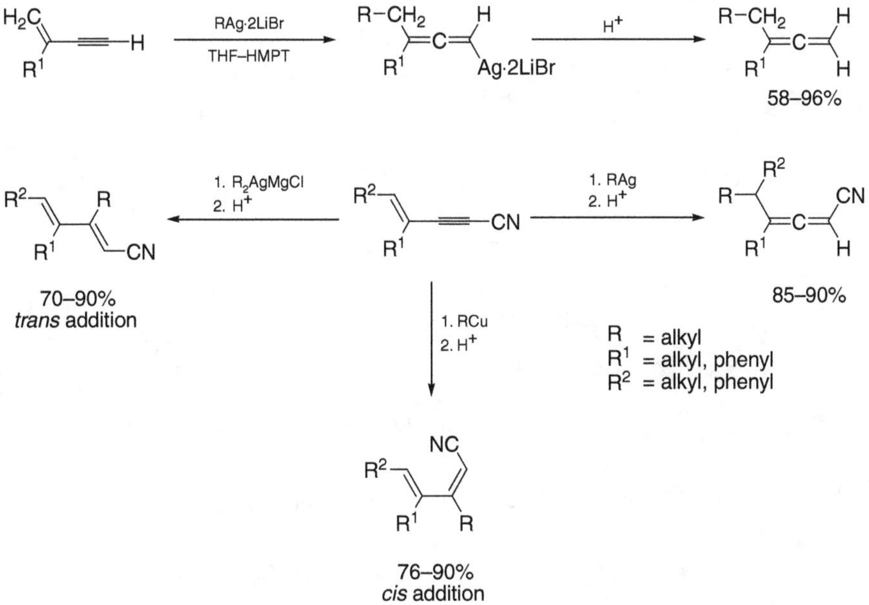

Scheme 1.4

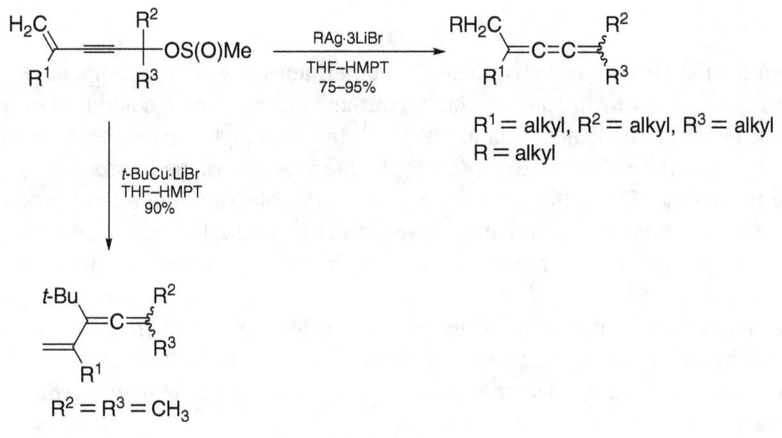

Scheme 1.5

Silver trinitromethane undergoes addition to adamantyl (Ad) halides,[31] methyl iodide, and trinitromethyl iodide (Scheme 1.7).[32]

1.2.2 Synthesis and Stability of Perfluoroalkylsilver Compounds

In comparison to the corresponding alkylsilver compounds, the perfluoroalkylsilver derivatives are significantly more stable and are able to participate in chemistry not

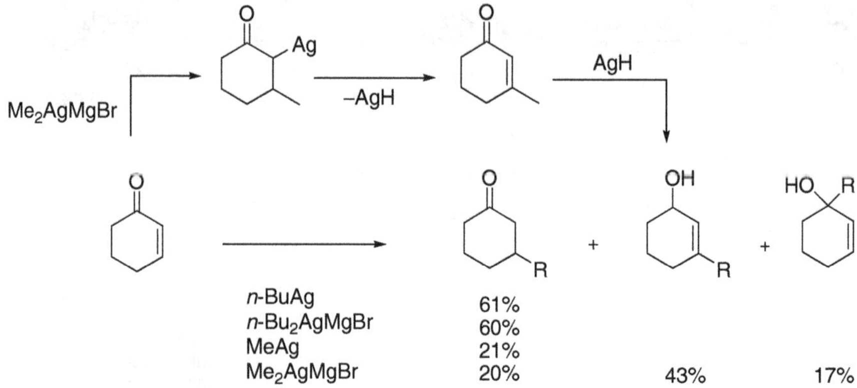

Scheme 1.6

$$AgC(NO_2)_3 + AdBr \longrightarrow AdC(NO_2)_3$$

$$AgC(NO_2)_3 + MeI \longrightarrow MeC(NO_2)_3$$

$$AgC(NO_2)_3 + IC(NO_2)_3 \longrightarrow (NO_2)_3CC(NO_2)_3$$

Scheme 1.7

known for the alkylsilver derivatives.[33] The perfluoroalkyl silver compounds were first synthesized by Miller and coworkers through the nucleophilic addition of silver fluoride to perfluoroalkenes (Scheme 1.8).[34,35] In an attempt to avoid the use of silver fluoride, Dyatkin and coworkers developed the reaction of perfluoroalkenes with silver trifluoroacetate in the presence of cesium or potassium fluoride to produce a variety of perfluoroalkyl silver derivatives.[36] Perfluoroalkylsilver compounds have also been synthesized by transmetallation of the corresponding cadmium reagents with silver nitrate.[37]

An interesting synthetic route to the perfluoroalkylsilver derivatives has been reported by Knunyants and coworkers with the production of perfluoroisopropylsilver in excellent yield via the decarboxylation of the silver salt of α-hydroperfluoroisobutyric acid in pyridine (Scheme 1.9).[38]

Tyrra has described the preparation of trifluoromethylsilver from Me_3SiCF_3 and silver fluoride in propionitrile (Scheme 1.10).[39] On treatment with PNPCl ($[PNP]^+ = Ph_3P{=}N{-}PPh_3{}^+$) trifluoromethylsilver was trapped as a mixture of argentates,

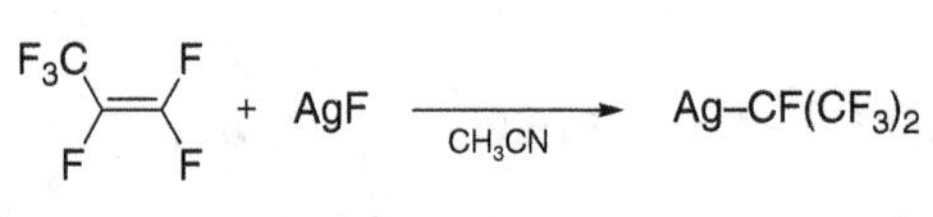

Scheme 1.8

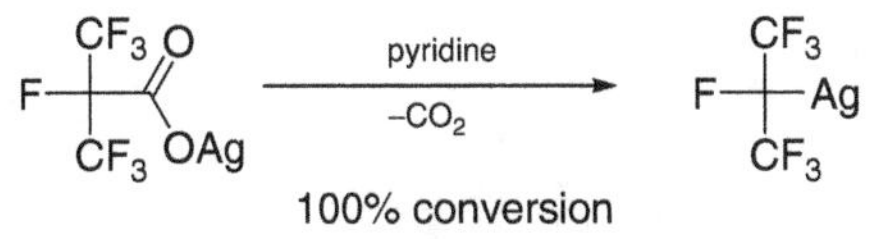

Scheme 1.9

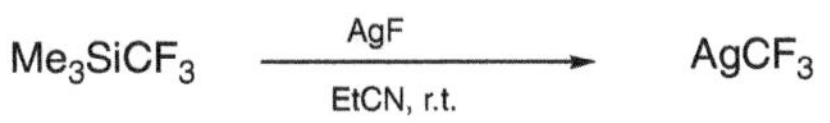

Scheme 1.10

[PNP][Ag(CF_3)$_2$] and [PNP][Ag(CF_3)Cl]. The latter was characterized by X-ray crystallography.[33,39]

1.2.3 Reactivity of Perfluoroalkylsilver Compounds

The known reactions of the perfluoroalkylsilver derivatives generally equate to nucleophilic substitution reactions and the oxidative perfluoroorganylation of group 12–16 elements.[39,40] Normant and coworkers have reported the reaction of both carbon dioxide and allyl bromide with perfluoroisopropylsilver (Scheme 1.11).[41] Heptafluoro-2-nitrosopropane has been prepared in good yield from perfluoroisopropylsilver by treatment with nitrosyl chloride.[42]

Dyatkin and coworkers have reported the reaction of perfluoroalkylsilver derivatives with iodine to give perfluoroalkyl iodides, and with elemental sulfur to give the perfluoroalkylthiosilver, which was subsequently alkylated with ethyl iodide to give the thioether (Scheme 1.12).[36]

Perfluoro-*tert*-butylmethyl sulfide was synthesized by treatment of the silver derivative with trifluoromethylsulfenyl chloride. The disulfide was also obtained from initial reaction with elemental sulfur, followed by treatment with the sulfenyl chloride (Scheme 1.13).[43]

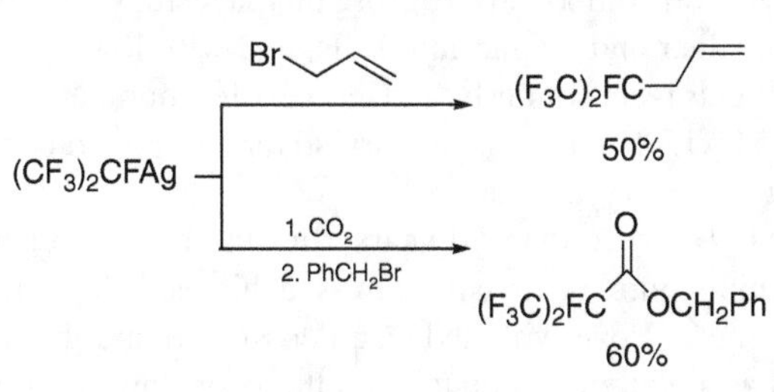

Scheme 1.11

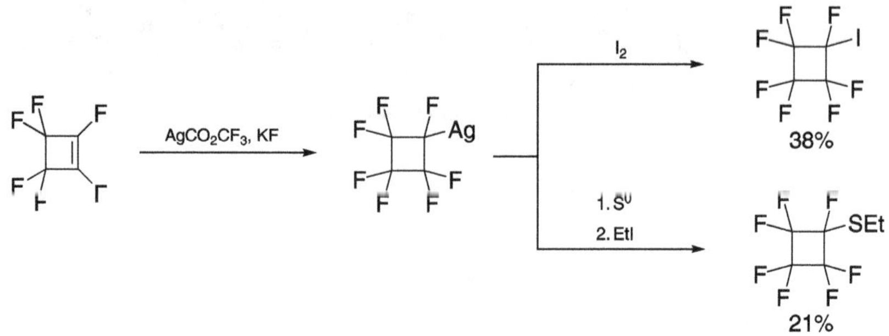

Scheme 1.12

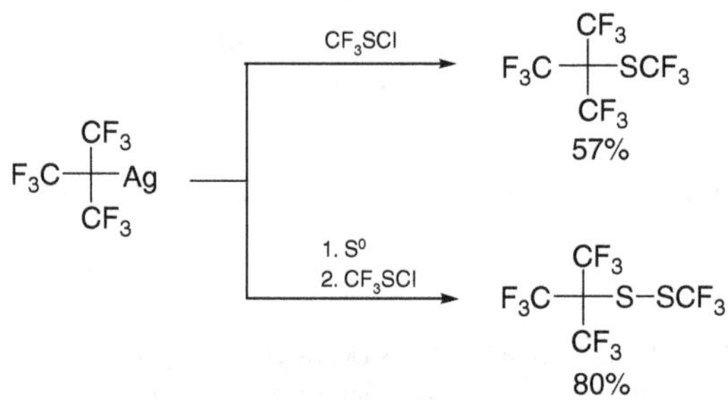

Scheme 1.13

1.3 C_{sp^2}-Ag

1.3.1 Synthesis and Stability of Arylsilver Compounds

The first synthesis of an arylsilver compound was reported by Krause in 1919, who isolated a compound with the formula $(PhAg)_2AgNO_3$ as a canary yellow powder from the reaction of silver nitrate with triphenylethyllead, which in the presence of "fuming HNO_3 . . . detonates with great violence and a flash of blinding white light."[44] Phenylsilver produced by this method decomposes thermally to a mixture of silver metal, benzene, and biphenyl.[6] In addition, a number of procedures were reported for the synthesis of arylsilver compounds from organomagesium compounds as mixtures, contaminated with silver and magnesium salts, and with low thermal stability.[45–47] Organolithium reagents react similarly, and in a very few number of cases in which the aryl was substituted with heteratoms, the arylsilver compound could be purified by means of extraction or recrystallization.[48–50]

It was only in 1972, more than 50 years after the first report by Krause, that a method for obtaining pure phenylsilver was achieved. Experimenting with the methodology of Krause, Beverwijk and Van der Kerk found that pure phenylsilver could be obtained as a white precipitate from the very slow addition of trialkylphenyllead or -tin compounds to ethanolic solutions of silver nitrate (Scheme 1.14).[51]

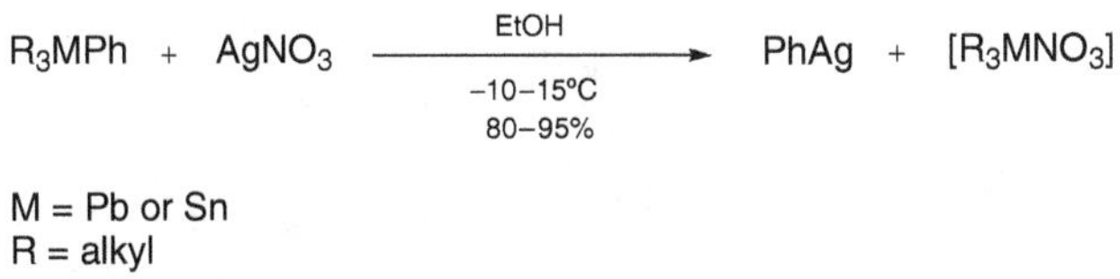

Scheme 1.14

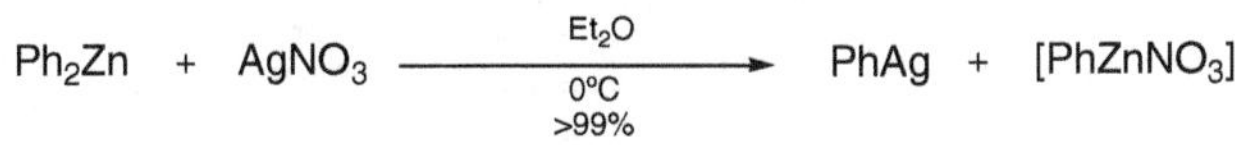

Scheme 1.15

A remarkable increase in the stability of pure phenylsilver is observed in comparison to that of the silver nitrate complex, with the compound surviving several days at room temperature under a nitrogen atmosphere.

Diarylzinc compounds react with silver salts to give arylsilver compounds of high purity and stability (Scheme 1.15).[52] Van der Kerk and coworkers synthesized phenylsilver and a number of methyl-substituted arylsilver compounds via this route, and found that *ortho*-methyl substitution significantly increased the thermal stability of the compound, as is the case for the corresponding arylcopper compounds (Table 1.5).[53]

As is the case for other organosilver compounds, arylsilver compounds are not monomeric. Cryoscopic molecular weight determination in benzene found phenylsilver to be polymeric and the methyl-substituted arylsilver compounds to be trimeric.[53] While cryoscopy of mesitylsilver, prepared from reaction of the corresponding Grignard reagent with silver chloride, gave a molecular weight corresponding to that of a dinuclear entity, determination of the structure by X-ray crystallography determined mesitylsilver to be tetrameric in the solid state.[54]

1.3.2 Reactivity of Arylsilver Compounds

In a very limited number of cases phenylsilver displays reactivity typical of moderately reactive organometallic compounds.[46] Gilman and coworkers reported the reaction of phenylsilver, prepared from phenylmagnesium iodide, with acetyl

TABLE 1.5. Temperatures of Decomposition of Arylsilver Compounds

Entry	R–Ag	Decomposition Temperature (°C)
1	Phenyl	74
2	2-Methylphenyl	91
3	3-Methylphenyl	82
4	4-Methylphenyl	78
5	2,6-Dimethylphenyl	170

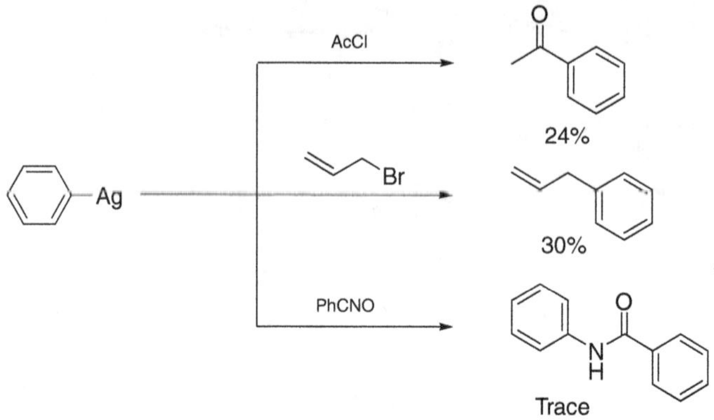

Scheme 1.16

chloride to give acetophenone in 24% yield, with allyl bromide to give allylbenzene in 30% yield, and with PhNCO to give a trace of benzanilide (Scheme 1.16).[46] The predominant product in all three cases was biphenyl.

1.3.3 Synthesis and Stability of Perfluoroarylsilver Compounds

As is general for organosilver compounds, perfluorophenylsilver is considerably more stable than phenylsilver.[55] The first example of a perfluoroarylsilver compound was reported by Miller and Sun in 1970.[49] Reaction of perfluorophenyllithium with silver trifluoroacetate, or of perfluorophenyl bromide with perfluoroisopropylsilver, gave perfluorophenylsilver as colorless crystals (Scheme 1.17). The compound was found to be stable up to 150°C, at which temperature it slowly dimerized to give perfluorobiphenyl and metallic silver. Since this initial report a number of syntheses of perfluorophenylsilver from perfluorophenyllithium followed.[56,57]

A new convenient synthesis of perfluorophenylsilver has been reported by Tyrra and coworkers. Treatment of trimethyl(perfluorophenyl)silane with silver fluoride in propionitrile gave perfluorophenylsilver in nearly quantitative yield (Scheme 1.18).[39,58] Perfluorophenylsilver synthesized by this method, isolated as a 1 : 1 adduct with propionitrile, has been characterized by X-ray crystallography.[58]

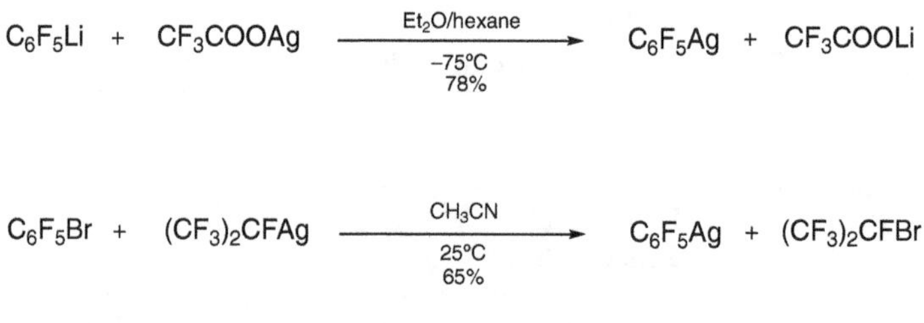

Scheme 1.17

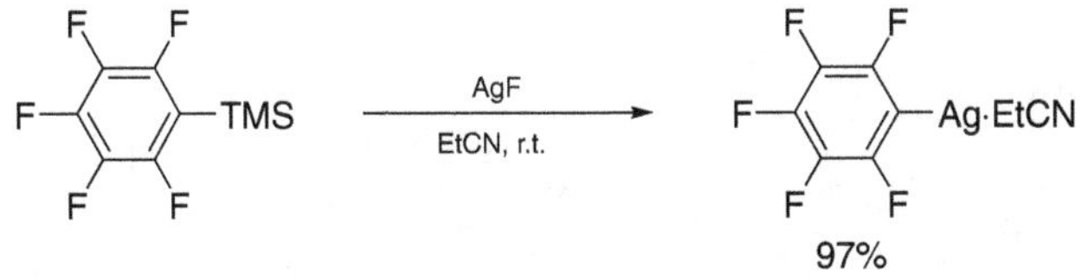

Scheme 1.18

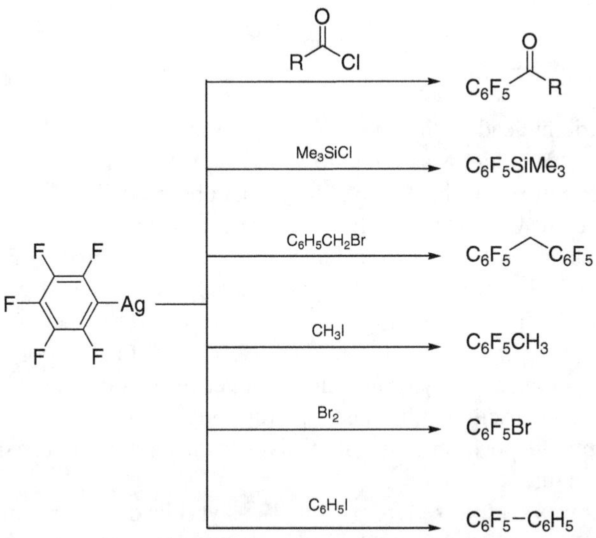

Scheme 1.19

1.3.4 Reactivity of Perfluoroarylsilver Compounds

The reactions of perfluorophenylsilver amount to the nucleophilic substitution of a variety of electrophiles and the oxidative perfluoroorganylation of group 12–16 elements (Scheme 1.19).[39,40,49,55,59,60]

1.3.5 Synthesis, Stability, and Reactivity of Alkenylsilver Compounds

Alkenysilver reagents have found very limited use in organic synthesis. This section focuses on the few methods where alkenylsilver reagents are synthesized with the intention of utilizing, or gaining further understanding of the nature of, the chemistry of these highly reactive species.

Alkenylsilver derivatives are relatively less thermally stable than alkynylsilver derivatives, but tend to be significantly more stable than the alkylsilver derivatives. The first investigation into the thermal stability of alkenylsilver reagents was reported by Glockling, who described the synthesis of isobut-1-enylsilver from the reaction of

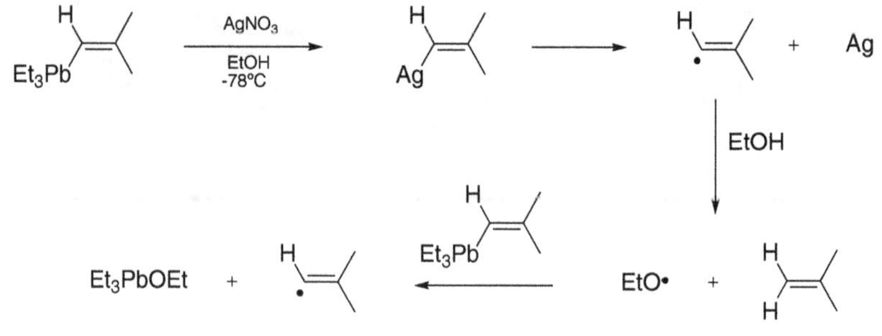

Scheme 1.20

isobut-1-enyltriethyllead with ethanolic silver nitrate at −78°C.[61,62] An orange precipitate resulted, which was found to slowly deposit silver metal on warming to −20°C, and significant isobutene evolution was observed above −10°C. The formation of isobutene was in fact catalytic in silver, suggesting the initiation of a chain reaction whereby an isobutenyl radical may abstract a proton from ethanol, followed by reaction of the incipient ethoxyl radical with isobut-1-enyltriethyllead (Scheme 1.20). In the absence of ethanol, the alkenylsilver reagent decomposed to give dimer and metallic silver upon gradual warming of the solution.

An increase in the stability of this alkenylsilver reagent can be effected through coordination. Both triethylamine and 2,2′-dipyridyl had a remarkable effect on stability, whereby deposition of metallic silver from solution occurred only slowly at room temperature.[62]

Styrylsilver was synthesized from triethyl-α-styryl lead as a deep red precipitate, and proved to be comparatively stable, with decomposition to silver requiring several days at room temperature.[5] Unlike isobut-1-enylsilver, the formation of dimer (1,4-diphenylbutadiene) was not observed; instead, an insoluble polymeric substance was formed.

Alkenylsilver compounds have been synthesized by treating vinyllithium reagents with silver salts. Both *cis*-1-propenylsilver and its corresponding tributylphosphine adduct were synthesized by this method for the purposes of investigating the outcome of the thermal decomposition of vinylsilver compounds. In both cases the product of dimerization predominated with almost complete retention of stereochemistry, thus suggesting that, as for the alkylsilver compounds, long-lived free propenyl radicals are not intermediates in this reaction (Scheme 1.21).[63,64] Kochi and Tamura report a similar finding whereby treatment of *cis*-propenylmagnesium bromide and methyl

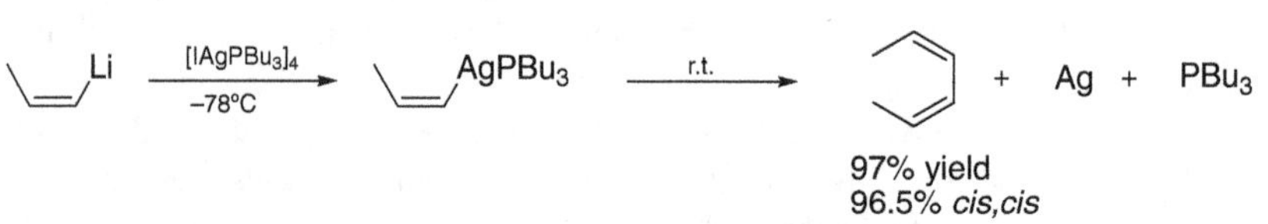

Scheme 1.21

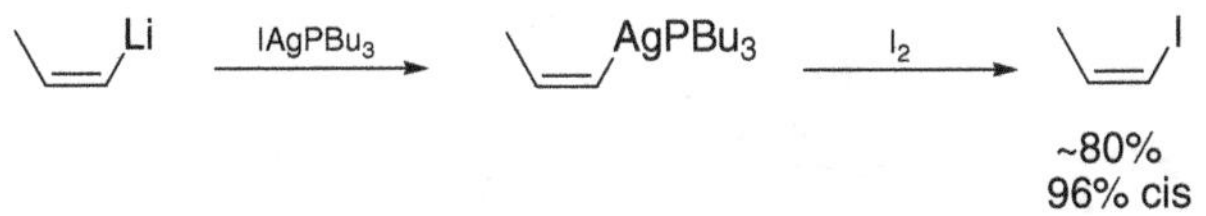

Scheme 1.22

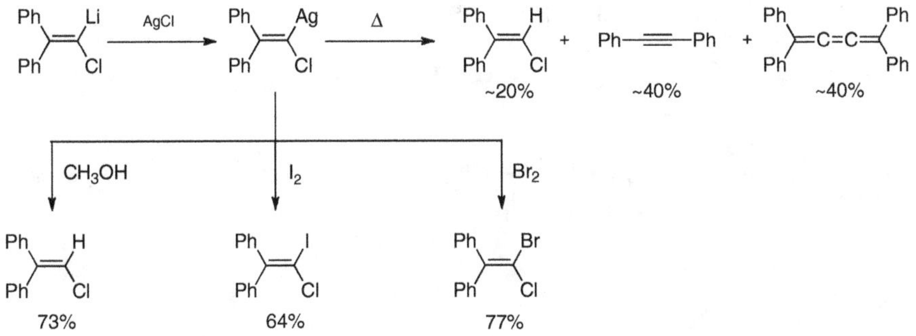

Scheme 1.23

bromide with catalytic amounts of silver(I) salt produced *cis*-but-2-ene in almost quantitative yield.[19]

The tributylphosphine adduct was found to react with iodine to give *cis*-1-iodopropene (Scheme 1.22).[64]

1-Chloro-2,2-diphenylvinylsilver has been produced at low temperatures by treating the lithiated derivative with silver chloride.[65] The compound decomposed at room temperature to give 1-chloro-2,2-diphenylethene, diphenylacetylene, and tetraphenylbutatriene. The silver derivative does not react with carbon dioxide, but gave the expected products of electrophilic substitution when treated with methanol, iodine, and bromine (Scheme 1.23).

1.3.6 Synthesis and Reactivity of Allenylsilver Compounds

Allenylsilver reagents have been prepared by the reaction of allenyllithium reagents with silver(I) salts,[66,67] and by the addition of alkylsilver reagents to conjugated enynes (see alkylsilver section 1.2.1 above), and have been added to a number of electrophiles in high yield (Scheme 1.24). In most cases, this resulted in allene contaminated with a small percentage of alkyne (<10%). Carbon disulfide reacted solely at the propargylic position to give β,γ-unsaturated γ-dithiolactones in excellent yield (Scheme 1.25),[16,68] whereas carbon dioxide reacted at the allenic position to produce allenyl carboxylic acids in good yield.[18] The allenylsilver compound did not react with methyl iodide under these conditions.[66,68]

Allenylsilver compounds appear to enter the palladium coupling catalytic cycle *via* transmetallation. It was found that *tert*-butyl allenylsilver, and the corresponding argentate derivative, underwent palladium-catalyzed coupling with iodobenzene to

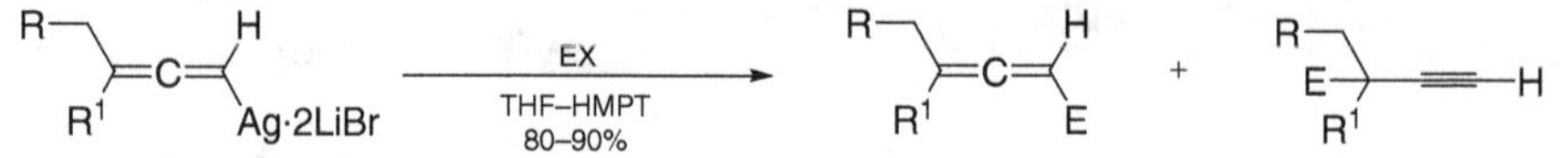

R = Bu, ^iPr, ^tBu R^1 = H or Me
EX = BrCN, I_2, MeS-SO_2Me, allyl bromide, CO_2, CS_2, Me_3SiCl, Me_3GeCl, Me_3SnCl, NBS, NCS

Scheme 1.24

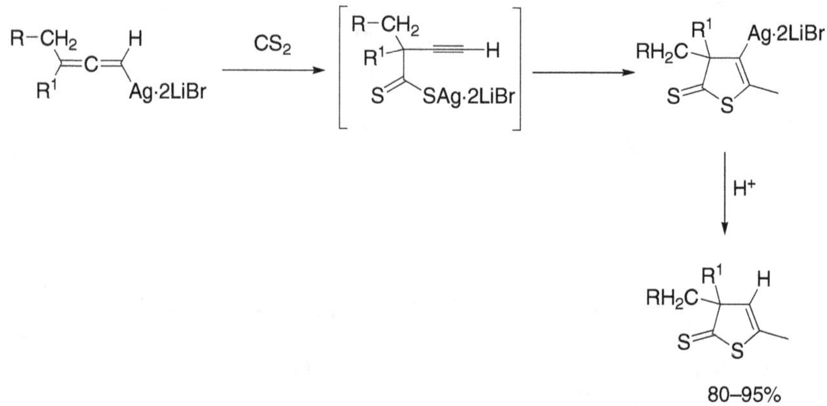

Scheme 1.25

give the aryl allene in good yield without any detectable amount of alkyne (Scheme 1.26).[66]

1.3.7 Synthesis of Perfluoroalkenylsilver Compounds

Perfluoroalkenylsilver compounds[55] have been synthesized by the treatment of perfluoroalkynes,[69] perfluoroallenes,[70] and 1-trimethylsilyl perfluoroalkenes[40] with silver fluoride; via the metathesis of lithium-substituted perfluoroalkenes with silver trifluoroacetate;[71] and by exchange between perfluorophenyl- or perfluoroisopropylsilver and the corresponding iodoperfluoroalkene (Scheme 1.27).[49]

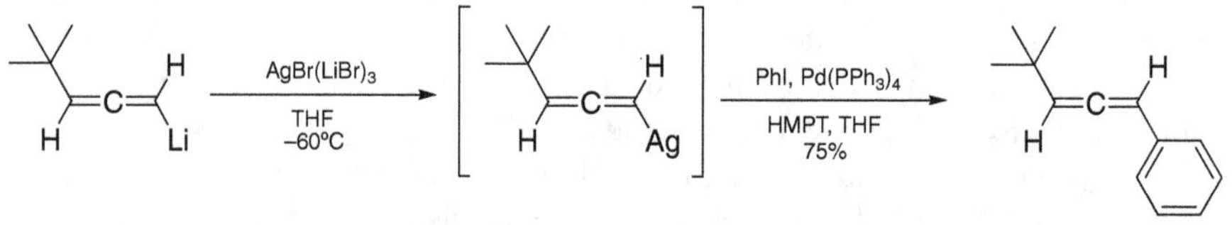

Scheme 1.26

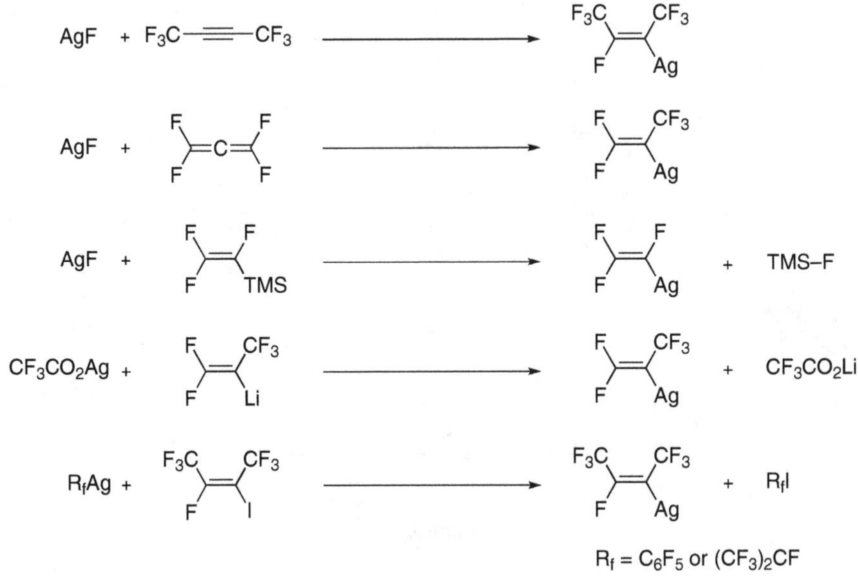

Scheme 1.27

1.3.8 Reactivity of Perfluoroalkenylsilver Compounds

Although finding relatively wider synthetic use than the alkenylsilver compounds, examples of reactions of the perfluoroalkenylsilver compounds are limited to a handful of reactions, including protolysis, bromination, dimerisation, methylation, and silylation (Schemes 1.28 and 1.29).[69,70]

1.3.9 Synthesis and Reactivity of Silver-Substituted Diazomethyl Compounds

Diazomethyl compounds are an extremely useful class of molecules for synthetic organic chemistry. Silver-substituted diazomethyl compounds have further increased the scope of these reagents. The thermally unstable argentodiazoesters are synthesized in situ by treating the corresponding diazoester with silver oxide. The resulting

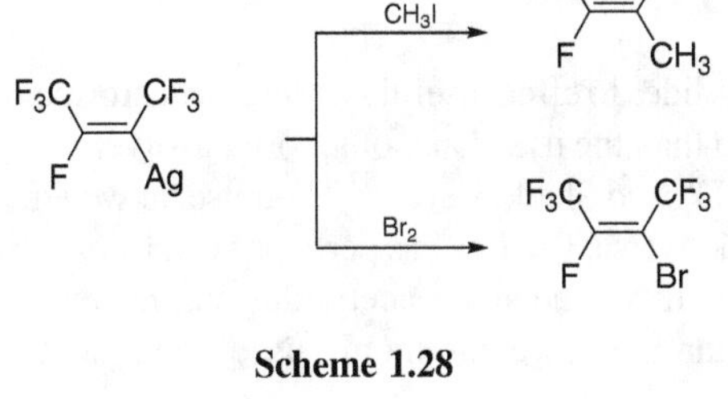

Scheme 1.28

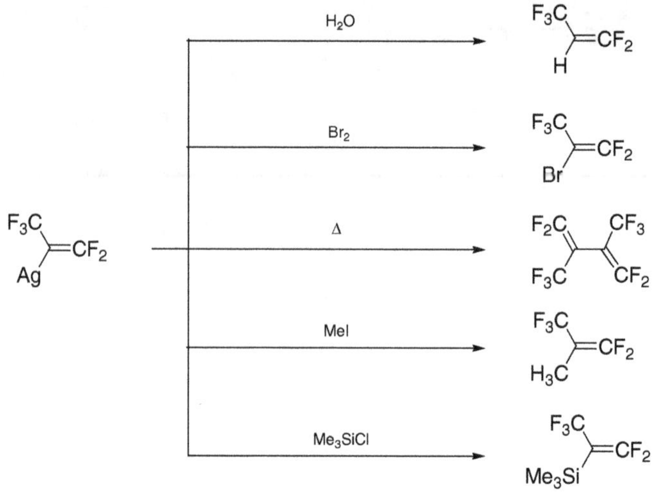

Scheme 1.29

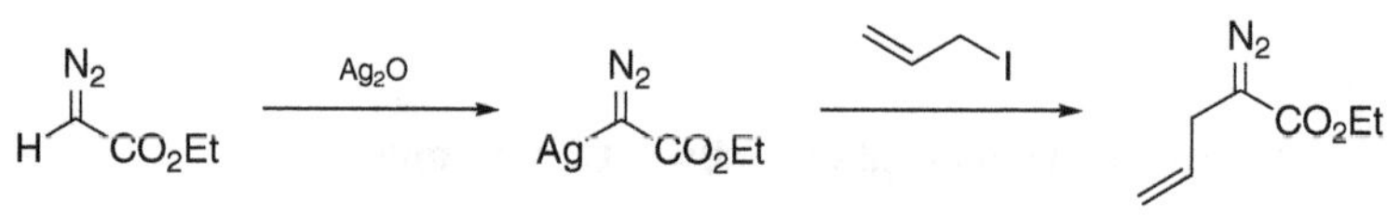

Scheme 1.30

argentodiazoesters may then easily participate in electrophilic substitution with S_N1 reactive alkyl bromides and iodides (Scheme 1.30).[72]

Silver(diazomethyl)phosphoryl compounds may also be synthesized by treatment of the corresponding diazophosphoryl compound with silver oxide. The thermally stable phosphoryl derivatives undergo electrophilic substitution with alkyl iodides[73] but, unlike the carbonyl-substituted derivatives, also undergo electrophilic diazoalkane substitution with a variety of Hückel aromatic salts (Scheme 1.31).[74–79]

Disilver diazomethane has also been synthesized. When treated with an alkyne, it produces the corresponding silver acetylide.[80]

1.4 C_{sp}-Ag

The use of silver acetylides to effect useful synthetic transformations has been perhaps more widely reported than the use of any other organosilver compound, as previously reviewed by Pale et al.[81] and Sladkov et al.[82] Their use, however, in comparison to that of other metal acetylides (Na, Cu, Li), has perhaps been limited by their relatively low reactivity and/or solubility. The silver acetylides are, however, generally comparatively more stable than the more commonly used metal acetylides, as they can be

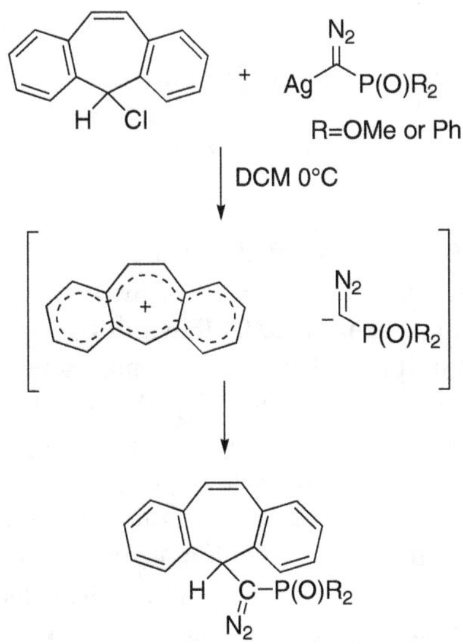

Scheme 1.31

stored for months without significant decomposition, and are prepared under relatively mild conditions.[83,84] These characteristics, in conjunction with the often unique chemistry displayed by this class of compounds, allow a number of novel transformations under conditions generally more suitable for sensitive substrates.

1.4.1 Synthesis of Silver Acetylides

Classically, silver acetylides have been prepared by the treatment of an alkyne in ammoniacal silver nitrate solution to give the silver acetylide as a precipitate, which may be easily collected by filtration and stored in the absence of light (Scheme 1.32).[83–86] A number of silver acetylides have been synthesized by this method, with a wide range of functionality. Other early methodology includes the synthesis of (1-butynyl)silver from the reaction of silver bromide with the corresponding Grignard reagent,[87] and of propynyl- and (phenylethynyl)silver from the corresponding potassium acetylide and silver nitrate in liquid ammonia.[88]

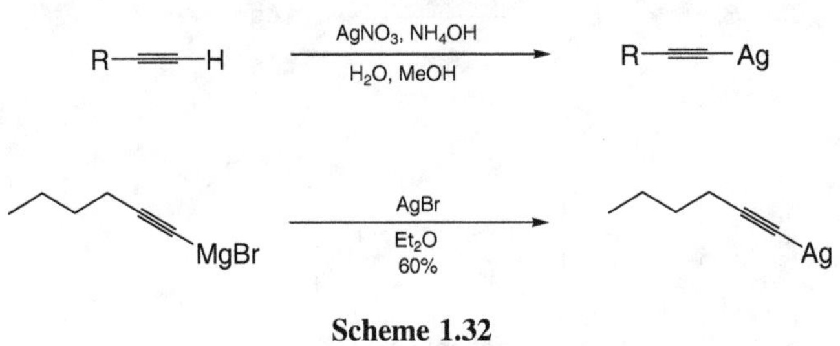

Scheme 1.32

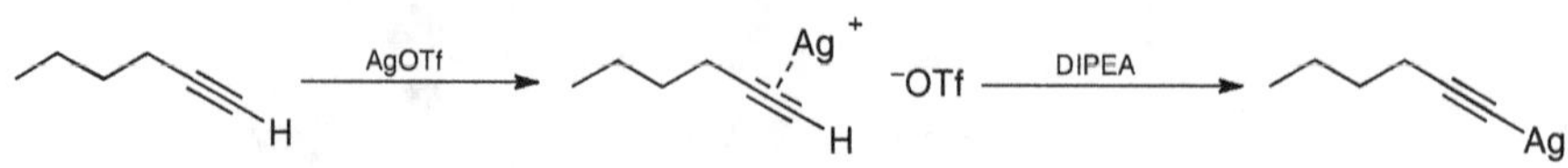

Scheme 1.33

Pale and coworkers have demonstrated that silver acetylides may be synthesized with a variety of silver salts in the presence of base in a number of solvents. By treating 1-hexyne in either deuterated benzene or DMF with silver triflate, the group was able to observe the formation of the π-alkyne–silver complex and subsequently the silver acetylide through the use of ^{1}H, ^{13}C, and ^{109}Ag NMR. The incipient π-alkyne–silver complex is rapidly deprotonated on addition of diisopropylethylamine, to give the silver acetylide as a white precipitate (Scheme 1.33).[89]

More recently, the well-known deprotection of 1-trimethylsilyl-1-alkynes with silver salts has been exploited for the isolation of silver acetylides (Scheme 1.34).[90] Treatment of 1-trimethylsilyl-1-alkynes with silver salts in alcohol gives a variety of silver acetylides under relatively very mild conditions. The reaction also proceeds through formation of the π-alkyne–silver complex, thus enabling nucleophilic attack by the solvent to give the silver acetylides. This methodology has the advantage of avoiding strongly basic conditions in the synthesis of sensitive substrates, and allows silver acetylide formation from acetylenic alcohols for which silver-catalyzed cyclization is a possibility.

Most silver acetylides are highly insoluble in standard organic solvents because of the formation of coordination polymers, and this characteristic can cause difficulties with characterization and limit reactivity. This can be overcome by the addition of coordinating donor molecules such as triphenylphosphine that perturb the formation of the polymeric structure.[91] In the experience of the authors, however, the addition of such ligands can decrease reactivity.

1.4.2 Reactivity of Silver Acetylides

1.4.2.1 Addition to Activated Carbonyls and Iminium Ions

The poorly nucleophilic silver acetylides are generally not considered reactive enough to add directly to unactivated carbonyl groups, although a number of

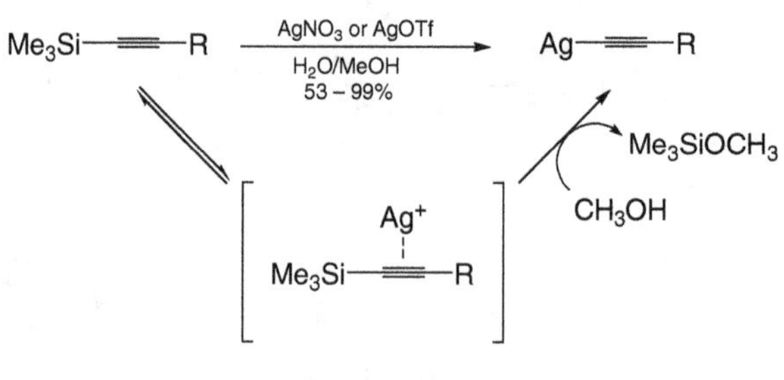

Scheme 1.34

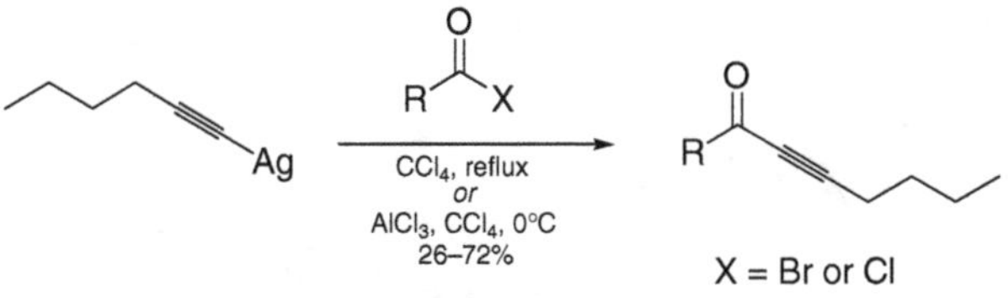

Scheme 1.35

examples of addition to activated carbonyls are known. The addition of silver acetylides to acid halides and carbon dioxide has been utilized in the synthesis of acetylenic ketones[92,93] and esters,[94,95] respectively. Davis and Schreiber reported the first addition of silver acetylides to acid halides to give ynones.[83] Silver acetylides bearing an alkyl chain were treated with an acid halide in either refluxing carbon tetrachloride, or at 0°C in the presence of aluminum trichloride to give ynones in moderate to good yield (Scheme 1.35). This methodology has been used by other groups to synthesize a number of acetylenic ketones under relatively mild conditions.[93]

This methodology was utilized effectively by Inanaga and coworkers in the total synthesis of methynolide.[96] Using the silver acetylide in this case allowed for the mild introduction of the acetylenic moiety, avoiding the use of strong base in the presence of the base sensitive ynone product (Scheme 1.36).

Even the somewhat menacing-looking peroxyl silver acetylides undergo addition to acid chlorides to give ketones (Scheme 1.37).[97]

Silver acetylides have been added to iminium salts to give propargylic amines. In the derivitization of cortanine hydrochloride for biological screening, a wide variety

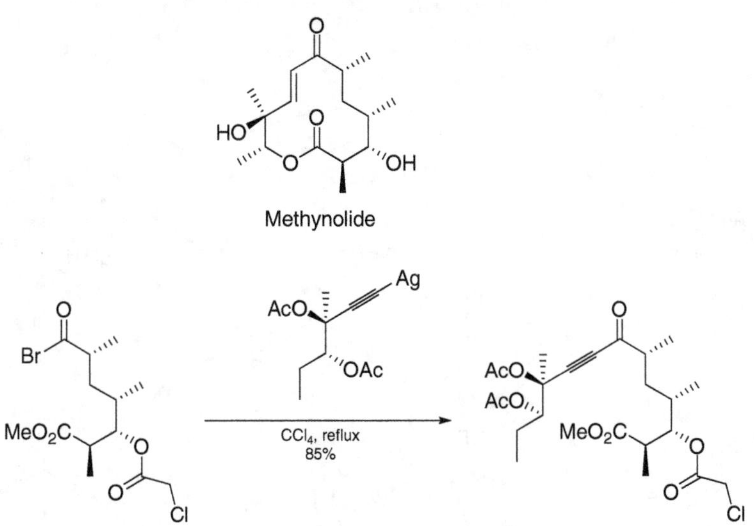

Scheme 1.36

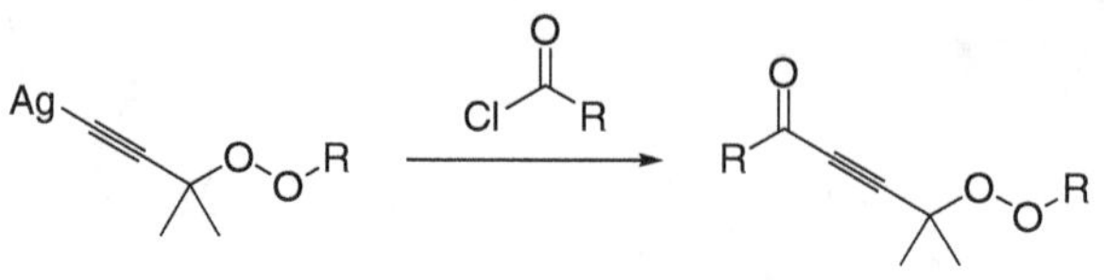

Scheme 1.37

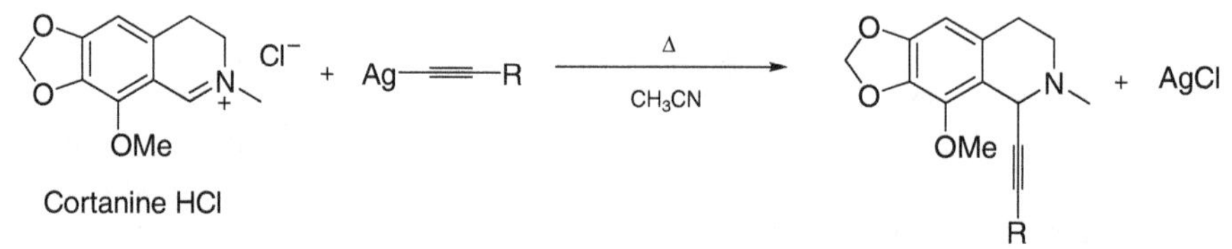

Scheme 1.38

of silver acetylides were added to the iminium salt to give alkynylated products in good yield (Scheme 1.38).[98,99] This same methodology was first used by the same group in the synthesis of noncyclic propargylic amines in a limited number of examples.[100]

β,γ-Alkynyl α-amino acid derivatives have been obtained from the silver-catalyzed addition of alkynes to α-iminoesters in good yield. It is likely that the π-alkyne–silver complex is deprotonated by the iminoester to give the silver acetylide in situ, which may then react with the activated iminium ion (Scheme 1.39).[101]

An enantioselective version of this reaction has been reported by Rueping et al.[102] Treatment of an α-iminoester and an alkyne with silver acetate and a binol phosphate derivative gave propargylic amines with the highest enantiomeric ratio (er) reported as 96 : 4. Although the proposed catalytic cycle invoked the in situ formation of the

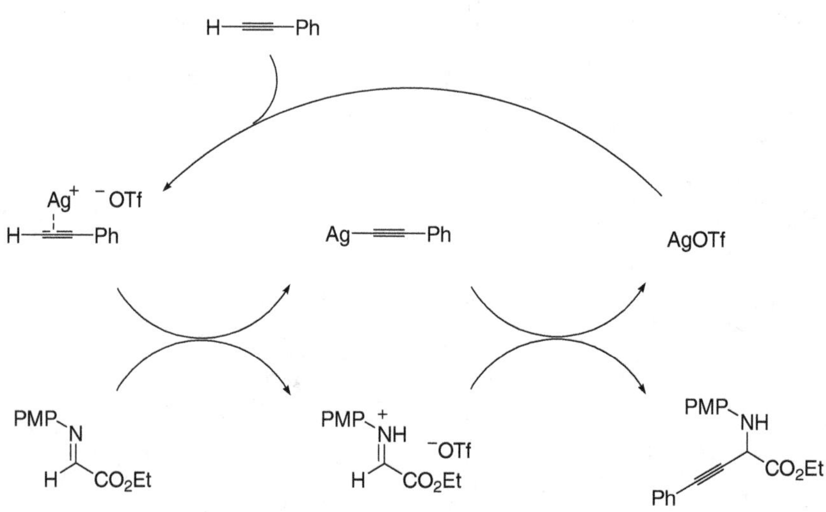

Scheme 1.39

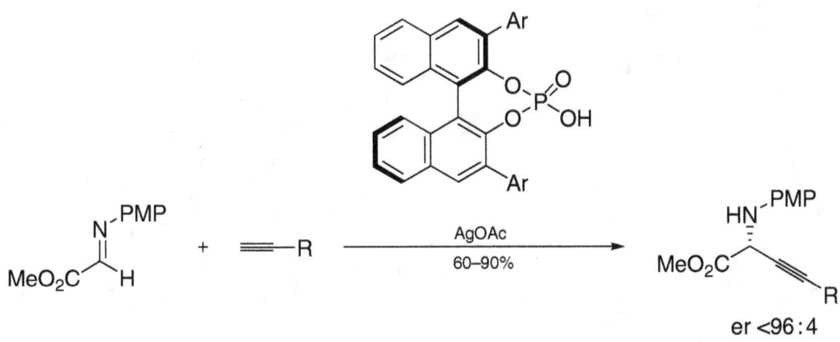

Scheme 1.40

silver acetylide and subsequent addition to an activated α-iminoester, the authors did not discount the possibility of participation of a chiral silver–phosphate complex (Scheme 1.40).

Li and coworkers have reported a novel silver-catalyzed, three-component coupling of an aldehyde, an amine, and an alkyne, and the first example of a direct addition of a silver acetylide to an aldehyde to produce propargylic alcohols (Scheme 1.41).[103–105] Treatment of an aldehyde with a silver halide and secondary amine gave propargylic amines in good yields in organic solvent, water, or ionic liquids. The proposed mechanism suggests addition of a silver acetylide to the iminium ion formed in situ.

Further screening of conditions revealed that the use of Cy_3PAgCl gave only propargylic alcohol rather than propargylic amine. The authors have postulated that, as silver acetylides are generally considered to be too stable to add directly to unactivated carbonyls, the electron-donating effect of the phosphine ligand activates the silver–carbon bond. The phosphine ligand in this reaction is thereby acting as a "chemoswitch," allowing for selection of either the propargylic amine or alcohol. The reaction is neither water nor air sensitive and provides a "greener" alternative to the corresponding Grignard reaction (Scheme 1.42).

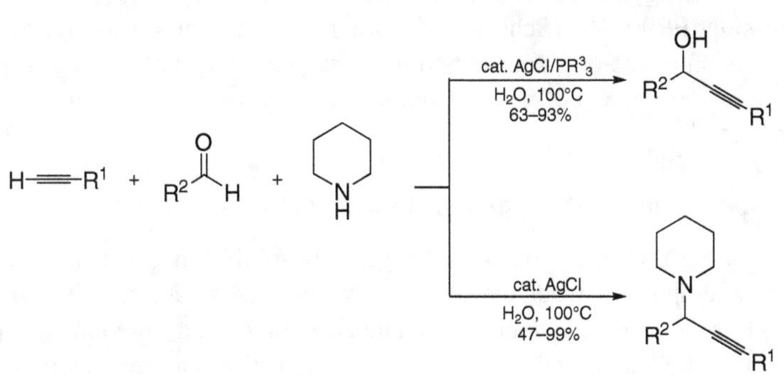

Scheme 1.41

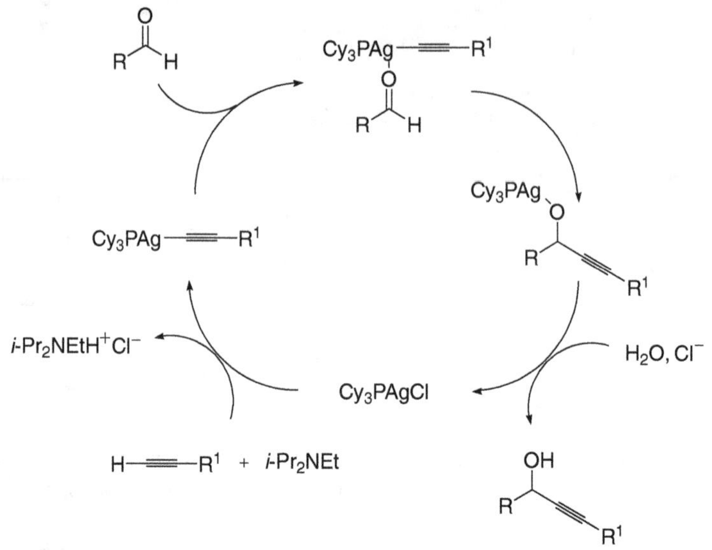

Scheme 1.42

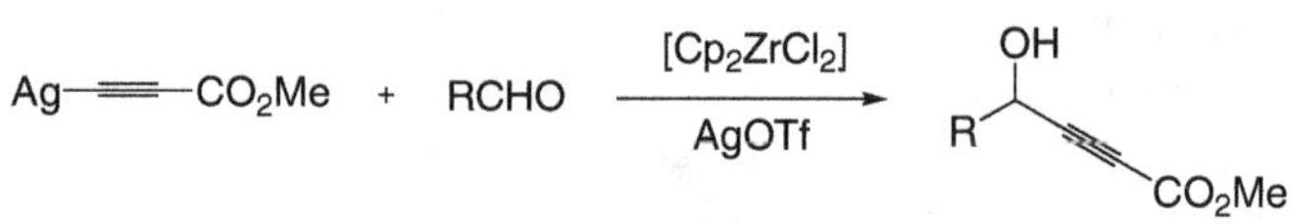

Scheme 1.43

Silver acetylides participate in the zirconium-catalyzed nucleophilic addition to carbonyls in the absence of base to produce functionalized γ-hydroxy-α,β-acetylenic esters.[106] Although this reaction involves a transmetallation rather than a direct addition of a silver acetylide to an unactivated aldehyde, its particular attraction of is the avoidance of strong bases to produce the reactive zirconium species (Scheme 1.43).

This methodology has been applied to the total synthesis of the potent anticancer natural product FR901464 (Scheme 1.44, where dr = diastereomeric ratio).[107]

An extension of the original methodology has seen the synthesis of propargylic alcohols in moderate to good yields from silver acetylides via zirconium acetylides (Scheme 1.45).[108]

1.4.2.2 Nucleophilic Substitution of Activated Heteroaromatics

With the finding that silver phenylacetylide was soluble in pyridine, Agawa and coworkers attempted to react the acetylide with benzoyl chloride in pyridine as solvent. While the expected ynone was isolated in poor yield, the predominant product of reaction was from initial acylation of the pyridine nitrogen to produce the electrophilic pyridinium salt, followed by addition of the silver acetylide to the 2

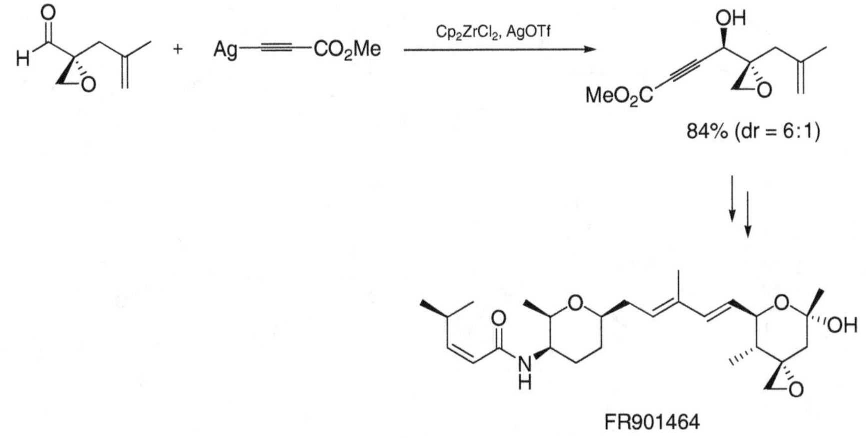

Scheme 1.44

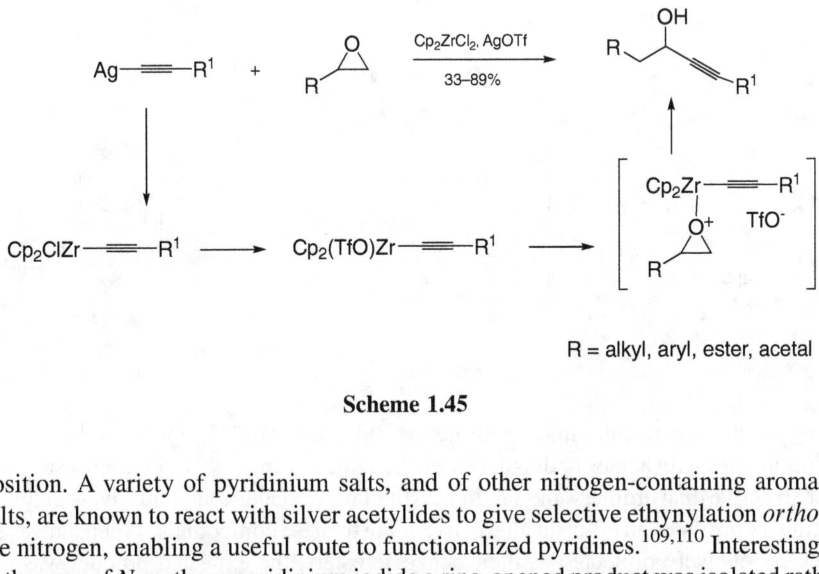

Scheme 1.45

position. A variety of pyridinium salts, and of other nitrogen-containing aromatic salts, are known to react with silver acetylides to give selective ethynylation *ortho* to the nitrogen, enabling a useful route to functionalized pyridines.[109,110] Interestingly, in the case of *N*-methoxypyridinium iodide a ring-opened product was isolated rather than the ethynylated pyridine. The formation of this product is rationalized because the methoxide anion is a poorer leaving group than the benzoate, thus allowing ring opening to compete with aromatization (Scheme 1.46).

1-Aminobenzimidazolium salts react with silver acetylides in refluxing DMF to give pyrazoles in moderate yield (Scheme 1.47).[111]

1.4.2.3 Reaction with Alkyl Halides

In the synthesis of nucleosides from their corresponding *C*-glycosyl derivatives, silver acetylides have been proved effective in introducing an acetylenic ester. The ribose

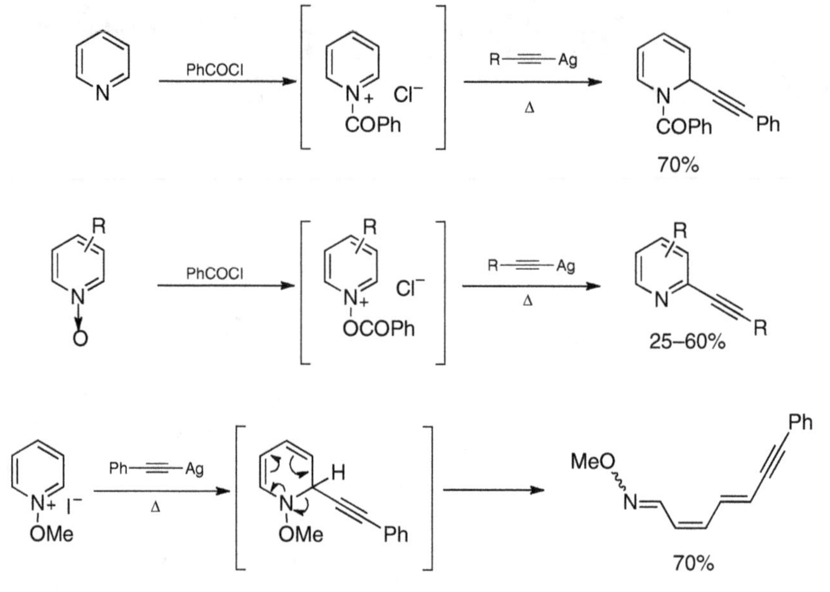

Scheme 1.46

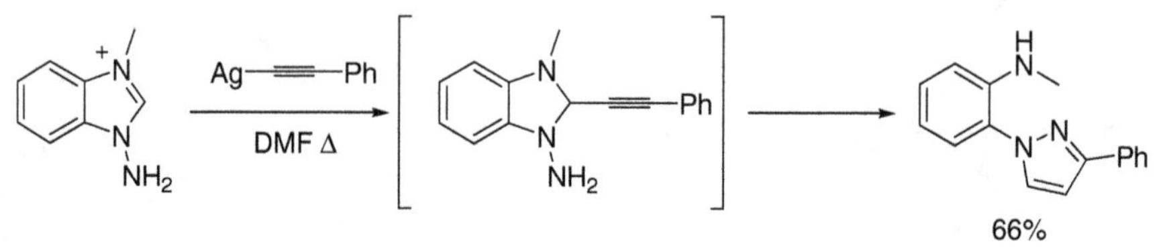

Scheme 1.47

moiety of these molecules may be further modified through 1,3-dipolar cycloaddition of the alkynes with azides or diazoalkanes (Scheme 1.48).[112,113] Acetylenic esters are more reactive than ordinary alkynes in 1,3-dipolar cycloadditions, and silver acetylide chemistry provides an efficient alternative route to these compounds. When, however, the alcohols were protected as their benzoate esters, the expected product was not

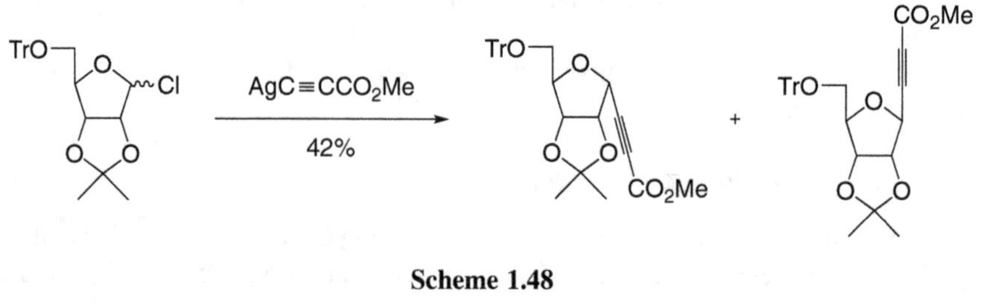

Scheme 1.48

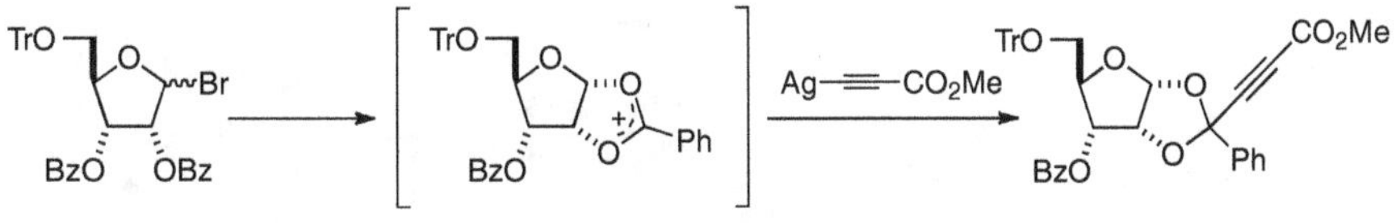

Scheme 1.49

obtained. The authors have invoked anchimeric assistance in displacement of the bromide by the benzoate ester, followed by attack on the incipient oxonium ion by the silver acetylide to give alkynylation alpha to the phenyl group (Scheme 1.49).

Direct bridgehead alkynylation of 1-iodoadamantane has been achieved by treatment with a variety of silver acetylides at reflux in *N*-methylmorpholine.[114] The methodology has been extended to the direct bridgehead substitution of methylbicyclo[2.2.2]octane and a carborate anion (Scheme 1.50).[115]

In an attempt to elucidate the mechanism, the reaction was performed in the presence of a radical trap (TEMPO=2,2,6,6-tetramethyl-1-piperidinoxyl), a reversible electron acceptor (dinitrobenzene), or a radical sensitizer (dimethoxybenzene). In all cases the proportion of coupled product was increased, suggesting that both radical and polar mechanisms contribute to the outcome of this reaction.

Primary alkyl halides can also be alkynylated by silver acetylides. Isabelle and coworkers reported the reaction of methyl iodide, ethyl iodide and d_3-methyl iodide with several silver acetylides to give disubstituted alkynes.[116] The authors preferred a non-radical-mediated mechanism for this reaction, as neither methane nor ethane, expected byproducts of a radical reaction, were observed.

1.4.2.4 Coupling Reactions

As is the case for allenylsilver reagents, silver acetylides undergo palladium-catalyzed coupling reactions. Ruitenberg and coworkers showed that (1-trimethylsilylethynyl)

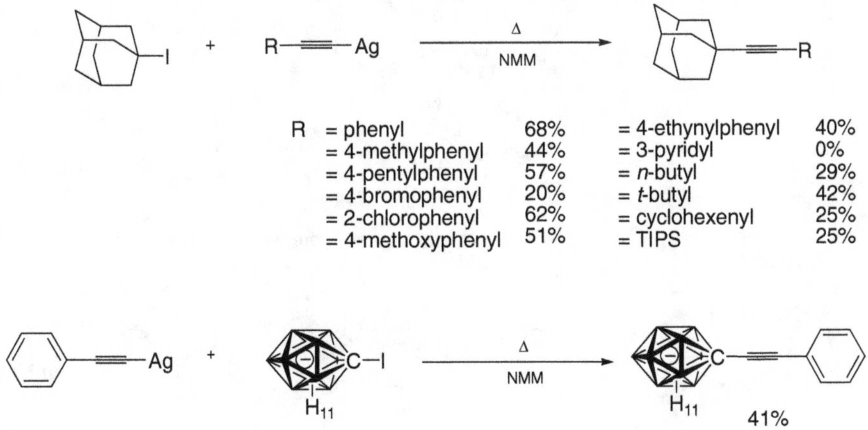

Scheme 1.50

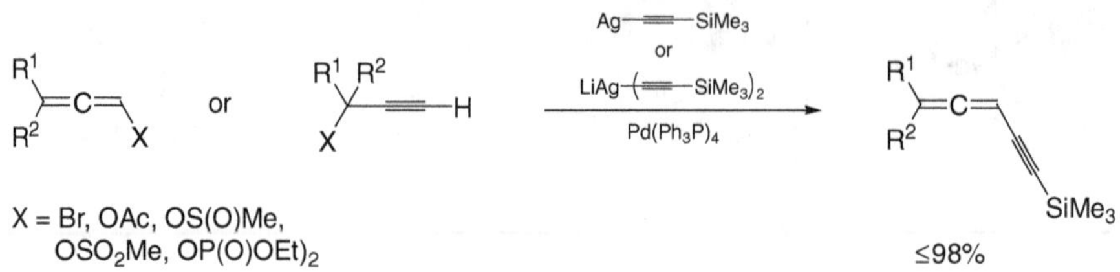

Scheme 1.51

silver, and the corresponding lithium argentate, react with 1-allenic or 2-propynic substrates in the presence of catalytic amounts of $Pd(Ph_3P)_4$ to give conjugated allenynes (Scheme 1.51).[117]

During attempts at the total synthesis of neocarzinostatin chromophore, Pale and coworkers sought to develop methodology for the coupling of epoxyalkynes with enol triflates. Standard Sonogashira conditions resulted in disappointing yields due to decomposition of the sensitive epoxyenyne moiety. It was found, however, that replacing copper(I) iodide with silver(I) salts led to the clean formation of product in at least comparable yield (Scheme 1.52 and Table 1.6).[118–120]

Because of the chemical similarities between silver and copper, silver acetylides were implicated in a catalytic cycle comparable to that proposed for the copper-catalyzed reaction, and the group has produced undeniable evidence for their participation. Authentic samples of silver acetylides were found to couple with enol triflates under the same conditions as for the catalytic reaction to produce enynes in

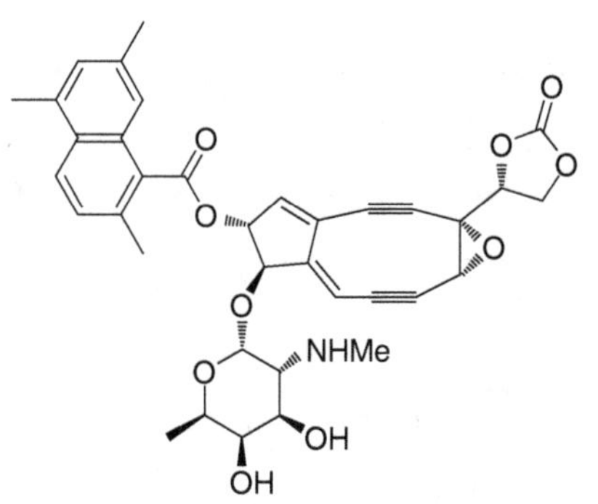

Neocarzinostatin chromophore

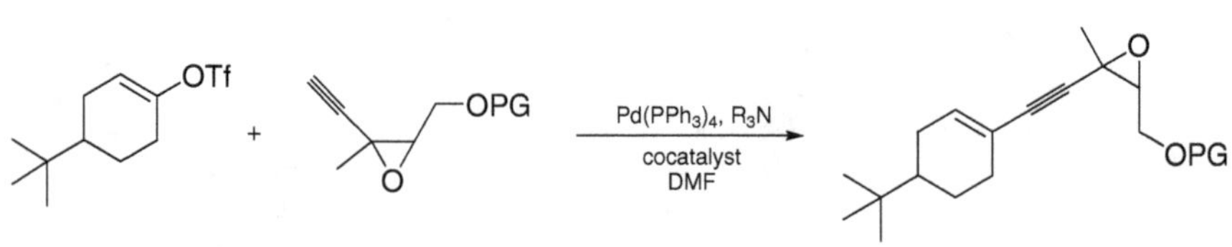

Scheme 1.52

TABLE 1.6. Silver-Catalyzed Sonogashira-Type Coupling

Entry	Cocatalyst	Catalyst Loading (mol%)	Time (h)	Yield (%)
1	CuI	10	1	44–54
2	AgI	10	24	51
3	AgI	20	20	78
4	$AgNO_3$	10	4	60
5	Ag_2CO_3	5	4	56

high yield.[121] Furthermore, ^{109}Ag NMR spectra of silver triflate in DMF-d_7 gave a signal at 80 ppm that shifted to 750 ppm on addition of 1-hexyne, and subsequently to 1050 ppm on addition of diisopropylethylamine. These values are consistent with initial formation of a silver π complex, followed by deprotonation and formation of the silver acetylide (Scheme 1.53).[89]

In 2006, Li and Wang reported the palladium-free, silver-catalyzed Sonogashira-type coupling of aryl halides and terminal alkynes. The reaction proceeds in high yield in the presence of catalytic silver iodide, triphenylphosphine, and potassium carbonate. Although the mechanism remains unclear, it is evident that the silver acetylide has a role, as the acetylide is formed on mixing of the reagents (Scheme 1.54).[122]

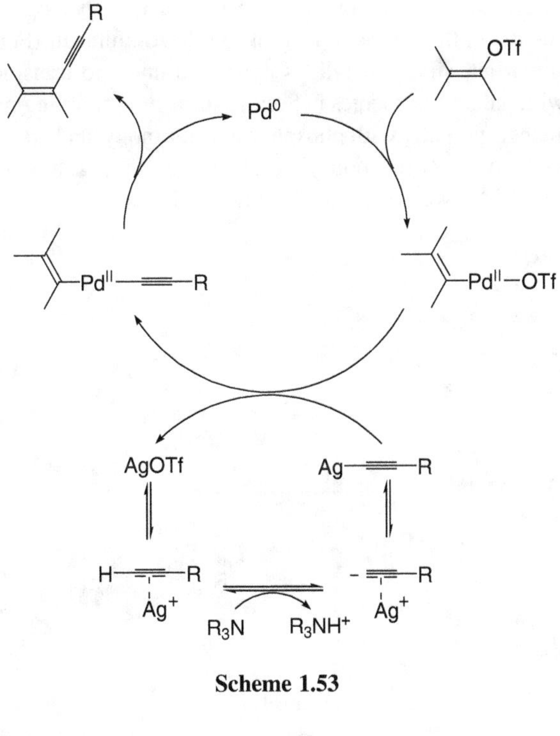

Scheme 1.53

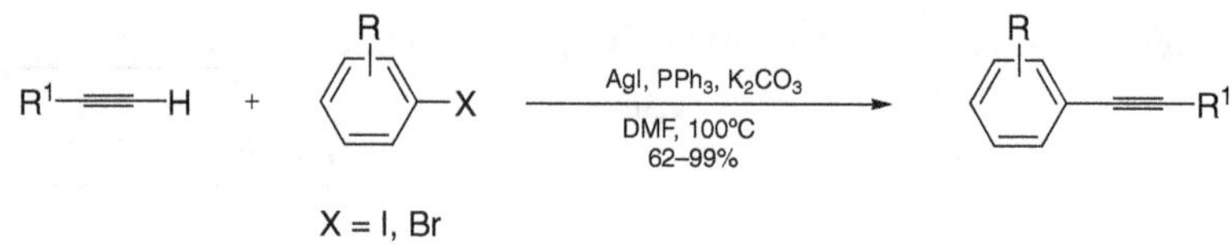

Scheme 1.54

Although catalytic amounts of Ag_2O have been found to promote the palladium-catalyzed coupling of aryl boronic acids and terminal alkynes, the authors in this case do not attribute the desired reactivity to the formation, and subsequent transmetallation, of a silver acetylide. Rather, it is proposed that the Ag_2O activates the alkynylpalladium complex to allow transmetallation from the boronic acid, and that any competing formation of the silver acetylide results in a homocoupling of the alkynes (Scheme 1.55).[123]

Silver acetylides undergo oxidative condensation in the presence of cupric chloride to give diynes.[124] Dimerization is a reaction common to a number of organosilver compounds, including both alkyl- and arylsilver derivatives.

1.4.2.5 Reactions with Non-carbon Electrophiles

Silver acetylides have also been used to form bonds with non-carbon centers to produce synthetically useful compounds. Sladkov and coworkers report the reaction of silver acetylides to arenediazonium chlorides to give areneazoethynes.[125,126] A necessary condition for the reaction was the presence of *para* or *ortho* electron acceptor substituents on the aromatic ring of the diazonium salt (Scheme 1.56).

As discussed, earlier, silver acetylides appear to undergo transmetallation with Cp_2ZrCl_2, and with palladium species in Sonogashira-type coupling reactions. Silver acetylides also react efficiently with phosphorous, antimony, and arsenic chlorides to give the ethynyl derivatives in good yield (Scheme 1.57). Ethynyl derivatives of mercury[127] and tin[128,129] were similarly synthesized.

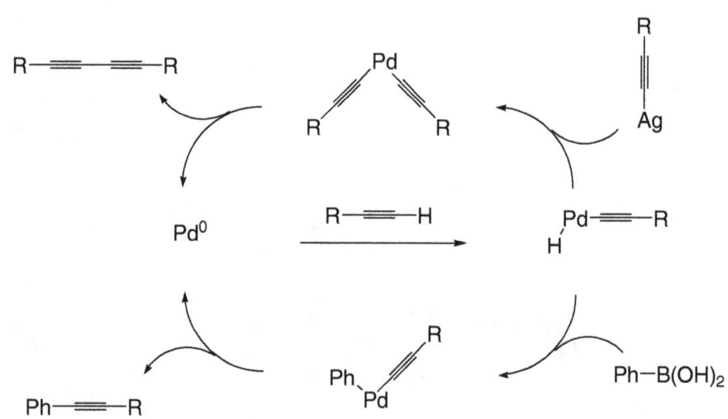

Scheme 1.55

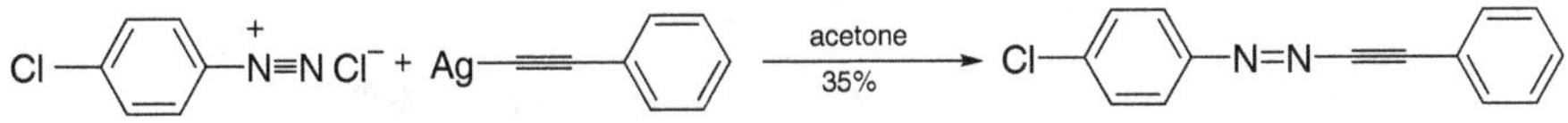

Scheme 1.56

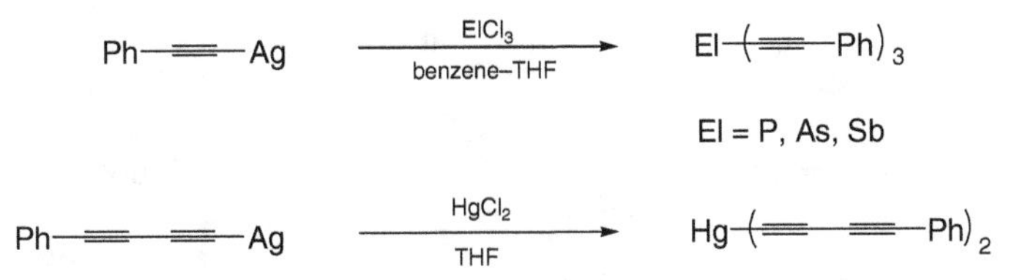

Scheme 1.57

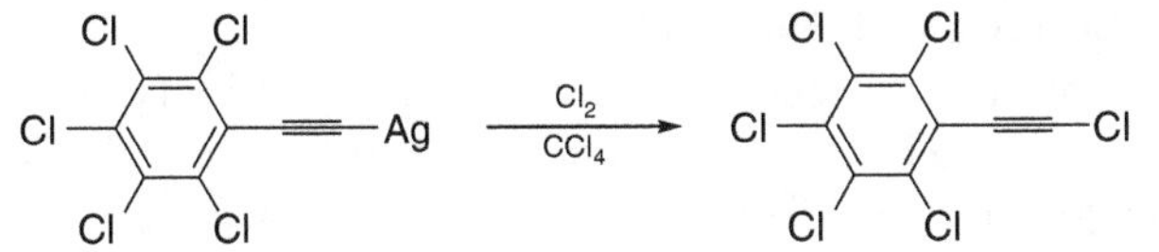

Scheme 1.58

Iodine, bromine, and chlorine react with silver acetylides to give haloalkynes. An early report of such a reaction was applied to the first synthesis of perchloropheny-lacetylene (Scheme 1.58).[130,131]

The silver acetylides of 2-hydroxy-2-ethynyladamantane and 4-*trans*-ethynyl-*cis*-2,6,6-trimethylbicyclo[3.1.1]heptan-4-ol react with iodine in chloroform to give the corresponding iodoalkynes under mild conditions in good yield.[132] Similarly, the bromoalkynes of various *n*-alkyl acetylenes have been synthesized through treatment of the corresponding silver acetylides with bromine.[133] Iodoform has been utilized in one example as an iodinating agent, whereby iodoform and silver phenylacetylide in refluxing dimethoxymethane gave 1-iodo-2-phenylacetylene in 70% yield.[127]

Alkynes bearing peroxyl functionality have also been facilely halogenated under these conditions (Scheme 1.59).[97]

The halogenation of alkynes via silver acetylides can be performed catalytically, utilizing a silver(I) salt and an electrophilic halogen source, such as *N*-bromosucci-nimide. Hofmeister and coworkers first developed this highly useful methodology in the synthesis of 17α-haloethynyl steroids. On treatment of the alkynes with catalytic silver nitrate and either *N*-bromosuccinimide or *N*-iodosuccinimide in acetone at room temperature, the corresponding halo derivatives were obtained in high yield,

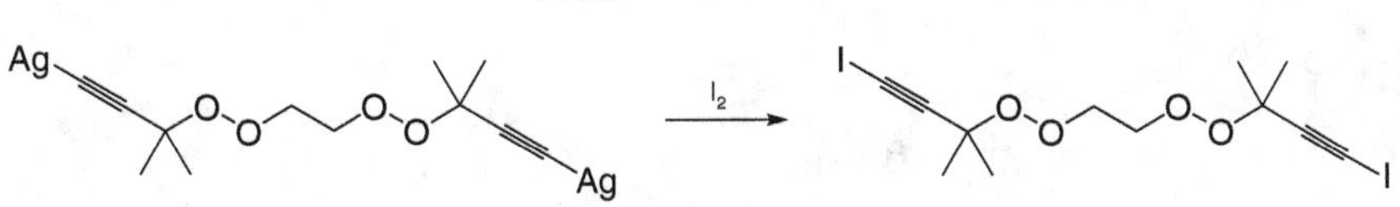

Scheme 1.59

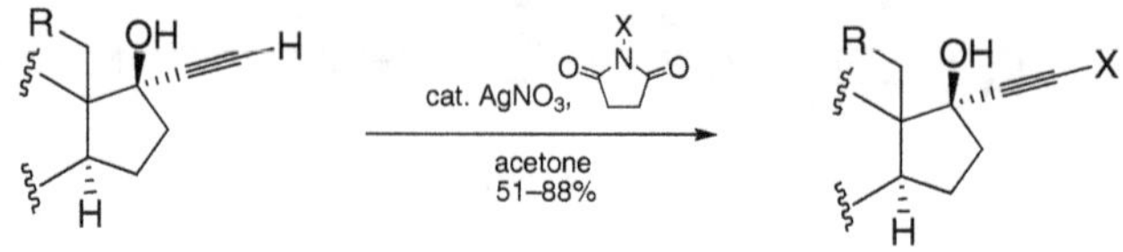

Scheme 1.60

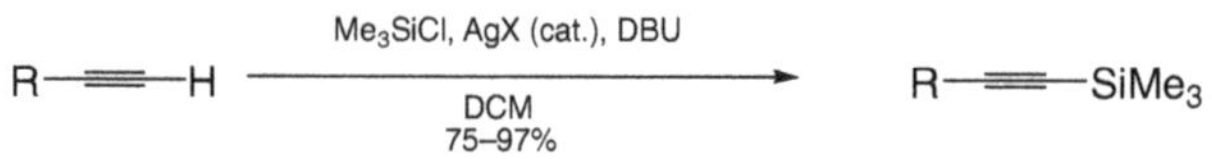

Scheme 1.61

although use of *N*-chlorosuccinimide failed to give the chloroalkyne (Scheme 1.60). Thus, this mild procedure obviated the need for strongly alkaline conditions and protection of functionality in the molecule.

The catalytic reaction has found general use in the synthesis of natural products and complex molecules, and has been reviewed exhaustively prior to this publication.[134]

In an analogous reaction, 1-trimethylsilyl-1-alkynes have been synthesized in good to excellent yield on treatment with trimethylsilyl chloride, catalytic amounts of silver salts, and DBU in refluxing dichloromethane (Scheme 1.61).[135]

1.4.2.6 *Fragmentation*

α-Acetylenic carbinols may undergo deethynylation to give ketones when treated with silver carbonate or silver oxide in DMSO. The reaction is promoted by intermolecular coordination between the silver of one silver acetylide molecule and the triple bond of another (Scheme 1.62).[136]

1.4.2.7 *Desilylation*

Perhaps the most widely reported use of silver acetylides is in the deprotection of 1-trimethylsilyl-1-alkynes. Schmidt and Arens first reported the treatment of 1-trimethylsilyl-1-alkynes with silver nitrate to give the silver acetylide in situ, which, when treated with potassium cyanide, is converted to the free alkyne (Scheme 1.63).[137] This very mild and selective, but toxic, means for the selective

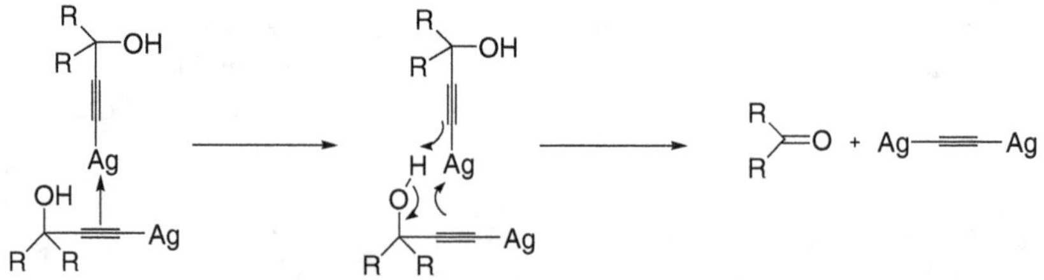

Scheme 1.62

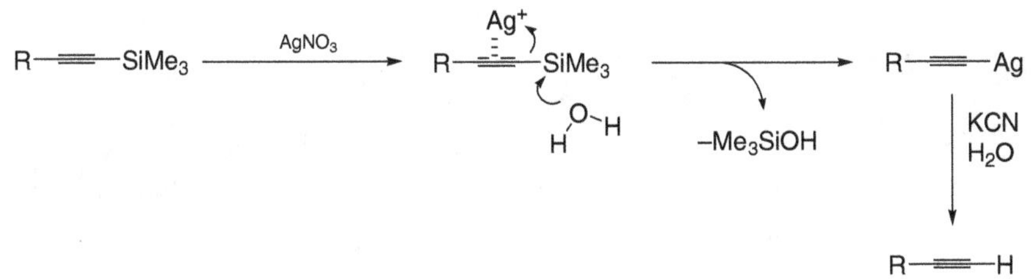

Scheme 1.63

desilylation of alkynes thus provides an alternative to strongly basic conditions and has been utilized in a number of natural product total syntheses.

In an important extension to this methodology, published in 2005, protodesilylation was effected with catalytic quantities of silver(I) salt in either acetone with a large excess of water,[138] or in a mixture of dichloromethane, methanol, and water.[90,139] Selectivity for deprotection of 1-trimethylsilyl-1-alkynes was very high in both cases, with the exception of substrates bearing silyl protected alcohols, which, under prolonged exposure to reaction conditions, gave varying percentages of free alcohol. This fall in selectivity is attributed to the formation of nitric acid during the catalytic cycle, and was remedied in one example by the addition of pyridine to the reaction mixture.

The usefulness of this specific and mild means of desilylation has been advantageously extended to include one-pot reaction of the formed silver acetylide. Catalytic desilylation/halogenation of 1-trimethylsilyl-1-alkynes has been utilized in a variety of syntheses, and, as for the halogenation of unprotected alkynes, is generally performed using a silver(I) salt in the presence of an electrophilic halogen source in acetone. A particularly elegant example of this methodology can be found in the hapten synthesis of Forsyth and coworkers, whereby the 1-trimethylsilyl-1-alkynes was cleanly converted in 95% yield to the iodoalkyne in the presence of other silyl protecting groups on the highly functionalized spiro-*N,O*-acetal core on treatment with $AgOCOCF_3$ and *N*-iodosuccinimide (Scheme 1.64).[140]

The versatility of silver-catalyzed desilylation has been further extended by its use in a one-pot catalytic desilylation/Sonogashira-type coupling reaction. In the total

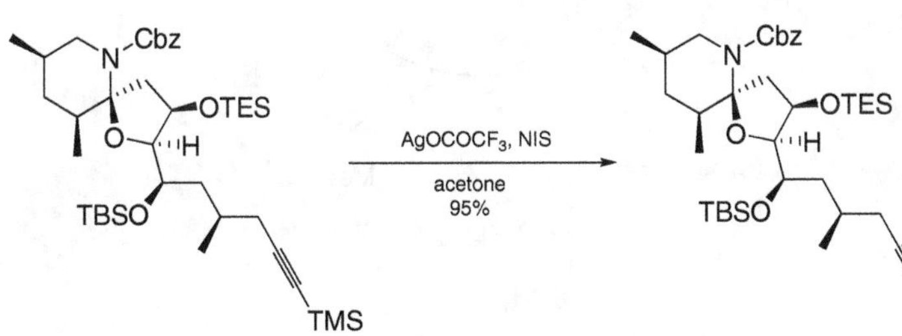

Scheme 1.64

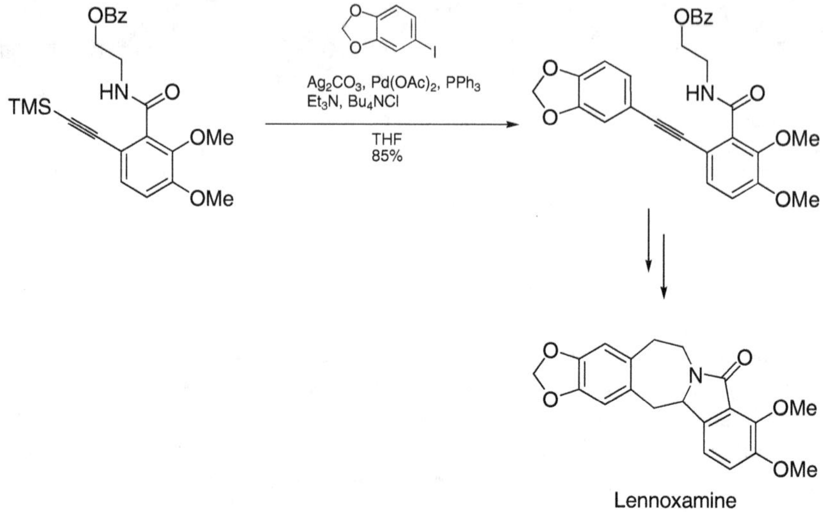

Scheme 1.65

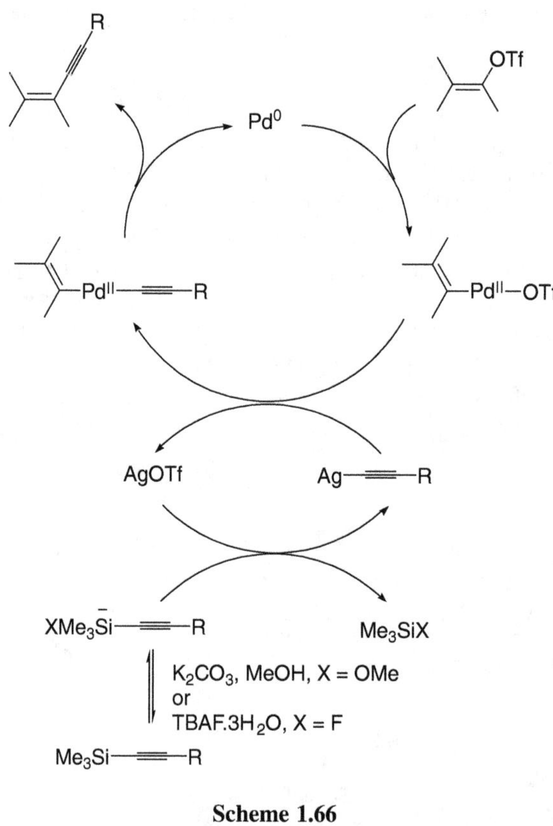

Scheme 1.66

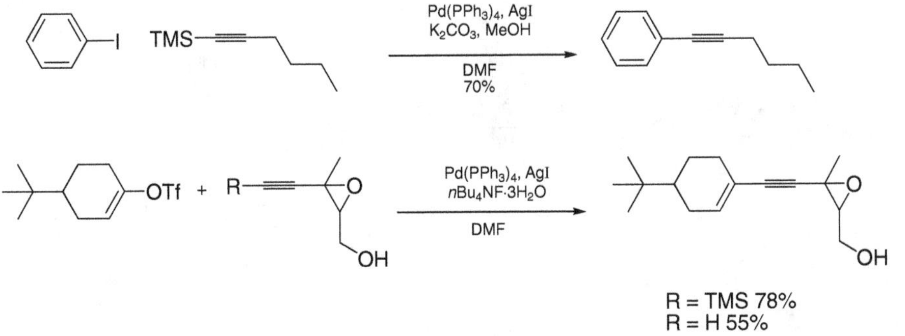

Scheme 1.67

synthesis of lennoxamine, Koseki and Nagasaka obtained a significantly improved yield of coupled product on treatment of the trimethylsilyl-protected alkyne with Ag_2CO_3, $Pd(OAc)_2$, PPh_3, Bu_4NCl, and triethylamine, in comparison to that obtained for standard Sonogashira conditions for the free alkyne (Scheme 1.65).[141] The reaction is, in this case, stoichiometric in silver, and although a mechanism is not given, the authors suggest the intermediacy of a silver acetylide. Development of the scope of this reaction showed that a variety of 1-trimethylsilyl-1-alkynes and aryl iodides were coupled under these conditions in good to excellent yield.[142]

Pale and coworkers provided the first example of combined desilylation/coupling catalytic for silver. They found that 1-trimethylsilyl-1-alkynes in the presence of tetrakis(triphenylphosphine)palladium, a silver(I) salt, and an activator (potassium carbonate in methanol, or TBAF·$3H_2O$) in DMF coupled with vinyl triflates and aryl iodides to give enynes good yields (Scheme 1.66).[143,144] Although silver(I) salt was not necessary for reaction when TBAF·$3H_2O$ was used for activation of the carbon–silicon bond, a small to significant improvement was observed for all reported

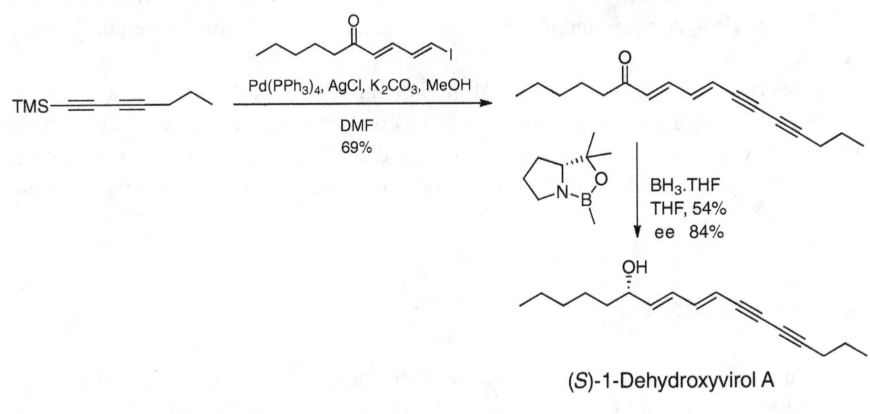

Scheme 1.68

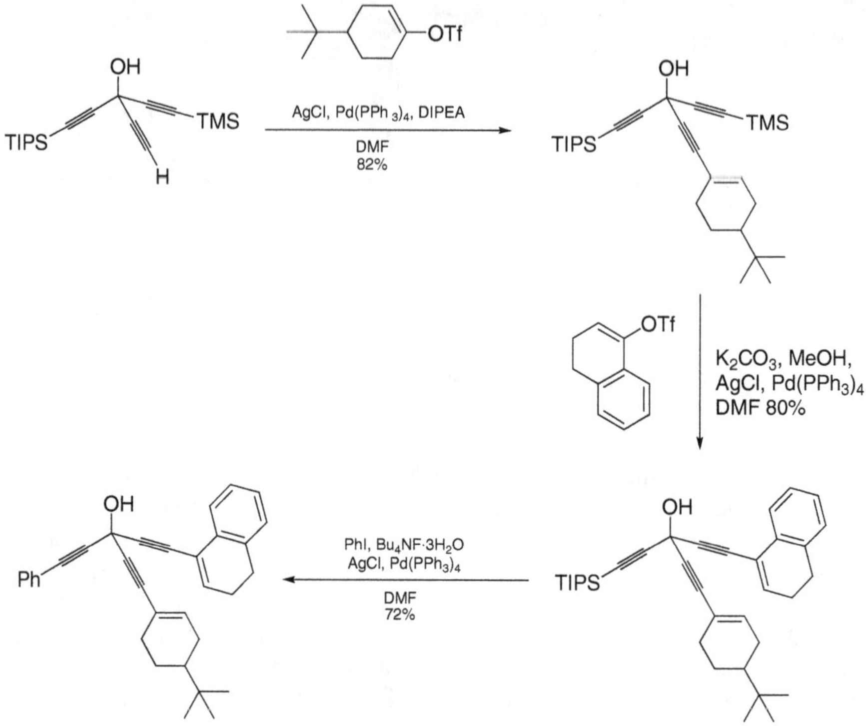

Scheme 1.69

cases on addition of silver(I) salt. Coupling was not observed in the absence of silver (I) salt when potassium carbonate in methanol was used.

The advantages of this methodology are particularly evident in the significantly higher yields obtained for sensitive substrates in comparison to that obtained for coupling of the free alkyne (Scheme 1.67).

The reaction conditions have been applied successfully to the total synthesis of (*S*)-1-dehydroxyvirol A by Fiandanese et al. (Scheme 1.68, where ee = enantiomeric excess).[145]

A powerful example of the utility of the Pale group methodology can be found in their selective coupling of molecules bearing a number of reactive alkynes. By careful selection of conditions, they selectively coupled each alkyne of a trialkyne, bearing one TMS group and one TIPS group, to vinyl triflates or aryl iodide (Scheme 1.69).[146,147]

1.5 CONCLUSION

In conclusion, it is clear that much scope remains for the development of methodology that exploits the unique chemistry of organosilver compounds. From the methods reviewed, it is evident that it is important to temper the reactivity of the organosilver compounds so as to find a balance between inertness and instability. There is no doubt that future success in this field may be found through the same balancing act.

REFERENCES

1. Silver-organische verbindungen, in *Gmelin Handbuch, Der Anoroganischer Chemie* Auflage 8, Teil B5, system nummer 61, RJ Meyer, EHE Pietsch, A Kotowski, M Beck-Goechring, K.-C Buschbeck, R. Keim. Ed: Springer-verlag, Berlin, 1975
2. Van Koten, G.; Noltes, J. G., Copper and silver, in *Comprehensive Organometallic Chemistry*, Vol. 2, Wilkinson, G.; Stone, R. G. A.; Abel, E. W., eds., Pergamon Press, Oxford, **1982**, pp. 709–763.
3. Schmidbaur, H.; Bayler, A.; Synthesis and uses of organosilver compounds, in *The Chemistry of Organic Derivatives of Gold and Silver*, Patai, S.; Rappoport, Z., eds., Wiley, Chichester, UK, **1999**, pp. 211–226.
4. Semerano, G.; Riccoboni, L., *Chem. Ber.* **1941**, *74*, 1089–1099.
5. Glockling, F.; Kingston, D., *J. Chem. Soc.* **1959**, 3001–3004.
6. Gilman, H.; Woods, L., *J. Am. Chem. Soc.* **1943**, *65*, 435–437.
7. Whitesides, G. M.; Bergbreiter, D. E.; Kendall, P. E., *J. Am. Chem. Soc.* **1974**, *96*, 2806–2813.
8. Whitesides, G. M.; Gutowski, F. D., *J. Org. Chem.* **1976**, *41*, 2882–2885.
9. Whitesides, G. M.; Panek, E. J.; Stedronsky, E. R., *J. Am. Chem. Soc.* **1972**, *94*, 232–239.
10. O'Hair, R. A. J., *Chem. Commun.* **2001**, *2002*, 20–21.
11. Joseph, L.; Gardner, J. H., *J. Org. Chem.* **1940**, *5*, 61–67.
12. Gardner, J. H.; Joseph, L.; Gollub, F., *J. Am. Chem. Soc.* **1937**, *59*, 2583–2584.
13. Gardner, J. H.; Borgstrom, P., *J. Am. Chem. Soc.* **1929**, *51*, 3375–3377.
14. Westmijze, H.; Kleijn, H.; Vermeer, P., *J. Organomet. Chem.* **1979**, *172*, 377–383.
15. Kleijn, H.; Westmijze, H.; Meijer, J.; Vermeer, P., *J. Organomet. Chem.* **1980**, *192*, 275–281.
16. Meijer, J.; Ruitenberg, K.; Westmijze, H.; Vermeer, P., *Synthesis* **1981**, 551–554.
17. Kleijn, H.; Westmijze, H.; Meijer, J.; Vermeer, P., *J. Organomet. Chem.* **1981**, *206*, 257–264.
18. Westmijze, H.; Kleijn, H.; Bos, H. J. T.; Vermeer, P., *J. Organomet. Chem.* **1980**, *199*, 293–297.
19. Tamura, M.; Kochi, J., *Synthesis* **1971**, 303–305.
20. Yamamoto, K.; Nakanishi, K.; Kumada, M., *J. Organomet. Chem.* **1967**, *7*, 197–202.
21. Nagano, T.; Hayashi, T., *Chem. Lett.* **2005**, *34*, 1152–1153.
22. Murphy, R.; Prager, R. H., *Aust. J. Chem.* **1976**, *29*, 617–626.
23. Murphy, R.; Prager, R. H., *Tetrahedron Lett.* **1976**, 463–464.
24. Brown, H. C.; Verbrugge, C.; Snyder, C. H., *J. Am. Chem. Soc.* **1961**, *83*, 1001.
25. Brown, H. C.; Snyder, C. H., *J. Am. Chem. Soc.* **1961**, *83*, 1002–1003.
26. Brown, H. C.; Hebert, N. C.; Snyder, C. H., *J. Am. Chem. Soc.* **1961**, *83*, 1001–1002.

27. Avasthi, K.; Ghosh, S. S.; Devaprabhakara, D., *Tetrahedron Lett.* **1976**, *17*, 4871–4874.
28. Chujo, Y.; Tomita, I.; Kozawa, Y.; Saegusa, T., *Macromolecules* **1993**, *26*, 2643–2644.
29. Kauffmann, T.; Huelsduenker, A.; Menges, D.; Nienaber, H.; Rethmeier, L.; Robbe, S.; Scherler, D.; Schricke, J.; Wingbermuehle, D., *Tetrahedron Lett.* **1990**, *31*, 1553–1556.
30. Wang, F. Q.; Khairallah, G. N.; Koutsantonis, G. A.; Williams, C. M.; Callahan, D. L.; O'Hair, R. A. J., *Phys. Chem. Chem. Phys.* **2009**, *11*, 4132–4135.
31. Luk'yanov, O. A.; Savost'yanova, I. A.; Gorelik, V. P.; Shlykova, N. I.; Tartakovskii, V. A., *Izvest. Akad. Nauk Ser. Khim.* **1992**, 1798–1803.
32. Hantzch, A.; Caldwell, K. S., *Chem. Ber.* **1906**, 2472–2478.
33. Tyrra, W., *Heteroat. Chem.* **2002**, *13*, 561–566.
34. Miller, W. T., Jr.; Burnard, R. J., *J. Am. Chem. Soc.* **1968**, *90*, 7367–7368.
35. Probst, A.; Raab, K.; Ulm, K.; Von Werner, K., *J. Fluorine Chem.* **1987**, *37*, 223–245.
36. Dyatkin, B. L.; Martynov, B. I.; Martynova, L. G.; Kizim, N. G.; Sterlin, S. R.; Stumbreviciute, Z.; Fedorov, L. A., *J. Organomet. Chem.* **1973**, *57*, 423–433.
37. Naumann, D.; Wessel, W.; Hahn, J.; Tyrra, W., *J. Organomet. Chem.* **1997**, *547*, 79–88.
38. Polishchuk, V. R.; Fedorov, L. A.; Okulevich, P. O.; German, L. S.; Knunyants, I. L., *Tetrahedron Lett.* **1970**, 3933–3936.
39. Tyrra, W., *J. Fluorine Chem.* **2001**, *112*, 149–152.
40. Tyrra, W.; Naumann, D., *J. Fluorine Chem.* **2004**, *125*, 823–830.
41. Dubot, G.; Mansuy, D.; Lecolier, S.; Normant, J. F., *J. Organomet. Chem.* **1972**, *42*, C105–C106.
42. Banks, R. E.; Dickinson, N.; Morrissey, A. P.; Richards, A., *J. Fluorine Chem.* **1984**, *26*, 87–92.
43. Rossman, D. I.; Muller, A. J.; Lewis, E. O., *J. Fluorine Chem.* **1991**, *55*, 221–224.
44. Krause, E.; Schmitz, M., *Chem. Ber.* **1919**, *52*, 2150–2164.
45. Krause, E.; Wendt, B., *Chem. Ber.* **1923**, *56*, 2064–2066.
46. Gilman, H.; Straley, J. M., *Recl. Trav. Chim. Pays-Bas* **1936**, *55*, 821–834.
47. Reich, R., *C. R. Chim.* **1923**, *177*, 322–324.
48. Wennerstrom, O., *Acta Chem. Scand.* **1971**, *25*, 2341–2349.
49. Miller, W. T., Jr.; Sun, K. K., *J. Am. Chem. Soc.* **1970**, *92*, 6985–6987.
50. Leusink, A. J.; Van Koten, G.; Noltes, J. G., *J. Organomet. Chem.* **1973**, *56*, 379–390.
51. Beverwijk, C. D. M.; Van der Kerk, G. J. M., *J. Organomet. Chem.* **1972**, *43*, C11–C12.
52. Boersma, J.; Des Tombe, F. J. A.; Weijers, F.; Van der Kerk, G. J. M., *J. Organomet. Chem.* **1977**, *124*, 229–233.
53. Hofstee, H. K.; Boersma, J.; Van der Kerk, G. J. M., *J. Organomet. Chem.* **1978**, *168*, 241–249.
54. Gambarotta, S.; Floriani, C.; Chiesi-Villa, A.; Guastini, C., *J. Chem. Soc., Chem. Commun.* **1983**, *19*, 1087–1089.
55. Burton, D. J.; Yang, Z. Y.; Morken, P. A., *Tetrahedron* **1994**, *50*, 2993–3063.
56. Smith, V. B.; Massey, A. G., *J. Organomet. Chem.* **1970**, *23*, C9–C10.
57. Uson, R.; Laguna, A.; Abad, J. A., *J. Organomet. Chem.* **1983**, *246*, 341–345.
58. Tyrra, W.; Wickleder, M., *Z. Anorg. Allg. Chem.* **2002**, *628*, 1841–1847.

59. Kremlev, M. M.; Tyrra, W.; Naumann, D.; Yagupolskii, Y. L., *J. Fluorine Chem.* **2005**, *126*, 1327–1331.
60. Tyrra, W.; Aboulkacem, S.; Hoge, B.; Wiebe, W.; Pantenburg, I., *J. Fluorine Chem.* **2006**, *127*, 213–217.
61. Glockling, F., *J. Chem. Soc.* **1955**, 716–720.
62. Glockling, F., *J. Chem. Soc.* **1956**, 3640–3642.
63. Whitesides, G. M.; Casey, C. P., *J. Am. Chem. Soc.* **1966**, *88*, 4541–4543.
64. Whitesides, G. M.; Casey, C. P.; Krieger, J. K., *J. Am. Chem. Soc.* **1971**, *93*, 1379–1389.
65. Koebrich, G.; Froehlich, H.; Drischel, W., *J. Organomet. Chem.* **1966**, *6*, 194–201.
66. Ruitenberg, K.; Kleijn, H.; Meijer, J.; Oostveen, E. A.; Vermeer, P., *J. Organomet. Chem.* **1982**, *224*, 399–405.
67. Meijer, J.; Ruitenberg, K.; Westmijze, H.; Vermeer, P., *Synthesis* **1981**, 551–553.
68. Westmijze, H.; Ruitenberg, K.; Meijer, J.; Vermeer, P., *Tetrahedron Lett.* **1980**, *21*, 1771–1772.
69. Miller, W. T., Jr.; Snider, R. H.; Hummel, R. J., *J. Am. Chem. Soc.* **1969**, *91*, 6532–6534.
70. Banks, R. E.; Haszeldine, R. N.; Taylor, D. R.; Webb, G., *Tetrahedron Lett.* **1970**, 5215–5216.
71. Morken, P. A.; Lu, H.; Nakamura, A.; Burton, D. J., *Tetrahedron Lett.* **1991**, *32*, 4271–4274.
72. Schoellkopf, U.; Rieber, N., *Angew. Chem. Int. Ed.* **1967**, *6*, 261.
73. Felcht, U.; Regitz, M., *Chem. Ber.* **1975**, *108*, 2040–2054.
74. Boehshar, M.; Heydt, H.; Regitz, M., *Tetrahedron* **1986**, *42*, 1815–1822.
75. Arenz, S.; Boehshar, M.; Regitz, M., *Chem. Ber.* **1986**, *119*, 1755–1765.
76. Bethaeuser, W.; Weber, B.; Heydt, H.; Regitz, M., *Chem. Ber.* **1985**, *118*, 1315–1328.
77. Regitz, M.; Weber, B.; Eckstein, U., *Liebigs Ann. Chem.* **1979**, 1002–1019.
78. Heydt, A.; Heydt, H.; Weber, B.; Regitz, M., *Chem. Ber.* **1982**, *115*, 2965–2980.
79. Eisenbarth, P.; Regitz, M., *Chem. Ber.* **1984**, *117*, 445–454.
80. Blues, E. T.; Bryce-Smith, D.; Irwin, J. G.; Lawston, I. W., *J. Chem. Soc. Chem. Commun.* **1974**, 466–467.
81. Halbes-Letinois, U.; Weibel, J. M.; Pale, P., *Chem. Soc. Rev.* **2007**, *36*, 759–769.
82. Sladkov, A. M.; Gol'ding, I. R., *Usp. Khim.* **1979**, *48*, 1625–1683.
83. Davis, R. B.; Scheiber, D. H., *J. Am. Chem. Soc.* **1956**, *78*, 1675–1678.
84. Royer, E. C.; Barral, M. C.; Moreno, V.; Santos, A., *J. Inorg. Nucl. Chem.* **1981**, *43*, 705–709.
85. Glaser, C., *Justus Liebigs Ann. Chem.* **1870**, *154*, 137–171.
86. Liebermann, C., *Justus Liebigs Ann Chem.* **1865**, *135*, 266–290.
87. Danehy, J. P.; Nieuwland, J. A., *J. Am. Chem. Soc.* **1936**, *58*, 1609–1610.
88. Nast, R.; Schindel, H., *Z. Anorg. Allg. Chem.* **1963**, *326*, 201–208.
89. Letinois-Halbes, U.; Pale, P.; Berger, S., *J. Org. Chem.* **2005**, *70*, 9185–9190.
90. Viterisi, A.; Orsini, A.; Weibel, J. M.; Pale, P., *Tetrahedron Lett.* **2006**, *47*, 2779–2781.
91. Bruce, M. I.; Humphrey, M. G.; Matisons, J. G.; Roy, S. K.; Swincer, A. G., *Aust. J. Chem.* **1984**, *37*, 1955–1961.

92. Yerino, L. V.; Osborn, M. E.; Mariano, P. S., *Tetrahedron* **1982**, *38*, 1579–1591.
93. Naka, T.; Koide, K., *Tetrahedron Lett.* **2003**, *44*, 443–445.
94. Fukue, Y.; Oi, S.; Inoue, Y., *J. Chem. Soc. Chem. Commun.* **1994**, 2091.
95. Tsuda, T.; Ueda, K.; Saegusa, T., *J. Chem. Soc. Chem. Commun.* **1974**, 380–381.
96. Inanaga, J.; Katsuki, T.; Takimoto, S.; Ouchida, S.; Inoue, K.; Nakano, A.; Okukado, N.; Yamaguchi, M., *Chem. Lett.* **1979**, *8*, 1021–1024.
97. Yuvchenko, A.; Moiseichuk, K. L.; Dikusar, E. A.; Zhukovskaya, N. A.; Ol'dekop, Y., *Zh. Obshch. Khim.* **1990**, *60*, 1587–1593.
98. Ukhin, L. Y.; Suponitskii, K. Y.; Kartsev, V. G., *Chem. Nat. Compd.* **2003**, *39*, 482–488.
99. Ukhin, L. Y.; Gol'ding, I. R.; Kartsev, V. G., *Chem. Nat. Compd.* **2004**, *40*, 156–159.
100. Ukhin, L. Y.; Komissarov, N.; Orlova, Z. I.; Tokarskaya, O. A.; Yanovskii, A. I.; Struchkov, Y. T., *Zh. Org. Khim.* **1987**, *23*, 1323–1325.
101. Ji, J. -X.; Au-Yeung, T. T. -L.; Wu, J.; Yip, C. W.; Chan, A. S. C., *Adv. Synth. Catal.* **2004**, *346*, 42–44.
102. Rueping, M.; Antonchick, A.; Brinkmann, C., *Angew. Chem. Int. Ed.* **2007**, *46*, 6903–6906.
103. Wei, C.; Li, Z.; Li, C. -J., *Org. Lett.* **2003**, *5*, 4473–4475.
104. Yao, X.; Li, C. -J., *Org. Lett.* **2005**, *7*, 4395–4398.
105. Li, Z.; Wei, C.; Chen, L.; Varma, R. S.; Li, C. -J., *Tetrahedron Lett.* **2004**, *45*, 2443–2446.
106. Koide, K.; Shahi, S. P., *Angew. Chem. Int. Ed.* **2004**, *43*, 2525–2527.
107. Albert, B. J.; Sivaramakrishnan, A.; Naka, T.; Czaicki, N. L.; Koide, K., *J. Am. Chem. Soc.* **2007**, *129*, 2648–2659.
108. Albert, B. J.; Koide, K., *J. Org. Chem.* **2008**, *73*, 1093–1098.
109. Nishiwaki, N.; Minakata, S.; Komatsu, M.; Ohshiro, Y., *Chem. Lett.* **1989**, *18*, 773–776.
110. Agawa, T.; Miller, S. I., *J. Am. Chem. Soc.* **1961**, *83*, 449–453.
111. Ukhin, L. Y.; Komissarov, N.; Orlova, Z. I.; Dolgopolova, N. A., *Zh. Obshch. Khim.* **1984**, *54*, 1676–1678.
112. De las Heras, F. G.; Tam, S. Y. K.; Klein, R. S.; Fox, J. J., *J. Org. Chem.* **1976**, *41*, 84–90.
113. Albrecht, H. P.; Repke, D. B.; Moffatt, J. G., *J. Org. Chem.* **1974**, *39*, 2176–2182.
114. Pouwer, R. H.; Williams, C. M.; Raine, A. L.; Harper, J. B., *Org. Lett.* **2005**, *7*, 1323–1325.
115. Pouwer, R. H.; Harper, J. B.; Vyakaranam, K.; Michl, J.; Williams, C. M.; Jessen, C. H.; Bernhardt, P. V., *Eur. J. Org. Chem.* **2007**, 241–248.
116. Isabelle, M. E.; Leitch, L. C., *Can. J. Chem.* **1958**, *36*, 440–448.
117. Ruitenberg, K.; Kleijn, H.; Westmijze, H.; Meijer, J.; Vermeer, P., *Recl. Trav. Chim. Pays-Bas* **1982**, *101*, 405–409.
118. Bertus, P.; Pale, P., *Tetrahedron Lett.* **1997**, *38*, 8193–8196.
119. Bertus, P.; Pale, P., *Tetrahedron Lett.* **1996**, *37*, 2019–2022.
120. Bertus, P.; Pale, P., *J. Organomet. Chem.* **1998**, *567*, 173–180.
121. Dillinger, S.; Bertus, P.; Pale, P., *Org. Lett.* **2001**, *3*, 1661–1664.
122. Li, P.; Wang, L., *Synlett* **2006**, *14*, 2261–2265.
123. Zou, G.; Zhu, J.; Tang, J., *Tetrahedron Lett.* **2003**, *44*, 8709–8711.
124. Sladkov, A. M.; Gol'ding, I. R., *Dokl. Akad. Nauk SSSR* **1971**, *200*, 132–133.

125. Sladkov, A. M.; Ukhin, L. Y., *Izvest. Akad. Nauk SSSR Ser. Khim.* **1964**, 1552–1553.
126. Sladkov, A. M.; Ukhin, L. Y.; Gorshkova, G. N., *Zh. Org. Khim.* **1966**, *2*, 1456–1459.
127. Gol'ding, I. R.; Sladkov, A. M., *Izvest. Akad. Nauk SSSR Ser. Khim.* **1972** (3), 529–530.
128. Le Quan, M.; Cadiot, P., *Bull. Soc. Chim. Fr.* **1965**, 35–44.
129. Johnson, O. H.; Holum, J. R., *J. Org. Chem.* **1958**, *23*, 738–740.
130. Ballester, M.; Castaner, J.; Riera, R.; Tabernero, I., *J. Org. Chem.* **1986**, *51*, 1413–1419.
131. Ballester, M.; Castaner, J.; Riera, R.; Tabernero, I.; Cornet, C., *Tetrahedron Lett.* **1977**, *27*, 2353–2354.
132. Dikusar, E. A.; Kozlov, N. G.; Koval'skaya, S. S.; Popova, L. A.; Moiseichuk, K. L., *Russ. J. Gen. Chem.* **2001**, *71*, 290–293.
133. Dikusar, E. A.; Zvereva, T. D.; Zhukovskaya, N. A.; Moiseichuk, K. L., *Russ. J. Gen. Chem.* **2001**, *71*, 917–920.
134. Yamamoto, Y., *Chem. Rev.* **2008**, *108*, 3199–3222.
135. Taniguchi, Y.; Inanaga, J.; Yamaguchi, M., *B. Chem. Soc. Jpn.* **1981**, *54*, 3229–3230.
136. Vitali, R.; Gladiali, S.; Gardi, R., *Gazz. Chim. Ital.* **1972**, *102*, 673–678.
137. Schmidt, H. M.; Arens, J. F., *Rec. Trav. Chim. Pay-B.* **1967**, *86*, 1138–1142.
138. Carpita, A.; Mannocci, L.; Rossi, R., *Eur. J. Org. Chem.* **2005**, 1859–1864.
139. Orsini, A.; Viterisi, A.; Bodlenner, A.; Weibel, J. M.; Pale, P., *Tetrahedron Lett.* **2005**, *46*, 2259–2262.
140. Forsyth, C. J.; Xu, J.; Nguyen, S. T.; Samdal, I. A.; Briggs, L. R.; Rundberget, T.; Sandvik, M.; Miles, C. O., *J. Am. Chem. Soc.* **2006**, *128*, 15114–15116.
141. Koseki, Y.; Nagasaka, T., *Chem. Pharm. Bull.* **1995**, *43*, 1604–1606.
142. Koseki, Y.; Omino, K.; Anzai, S.; Nagasaka, T., *Tetrahedron Lett.* **2000**, *41*, 2377–2380.
143. Halbes, U.; Pale, P., *Tetrahedron Lett.* **2002**, *43*, 2039–2042.
144. Bertus, P.; Halbes, U.; Pale, P., *Eur. J. Org. Chem.* **2001**, 4391–4393.
145. Fiandanese, V.; Bottalico, d.; Cardellicchio, C.; Marchese, G.; Punzi, A., *Tetrahedron* **2005**, *61*, 4551–4556.
146. Halbes-Letinois, U.; Vasiliev, A.; Pale, P., *Eur. J. Org. Chem.* **2005**, 2828–2834.
147. Halbes-Letinois, U.; Pale, P., *J. Organomet. Chem.* *687*, **2003**, 420–424.

2

CYCLOADDITION REACTIONS

ALEX M. SZPILMAN
Schulich Faculty of Chemistry, Technion—Israel Institute of Technology, Haifa, Israel

ERICK M. CARREIRA
Laboratory of Organic Chemistry, ETH Zürich, Switzerland

Silver in Organic Chemistry Edited by Michael Harmata

2.1 INTRODUCTION

In the 2001 book *Cycloaddition Reactions in Organic Synthesis*,[1] the use of silver salts as mediators or catalysts was conspicuous only by its absence. The catalytic effects of silver were briefly mentioned in the chapter on [3 + 2] cycloadditions. It is refreshing that in 7 years the field has progressed so far as to warrant a number of reviews, each covering various aspects of silver-mediated synthesis.[2–5] This chapter seeks to afford the reader with a comprehensive survey of the field from its inception to the end of 2008.

2.2 [2+2] CYCLOADDITIONS

In one of the first contributions to the field, Friedman reported on the ability of silver to influence the reactivity of benzyne toward benzene (Scheme 2.1). In the absence of silver, benzyne **2**, generated in situ from **1**, reacted with benzene in a [4+2] manner to afford adduct **3** in addition to small amounts of dimerization product **4**.[6] However, in the presence of substochiometric amounts of silver ions, the course of the reaction was altered, leading to formation of biphenyl **5** and benzocycloctatriene **6**. A contingent of other metal ions including Tl^+, Cu^+, Cu^{2+}, Hg_2^{2+}, and Hg^{2+} were unable to influence product formation in a similar manner.

On the basis of these observations, Friedman[6] and Paquette[7] proposed the mechanism shown in Scheme 2.2. Thus, silver was suggested to activate benzyne for electrophilic attack to afford silver intermediate **8**. Intramolecular proton transfer would account for the formation of **5**, while cyclization would result in formation of **9** with subsequent $6e^-$ disrotary electrocyclic ring opening to afford **6**.

Masamune reported that reaction of **10** with $AgClO_4$ in the presence of 3 equiv of maleic anhydride resulted in CO evolution and formation of cycloadduct **11** in 85%

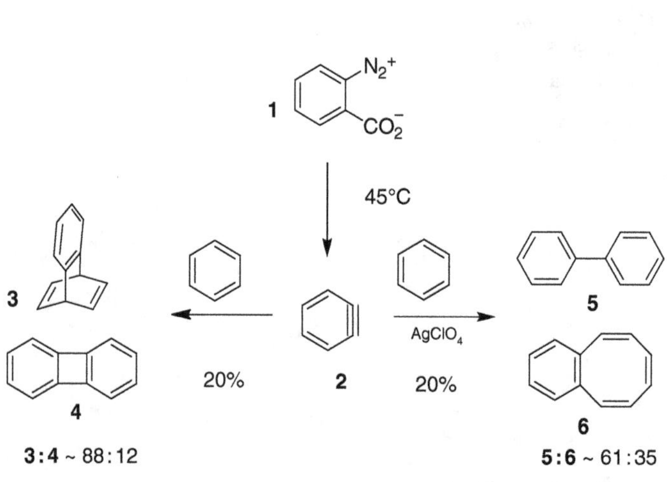

Scheme 2.1

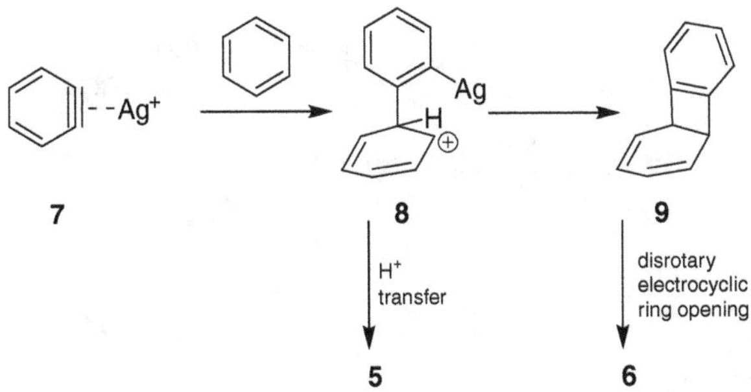

Scheme 2.2

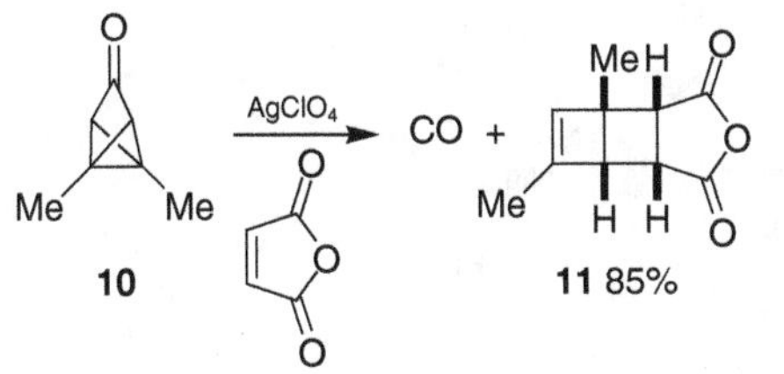

Scheme 2.3

yield (Scheme 2.3).[8] While no mechanistic studies were performed, it was postulated that ring opening takes place by an ionic mechanism with possible formation of 1,3-dimethylcyclobutadiene. In support of this hypothesis, in situ generation of free 1,3-dimethylcyclobutadiene, from its stable tricarbonyliron complex, in the presence of maleic anhydride leads to the formation of **11** in high yield.

Kozmin reported that $AgNTf_2$ is an efficient catalyst for [2+2] cycloaddition of TIPS alkynyl ethers **12** with electrophilic alkenes (Scheme 2.4).[9] The finding arose serendipitously from an attempt to generate $TIPSNTf_2$ in situ. Notably, $AgNTf_2$ was superior to a broad selection of standard Lewis acid (e.g., $BF_3{\cdot}OEt_2$ and $TiCl_4$) and other silver salts (e.g., AgOTf). The reaction is of a fairly wide scope with respect to the alkene **13**. A range of alkenes with electron-withdrawing substituents are able to participate in the reaction, including acrylonitrile, methyl acrylate, and cyclohexenone. The substituent on the alkyne can be either alkyl or phenyl. Mechanistic studies involving deuterium labeling and NMR spectroscopy indicated that the mechanism is stepwise and is initiated by the complexation of $AgNTf_2$ to alkyne **12**.

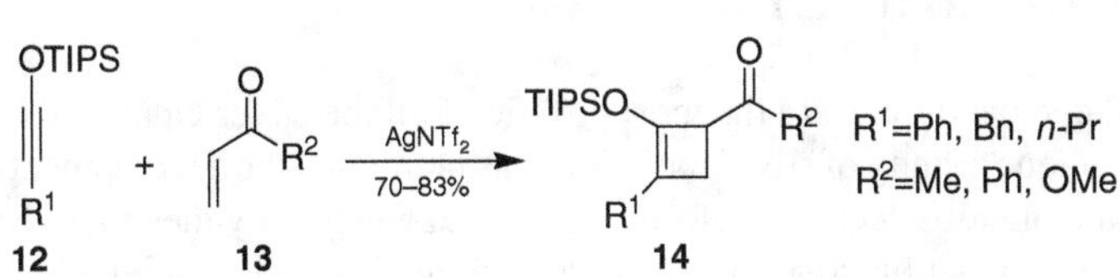

Scheme 2.4

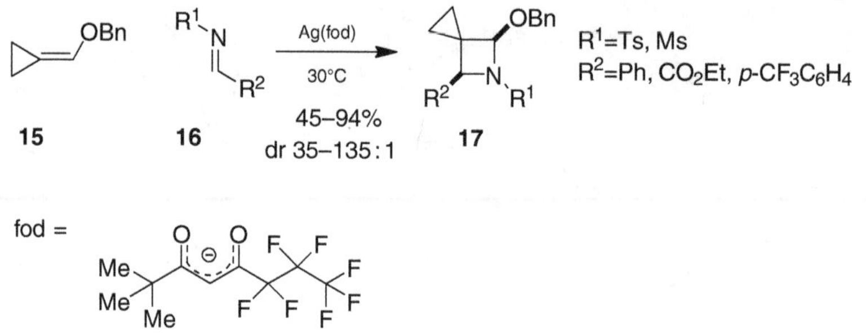

Scheme 2.5

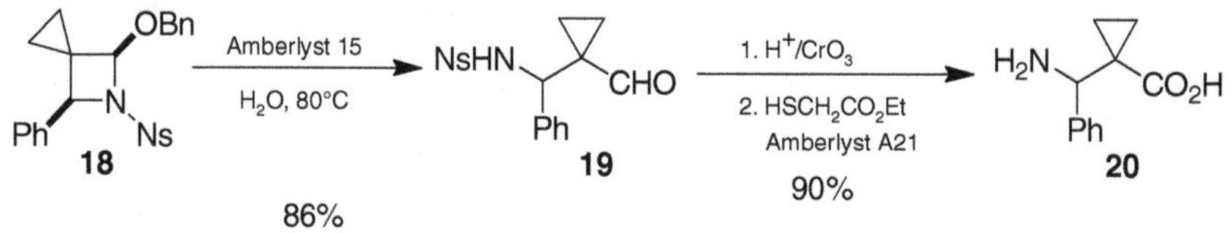

Scheme 2.6

The Nakamura group reported the [2+2] cycloaddition of alkoxymethylenecyclopropanes to imines (Scheme 2.5).[10] They identified Ag(fod) from an assortment of metal salts including gold ($AuBr_3$), palladium [$Pd(OAc)_2$], and other silver species [Ag(acac)].

Interestingly, the uncatalyzed reaction takes place at 80°C and is a rare case of a thermal [2+2] cycloaddition. However, the catalytic process offers advantages in terms of higher *cis*/*trans* ratios. It should be noted that the *trans* isomer can be isomerized into the thermodynamically favored *cis* product by extended heating to 80°C in acetonitrile. The utility of the products was demonstrated by the synthesis of amino acid **20** (Scheme 2.6).

In summary, it is clear that silver possesses unique abilities to activate both olefinic species and small strained rings for [2+2] cycloaddition reactions. Nevertheless, reports of silver-mediated [2+2] cycloadditions remain rare.

2.3 [3+2] CYCLOADDITIONS

The [3+2] cycloadditions are the most prolific of all the silver-catalyzed cycloadditions. The unique affinity of silver for imines has facilitated the development of highly efficient and enantioselective cycloadditions of azomethine ylides to alkenes. Judicious choice of reaction conditions is crucial in achieving high yields for different substitution patterns.

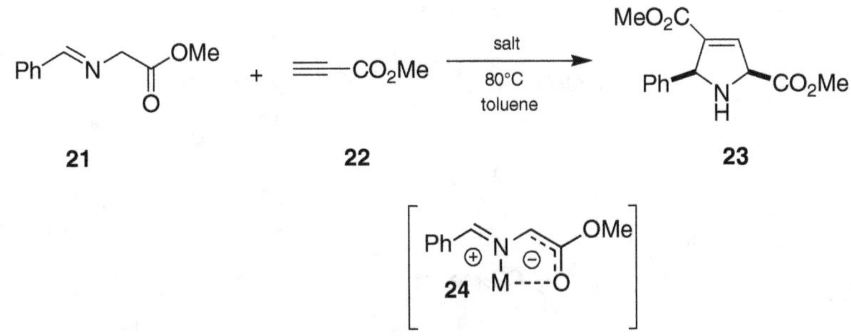

Scheme 2.7

2.3.1 [3+2] Cycloadditions of Azomethine Ylides

2.3.1.1 Discovery and Development of the Silver-Catalyzed [3+2] Cycloaddition of Azomethine Ylides

The first example of the title process was reported by the Grigg group in 1982.[11] This work examined the cycloaddition of Schiff base **21**, generated from benzaldehyde and glycine, with methyl propiolate **22** in the presence of a metal salt incorporating a Lewis acidic metal and a basic counteranion. The uncatalyzed reaction occurred in d_8-toluene at 80°C to afford product **23** in 94% yield and with a starting material half-life ($t_{1/2}$) of 38 h (Scheme 2.7; Table 2.1, entry 1).[12] Silver acetate was able to catalyze the reaction, affording the product in 95% yield (entry 3). In contrast, Brønsted acid was found to be detrimental to product formation (entry 2). Zinc acetate bishydrate and lithium acetate bishydrate were also efficient catalysts for the reaction (entries 4 and 5).

Solvent effects were found to be substantial for this [3+2] cycloaddition. Thus, the half-life of the thermal reaction of **21** with maleic anhydride at 105°C was reduced to 59 min in d_3-$MeNO_2$ compared with 120 min in d_8-toluene. The authors proposed the key notion that a reactive, conformationally rigid metalated azomethine species **24** is responsible for the high diastereoselectivity observed in the reaction.

A second pivotal contribution from the Grigg group asserted the positive influence of the use of amine bases in combination with Lewis acids to catalyze the thermal

TABLE 2.1. Effects of Acid on the Cycloaddition of Iminoester (21) and Methyl Propiolate (22)

Entry	Lewis Acid[a]	$t_{1/2}$ of **21** (h)	Yield (%)[b]
1	None	38	94
2	CH_3COOH	1.8	0
3	AgOAc	3.25	95
4	$Zn(OAc)_2{\cdot}2H_2O$	3.0	88
5	$LiOAc{\cdot}2H_2O$	5.5	93
6	$Mg(OAc)_2$	8.75	0

[a] Run with 1.5 equiv.

[b] Determined by NMR with hexamethyl benzene as internal standard.

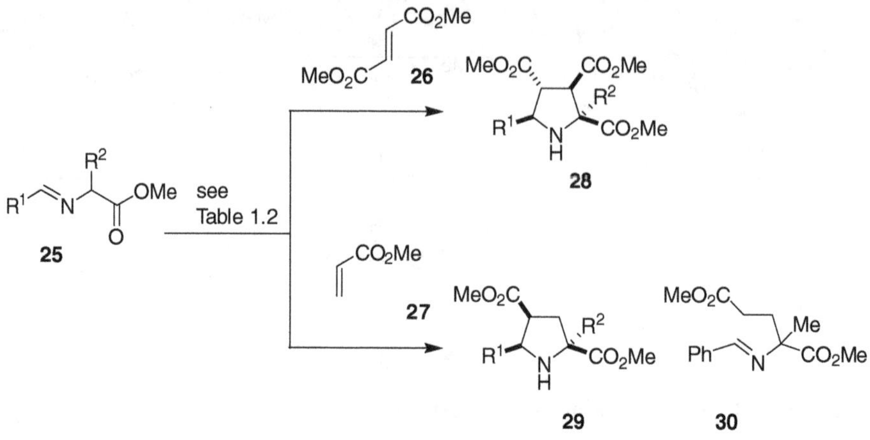

Scheme 2.8 Optimization of the dipolar cycloaddition of iminoesters. See Table 2.2 for details.

reaction.[13] An important finding was that a combination of a metal salt and an amine base would minimize stereomutation[14] of the putative dipole, and thus improve diastereoselectivity. For this type of protocol it was found that silver acetate is vastly superior to both lithium bromide and zinc bromide.

A range of solvents were investigated. This screen confirmed the earlier finding that cycloaddition rates in polar solvents are substantially higher than in nonpolar solvents. Acetonitrile was found to be the solvent of choice. Thus, the reaction of aromatic iminoesters **25** with dipolarophiles **26** and **27** took place in acetonitrile in the presence of silver acetate and triethylamine (Scheme 2.8). The reaction was completed within 0.5–2.5 h at room temperature to afford the pyrrolidine products **28** and **29**, respectively, in good yield. In contrast, lithium bromide catalysis resulted in formation of conjugate addition byproducts of type **30**, which were isolated in fair yields (35–45%). A key finding was that substochiometric amounts of silver acetate could be used, without significantly lowering the yield (Table 2.2, entry 8).

TABLE 2.2. Influence of Metal Salts and Solvent on the Reaction Shown in Scheme 2.8

Entry	R^1	R^2	Dipolarophile	Lewis Acid	Solvent	Time (h)	Yield (%)
1	Ph	Me	**27**	LiBr (1.5)	MeCN	0.5	55[b]
2	Ph	Me	**27**	AgOTs (1.5)	THF	48	81
3	Ph	Me	**27**	AgOAc (1.5)	MeCN	0.5	80
4	Ph	Me	**26**	AgOAc (1.5)	THF	36	90
5	Ph	Me	**26**	AgOAc (1.5)	MeCN	0.67	86[c]
6	2-Naphthyl	H	**27**	AgOAc (1.5)	THF	16	87
7	2-Naphthyl	H	**27**	AgOAc (1.5)	MeCN	0.5	100
8	2-Naphthyl	H	**27**	AgOAc (0.15)	MeCN	2.5	87[d]

[a] Run with 1.0 equiv of triethylamine unless otherwise noted.
[b] With 45% of **30**.
[c] 7:1 mixture of diastereoisomers.
[d] 0.1 equiv triethylamine.

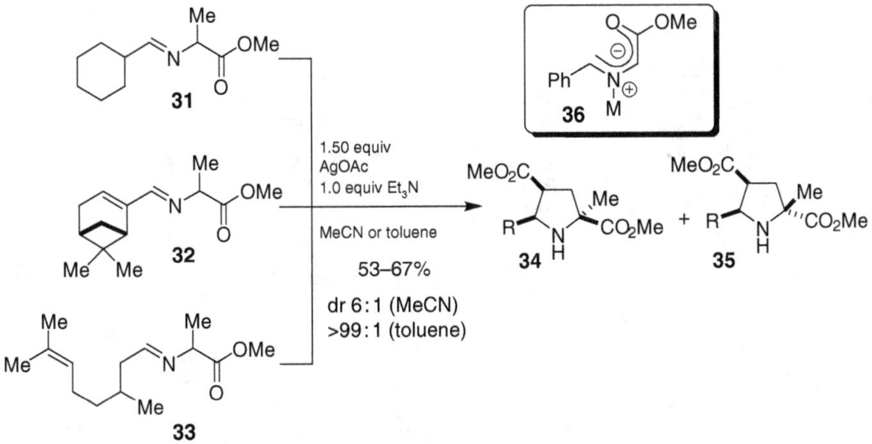

Scheme 2.9

It proved possible to expand the scope of this cycloaddition protocol to aliphatic iminoesters (Scheme 2.9).[15] For aliphatic iminoesters like **31** and **33**, the existence of enamine-to-imine tautomerization during the reaction can lead to the formation of byproducts. However, the reaction of aldimines **31–33** under the silver acetate/triethylamine/acetonitrile protocol led to the formation of the products in 53–67% yield. In the case of chiral iminoester **32**, stereocontrol was poor (30% de). Lithium bromide also catalyzed the reaction, but at lower rates and produced compounds of type **30** (see Scheme 2.8) as the main product. Paradoxically, although reaction rates were lower, stereoselectivity was higher in nonpolar solvents. Grigg suggested that this was because product formation through azomethine intermediate could take place through either conformation **24** or **36**. Since the latter has greater separation of charge, its contribution should be diminished in nonpolar solvents.

As has been eluded to in the preceding text, the choice of the metal is crucial. Lithium, zinc, or other metal salts are often capable of catalyzing a specific azomethine cycloaddition reaction. However, silver salts frequently provide the products in higher yields. Furthermore, the silver salt anion is of crucial importance.[16] Silver acetate (429.8 CHF/100 g, 99% purity Acros) is usually the salt of choice. For example, the reaction of naphthyl iminoester **25** ($R^1 = H$) with methyl acrylate catalyzed by 1.5 equiv of silver acetate and 1 equiv of triethylamine in acetonitrile affords the product in quantitative yield within 0.5 h. In contrast, catalysis by silver carbonate and silver tartrate affords the product in 67% and 61%, respectively, at considerably longer reaction times. On the other hand, silver nitrate (253.3 CHF/500 g, 99.9% purity, ABCR), is a more cost-efficient alternative and competitive in terms of yields (87%). It should be noted that no asymmetric induction could be observed in the silver tartrate–catalyzed reactions. Grigg has also reported that zeolite-bound Ag_2O, CuCl, CuCN, $NiCl_2$(dppe), and $NiCl_2(PPh_3)_2$ catalyzes the reaction.[17] In the case of silver oxide-impregnated zeolites, imine- and DBU-dependent leaching appears to be responsible for the catalytic activity.

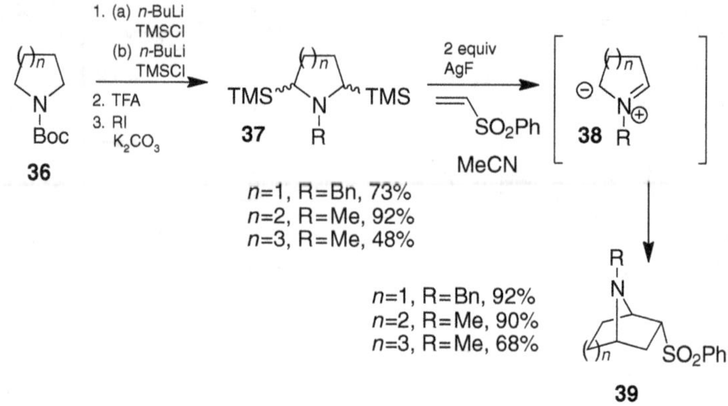

Scheme 2.10

The Pandey group has developed a silver fluoride–promoted desilylation of tertiary bis(silyl)amines as an interesting alternative method to access azomethine ylides (Scheme 2.10).[18] Notably, this method allows the generation of nonstabilized azomethine ylides under essentially neutral conditions. The starting materials are prepared by a three-step process, sometimes coupled into a single operation. For example, Boc-protected pyrrolidine **36** can be sequentially deprotonated and silylated twice in a one-pot reaction (Scheme 2.10). Removal of the Boc group and alkylation of the free amine leads to bis(silyl)amine **37**. When this compound is treated with 2 equiv of silver fluoride in the presence of phenyl vinyl sulfone, rapid formation of products **39** as single stereoisomers results.

During the reaction silver(0) deposits on the sides of the reaction vessel. The sequence of events leading to the azomethine ylide is unclear. However, evidently single-electron transfer (SET) from the amine to silver takes place either prior or subsequent to fluoride-enabled silyl cleavage. This process is repeated with a second equivalent of silver fluoride resulting in the formation of **38** either in free form or more likely as its silver complex. The scope of the method was expanded to the synthesis of bicylic systems exemplified here by tropinone **42** (Scheme 2.11).[19] Pandey has also extended the protocol to the synthesis of tricycloalkanes[20] and applied it to a total synthesis of the poisonous frog alkaloid epibatidine.[21]

2.3.1.2 *Auxiliary-Based Asymmetric [3+2] Cycloadditions*

As mentioned above, Grigg et al. had performed a brief reconnaissance into chiral silver complexes as inducers of chirality in the [3+2] cycloaddition.[16] This work

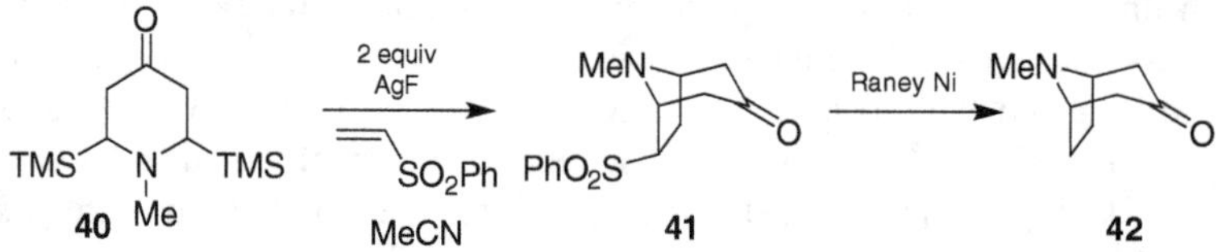

Scheme 2.11

focused on silver salts similar in nature to silver acetate, namely, silver tartrate and silver *O*-benzoyltartrate. While these reagents were able to promote the cycloaddition reaction, no asymmetric induction was observed. Asymmetric induction in [3+2] cycloaddition reactions by the action of a chiral metal complex had, however, already been achieved using superstochiometric amounts of a cobalt–ephedrine complex (see discussion below).[22] However, in the early 1990s the advent of practical asymmetric catalysis using substochiometric amounts of chiral silver complexes was still an elusive goal. As has historically been the case,[23] the focus of earlier studies shifted to the use of auxiliaries to induce stereoselective cycloadditions of azomethine ylides and alkenes.

As aliphatic chiral iminoesters gave poor diastereoselectivity in the [3+2] cycloaddition (Scheme 2.9),[15] the appendage of the chiral steering group to the dipolarophile component was attempted. Menthyl acrylate (**44**) proved promising in this regard (Scheme 2.12).[24] However, menthyl acrylate (**44**) proved much less reactive than methyl acrylate (**27**). Thus, cycloaddition of **43** with menthyl acrylate required 8 h to reach full conversion compared with 1 h for methyl acrylate. Accompanying the long reaction times were low yields of cycloadduct (36% in the specific case) and the formation of byproducts. Accordingly, efforts were directed at increasing the rate of cycloaddition. The pK_a of arylaldimines of α-amino esters is 17–19.5. It was argued that stronger bases should increase the rate of cycloaddition.[24] Indeed, the use of DBU in place of triethylamine had already proved itself effective in increasing the yield in lithium salt–catalyzed reactions.[15]

The rate and yield of the cycloaddition of **43a** to menthyl acrylate **44** in toluene were found to be directly related to the basicity of the base used. Thus, yield and rate

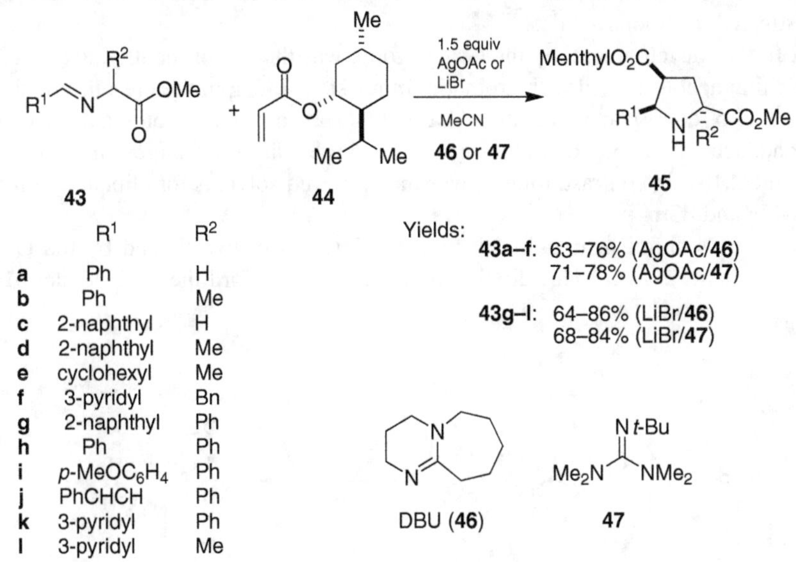

	R^1	R^2
a	Ph	H
b	Ph	Me
c	2-naphthyl	H
d	2-naphthyl	Me
e	cyclohexyl	Me
f	3-pyridyl	Bn
g	2-naphthyl	Ph
h	Ph	Ph
i	*p*-MeOC$_6$H$_4$	Ph
j	PhCHCH	Ph
k	3-pyridyl	Ph
l	3-pyridyl	Me

Scheme 2.12

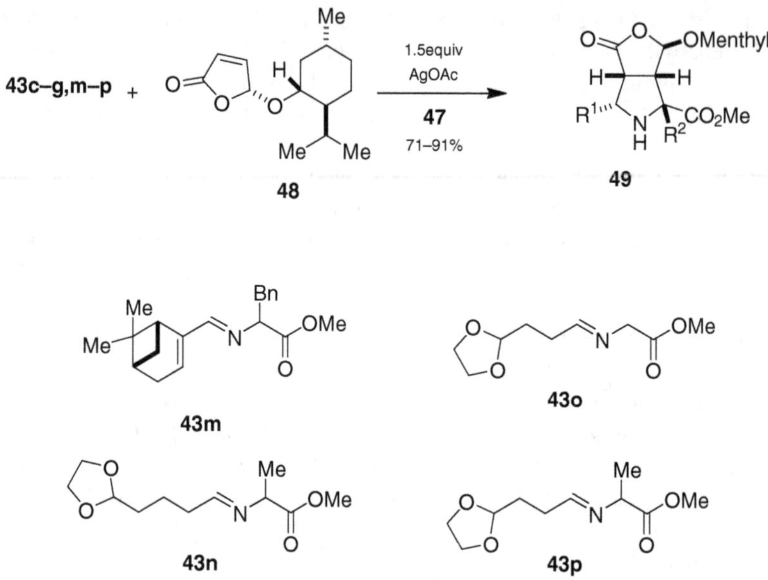

Scheme 2.13

increased in the following order: **47** (pK_a of the conjugate acid 14.6 in DMSO) > DBU **46** (pK_a 12) > Et_3N (pK_a 9.0). The cycloaddition of **43a** with menthyl acrylate was completed in 1 h using **47** as the base and afforded the product as a single diastereomer in 75% yield. For aryl iminoesters **43g–l** bearing bulky R^2 substituents, lithium bromide was the catalyst of choice. Unsaturated chiral amides were also investigated as dipolarophiles.[25]

A further development, by the Grigg group, was the use of menthyl acetal **48**.[26] This chiral acetal reacted with aromatic iminoesters **43c–g** in the presence of silver acetate (1.5 equiv) and guanidine base **47** (1.2 equiv) in acetonitrile, to give cycloadducts **49** in good yields and as a single diastereoisomer in each case (Scheme 2.13). In contrast, toluene was the preferred solvents for aliphatic iminoesters **43e** and **43m–p**.

Hiemstra's chiral hemiaminal lactam **50**[27] was also investigated by the Grigg group (Scheme 2.14).[26] This dipolarophile reacted to afford the cycloadduct **51** in

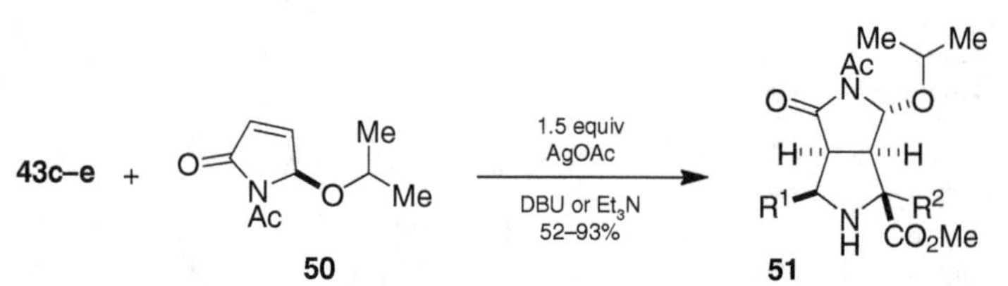

Scheme 2.14

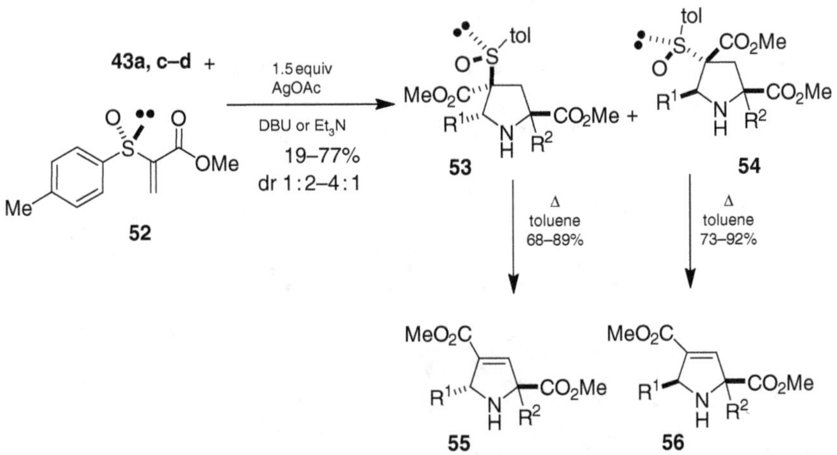

Scheme 2.15

moderate to excellent yields. Dichloromethane proved the best solvent for this reaction.

The Ruano group pioneered the use of sulfoxide-based auxiliaries.[28] Methyl *S-para*-tolylsulfinyl acrylate **52** underwent reaction with iminoesters **43a, c–d** using Grigg's catalyst system (AgOAc, DBU) in acetonitrile and THF (Scheme 2.15). Silver acetate catalyst loadings ranged from 7.5 mol% to 1.5 equiv. In general, good yields were obtained but diastereoselectivity was highly variable. Interestingly, the ratios of diastereoisomers were highly dependent on the identity of the solvent and base, but varied little as a function of catalyst loading or base/catalyst ratios. The auxiliary was removed by thermally induced *syn* elimination of *para*-toluenesulfenic acid, which took place to produce the corresponding 2-carboxy-2,3-pyrrolidines **55** and **56** in high yield.

The chemistry of sulfoxide **57** has been studied in the Diels–Alder reaction with dienes.[29] Ruano et al. studied its reaction with iminoesters **43a, c–d** under a variety of conditions (Scheme 2.16).[30] Silver-catalyzed cycloaddition at 0°C or room temperature afforded 1 : 1 mixtures of diastereomers **58** and **59** in variable yields. In contrast, deprotonation in the absence of silver using LiHMDS at –78°C afforded product **58** in 98 : 2 dr and 89–93% yield.[30]

The authors suggested a stepwise mechanism in which precomplexation of the sulfoxide to the lithium azaenolate would take place, thus allowing conjugate addition to follow.[30] This notion was based on Posner's[31] and Paquette's[32] earlier work on conjugate additions of nucleophiles to **57**. As before, thermal elimination of toluene sulfenic acid led to the conjugated products (not shown) in 92–93% yields for R^2 = methyl.[30] However, for R^2 = H thermal elimination of arylsulfenic acid did not afford any dihydropyrrole product but rather led to the formation of pyrroles. Treating **58** or **59** with Raney nickel at low temperature led to the rapid formation of **60** and **61**, respectively, in 82–86% yield.

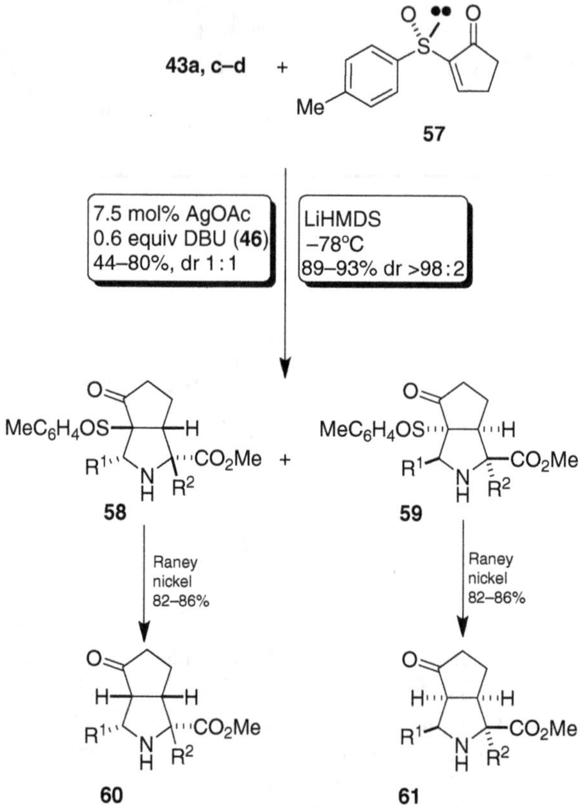

Scheme 2.16

Nájera and Sansano et al. studied the reaction of acryl esters **63** of *R*- and *S*- lactate (only *S* shown; see Scheme 2.17).[33] The reaction transpired at room temperature catalyzed by 10 mol% silver acetate in the presence of either 10 mol% KOH or triethylamine. Both methods afforded the *endo* products in moderate to good yields and excellent stereoselectivity. The auxiliary was cleaved by the action of aqueous base. In collaboration with the Cossio group, these authors have applied the method to synthesize a number of hepatitis C RNA polymerase inhibitors.[34]

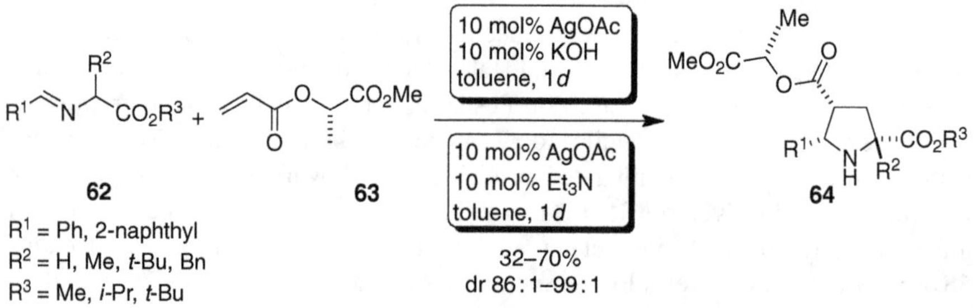

Scheme 2.17

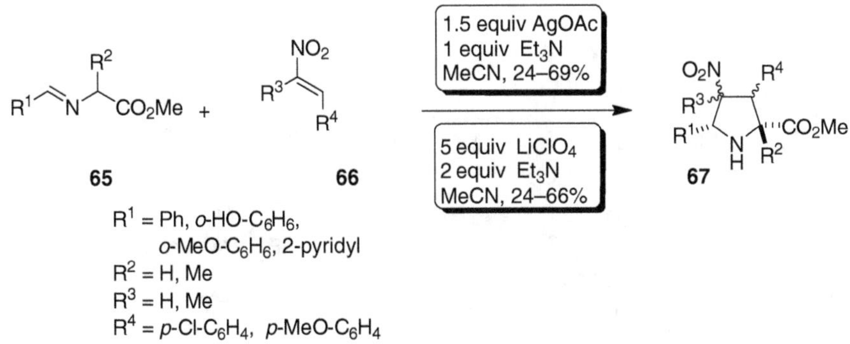

Scheme 2.18

Cossio and Linden et al. studied the reaction of aryl iminoester-derived azomethine ylides **65** with 2- and 1,2-disubstituted nitroalkenes **66** using silver acetate and triethylamine in acetonitrile (Scheme 2.18).[35] The product 3-nitropyrrolidines were formed in yields of 24–69% with excellent regioselectivity. These findings mirrored earlier work by Toke et al.[36] Unfortunately, diastereoselectivity was generally in the 1 : 1–4 region. Only in two cases where yields were low were the products formed in a 2 : 98 ratio. Lithium perchlorate was able to catalyze the reaction as well with similar yields. Remarkably, the diastereoselectivity with lithium perchlorate in some cases resembled that obtained with silver acetate and in others was reversed. Cossio also reported the stereoselective reaction of phenyl isocyanates with azomethine ylides **65** under the same conditions.[37] These reactants combined to give the expected products as single stereoisomer in 30–50% yield.

While this initial excursion into the chemistry of nitroalkenes was disappointing, it proved possible to turn the tables by the use of a propionate aldol auxiliary attached at the 2 position of the nitroalkenes, namely, compounds **69** (Scheme 2.19).[38] These chiral nonracemic nitroalkenes reacted stereospecifically with metalated azomethine ylides to afford the corresponding product **70** in 60–92% yield. Only 10 mol% of silver acetate was required to achieve complete conversion within 5 h at room

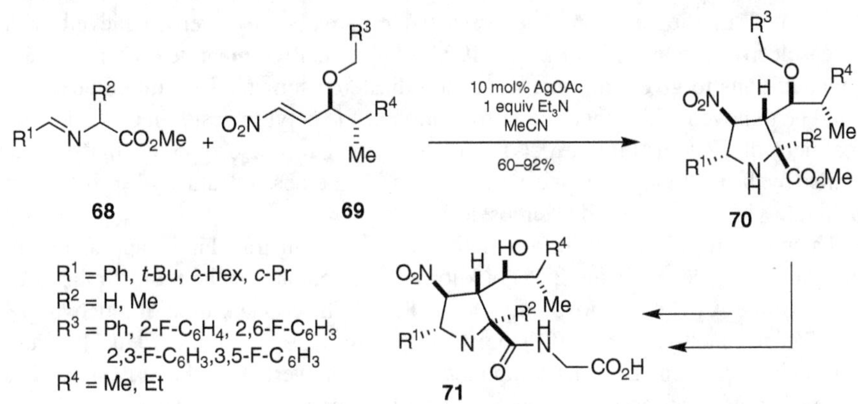

Scheme 2.19

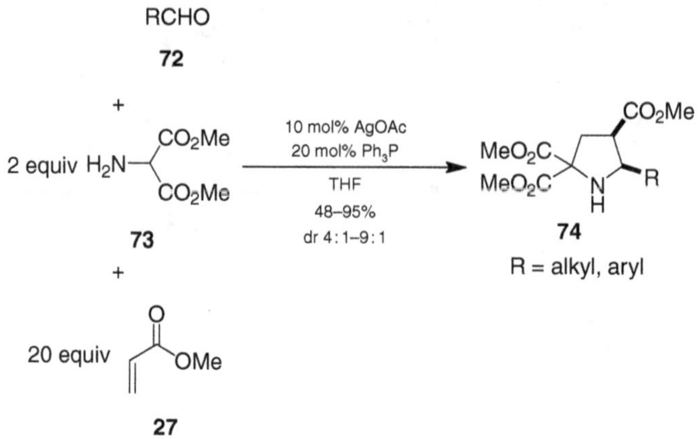

Scheme 2.20

temperature. A full equivalent of triethylamine served as the base. Diastereoselectivity was better than 99 : 1 when R^3 was an aryl group and ranged from 99 : 1 to 91 : 9 for aliphatic iminoesters. Compounds **70** were converted into a series of integrin inhibitors **71** by hydrolysis of the methyl ester function and amide formation to glycine. These compounds were shown to inhibit adhesion of cancer cells to microvascular endothelium in vitro and metastasis in vivo.

The Garner group has developed a silver acetate/triphenylphosphine (10 mol%) catalyzed one-pot three-component coupling (Scheme 2.20) of aldehydes **72**, methyl acrylate (**27**), and dimethyl-2-aminomalonate (**73**, 2 equiv).[39] Notably, both aliphatic and aromatic aldehydes may participate in the reaction and yields up to 95% can be achieved.

A breakthrough was achieved by replacing 2-aminomalonate with glycyl camphorsultam **75** (Scheme 2.21).[40,41] This auxiliary allowed the one-pot three-component coupling of dimethyl maleate (**76**) and a variety of aliphatic aldehydes. Following the findings of the Zhou group (see text below), no amine base was used. Furthermore, in addition to dimethyl maleate (**76**), a variety of other dipolarophiles, including methyl acrylate (**27**), *N*-phenylmaleimide, and phenyl vinyl sulfone, participated in the reaction. Yields in the range 59–94% were achieved with stereoselectivity ranging from 7 : 1 to 19 : 1. Only dimethyl fumarate (**26**) underwent cycloadditions to give a mixture of several diastereoisomers. The sultam auxiliary could be removed with concomitant formation of the octyl thioester in 68% yield by treatment of **77** with lithium octyl thiolate. Garner has also reported that imines formed by the reaction of glycine with Garner's (serine-derived) aldehyde proceeds in moderate yields and good diastereoselectivity.[42]

The power of this cycloaddition method is clearly illustrated in its application to the synthesis of **80** (Scheme 2.22), a known intermediate for the total synthesis of cyanocycline A (**81**) and bioxalomycin β2 (**82**).[43] The condensation of aldehyde **78** and **75** in neat methyl acrylate (**27**) took place to afford the pyrrolidine **79** in 74% yield as a single stereoisomer.[44] This compound was then converted into **80** in 12 steps. The method has also been applied to the synthesis of a number of analogs of **81** and **82** with

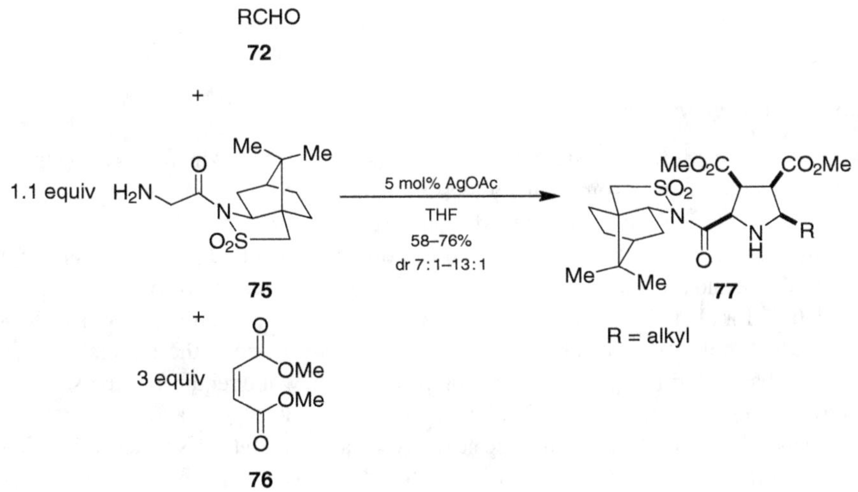

Scheme 2.21

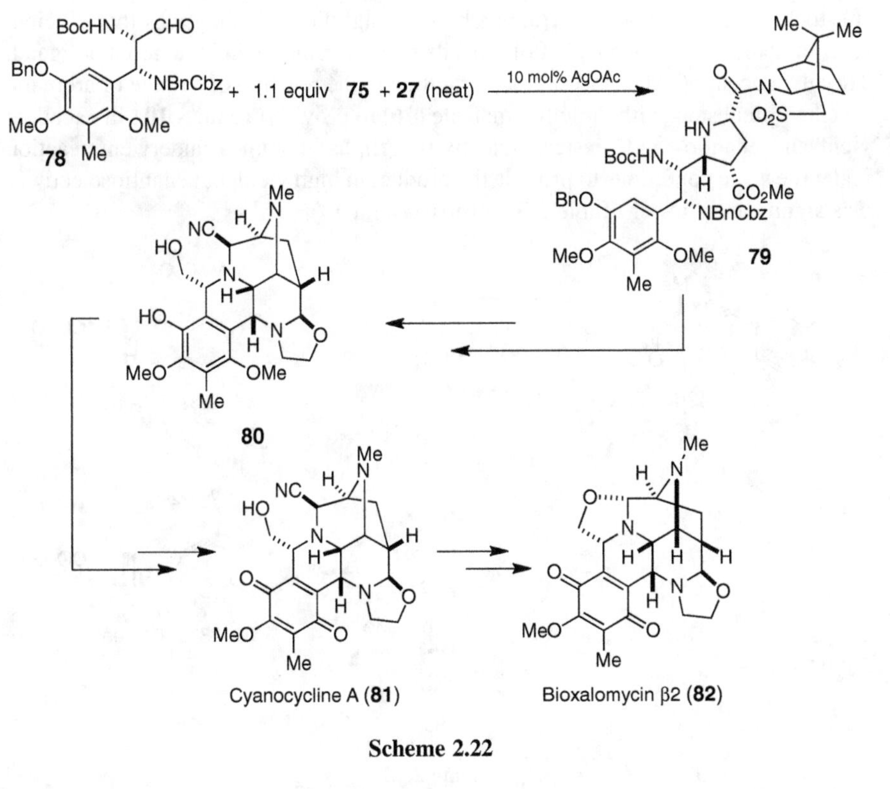

Scheme 2.22

notable biological properties.[45] The broad range of aldehydes that may participate in the reaction marks this method as one of the most generally useful to date.

2.3.1.3 *Catalytic Asymmetric [3+2] Cycloadditions*

As noted above, the first use of a chiral silver salt (silver tartrate) to induce asymmetry in the cycloaddition of azomethine ylides with dipolarophiles was unsuccessful.[16] However, the Grigg group, undaunted by this setback, tested chiral bisphosphine **83** as a ligand in the silver triflate–catalyzed reaction of **43d** with dipolarophiles phenyl vinyl sulfone and methyl acrylate **27** (Scheme 2.23).[46] This reaction proceeded in good yield, 64% for phenyl vinyl sulfone and 83% for methyl acrylate, in both cases in 70% ee. The preliminary nature of these findings was overshadowed by the authors' success with a cobalt chloride–ephedrine (**85**) complex, which, while requiring superstochiometric amounts of catalyst, provided the product **86** in 84% yield and 96% ee.[22]

It was to be almost 8 years until the next silver-based catalytic asymmetric reaction was reported. In their pioneering work,[47] the Zhang group screened a number of privileged phosphine ligands including BINAP (**87**), the Trost ligand (**88**), Duphos (**89**), and PennPhos (**90**) in the reaction of iminoester **103** with dimethyl maleate (**76**) (see Fig. 2.1, Scheme 2.24, and Table 2.3). However, the best results were achieved with the FAP ligands **91** and **92**. Of these, the bulkier xylyl congener **92** provided the best enantioselectivity.

The cycloaddition reaction was performed at room temperature and employed substochiometric amounts of Hünig's base. A significant finding was that efficient reaction took place with only 3 mol% of silver acetate and a similar amount of ligand. The silver acetate/FAP catalyst system facilitated the reaction of a range of aromatic glycine iminoesters with dimethyl maleate (**76**) to provide products **104** in excellent yield and enantio- and diastereoselectivity. Aliphatic imines underwent reaction under the same conditions to provide the product in high yield, but enantioselectivity was significantly lower (Table 2.3, entries 12 and 13).

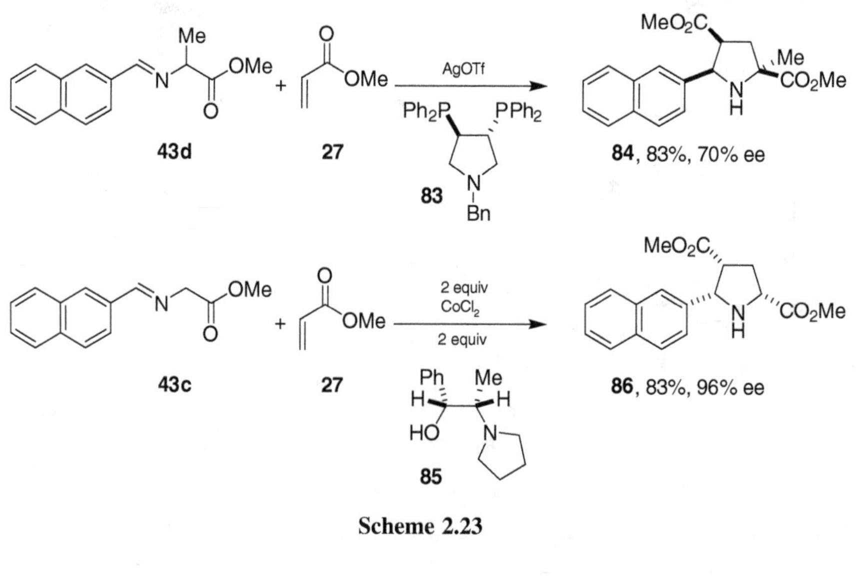

Scheme 2.23

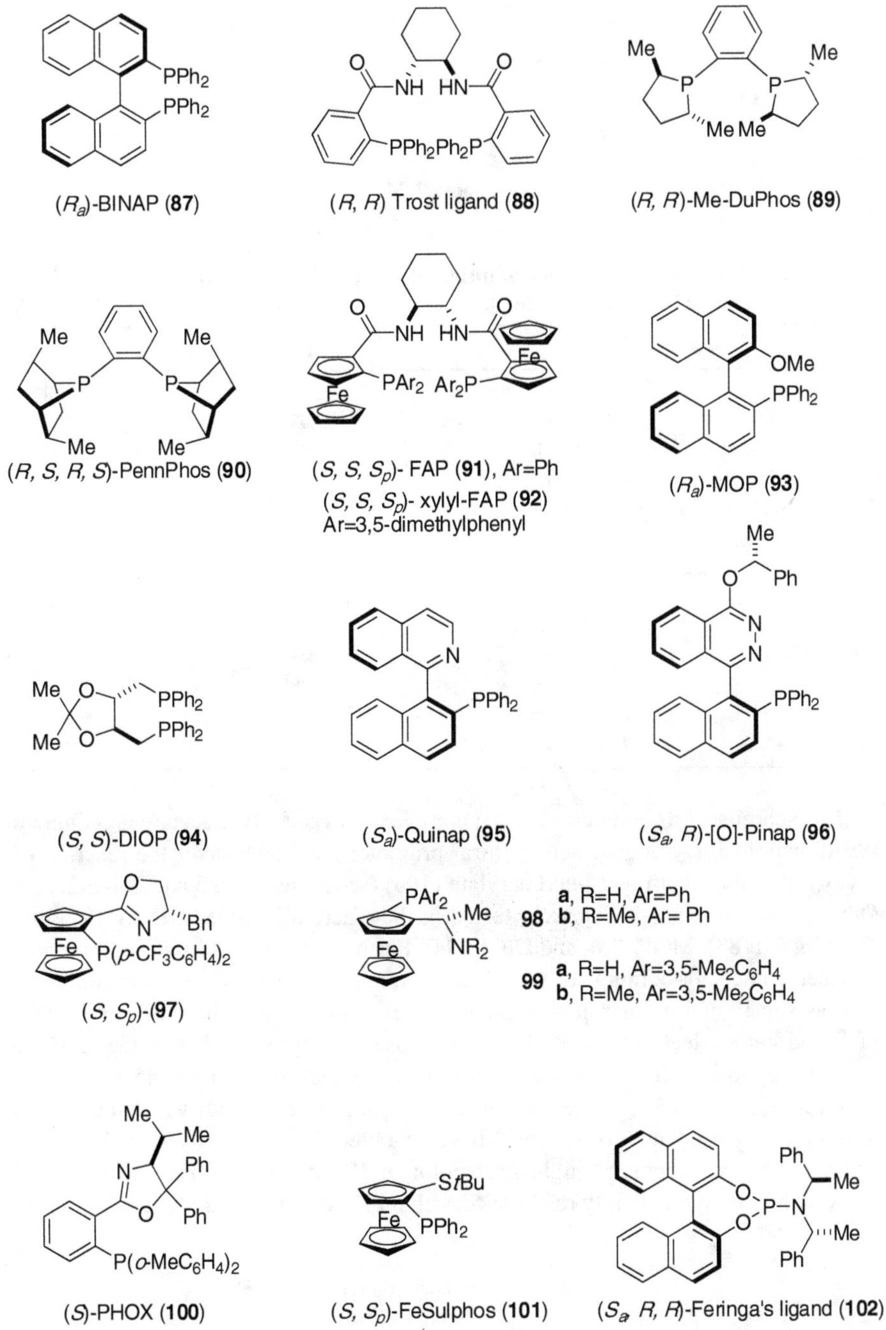

Figure 2.1 Chiral ligands used in catalytic asymmetric [3 + 2] cycloadditions.

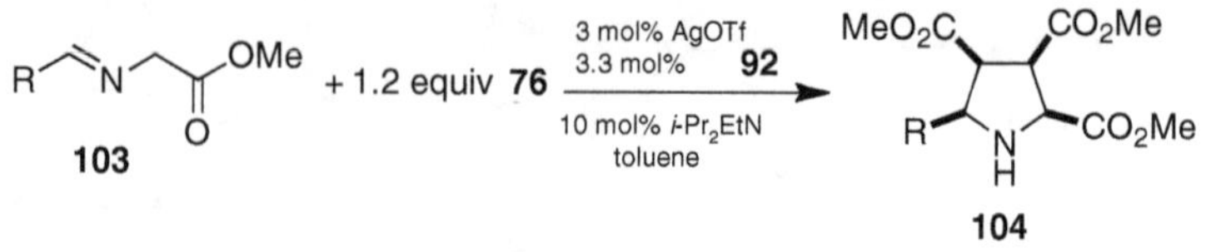

Scheme 2.24

TABLE 2.3. Enantioselective Cycloaddition of Iminoesters (103) and Dimethyl Maleate (76) (Scheme 2.24)

Entry	R	Yield (%)	ee (%)
1	Ph	87	87
2	*p*-Me-C_6H_4	93	88
3	*p*-MeOC_6H_4	98	92
4	*p*-Cl-C_6H_4	96	92
5	*p*-F-C_6H_4	96	90
6	*p*-CN-C_6H_4	90	96
7	*o*-Xl-C_6H_4	96	86
8	*o*-Me-C_6H_4	97	90
9	1-Naphthyl	73	85
10	2-Naphthyl	98	97
11	3-Pyridyl	98	84
12	*i*-Pr	82	70
13	*c*-Hex	82	81

The Schreiber group reported that a complex between silver acetate and Quinap (**95**) functioned as a highly active chiral promoter and catalyst for the reaction of arylglycyl imines with *tert*-butyl acrylate (**106**) (see Scheme 2.25 and Table 2.4).[48] Other phosphine ligands that were tested in the reaction included BINAP (**87**), the Trost ligand (**88**), MOP (**93**), and DIOP (**94**). Further, screening of starting materials revealed some shortcomings of the system. Iminoesters bearing an α substituent as well as substituted dipolarophiles underwent reaction to give the products in high yield and stereoselectivity but moderate enantioselectivity. Nevertheless, the catalytic system was notable for its ability to promote cycloaddition even at –45°C.

While Quinap (**95**) is an excellent and versatile ligand from which numerous developments in asymmetric catalysis have benefited,[49] its synthesis is cumbersome, and hence its price remains high (*R* enantiomer 100 mg, 337 CHF). Carreira et al. showed that the synthetically readily accessible PINAP ligand class is as versatile as

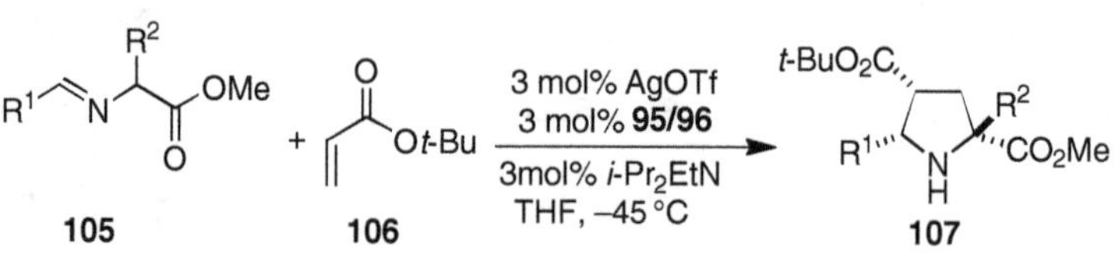

Scheme 2.25

TABLE 2.4. Yield and Enantioselectivity for Formation of 107 Using Ligand 95 (Scheme 2.25)

Entry	R^1	R^2	Yield (%)	ee (%)
1	*p*-$MeOC_6H_4$	H	93	95
2	*p*-Br-C_6H_4	H	89	95
3	*p*-NC-C_6H_4	H	92	96
4	2-Naphtyl	H	89	94
5	2-Toluyl	H	95	89
6	Ph	Me	98[a]	80
7	Ph	Bn	93[b]	77

[a] Reaction performed with 10 mol% catalyst loading at −20°C for 24 h.
[b] Reaction performed with 10 mol% catalyst loading at −20°C for 48 h.

TABLE 2.5. Yield and Enantioselectivity for Formation of 107 Using Ligand 96 (Scheme 2.25)

Entry	R^1	R^2	Yield (%)	ee (%)
1	*p*-$MeOC_6H_4$	H	88	92
2	*p*-Br-C_6H_4	H	94	95

Quinap (**95**) in a number of mutually unrelated transition-metal-catalyzed asymmetric reactions.[50] Thus, **105** added to *tert*-butyl acrylate **106** using silver acetate and PINAP ligand **96**, under the conditions described in Scheme 2.25, to provide the desired product **109** in excellent enantioselectivity and yield (see Scheme 2.6 and Table 2.5). Notably, the reaction required only 3 mol% of catalyst for complete conversion at –40°C within 36 h.

The Zhou group reported the use of silver acetate and *N,P*-ferrocenyl ligand **97** in the azomethine cycloaddition with dimethyl maleate (**76**) as shown in Scheme 2.27

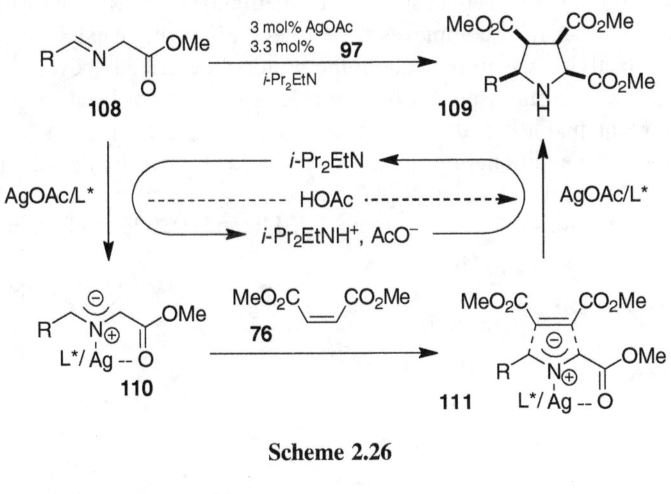

Scheme 2.26

TABLE 2.6. Yields and Enantioselectivity for Formation of 109 (Scheme 2.27)

Entry	R	Yield (%)	ee (%)
1	Ph	85	97
2	*p*-MeOC_6H_4	94	98
3	*p*-Cl-C_6H_4	99	97
4	*p*-F-C_6H_4	96	97
5	*p*-CN-C_6H_4	91	97
6	*o*-Cl-C_6H_4	98	97
7	*o*-Me-C_6H_4	99	98
8	1-Naphthyl	85	98
9	2-Naphthyl	95	98
10	3-Pyridyl	76	93
11	*i*-Pr	56	88

(see Table 2.6).[51] Zhou proposed the mechanism shown in Scheme 2.26, based on the earlier proposals by Grigg and Zhang.[46,47]

Zhou proposed that silver acetate should be able to catalyze the reaction without the commonly reported use of strong amine bases (indeed, this had already been shown using excess silver acetate in the pioneering studies of Grigg[12]). Thus, acetate/acetic acid was proposed for its ability to function as a proton shuttle for the reaction. This proposal was borne out by experiment. Silver acetate/**97** was able to efficiently catalyze the reaction of arylglycyl iminoester **108** with dimethyl maleate (**76**) at low temperature in ether without the addition of an external base (Scheme 2.27, Table 2.6). High yields and enantioselectivity were achieved. Notably, the reaction time was significantly reduced compared to earlier work. As a corollary, the authors investigated the significance of the anion in the silver salt. As predicted, a range of silver salts with basic counteranions catalyzed the reaction, but silver triflate, for instance, was unable to catalyze the reaction unless Hünig's base was added as an external base.

Zhou also reported a series of related *P*,*S*-ferrocenyl ligands and their use in the [3+2] cycloaddition of aryl-substituted azomethine ylides with *N*-phenylmaleimide.[52] While these silver complexes were able to efficiently catalyze the reaction, the enantioselectivity was lower than in the protocol described above.

Although mechanistic aspects of these reactions have not been investigated in depth, important findings and clues emanate from the many reports summarized in this chapter. An additional intriguing finding was reported by the Zhou group in 2007.[53] While screening a series of *N*,*P*-ferocenyl ligands of types **98** and **99**, they observed that reactions involving bismethylated ligands **98b** led to formation of the

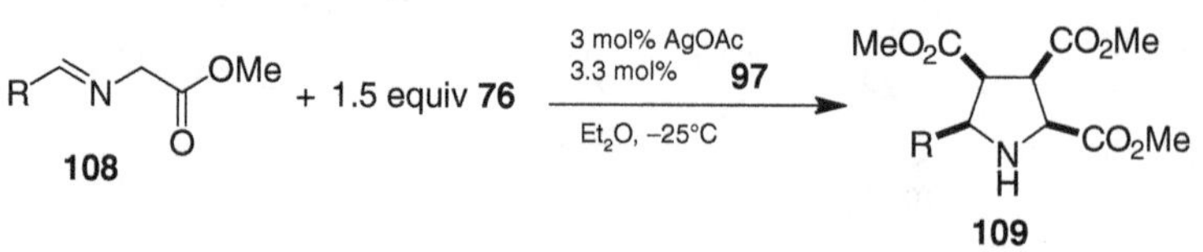

Scheme 2.27

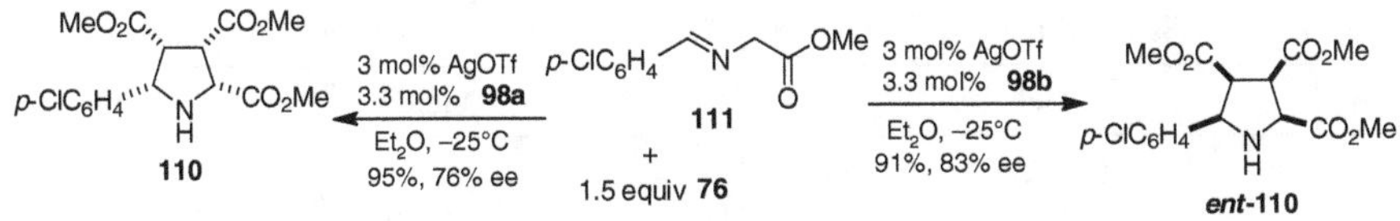

Scheme 2.28

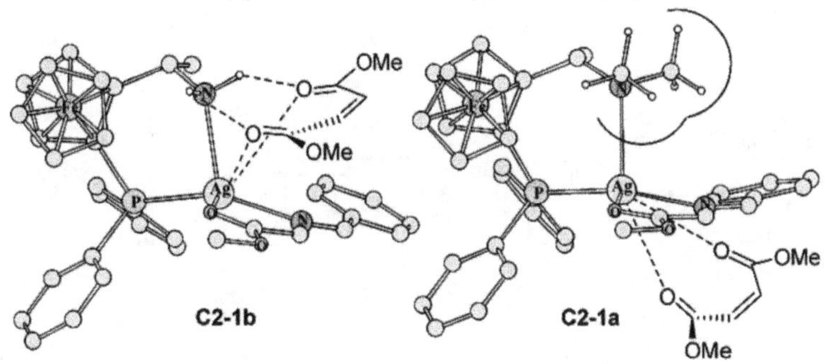

Figure 2.2 Models showing effects of presence or absence of H bonding on silver/ligand complex and dipolarphile. (Taken from ref. 53 with permission.)

products *ent*- **110** in high yield and enantioselectivity (Scheme 2.28). Curiously, when ligand **98a** (bisdesmethyl **98b**) was used, the same product was formed in high yield and enantioselectivity, but with the opposite absolute configuration. Clearly the ability of the ligand **98a** to form additional hydrogen bonds led to fundamental changes in the structure of the silver azomethine complex, resulting in formation of the enantiomer of the product formed with ligand **98b**. This trend proved to be consistent even with other ligands of the same type (e.g., **99a** vs. **99b**).

Computational studies yielded the models shown in Figure 2.2. These models illustrate how the existence or absence of hydrogen bonding between the silver/ligand complex and the dipolarphile may lead to selection of opposite faces in the reaction with the silver-bound azomethine ylide derived from **111**.

While virtually all of the research described above has focused on the intermolecular cycloaddition of azomethine ylides, the intramolecular process holds considerable promise for the synthesis of polycyclic natural products. The Pfaltz group reported an intramolecular catalytic asymmetric cyclization of aryl iminoesters **112** using a complex of silver acetate with PHOX type ligand **100** (Scheme 2.29,

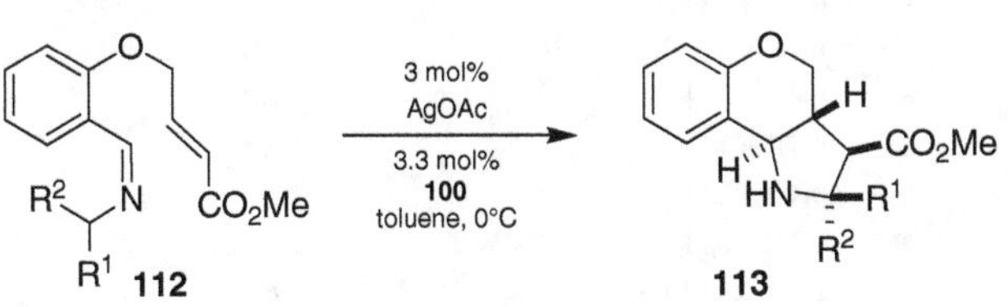

Scheme 2.29

TABLE 2.7. Yields and Enantioselectivity for Formation of 113 (Scheme 2.29)

Entry	R^1	R^2	Yield (%)	ee (%)
1	CO_2Me	H	74	96
2	CO_2tBu	H	66	99
3	CO_2Me	Me	61	96
4	2-Pyridyl	H	70	83

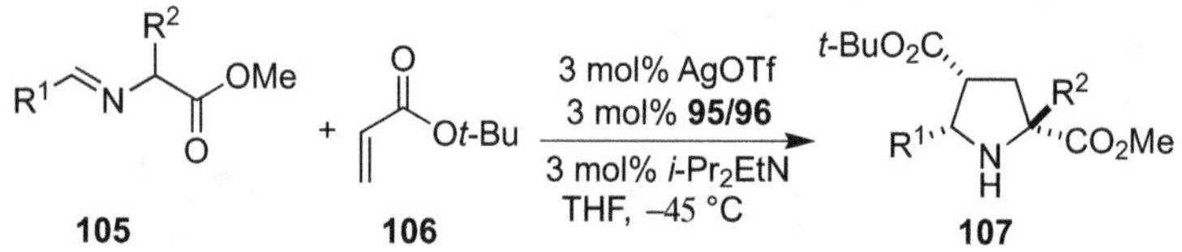

Scheme 2.30

Table 2.7).[54] The tricyclic products **113** were formed in good yield and excellent enantioselectivity. Interestingly, the PHOX–silver complex performed less well in intermolecular cycloadditions.

Nájera reported that silver perchlorate forms a catalytically active complex with (*S*)-BINAP (**87**).[55] This complex efficiently catalyzes the reaction of aryl iminoesters **108** with *N*-methylmaleimide **114** in the presence of the external base triethylamine (Scheme 2.30, Table 2.8). High enantioselectivity is obtained in this reaction. As a consequence of the low solubility of the silver BINAP complex in the reaction medium toluene, it can be quantitatively recovered from the reaction mixture by simple filtration. The recovered catalyst could be recycled at least 4 times with no apparent loss in activity. Carretero reported that *N*-phenylmaleimide also reacted with

TABLE 2.8. Yields and Enantioselectivity for Formation of 115 (Scheme 2.30)

Entry	R	Yield (%)	ee (%)
1	Ph	90	>99
2	Ph	90	>99
3	2-Naphthyl	89	99
4	*o*-Cl-C_6H_4	82	82
5	*p*-Me-C_6H_4	88	86
6[a]	*p*-Me-C_6H_4	88	99
7	*p*-MeO-C_6H_4	85	80
8	*p*-Cl-C_6H_4	87	64
9[a]	*p*-Cl-C_6H_4	87	98
10	2-Thienyl	87	90

[a] Reaction performed with 5 mol% DBU instead of triethylamine.

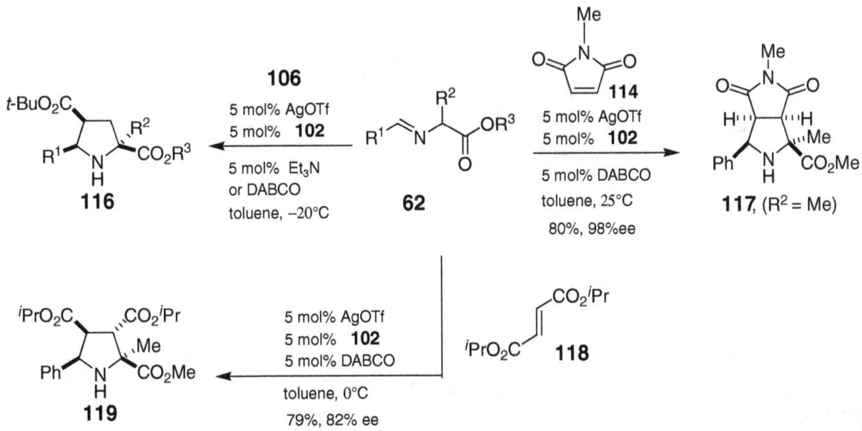

Scheme 2.31 Cycloaddition of **62** with various dipolarophiles.

a complex formed from silver acetate and Fesulphos ligand (**101**) and **108** to give products in up to 89% ee. However, in this case copper salts performed better.[56]

Nájera et al. have shown that Feringa's phosphoramidite ligand **102** is superior to other ligands in the cycloaddition of aryl iminoesters **62** with *tert*-butyl acrylate (**106**), *N*-methylphthalimide **114**, and diisopropyl fumarate (**118**).[57] As shown in Scheme 2.31 and Table 2.9, the products **116** and **117** were formed in excellent yields and exquisite *endo* selectivity.

Most of the examples given in the preceding text relied on silver phosphine complexes as catalysts. The Jørgensen group examined the use of cinchona-derived ligands in the silver fluoride–catalyzed reaction of aromatic iminoesters **103** with methylacrylate (**27**). Yields were excellent, but enantioselectivities were only in the 41–73% range.[58]

While significant progress has been achieved in catalytic asymmetric cycloaddition of aryl iminoesters with dipolarophiles, the field is still in its infancy. Notably,

TABLE 2.9. Yields and Enantioselectivities for Synthesis of 116 (Scheme 2.31)

Entry	R^1	R^2	R^3	Yield (%)	ee (%)
1[a]	*o*-Me-C_6H_4	H	Me	83	98
2[a]	*o*-Cl-C_6H_4	H	Me	90	>98
3[a,b]	Ph	H	*i*-Pr	81	98
4[c]	*p*-MeOC_6H_4	H	*i*-Pr	79	98
5[c]	Ph	Me	Me	78	91
6[c]	2-Thienyl	Me	Me	77	92
7[c]	2-Thienyl	*t*-Bu	Me	70	82
8[c]	Ph	Bn	Me	77	98

[a] DABCO as base.
[b] Performed at 0°C.
[c] Et_3N as base.

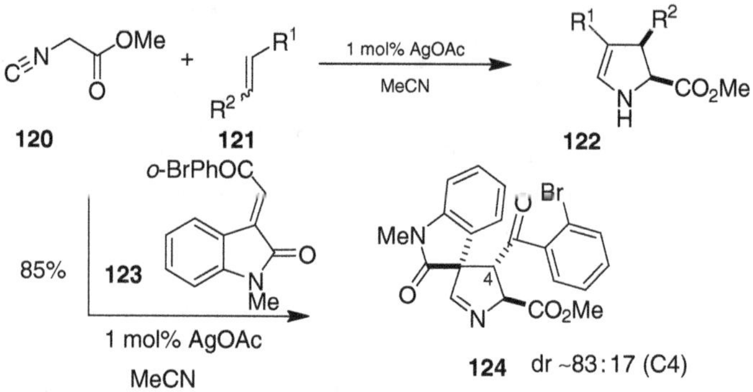

Scheme 2.32

the lack of protocols capable of leading to asymmetric cycloaddition of aliphatic iminoesters marks this as an important direction for future research. Additionally, the application of intramolecular cycloaddition may provide unique solutions in the synthesis of complex molecules. Also understanding of the mechanisms involved is still lacking, although it is likely that advances in this area would catalyze further developments.

2.3.1.4 Selected Applications and Extensions of Azomethine [3+2] Cycloadditions

Work from the Grigg group since the late 1990s has focused on extensions of the chemistry detailed above. In one effort, it was shown how the unique reactivity of isocyanides allows them to function as azomethine ylide precursors. The reaction of isocyanide **120** with dipolarophiles **121** proceeds in the presence of silver acetate to afford dehydropyrroles **122** in good yield (Scheme 2.32, Table 2.10). Reaction with dimethyl maleate (**76**) afforded the corresponding product in excellent yield and diastereoselectivity. In contrast, its stereoisomer, dimethyl fumarate (**26**), furnished a 6 : 4 mixture of isomers under identical conditions.[59] The authors proposed a stepwise mechanism for this reaction (Scheme 2.33).

Tepe has reported the synthesis of the structurally related Δ^1-pyrrolines (Scheme 2.34, Table 2.11).[60] These structures are found in numerous natural products.

TABLE 2.10. Yields of Compound 122 (Scheme 2.32)

Entry	R^1	R^2	Yield (%)	dr
1	CO_2Me	H	67	—
2	CN	H	72	—
3	CHO	H	73	—
4	$COCH_3$	H	82	—
5	CO_2Me	CO_2Me (*cis*)	65	96 : 4
6	CO_2Me	CO_2Me (*trans*)	88	63 : 37

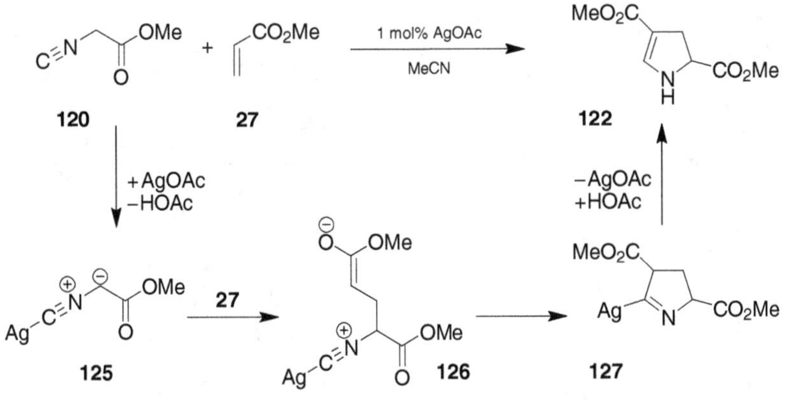

Scheme 2.33

This cycloaddition takes place to give the *exo* cycloadduct, rather than the more common *endo* cycloadduct, as the major product. Presumably this is a consequence of the intermediate munchnone **131** being locked in an *anti* orientation as compared to the commonly favored *syn* orientation (as in **24**).

Grigg has also shown that *trans*-substituted nitroolefins react with azomethine ylides in good yield (Scheme 2.35).[61] Although *trans*-disubstituted dipolarophiles often give a mixture of *endo* and *exo* products, the diastereoselectivity was generally excellent except for (*E*)-2-methoxy-4-(2-nitrovinyl)phenol, which reacted with 2-naphthylmethyliminoglycyl ester **103** to give the product in only 42% yield and 2 : 1 *endo* : *exo* diastereoselectivity. Low diastereoselectivity also resulted from the cycloaddition of these nitroolefins with homoserine-lactone-derived imines.[61]

The scope of the silver-catalyzed [3+2] cycloaddition is not limited to amino acid–derived iminoesters. Grigg has reported the use of aminophosphane-derived azomethine ylides in the cycloaddition to afford the corresponding phosphonate-substituted pyrrolidines (Scheme 2.36).[62] The diastereoselectivity of these cycloadditions is not inferior to those of standard amino acid–derived imines.

The incorporation of fluorine is one of the most commonly employed strategies in lead optimization in medicinal chemistry. Franchini has reported that *trans* methyl 3-fluoroacrylate (**137**) undergoes reaction with glycine menthyl ester– derived imines **136** and **139** to afford fluorinated prolines **138** and **140** in good yield and diastereoselectivity (Scheme 2.37).[63]

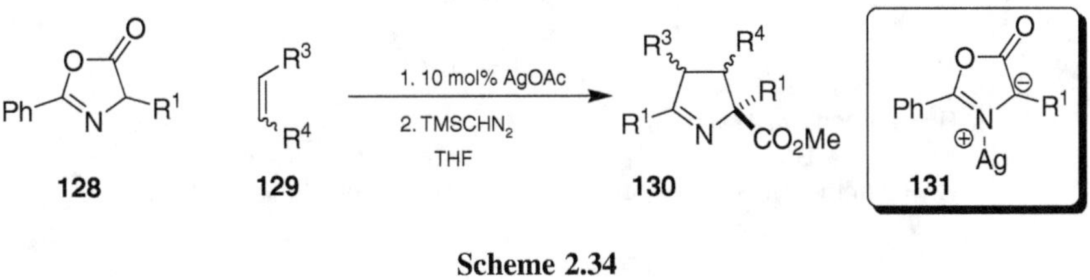

Scheme 2.34

TABLE 2.11. Yields of Δ^1-Pyrrolines

Entry	R^1	Dipolarophile (**129**)	Product (**130**)	Yield (%)
1	Me	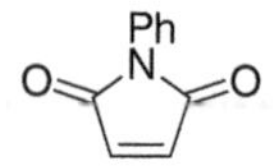	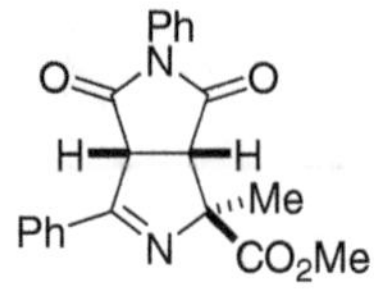	78
2	Me	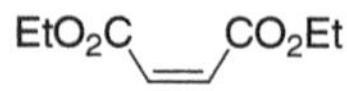	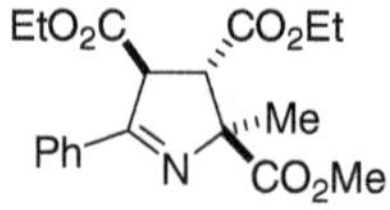	75
3	Me	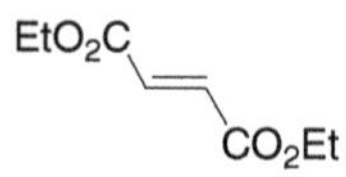	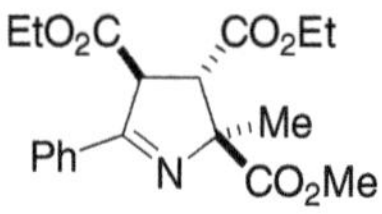	75[a]
4	Me	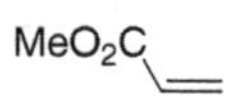	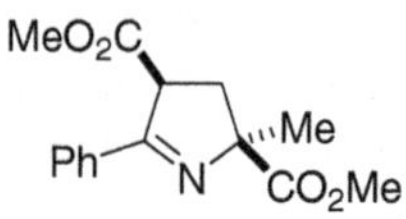	95[b]
5	Ph	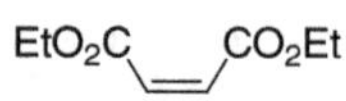	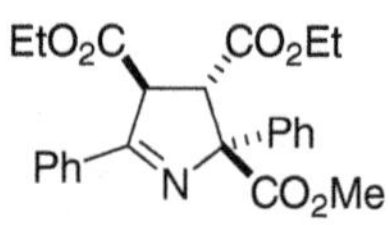	15

[a] 2 : 1 mixture of exo/endo epimers.
[b] 1 : 1 mixture of regioisomers.

Raghunathan has disclosed a synthesis of spiropyrrolidines **143** based on the Grigg triethylamine protocol (Scheme 2.38).[64] The structurally complex spirocycles are formed in a single step, in high yield from simple starting materials. This protocol was extended to the synthesis of β-lactam spirocyclic compounds.[65]

Alcaide studied the silver acetate/triethylamine-catalyzed reaction of chiral β-lactam iminoesters **144** with methyl acrylate (**27**) as shown in Scheme 2.39.[66] These

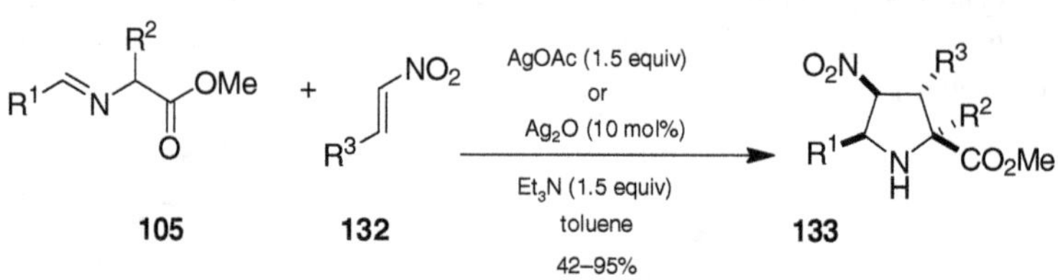

R^1 = 2-naphthyl, 4-biphenyl
R^2 = H, Me, Bn
R^3 = 9-anthracyl, 3-indolyl, 4-HO-3-MeOC_6H_3, 2-furyl, 2-thienyl, 3-pyridyl

Scheme 2.35

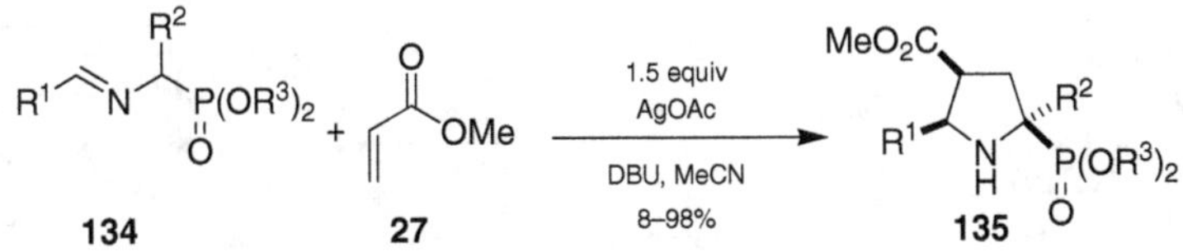

R^1 = 2-naphthyl, 2-thienyl, *c*-Hex, *p*-$NO_2C_6H_4$, *o*-IC_6H_4, 2-pyridyl
R^2 = H, Me, Ph, Bn, CH_2OH, CH_2OTBS
R^3 = Me, Et

Scheme 2.36

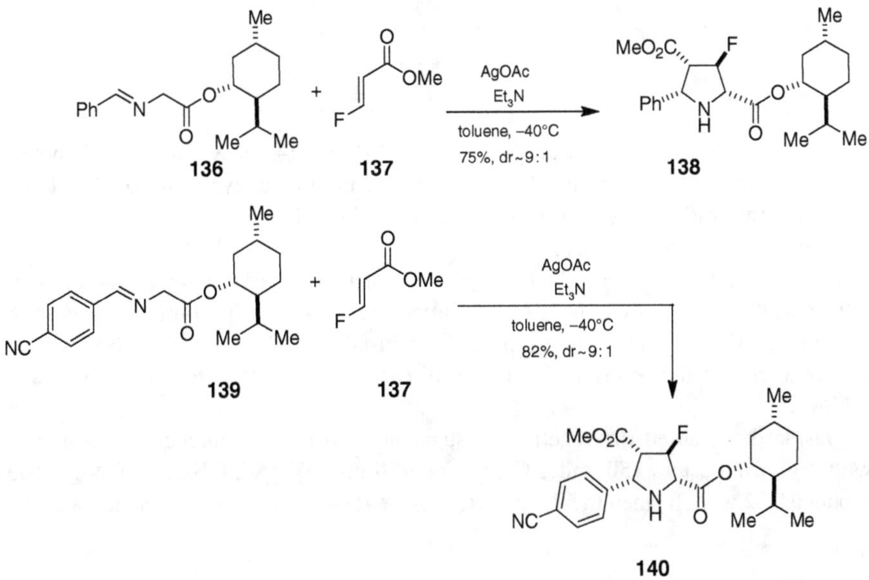

Scheme 2.37

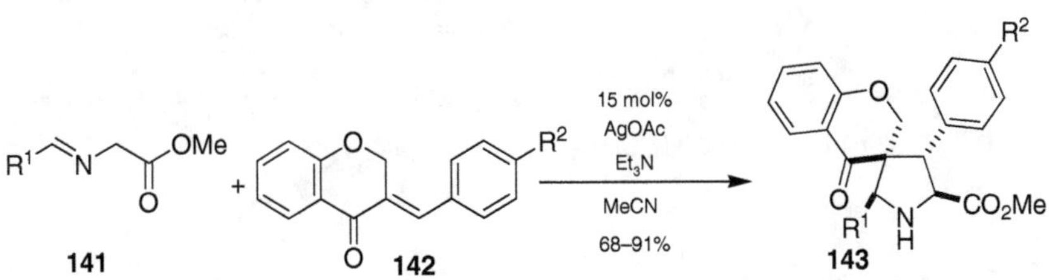

R^1 = Ph, *p*-ClC_6H_4, *p*-MeO-C_6H_4
R^2 = H, Cl, OMe, Me, NO_2

Scheme 2.38

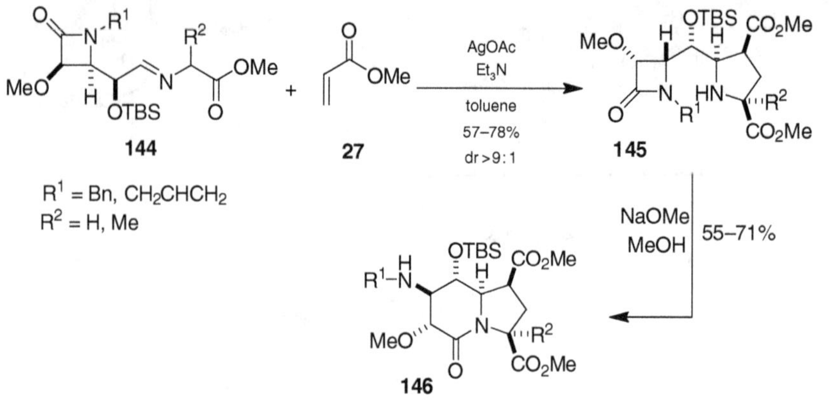

Scheme 2.39

reactions proceeded to afford the pyrrolidines **145** in 57–78% yield and with better than 9 : 1 diastereoselectivity. The compounds could be cyclized to afford the corresponding indolizidinones **146** in moderate yield. The indolizidinone moiety is another common motif in natural products.

Pätzel achieved a stereoselective synthesis of methylcarboxy-substituted pyrrolidines **149** by reaction of phenyl iminoester **147** with chiral enones **148** (Scheme 2.40).[67] Silver acetate (15 mol%) and DBU (1.2 equiv) in THF was used as promoters for the reaction. In almost all cases, selectivity better than 95 : 5 was achieved.

Bashiardes studied the reaction of sugar-derived chiral nonracemic pyran **151** with aryl iminoesters **150** under Grigg's conditions (AgOAc, DBU, MeCN).[68] The products **152** were formed in 59–66% yield as single diastereoisomers (Scheme 2.41).

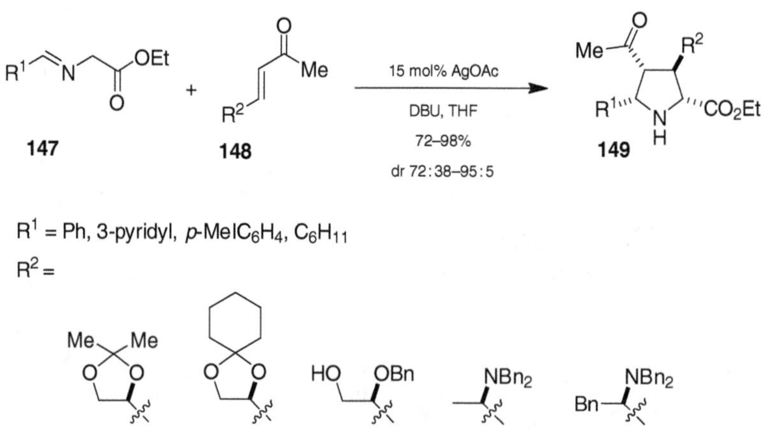

Scheme 2.40

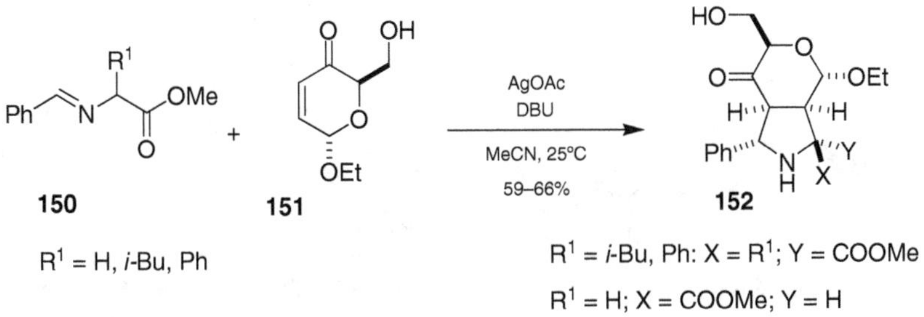

Scheme 2.41

Grigg has pioneered the coupling of two or more reaction protocols into tandem or sequential one-pot procedures. These efforts have led to successful protocol for a silver-catalyzed azomethine ylide cycloaddition and a subsequent rhodium-catalyzed [2+2+2] cycloisomerization,[69] as well as a tandem Heck–azomethine ylide cycloaddition.[70] The Grigg group has also developed strategies for the rapid assembly of complex polycyclic structures by running an azomethine ylide cycloaddition reaction in sequence with a further cyclization reaction. For example, azomethine cycloaddition followed by a Pictet–Spengler cyclization gives access to a variety of bridged tri-, tetra-, penta-,[71] and hexacyclic[72] indole derivatives. This protocol has also been carried out on solid phase by employing Wang resin-linked acrylate ester as the dipolarphile in the [3+2] cycloaddition.[73] After the Pictet–Spengler step the products were cleaved of the solid phase using TFA.[74]

As a part of a program directed toward the synthesis of the potent topisomerase I inhibitors, the lamellarins (e.g., **153** and **154**), Porco has reported the silver triflate–catalyzed tandem cycloisomerization–azomethine ylide cycloaddition of **155** (Scheme 2.42).[75] The postulated mechanism of this intriguing and highly efficient process is shown in Scheme 2.43. Silver-catalyzed addition of the imine nitrogen to the alkyne results, on subsequent deprotonation, in the formation of an azomethine ylide **160**. This ylide participates in [3+2] cycloaddition with the alkyne component leading to formation of a dehydropyrrole **161**. Finally, oxidation by adventitious oxygen leads to formation of the product **162**.

2.3.2 Other [3+2] Cycloadditions

Billups reported the cycloaddition of benzocyclopropene **163** to 1,3-butadiene (Scheme 2.44). The reaction proceeded at 0°C with only 1 mol% of $AgBF_4$ to afford product **165** in 64% yield.[76] Under these conditions, compound **163** also reacted in a [3+2] fashion with alkenes, allenes, and disubstituted alkynes but in much lower yields (11–25%) of [3+2] cycloaddition products. A phenylsilver species has been invoked as an intermediate. The reaction probably proceeds through a cationic mechanism.

Saito reported the extension of Billups' tandem silver-catalyzed ring-opening cycloaddition reaction methodology to addition to imines (Scheme 2.45).[77] Naphtho-[*b*]cyclopropene **167** added to substituted aryl imines **168** at room temperature in the

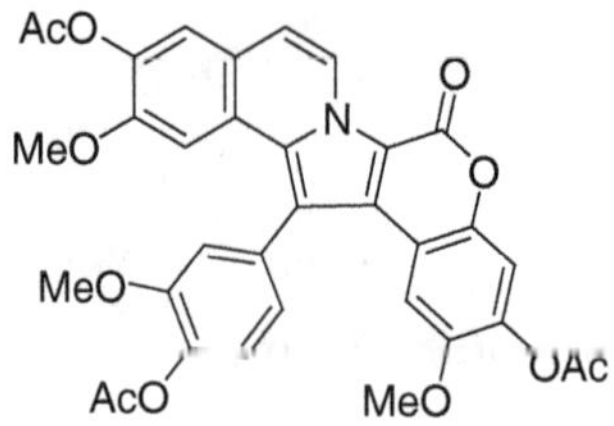

Lamellarin D-triacetate (**153**)

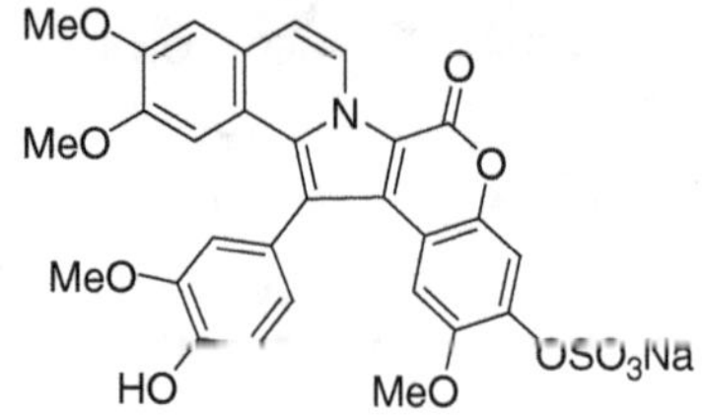

Lamellarin α20-sulfate (**154**)

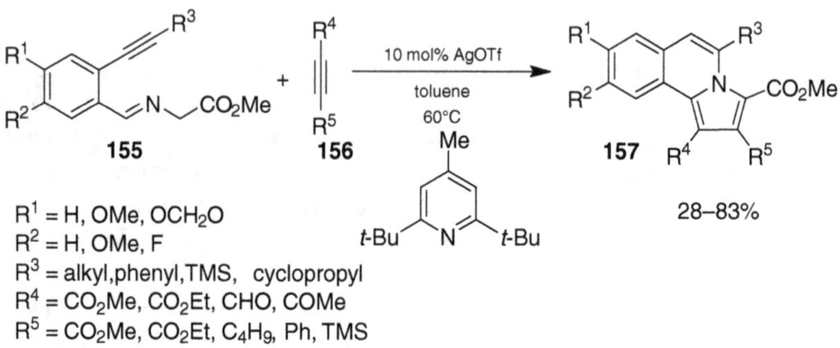

Scheme 2.42

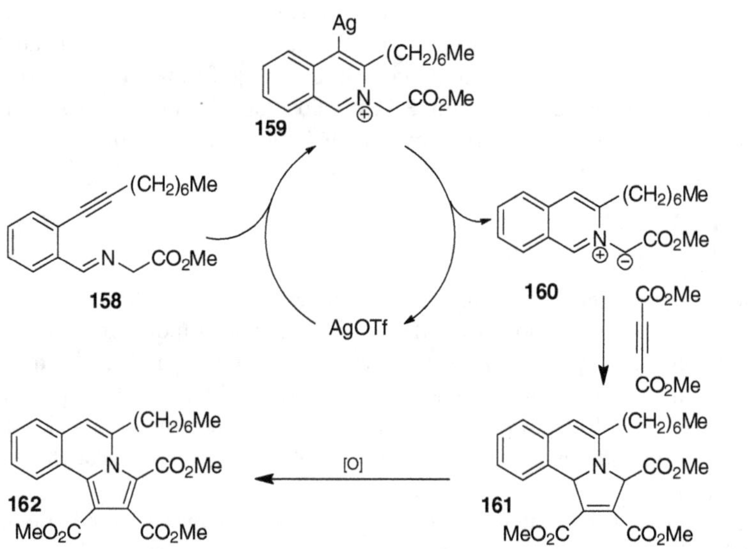

Scheme 2.43

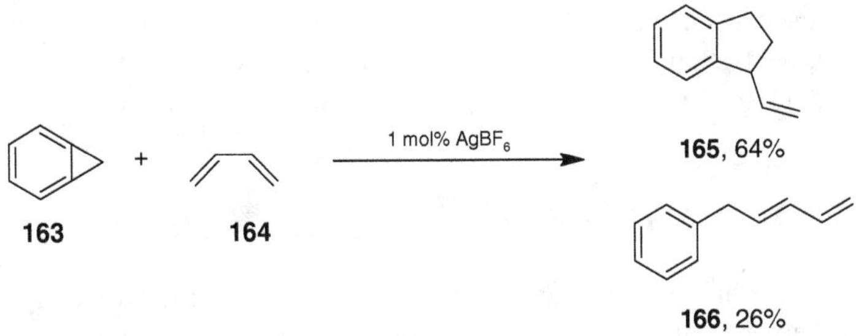

Scheme 2.44

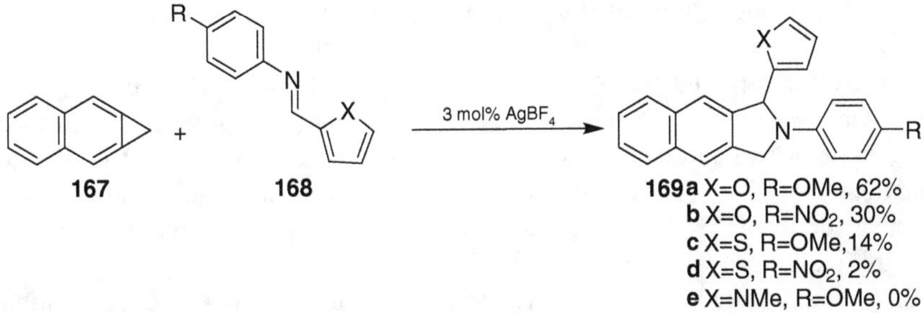

Scheme 2.45

presence of 3 mol% $AgBF_4$ to afford cycloadducts **169** in 0–62% yield. The best yields were achieved for electron-rich aryliminyl furans. Notably, no reaction took place under thermal conditions. Catalysis of the reaction by $Yb(fod)_3$ could also be achieved, but this system required 60°C to proceed. Yields were moderate.

2.4 [3+3] CYCLOADDITIONS

Saito also investigated the reaction of naphtho[*b*]cyclopropene **167** with troponone-derived imines **170** (Scheme 2.46). Reaction took place to afford tetracyclic dihydroisoquinolines **171** in modest yields.[78] However, the preferred product was the simple addition product **172**. The yields of cycloadducts **171** were generally only 8–15%.

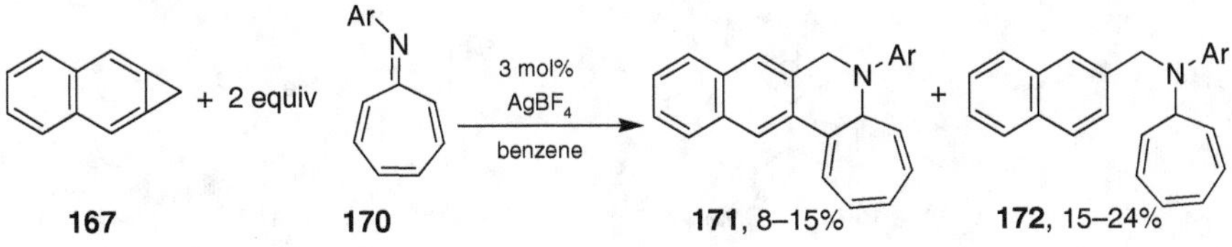

Scheme 2.46

The proposed mechanism involves silver-catalyzed attack by the imine function on the cyclopropyl system. This is followed by conjugate addition of the intermediate silver species on the tropone ring system and subsequent isomerization to afford **171**.

2.5 [4+2] CYCLOADDITIONS

In an early contribution to the field, Jendralla showed that the inherently unstable *trans*-cycloheptene could be stabilized by the addition of silver salts ($AgClO_4$ and AgOTf) as 1 : 1 adduct **175** (see Scheme 2.47).[79] This complex undergoes facile additions to dienes **176–178** to give the [4+2] cycloadducts **179–181** in good yields. Interestingly, the complex could be dissociated by addition of *trans*-cyclooctene. This is due to the higher affinity of silver for trans-cyclooctene, with which it forms 3 : 1 adducts.[79]

Mukaiyama reported that 4–10 mol% of $Ph_2Sn{=}S$ and $AgClO_4$ catalyzed the Diels–Alder reaction between but-3-ene-2-one (**183**) or (*E*)-1-phenylbut-2-en-1-one (**182**) and unactivated dienes **184–186** to afford the products **187–192** in 57–95% yield (Scheme 2.48).[80] The reaction with cyclic dienes **185** and **186** was highly *endo*-selective (dr 99 : 1). Naphthalene-1,4-dione also participated in the reaction with similar yields.

Silver-impregnated solids may also serve as catalysts in [4+2] cycloadditions. Thus, Mayoral reported that silver ion–exchanged montmorillite K10 functioned as a catalyst for the reaction of methyl 2-cyanocinnamate and cyclopentadiene **185** (Scheme 2.49).[81] The use of such heterogeneous catalysts may enhance catalyst recovery, an important consideration for industrial applications.

With the relative air and water stability of silver phosphine complexes in mind, the Frost and Weller groups reported the use of a silver(I) carborane triphenylphosphine complex as a catalyst for the aza–Diels-Alder reaction.[82,83] These unique catalyst complexes were able to catalyze the reaction of Danishefsky's diene (**195**) with **196** in

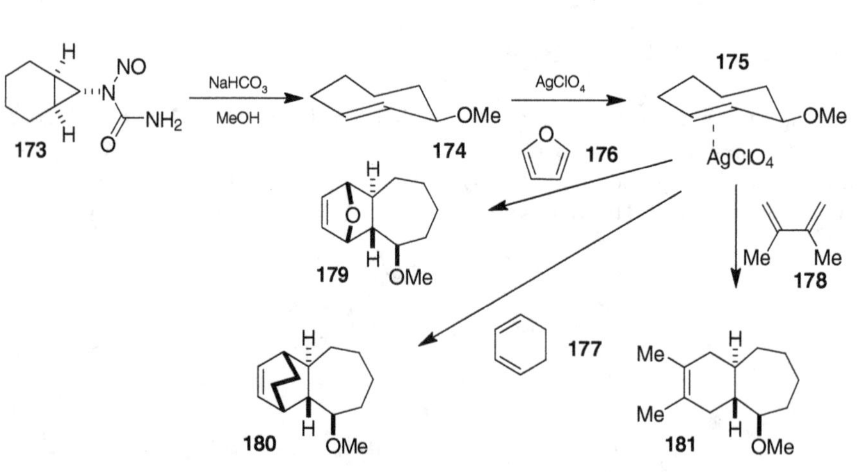

Scheme 2.47

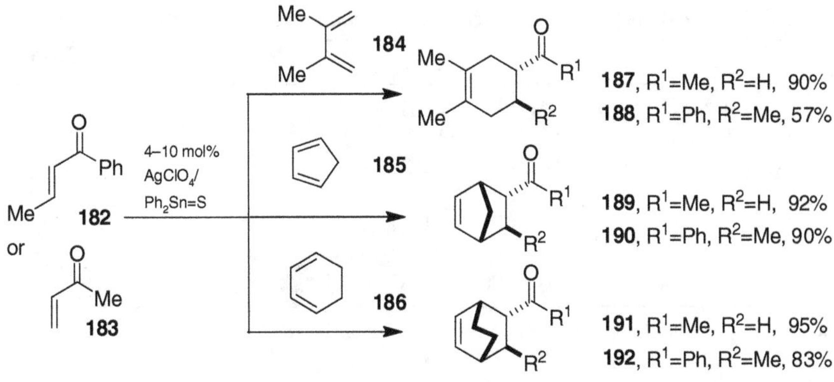

Scheme 2.48

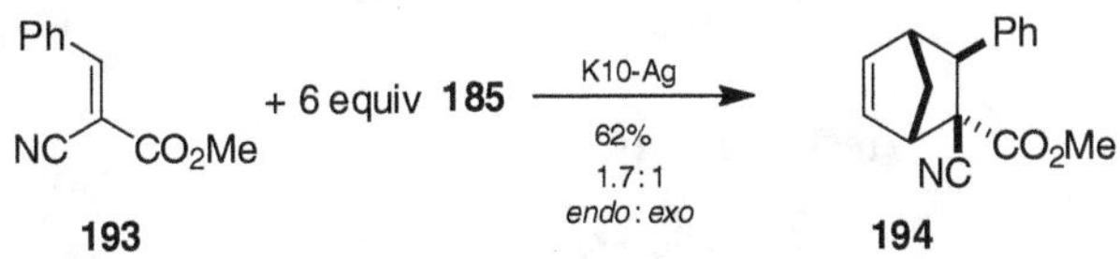

Scheme 2.49

CD_2Cl_2 at room temperature at low catalyst loadings (0.1 mol%) (Scheme 2.50, Table 2.12). Interestingly, catalytic activity was dependent on the presence of 50 mol% water. No reaction took place in its absence. This method was also studied using water as the solvent. The complex $(PPh_3)Ag(CB_{11}H_6Br_6)$ catalyzed the reaction with a turnover frequency (TOF) of 4000 h^{-1}. Thus, it was more active than $(PPh_3)Ag(CB_{11}H_{12})$. Both catalysts were more active than $AgOTf(Ph_3P)$, $AgClO_4(Ph_3P)$ and $AgBF_4(Ph_3P)$.

Complexation of the three silver salts $Ag(CB_{11}H_{12})$, $Ag(CB_{11}H_6Br_6)$, and Ag(OTf) to polymer bound triphenylphosphine also yielded active catalyst systems. The polymer-bound catalyst could be recycled 3 times with no loss of activity. Dimeric complexes [e.g., $[Ag(PPh_3)_2(CB_{11}H_{12})]_2$] were significantly poorer catalysts.

Kobayashi examined the efficacy of various silver salts (e.g., AgOTf, $AgClO_4$, Ag_2CO_3) as catalysts for the aza-Diels–Alder cycloaddition of **195** to imines formed in situ in a 9 : 1 THF/water mixture (Scheme 2.51).[84] The silver salts with basic anions (e.g., Ag_2CO_3) were unable to catalyze the reaction, but the silver salts with nonbasic

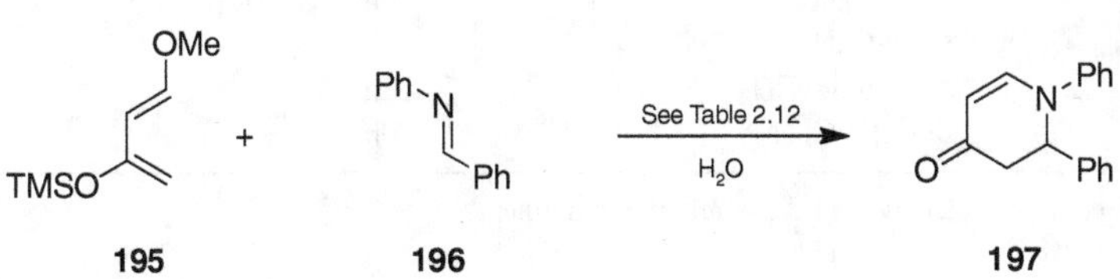

Scheme 2.50

TABLE 2.12. Yields of 197 Using Various Silver Salts (Scheme 2.50)[a]

Entry	Catalyst	Yield (%)
1	$[Ag(PPh_3)(BF_4)]$	35
2	$[Ag(PPh_3)(OTf)]$	70
3	$[Ag(PPh_3)(ClO_4)]$	90
4	$[Ag(PPh_3)(CB_{11}H_{12})]$	98
5	$[Ag(PPh_3)_2(CB_{11}H_{12})]$	99
6	$[Ag(PPh_3)(CB_{11}H_6Br_6)]$	99
7	$[Ag(PPh_3)_2(CB_{11}H_6Br_6)]$	85

[a] Reactions run in water for 60 min isolated yields.

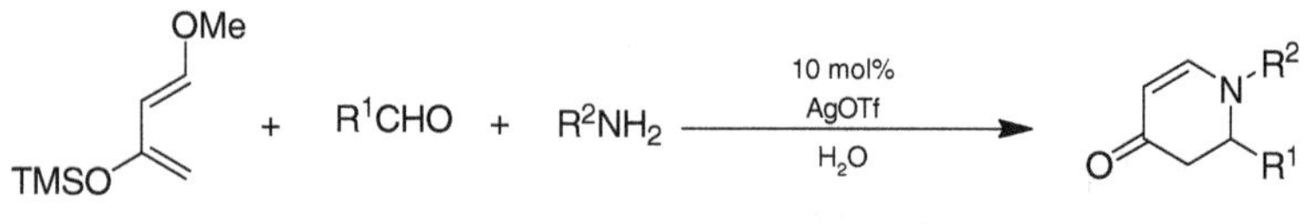

Scheme 2.51

anions (e.g., AgCl) all promoted the reaction to some extent; AgOTf was superior to all other potential catalysts examined.

Further studies revealed that a 10 mol% loading of AgOTf was sufficient to catalyze the reaction of Danishefsky's diene (**195**) with a variety of aromatic phenylimines bearing electron-withdrawing/donating substituents, in 57–92% yield within 2–3 h. These studies were carried out in water. Electron-poor imines generally required the use of 3 equiv of diene **195** to obtain satisfactory yields. The method was extended to a one-pot three-component protocol with in situ formation of the imine from the aniline and 1.5 equiv of the aldehyde (Scheme 2.51, Table 2.13). Because of

TABLE 2.13. Yields of 200 for Several Aldehydes (Scheme 2.51)[a]

Entry	R^1	R^2	Yield (%)
1	Ph	Ph	63
2[b]	Ph	Ph	80
3[b,c]	Ph	*p*-BrC_6H_4	90
4[b]	Ph	*o*-$MeOC_6H_4$	56
5	*c*-C_6H_{12}	Ph	70
6[b]	*c*-C_6H_{12}	Ph	51
7[b]	$PhCH_2CH_2$	Ph	53
8	*i*-Pr-CH_2	Ph	72

[a] With 1.5 equiv of aldehyde and diene relative to amine.
[b] With 10 mol% Triton X-100 added.
[c] With 3 equiv diene.

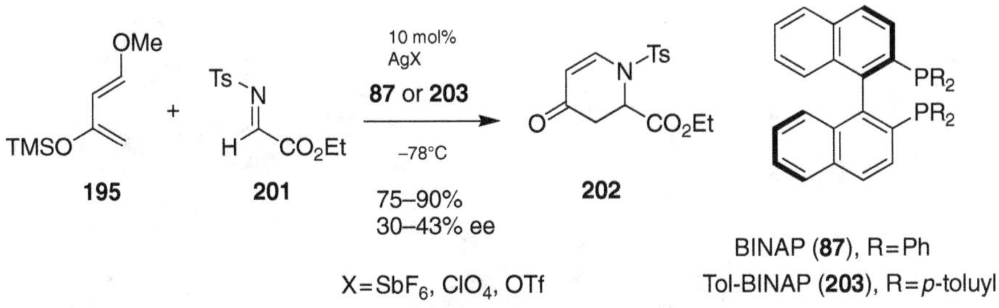

Scheme 2.52

the low solubility of the reagents in water, the reaction mixture was heterogenous. The complexity of the physical state of the mixture and its dependence on reactant structure is evident from the fact that in some cases the addition of 10 mol% Triton X-100, a nonionic surfactant, improved yields for aromatic aldehydes while in the case of cyclohexane carbaldehyde the yield was reduced (compare, e.g., entries 1 and 2 with entries 5 and 6 in Table 2.13).

Attempts to develop enantioselective protocols for the aza-Diels–Alder reaction were reported simultaneously with those described above. A first contribution in this area was the report by the Jørgensen group,[85] who studied the influence of salts of copper, silver, palladium, and zinc. Copper(I) perchlorate provides optimal yields and enantioselectivity, but complexes of BINAP (**87**) and Tol-BINAP (**203**) with $AgSbF_6$, AgOTf, and $AgClO_4$ were able to catalyze the reaction, albeit with low enantioselectivity (Scheme 2.52).

In contrast, the Hoveyda and Snapper groups reported the efficient catalysis of silver acetate complexes of imine **204**.[86] These authors first established the superior performance of *N-ortho*-methoxyaniline-substituted imines in the reaction with Danishefsky diene **195**. Because of the modular nature of ligand **206**, optimization of its structure could be achieved by surveying a combinatorial library of amino acid derivatives. The *l*-isoleucine *para*-methoxyaniline conjugate **206** proved the best. This ligand facilitated full conversion within 2–3 h with high yield and enantioselectivity (Scheme 2.53).

Through conjugation to a Wang resin, the ligand could be immobilized as its benzyl congener **207**. A complex between **207** and AgOAc (5 mol%) was able to catalyze the reaction between **204** (Ar = 2-naphthyl) and Danishefsky's diene **195** in 96% yield and 86% ee. This drop in enantiomeric excess, compared to the reaction featuring silver acetate/**206** complex, accurately reflects the drop in enantioselectivity observed by using the non-resin-bound analog of **207** and so appears to be unrelated to the solid phase. The resin-bound catalyst AgOAc/**207** complex could be recycled at least 5 times with no drop in activity or stereoselection.

The group of Yamamoto reported the catalytic enantioselective hetero-Diels–Alder reactions of azo compound **209** and dienes **208** (Scheme 2.54).[87] In a ligand screening the use of BINAP (**87**) gave higher conversion and enantioselectivity than both Segphos (**211**) and Difluorphos (**212**). Interestingly, the optimal silver

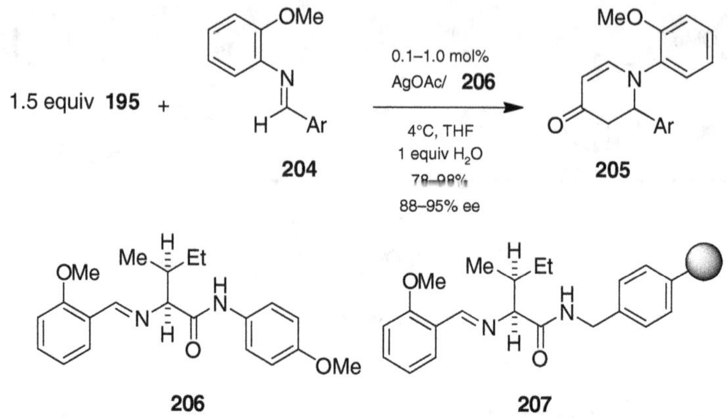

Scheme 2.53

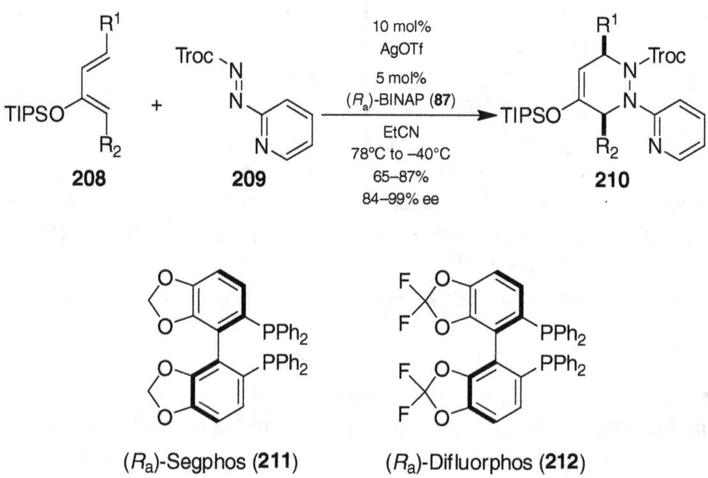

Scheme 2.54

triflate–to-BINAP ratio was 2 : 1. This would indicate the presence of a bimetallic catalyst exhibiting cooperative effects of two neighboring silver atoms.

2.6 CONCLUDING REMARKS

Silver salts and phosphine complexes have proved themselves as catalysts and mediators of a range of cycloadditions. No doubt only the tip of the iceberg has been uncovered regarding the possible applications of silver-mediated cycloadditions.

REFERENCES

1. Kobayashi, S.; Jørgensen, K.A., eds., Cycloaddition Reactions in Organic Synthesis, Wiley-VCH, **2001**.
2. Pellissier, H., *Tetrahedron* **2007**, *63*, 3235–3285.
3. Alvarez-Corral, M.; Munoz-Dorado, M.; Rodriguez-Garcia, I., *Chem. Rev.* **2008**, *108*, 3174–3198.
4. Calaza, M. I.; Cativiela, C., *Eur. J. Org. Chem.* **2008**, 3427–3448.
5. Husinec, S.; Savic, V., *Tetrahedron: Asymmetry* **2005**, *16*, 2047–2061.
6. (a) Friedman, L., *J. Am. Chem. Soc.* **1967**, *89*, 3071–3073; (b) Friedman, L.; Lindow, D. F., *J. Am. Chem. Soc.* **1968**, *90*, 2324–2328.
7. Paquette, L. A., *Chem. Commun.* **1971**, 1076–1077.
8. Ona, H.; Sakai, M.; Suda, M.; Masamune, S., *Chem. Commun.* **1973**, 45–46.
9. Sweis, R. F.; Schramm, M. P.; Kozmin, S. A., *J. Am. Chem. Soc.* **2004**, *126*, 7442–7443.
10. Nakamura, I.; Nemoto, T.; Yamamoto, Y.; de Meijere, A., *Angew. Chem. Int. Ed.* **2006**, *45*, 5176–5179.
11. Grigg, R.; Gunaratne, H. Q. N., *Chem. Commun.* **1982**, 384–386.
12. Grigg, R.; Gunaratne, H. Q. N.; Sridharan, V., *Tetrahedron* **1987**, *43*, 5887–5898.
13. Barr, D. A.; Grigg, R.; Gunaratne, H. Q. N.; Kemp, J.; Mcneekin, P.; Sridharan, V., *Tetrahedron* **1988**, *44*, 557–570.
14. Grigg, R.; Kemp, J., *Tetrahedron Lett.* **1980**, *21*, 2461–2464.
15. Grigg, R.; Montgomery, J.; Somasunderam, A., *Tetrahedron* **1992**, *48*, 10431–10442.
16. Amornraksa, K.; Barr, D. A.; Donegan, G.; Grigg, R.; Ratananukul, P.; Sridharan, V., *Tetrahedron* **1989**, *45*, 4649–4668.
17. Grigg, R.; Cooper, D. M.; Holloway, S.; McDonald, S.; Millingtona, E.; Sarkera, M. A. B., *Tetrahedron* **2005**, *61*, 8677–8685.
18. Pandey, G.; Lakshmaiah, G., *Tetrahedron Lett.* **1993**, *34*, 4861–4864.
19. Pandey, G.; Lakshmaiah, G.; Ghatak, A., *Tetrahedron Lett.* **1993**, *34*, 7301–7304.
20. Pandey, G.; Sahoo, A. K.; Bagul, T. D., *Org. Lett.* **2000**, *2*, 2299–2301.
21. Pandey, G.; Laha; Joydev K.; Lakshmaiah, G., *Tetrahedron* **2002**, *58*, 3525–3534.
22. Allway, P.; Grigg, R., *Tetrahedron Lett.* **1991**, *32*, 5817–5820.
23. Carreira, E. M.; Kværnø, L., *Classics in Stereoselective Synthesis*, Wiley-VCH, **2009**.
24. (a) Barr, D. A.; Dorrity, M. J.; Grigg, R.; Malone, J. F.; Montgomery, J.; Rajviroongit, S.; Stevenson, P., *Tetrahedron Lett.* **1990**, *31*, 6569–6572; (b) Barr, D. A.; Dorrity, M. J.; Grigg, R.; Hargreaves, S.; Malone, J. F.; Montgomery, J.; Redpath, J.; Stevenson, P.; Thornton-Pett, M., *Tetrahedron.* **1995**, *51*, 273–294.
25. (a) Nyerges, M.; Bendell, D.; Arany, A.; Hibbs, D. E.; Coles, S. J.; Hursthouse, M. B.; Groundwater, P. W.; Meth-Cohn, O., *Synlett* **2003**, 947–950; (b) Nyerges, M.; Bendell, D.; Arany, A.; Hibbs, D. E.; Coles, S. J.; Hursthouse, M. B.; Groundwater, P. W.; Meth-Cohn, O., *Tetrahedron* **2005**, *61*, 3745–3753.
26. Cooper, D. M.; Grigg, R.; Hargreaves, S.; Kennewell, P.; Redpath, J., *Tetrahedron* **1995**, *51*, 7791–7808.
27. Koot, W.-J.; Hiemstra, H.; Speckamp, W. N., *J. Org. Chem.* **1992**, *57*, 1958–1961.
28. Ruano, J. L. G.; Tito, A.; Peromingo, M. T., *J. Org. Chem.* **2002**, *67*, 981–987.

29. (a) Alonso, I.; Carretero, J. C.; Ruano, J. L. G., *Tetrahedron Lett.* **1989**, *30*, 3853–3856; (b) Alonso, I.; Carretero, J. C.; Ruano, J. L. G.; Cabrejas, M. L. M.; Lopez-Solera, I.; Raithby, P. R., *Tetrahedron Lett.* **1994**, *35*, 9461–9464.
30. Ruano, J. L. G.; Tito, A.; Peromingo, M. T., *J. Org. Chem.* **2003**, *68*, 10013–10019.
31. (a) Posner, G. H.; Hulce, M.; Mallamo, J. P., *J. Org. Chem.* **1981**, *46*, 5244–5246; (b) Posner, G. H.; Mallamo, J. P.; Miura, K., *J. Am. Chem. Soc.* **1981**, *103*, 2886–2888; (c) Posner, G. H.; Asirvatham, E., *J. Org. Chem.* **1985**, *50*, 2589–2591. (d) Posner, G. H., *Acc. Chem. Res.* **1987**, *20*, 72–78.
32. Paquette, L. A.; Tae, J.; Arrington, M. P.; Sadoun, A. H., *J. Am. Chem. Soc.* **2000**, *122*, 2742–2748.
33. Nájera, C.; De Gracia Retamosa, M.; Sansano, J. M., *Tetrahedron Asym.* **2006**, *17*, 1985–1989.
34. Nájera, C.; De Gracia Retamosa, M.; Sansano, J. M.; De Cozar, A.; Cossio, F. P., *Eur. J. Org. Chem.* **2007**, 5038–5049.
35. Ayerbe, M.; Arrieta, A.; Cossio, F. P.; Linden, A., *J. Org. Chem.* **1998**, *63*, 1795–1805.
36. Nyerges, M.; Rudas, M.; Toth, G.; Herenyi, B.; Kadas, I.; Bitter, I.; Toke, L., *Tetrahedron* **1995**, *51*, 13321–13330.
37. Erkizia, E.; Aldaba, E.; Vara, Y.; Arrieta, A.; Gornitzka, H.; Cossio, F. P., *Arkivoc* **2005**, 189–199.
38. Zubia, A.; Mendoza, L.; Vivanco, S.; Aldaba, E.; Carrascal, T.; Lecea, B.; Arrieta, A.; Zimmerman, T.; Vidal-Vanaclocha, F.; Cossio, F. P., *Angew. Chem. Int. Ed.* **2005**, *44*, 2903–2907.
39. Garner, P.; Kaniskan, H. U., *J. Org. Chem.* **2005**, *70*, 10868–10871.
40. Garner, P.; Kaniskan, H. U.; Hu, J.; Youngs, W. J.; Panzner, M., *Org. Lett.* **2006**, *8*, 3647–3650.
41. Vandewalle, M.; Van Der Eycken, J.; Oppolzer, W.; Vullioud, C., *Tetrahedron* **1986**, *42*, 4035–4043.
42. Garner, P.; Kaniskan, H. U., *Tetrahedron Lett.* **2005**, *46*, 5181–5185.
43. For the conversion of structure 80 into cyanocycline A, see Fukuyama, T.; Li, L.; Laird, A. A.; Frank, R. K., *J. Am. Chem. Soc.* **1987**, *109*, 1587–1589. For conversion of cyanocycline A into bioxolamycin β2, see Fukuyama, T., *Adv. Heterocycl. Chem.* **1992**, *2*, 189–249.
44. Kaniskan, H. U.; Garner, P., *J. Am. Chem. Soc.* **2007**, *129*, 15460–15461.
45. Kahsai, A. W.; Cui, J.; Kaniskan, H. U.; Garner, P. P.; Fenteany, G., *J. Biol. Chem.* **2008**, *283*, 24534–24545.
46. Grigg, R., *Tetrahedron: Asymmetry* **1995**, *6*, 2475–2486.
47. (a) Longmire, J. M.; Wang, B.; Zhang, X., *Tetrahedron Lett.* **2000**, *41*, 5435–5439; (b) You, S. -L.; Hou, X. -L.; Dai, L. -X; Gao, B. -X.; Sun, J., *Chem. Commun.* **2000**, 1933–1934.
48. Xiaodong Li, C. C.; Schreiber, S. L., *J. Am. Chem. Soc.* **2003**, *125*, 10174–10175.
49. Kocovsky, P.; Vyskocil, S.; Smrcina, M., *Chem. Rev.* **2003**, *103*, 3213–3245.
50. Knöpfel, T. F.; Aschwanden, P.; Ichikawa, T.; Watanabe, T.; Carreira, E. M., *Angew. Chem. Int. Ed.* **2004**, *43*, 5971–5973.
51. Zeng, W.; Zhou, Y.-G., *Org. Lett.* **2005**, *7*, 5055–5058.

52. Zeng, W.; Zhou, Y.-G., *Tetrahedron Lett.* **2007**, *48*, 4619–4622.
53. Zeng, W.; Chen, G.-Y.; Zhou, Y.-G.; Li, Y.-X., *J. Am. Chem. Soc.* **2007**, *129*, 750–751.
54. Stohler, R.; Wahl, F.; Pfaltz, A., *Synthesis* **2005**, 1431–1436.
55. Nájera, C.; De Gracia Retamosa, M.; Sansano, J. M., *Org. Lett.* **2007**, *9*, 4025–4028.
56. Cabrera, S.; Arrayas, R. G.; Martin-Matute, B.; Cossio, F. P.; Carretero, J. C., *Tetrahedron* **2007**, *63*, 6587–6602.
57. Nájera, C.; De Gracia Retamosa, M.; Sansano, J. M., *Angew. Chem. Int. Ed.* **2008**, *47*, 6055–6058.
58. Alemparte, C.; Blay, G.; Jørgensen, K. A., *Org. Lett.* **2005**, *7*, 4569–4572.
59. Grigg, R.; Lansdell, M. I.; Thornton-Pett, M., *Tetrahedron* **1999**, *55*, 2025–2044.
60. Peddibhotla, S.; Tepe, J. J., *J. Am. Chem. Soc.* **2004**, *126*, 12776–12777.
61. Grigg, R.; Kilner, C.; Sarker, M. A. B.; De la Cierva, C. O.; Dondas, H. A., *Tetrahedron* **2008**, *64*, 8974–8991.
62. Dondas, H. A.; Durust, Y.; Grigg, R.; Slatera, M. J.; Sarkera, M. A. B., *Tetrahedron* **2005**, *61*, 10667–10682.
63. Bonini, B. F.; Boschi, F.; Franchini, M. C.; Fochi, M.; Fini, F.; Mazzanti, A.; Ricci, A., *Synlett* **2006**, 543–546.
64. Subramaniyan, G.; Raghunathan, R., *Tetrahedron* **2001**, *57*, 2909–2913.
65. Subramaniyan, G.; Raghunathan, R.; Castro, A. M. M., *Tetrahedron* **2003**, *59*, 335–340.
66. Alcaide, B.; Almendros, P.; Redondo, M. C.; Ruiz, M. P., *J. Org. Chem.* **2005**, *70*, 8890–8894.
67. (a) Pätzel, M.; Galley, G., *Tetrahedron Lett.* **1993**, *34*, 5707–5710; (b) Galley, G.; Liebscher, J.; Pätzel, M., *J. Org. Chem.* **1995**, *60*, 5005–5010.
68. Bashiardes, G.; Cano, C.; Mauzé, B., *Synlett* **2005**, 587–590.
69. Grigg, R.; Sridharan, V.; Wang, J.; Xu, J., *Tetrahedron* **2000**, *56*, 8967–8976.
70. Grigg, R.; Millington, E. L.; Thornton-Pett, M., *Tetrahedron Lett.* **2002**, *43*, 2605–2608.
71. Dondas, H. A.; Duraisingham, J.; Grigg, R.; Maclachlan, W. S.; Macpherson, D. T.; Thornton-Pett, M.; Sridharan, V.; Suganthan, S., *Tetrahedron* **2000**, *56*, 4063–4070.
72. Grigg, R.; Thornton-Pett, M.; Yoganathan, G., *Tetrahedron* **1999**, *55*, 8129–8140.
73. Dondas, H. A.; Grigg, R.; Maclachlan, W. S.; Macpherson, D. T.; Markandu, J.; Sridharan, V.; Suganthan, S., *Tetrahedron Lett.* **2000**, *41*, 967–970.
74. Blaney, P.; Grigg, R.; Rankovic, Z.; Thornton-Pett, M.; Xu, J., *Tetrahedron* **2002**, *58*, 1719–1737.
75. Su, S.; Porco, J. A., *J. Am. Chem. Soc.* **2007**, *129*, 7744–7745.
76. Billups, W. E.; Chow, W. Y.; Smith, C. V., *J. Am. Chem. Soc.* **1974**, *96*, 1979–1980.
77. Saito, K.; Ono, K.; Ohkita, M.; Fukaya, M.; Ono, K.; Kondo, Y., *Heterocycles* **2002**, *60*, 773–778.
78. Saito, K.; Ono, K.; Ito, N.; Tada, N.; Ando, S., *Heterocycles* **2002**, *57*, 235–240.
79. Jendrella, H., *Angew. Chem. Int. Ed. Engl.* **1980**, *19*, 1032–1033.
80. Mukaiyama, T.; Watanabe, K.; Shiina, I., *Chem. Lett.* **1995**, *24*, 1–2.
81. Fraile, J. M.; Garcia, J. I.; Mayoral, J. A.; Pires, E.; Tarnai, T.; Figueras, F., *Appl. Catal. A* **1996**, *136*, 113–123.

82. Hague, C.; Patmore, N. J.; Frost, C. G.; Mahon, M. F.; Weller, A. S., *Chem. Commun.* **2001**, 2286–2287.

83. Patmore, N. J.; Hague, C.; Cotgreave, J. H.; Mahon, M. F.; Frost, C. G.; Weller, A. S., *Chem. Eur. J.* **2002**, *8*, 2088–2098.

84. Loncaric, C.; Manabe, K.; Kobayashi, S., *Adv. Synth. Catal.* **2003**, *345*, 475–477.

85. Yao, S.; Johannsen, M.; Hazell, R. G.; Jørgensen, K. A., *Angew. Chem. Int. Ed.* **1998**, *37*, 3121–3124.

86. Josephsohn, N. S.; Snapper, M. L.; Hoveyda, A. H., *J. Am. Chem. Soc.* **2003**, *125*, 4018–4019.

87. Kawasaki, M.; Yamamoto, H., *J. Am. Chem. Soc.* **2006**, *128*, 16482–16483.

3

SIGMATROPIC REARRANGEMENTS AND RELATED PROCESSES PROMOTED BY SILVER

JEAN-MARC WEIBEL, AURÉLIEN BLANC, AND PATRICK PALE

Laboratory for Organic Synthesis and Reactivity, Institute of Chemistry, University of Strasbourg, France

Silver in Organic Chemistry Edited by Michael Harmata

3.1 INTRODUCTION

Sigmatropic shifts represent a large class of reactions involving the migration of at least one sigma bond. Therefore, such migrations lead to skeletal rearrangements of the carbon frame within the molecule undergoing this reaction.

Sigmatropic rearrangements usually involve σ bonds adjacent to a π system or a σ bond included in a strained system. As other transition metals, but with specific properties due to its d^{10} electronic configuration, *f* orbitals and a relativistic effect,[1] silver easily interacts with such systems. Silver salts have thus been explored as catalysts to facilitate and promote sigmatropic rearrangements.

3.2 WOLFF AND ARNDT–EISTERT REARRANGEMENTS AND RELATED REACTIONS

The Wolff and Arndt–Eistert rearrangements are probably among the earliest known reactions promoted by silver ions.[2,3] Discovered at the turn of the nineteenth/twentieth century, the Wolff rearrangement allows the transformation of α-diazoketones to carboxylic acids,[4] while the Arndt–Eistert rearrangement is a similar sequence also leading to carboxylic acids, but including the preparation of α-diazoketones from a shorter acid chloride (Scheme 3.1).[5]

Numerous conditions have been developed for this transformation, but reproducible yields have usually been obtained by mixing a silver salt with a coreagent, such as silver nitrate associated with wet ammonia, silver oxide with triethylamine or sodium thiosulfate, and silver benzoate with triethylamine. Nonbasic conditions have also been described by Koch and Podlech using silver trifluoroacetate deposited on silica.[6] These modifications have been developed for the homologation of Fmoc-protected amino acids.

Mechanistic studies suggested that the in situ reduction of silver salts to silver nanoclusters (Ag_n) produces the true reagent in the Wolff rearrangement.[7] As a result, such reactions have been improved by directly using silver nanoclusters.[8]

The Wolff rearrangement and the Arndt–Eistert homologation sequence are very useful in organic synthesis. One of the most popular applications involves amino acids. An interesting example has been described as a key reaction in the synthesis of a ^{14}C-labeled amino acid used for deciphering the biosynthesis of penicillin N from glutamic acid (Scheme 3.2).[9] This rearrangement proceeds without racemization and can thus be applied in peptide synthesis.

Indeed, applied to α-amino acids, the Arndt–Eistert rearrangement leads to β-amino acids with retention of the configuration of the migrating group. Such

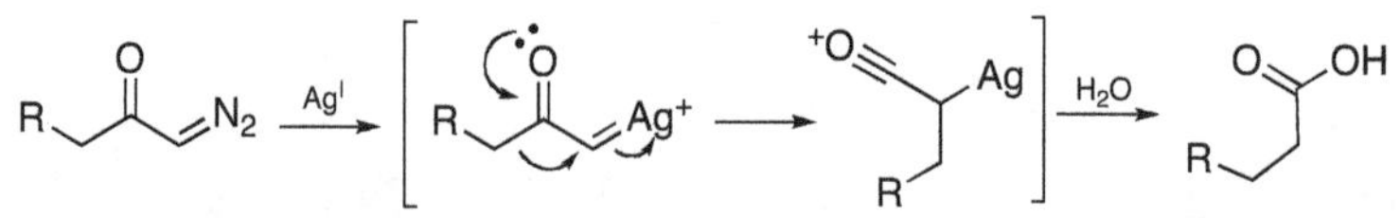

Scheme 3.1

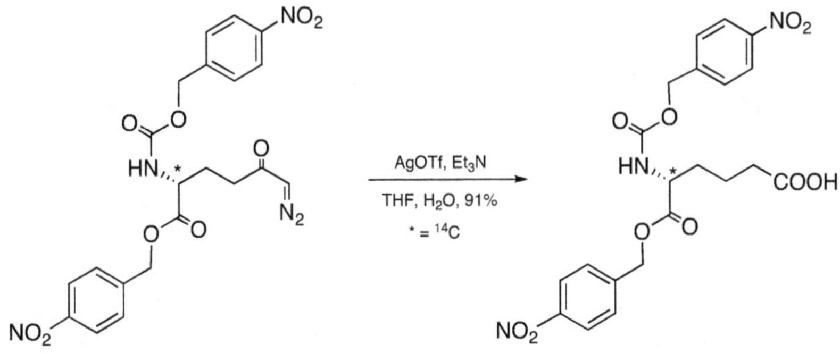

Scheme 3.2

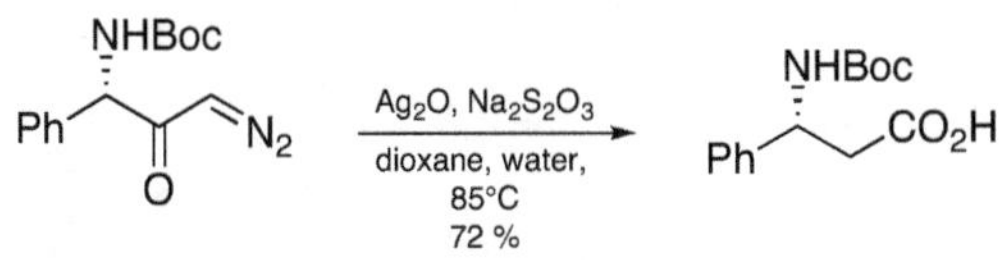

Scheme 3.3

compounds can be incorporated in nonproteinogenic peptides, inducing turns and other special secondary structures (Scheme 3.3).[10]

Since ketene is probably the intermediate of the Wolff rearrangement, the choice of solvents dictates the nature of the product. Indeed, water gave carboxylic acids, whereas alcohols or amines led to esters and amides, respectively. These combinations have been applied to the synthesis of more complex molecules. For example, the total synthesis of carbonolide B, a 16-membered macrolide antibiotic, relied on Arndt–Eistert homologation. In this sequence, a protected furanuronic acid was transformed to the corresponding α-diazoketone, which was then converted to its homologous carboxylic ester. The reaction was achieved using catalytic amounts of silver benzoate and excess of triethylamine in methanol (Scheme 3.4).[11]

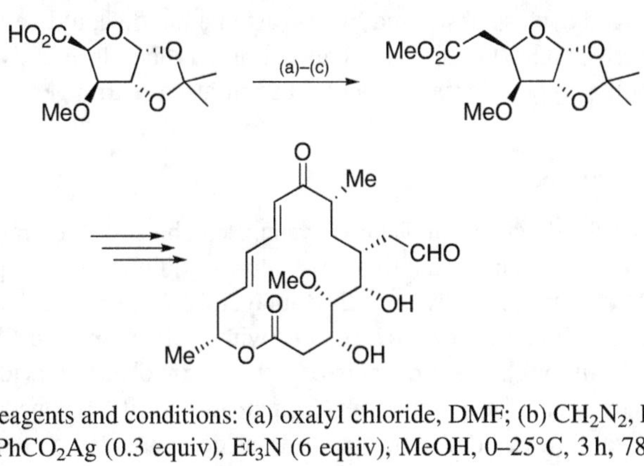

Scheme 3.4 Reagents and conditions: (a) oxalyl chloride, DMF; (b) CH_2N_2, Et_2O, 76% over two steps; (c) $PhCO_2Ag$ (0.3 equiv), Et_3N (6 equiv), MeOH, 0–25°C, 3 h, 78%.

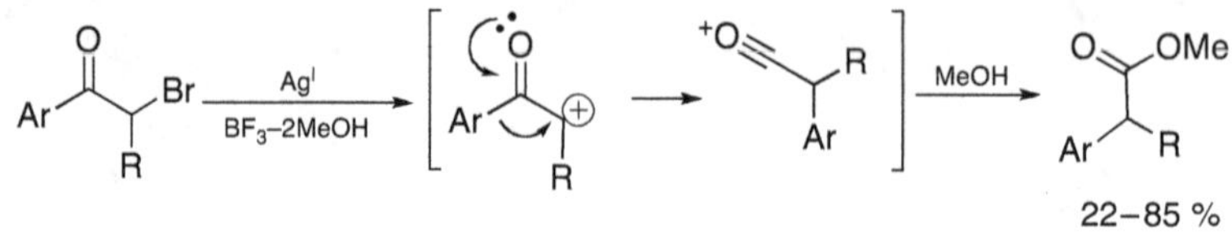

Scheme 3.5

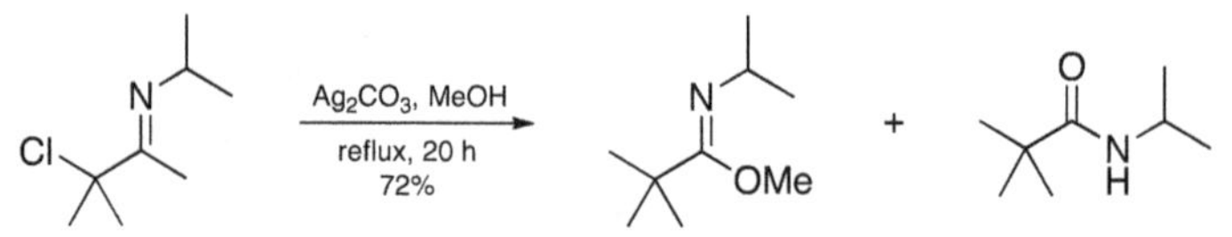

Scheme 3.6

Looking for a new route toward arylalkanoic acids, Giordano et al. observed that silver-assisted methanolysis of primary and secondary α-bromoalkyl aryl ketones led to a mixture of substitution and rearrangement products. Following further studies, these authors developed conditions that mainly provided the rearrangement product, an arylalkanoic ester. The yields were better with electron-rich aromatic groups and with alkyl substituents, suggesting a mechanism similar to the Arndt–Eistert rearrangement (Scheme 3.5).[12]

The silver-induced reactions of α-haloimines have been compared to those of α-haloketones (Scheme 3.6).[13] These silver-assisted reactions of α-haloimines have been interpreted in terms of the α-imidoylcarbenium intermediates, which rearrange in a fashion analogous to that of the corresponding ketones.

3.3 RING REARRANGEMENTS

Strained or bridged cycles substituted by at least one halide could be rearranged on treatment with silver salts. On the other hand, strained σ bonds directly interact with silver ions, leading to bond breakage and thus initiating rearrangement.

3.3.1 Halogenoamines

N-Chloramines embedded within bicyclic systems can be easily rearranged into aza-bridged and enlarged cyclic derivatives. *N*-Chloroisoquinuclidines, specifically, 2-azabicyclo[2.2.2]octane derivatives, or 2-azabicyclo[2.2.1]heptane derivatives gave 1-azabicyclo[3.2.1]heptanes on treatment with silver nitrate in methanol (Scheme 3.7).[14] Although the mechanism probably involved chloride abstraction by silver and iminium formation after sigmatropic shift, a concerted pathway catalyzed by silver ions could also be operative, as demonstrated by the preservation

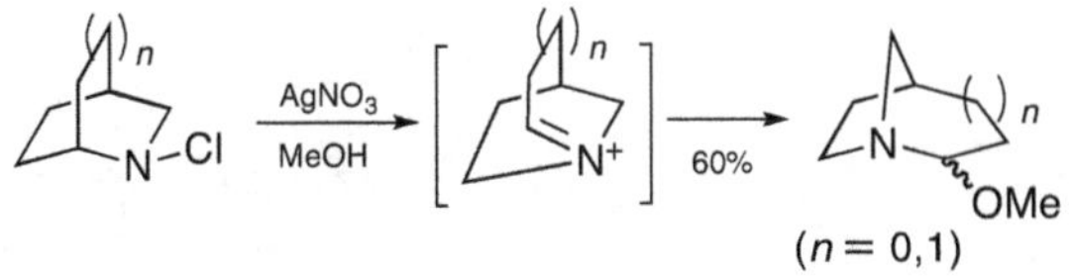

Scheme 3.7

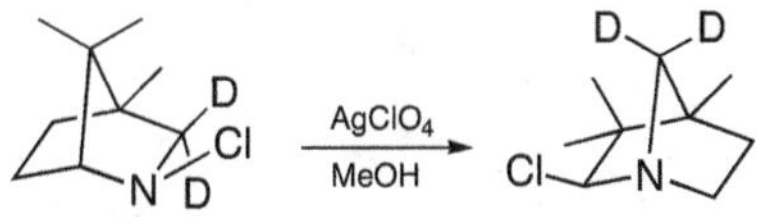

Scheme 3.8

of chloride in the rearrangement of deuterated *N*-chloro-2-azabicyclo[2.2.1]heptane derivatives in the presence of silver perchlorate (Scheme 3.8).[15]

Further studies revealed that the participation of both σ and π electrons played an important role in controlling the stereochemistry of such rearrangements.[16,17]

Of particular interest were the 7-aza analogs, for which the stereochemistry at the nitrogen atom is well defined, due to a higher inversion barrier relative to other amino derivatives. For such compounds, the silver-promoted rearrangement of each stereoisomer at the nitrogen atom (invertomers) proved to be dependent on the configuration at nitrogen. Thus, the *syn* invertomer of 7-azabenzonorbornadiene gave the normal solvolysis product, resulting from π participation of the alkene moiety in chloride abstraction by silver ion followed by *anti* addition of methanol (Scheme 3.9, top). In contrast, the *anti* isomer led to a rearranged product, a 1-azabicyclo[3.2.0]hept-3-ene derivative. The latter arose through π participation of the benzene unit in chloride loss, followed by methanol opening of the thus-formed multibridged intermediate at the position that regenerated benzene aromaticity (Scheme 3.9, bottom).[18–20]

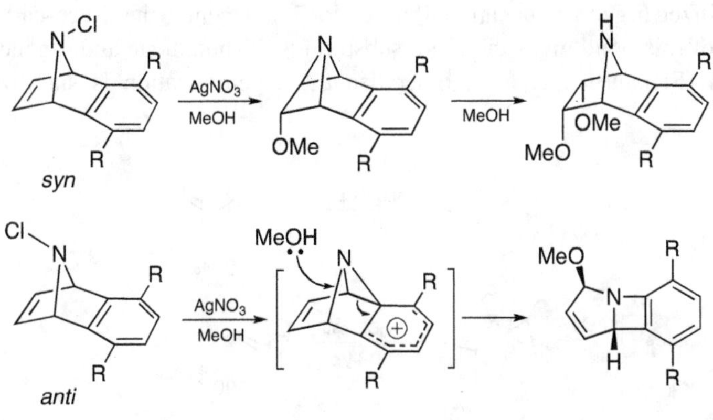

Scheme 3.9

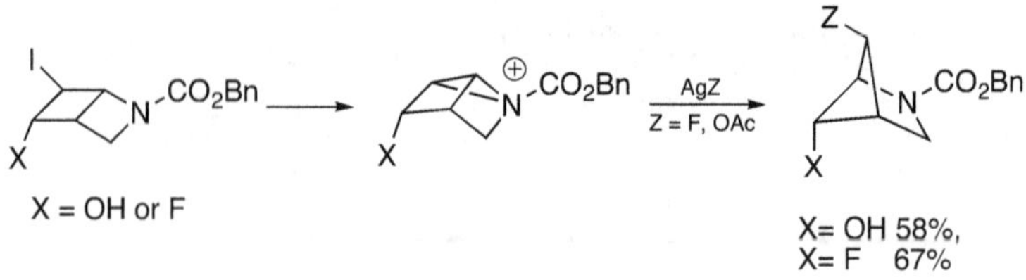

Scheme 3.10

In a similar way, 6-*exo*-iodo or bromo-2-azabicyclo[2.2.0]hexanes rearranged into 5,6-difunctionalized 2-azabicyclo[2.1.1]hexanes on treatment with silver or mercury salts. An aziridinium intermediate was probably formed in the presence of silver ions. Opening of this aziridinium then occurred in a stereocontrolled manner by addition of the silver counterion (Scheme 3.10).[21]

3.3.2 Cyclopropane Derivatives

Strained cyclic systems such as cyclopropane derivatives proved sensitive to silver ions. Indeed, silver usually induced ring opening of cyclopropanes, leading to argento cationic intermediates, which evolved further depending on substituents. Most, if not all, of these ring rearrangements resulted from initial formation of a σ complex of the cyclopropane with silver ion.

Bicyclo[1.1.0]butanes are typical of such behavior. They can be considered as joined cyclopropanes and as such, readily react in the presence of catalytic amount of silver salts. These compounds usually lead to 1,3-butadienes (Scheme 3.11).[22–24]

Screening and kinetic investigations of these rearrangements suggested that the argento cationic intermediate formed on interaction of silver ion with one cyclopropane unit could further rearrange through ring opening of the second cyclopropane unit, but that the latter depended on the nature of the substituent. With alkyl or aryl substituents able to stabilize cations, opening seems to proceed in order to form the most stabilized organosilver cation. If the cation is adjacent to the silver–carbon bond, cleavage of this bond gives the more substituted 1,3-butadiene and regenerates the silver ion (Scheme 3.12, top); otherwise the argento cation is stabilized by a

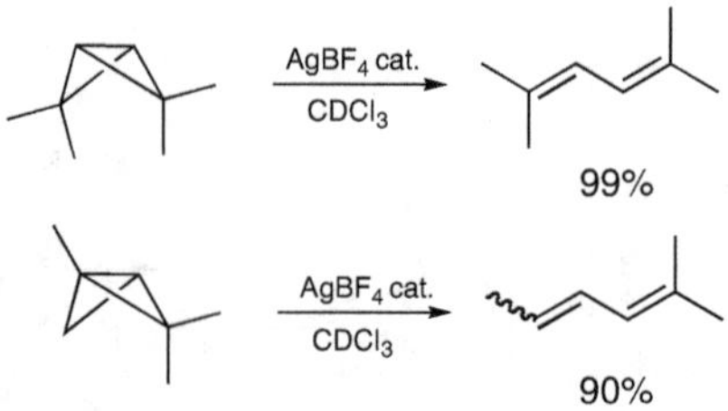

Scheme 3.11

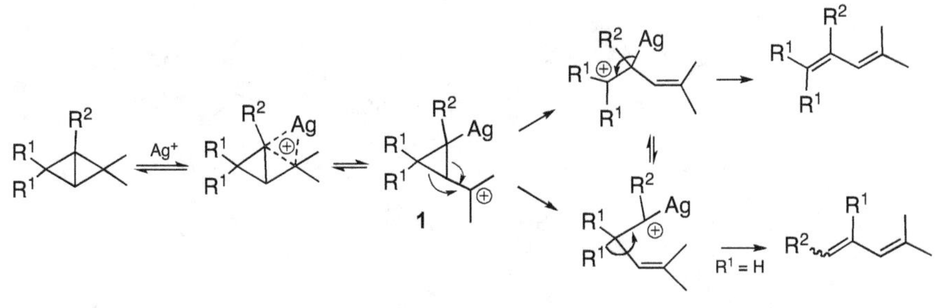

Scheme 3.12

sigmatropic shift, especially H shifts, leading to another 1,3-butadiene (Scheme 3.12, bottom).

It is worth noting that both secondary argento cation intermediates could be in equilibrium through either an H or vinyl shift, depending on the nature of the R^1 group (Scheme 3.12).

Parallel studies by Masamune et al. led nevertheless to the conclusion that the argento carbonium ion **1** proposed by Paquette et al. may not necessarily be involved in such Ag^I-catalyzed rearrangements.[25,26]

Such rearrangements have been exploited to homologate cyclic dienes, to prepare bridged annulenes and cyclohepta-1,3,5-trienes (Scheme 3.13).[27,28]

Interestingly, combining such fragmentation and rearrangement with hyperconjugative stabilization of the cyclopropylcarbinyl cation when a silyl or comparable substituent at the adequate position allowed the control of reaction course. Thus, 1-substituted norcarenes were obtained mainly after treatment of silylated or germylated tricyclo[4.1.0.0^{2,7}]heptanes with a catalytic amount of silver perchlorate (Scheme 3.14). The "classical" rearrangement product, a 1,3-cycloheptadiene in this case, was nevertheless observed.[29]

This silver-catalyzed ring rearrangement could be explained by preferential interaction of the silver cation with the most substituted edge of the cyclopropane to form an argento cation. In the latter, the presence of C–Si or Ge bond with the right orientation stabilized the cation sufficiently so that it classically evolved through β-H

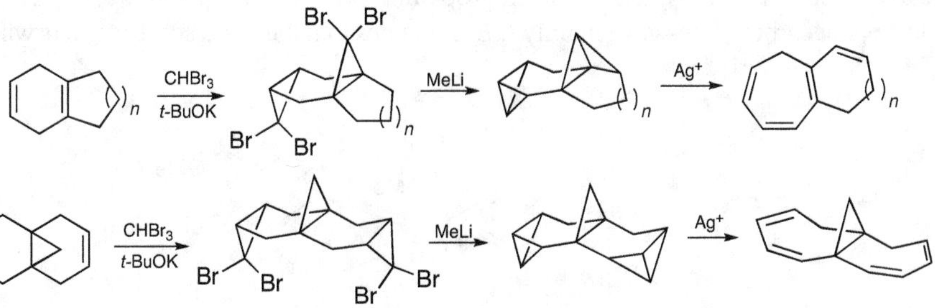

Scheme 3.13

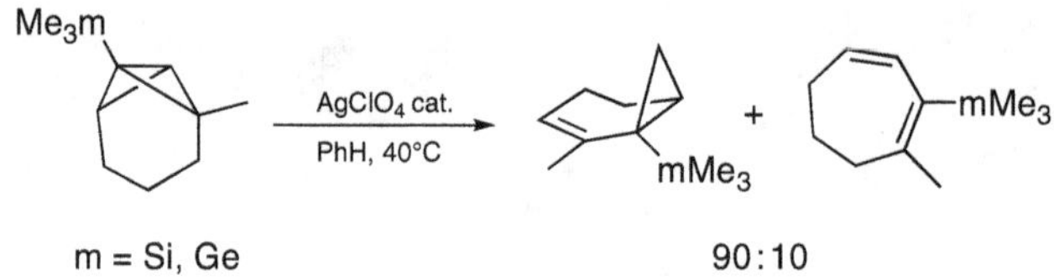

Scheme 3.14

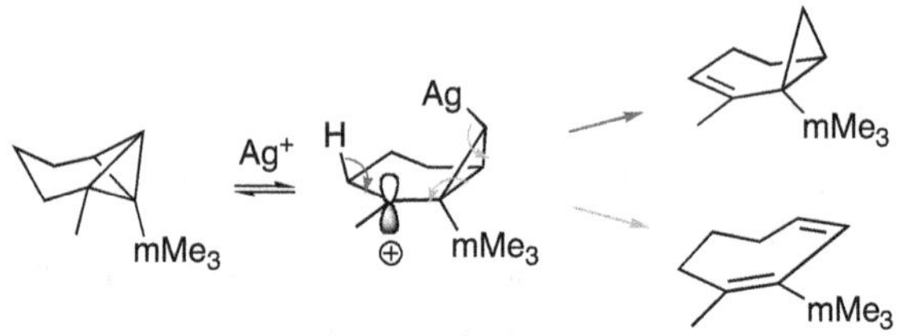

Scheme 3.15

elimination (Scheme 3.15, top), leading to the norcarene derivative. Nevertheless, the argentocyclopropyl cation could fragment as mentioned above and thus give a diene derivative (Scheme 3.15, bottom),

The highly strained cycloproparenes behave in the same way in the presence of silver salts. For example, the simplest member of this family gave the benzyl ether in a protic, nucleophilic solvent within a few minutes, while in an anhydrous and nonnucleophilic solvent, such as chloroform, dimerization occurred (Scheme 3.16).[30]

The even more strained alkylidene cycloproparenes gave rise to the same kind of σ-complex intermediate with silver ion. In the presence of alcohol, trapping of this intermediate occurred, leading to alkoxystyrene derivatives. Water could also act in the same way, yielding arylmethylketones after keto–enol equilibration. However, if a proton was present on the alkylidene moiety, H shift occurred, leading to an arylalkyne. No dimerization was observed in this case, probably due to steric constraints in such a process (Scheme 3.17).[31]

As one would expect because of the increasing strain, cyclopropenes readily reacted with silver ions, leading to ring-opening products. For example, dialkylcyclopropenecarboxylates gave mainly *E,E*-dienoates together with some isomers when

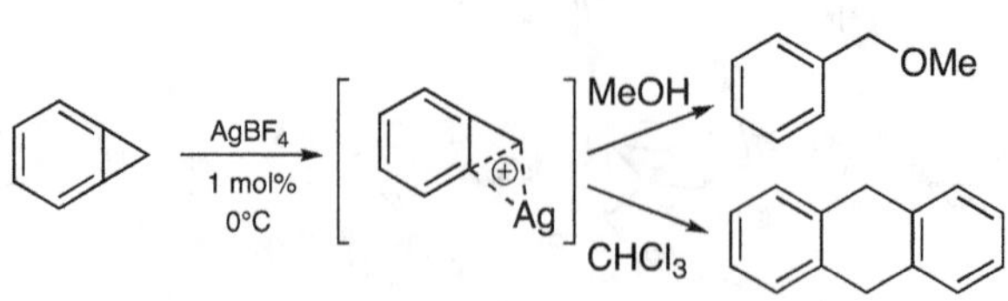

Scheme 3.16

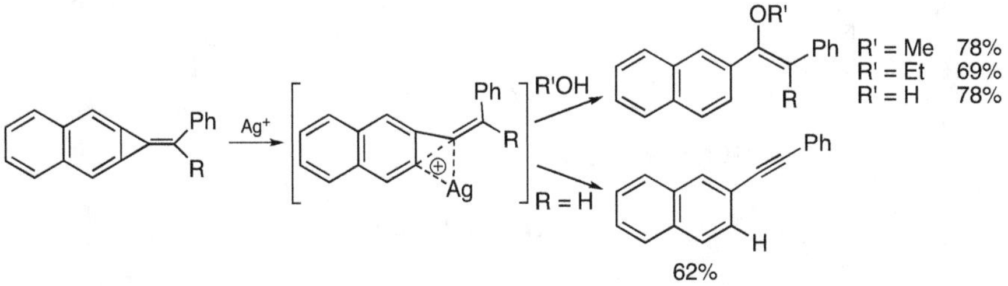

Scheme 3.17

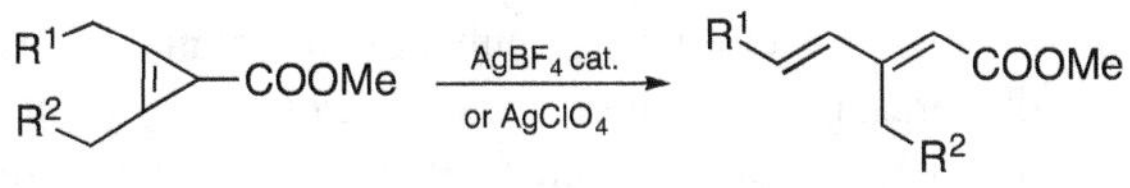

Scheme 3.18

submitted to a catalytic amount of silver perchlorate or tetrafluoroborate (Scheme 3.18).[32]

3,3′-Ethylenebicyclopropenyls have also been investigated as substrates for silver rearrangement because they would have elegantly led to Dewar benzenes by silver-catalyzed rearrangement of the bicyclopropenyl moiety. Unfortunately, treatment of such compounds with catalytic amounts of silver perchlorate in dioxane or deuterated chloroform afforded only polymeric materials. Replacing the solvent with methanol or MeOD induced the opening of one or both cyclopropene rings (Scheme 3.19).[33]

The mechanistic aspects of the silver(I)-promoted rearrangement of cyclopropene derivatives have been investigated, confirming preferential attack of Ag^+ on the σ bond to give an argentocarbenium ion. This intermediate is trapped by methanol, leading to a vinylsilver intermediate that is deuterated (Scheme 3.20).[33,34]

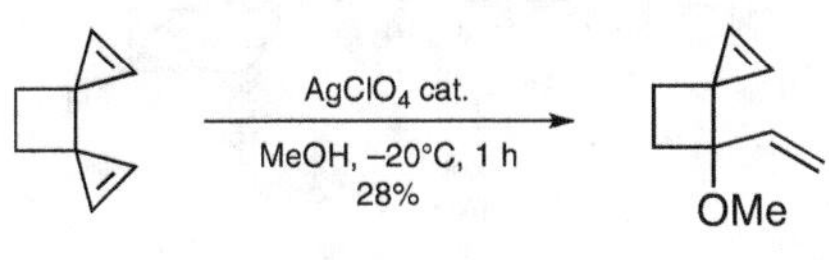

Scheme 3.19

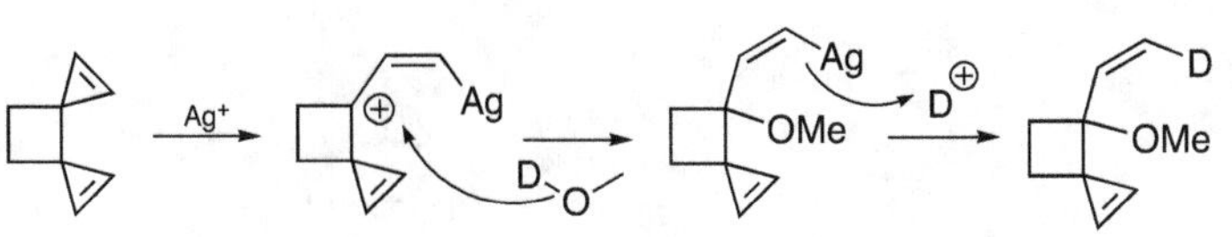

Scheme 3.20

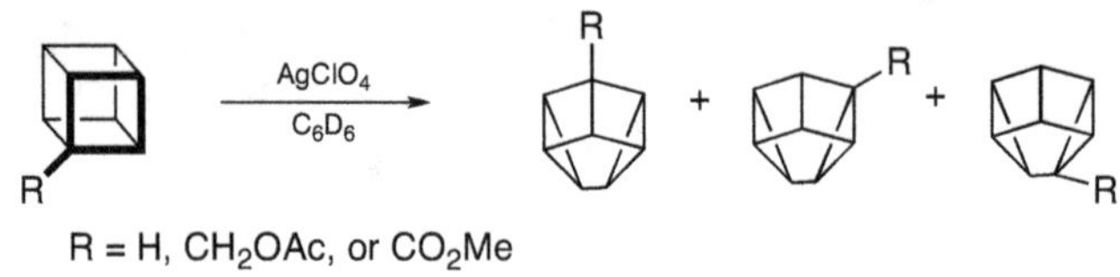

Scheme 3.21

3.3.3 Cubane Derivatives

The skeleta of other strained systems such as cubanes and related compounds can also be reorganized in the presence of silver salts. Rearrangements of cubanes were described in the early 1970s. Eaton et al.[35] showed that mono- or disubstituted cubanes gave a mixture of polycyclic regioisomers (Scheme 3.21). No yields were given.

Homocubanes were studied or used in synthesis more extensively. Paquette et al.[36,37] showed that the rates of homocubane isomerization to norsnoutane in benzene with silver perchlorate followed a second-order catalytic rate law (Scheme 3.22). Kinetic data suggested that a single site was involved in the silver interaction with homocubane.

This new reorganization of homocubyl-caged systems led to highly complex carbon frameworks. Heteroatoms could also be incorporated in the molecular skeleton (Scheme 3.23).[38]

Unfortunately, this reaction did not always lead to norsnoutane structures. Indeed, treatment of a 9-hydroxyhomocubane ester with silver perchlorate in refluxing benzene did not afford the expected norsnoutane derivative, but led instead to a

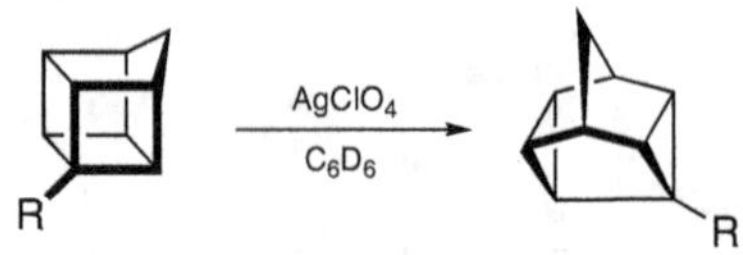

Scheme 3.22

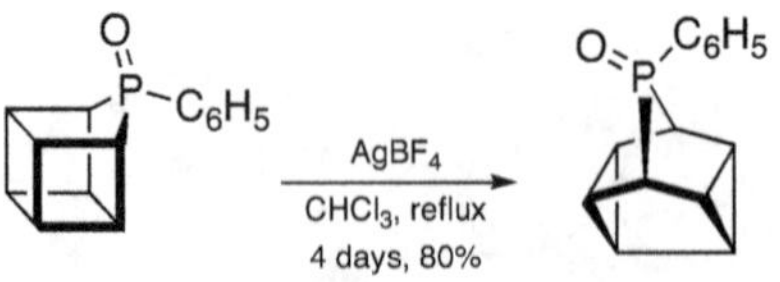

Scheme 3.23

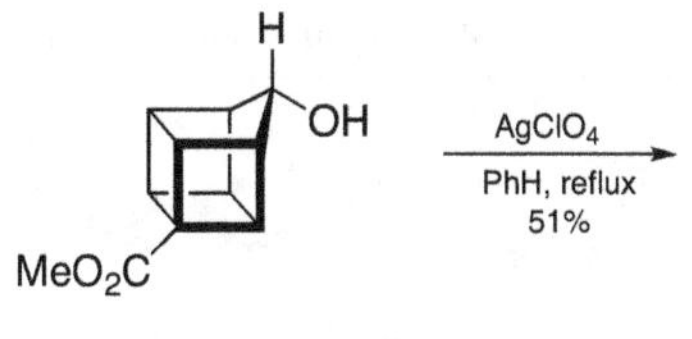

Scheme 3.24

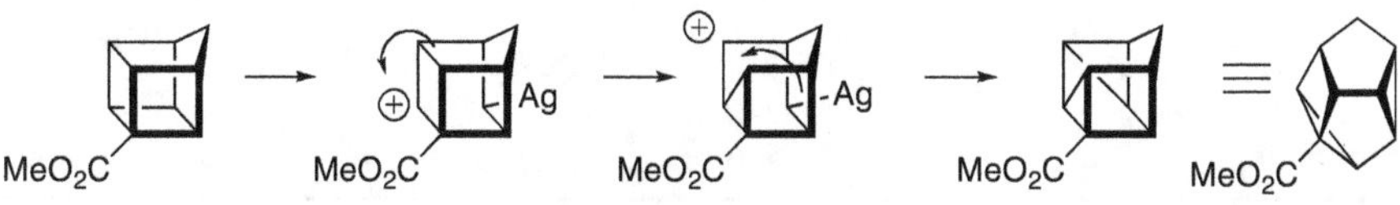

Scheme 3.25

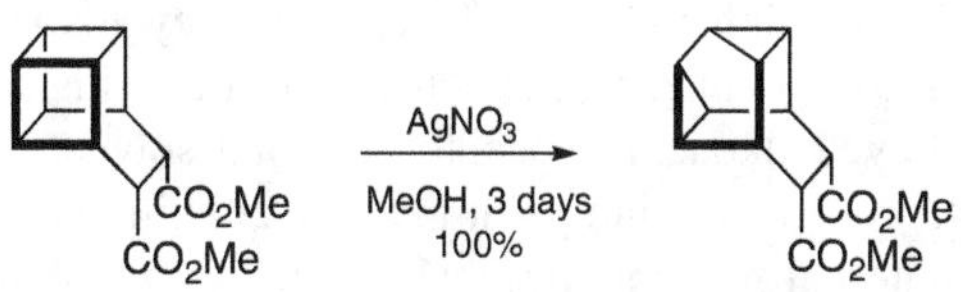

Scheme 3.26

different polycylic framework (Scheme 3.24). The course of the rearrangement is probably dictated by the stereoelectronic effect of the substituents.[39]

As for the cyclopropane rearrangement, the reaction probably involved the formation of a cationic silver cationic species followed by regioselective σ-bond migration, leading preferentially to the more stable carbocation intermediate that is finally intramolecularly trapped by the organosilver species (Scheme 3.25).[39]

Bishomocubanes have also been used as substrates for silver-catalyzed rearrangements (Scheme 3.26).[40]

3.3.4 Halogenocyclopropane Derivatives

Although usually described as driven by halide abstraction by silver ion, dibromocyclopropanes readily rearranged in the presence of silver salts in a process similar to those mentioned above. Such rearrangements usually led to ring expansion products.

E-Cyclooctene can easily be converted to its dibromocyclopropyl derivative, and the latter reacted with silver perchlorate in protic solvents, efficiently leading to the corresponding alkoxybromocyclononenes (Table 3.1).[41] Stereochemical studies starting from (−)-*E*-cyclooctene suggested the intermediate formation of a bromoallyl cation with a well-defined conformation, to which the nucleophile added (Table 3.1).[42,43]

TABLE 3.1. Silver-Catalyzed Formation of Alkoxybromocyclononene

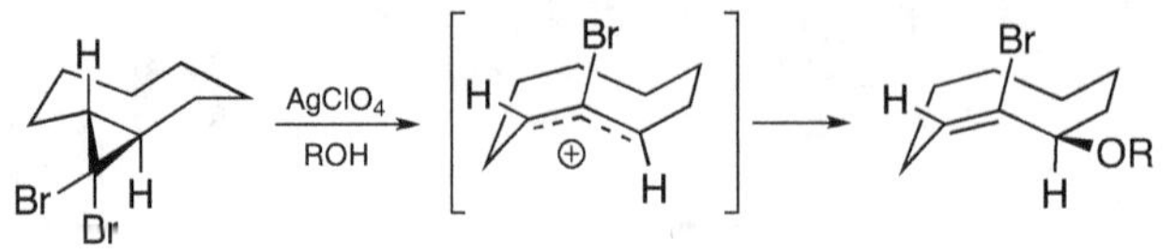

R	Time (min)	Yield (%)
H	10	82
Me	10	>99
Et	240	92
Ac	30	80

Complementary investigations revealed that the stereoselectivity of this ring expansion is indeed very high in favor of the *E*-(*cis*)-cycloalkenes when starting from 7- to 9-membered cycloalkenes, but is reversed if the ring size increases (from 10- to 13-membered cycloalkenes), especially in polar solvents.[44]

Combining cyclopropane opening with a nucleophilic attack could also be performed with an internal nucleophile, thus providing a way to form new rings. An interesting example was described by Danheiser et al.[45] in the mid-1980s. Dibromocyclopropanes carrying tethered alcohol or carboxylic acid functional group were efficiently converted to bromovinyl tetrahydrofurans or pyrans as well as to the corresponding lactones on treatment by silver salts (Scheme 3.27).

A very interesting extension of these bromocyclopropane opening reactions mediated by silver salts has been described by Collet et al.[46] Quinolines and isoquinolines could be homologated through a two-step process, including a silver-catalyzed ring expansion. These heterocycles were first alkylated by the anion derived from bromoform after *N*-benzylation. The resulting tribromomethyl derivatives were then treated with silver nitrate in a water/methanol solution (Scheme 3.28). The silver ion probably promoted the intermediate formation of a dibromoaziridinium ion. In the quinoline series, methanol added at the vinylic position and opened the aziridinium ion in an S_N2' reaction, while in the isoquinoline series, the aziridinium ion opening resulted from a direct S_N2 reaction. In both cases, hydrolysis led to the

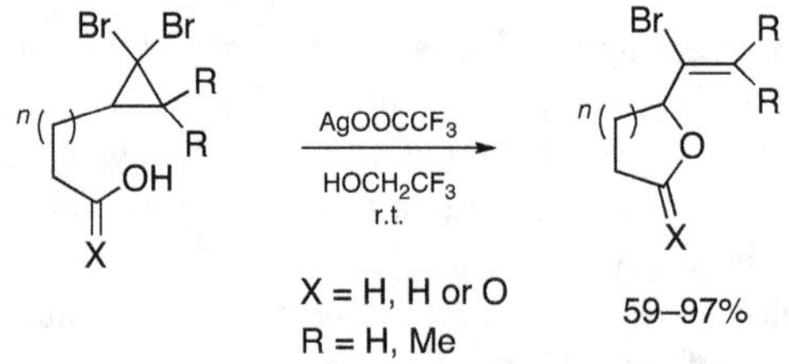

Scheme 3.27

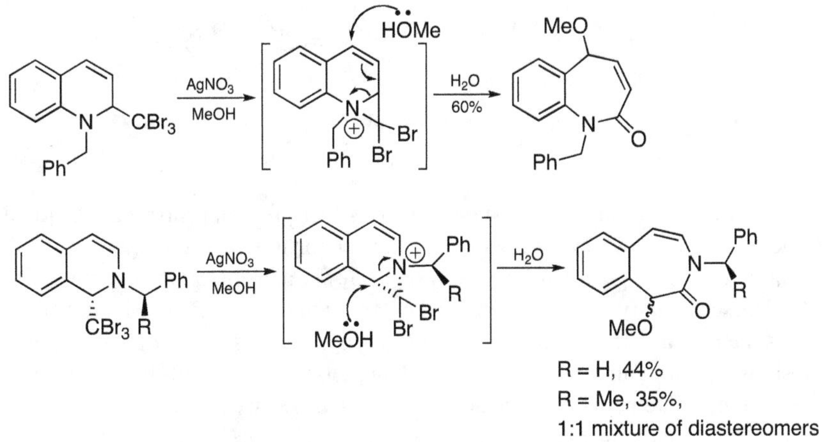

Scheme 3.28

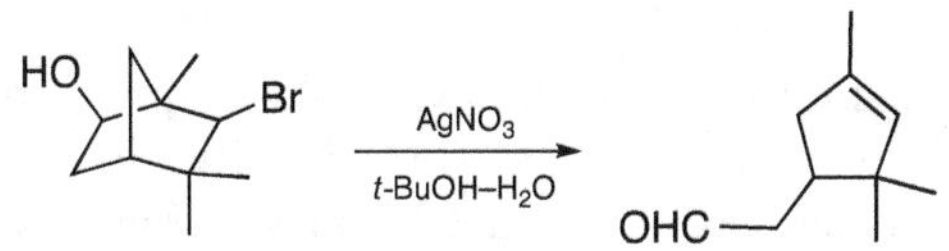

Scheme 3.29

corresponding seven-membered benzolactams. However, the same silver-catalyzed reaction applied to an optically pure *N*-phenethylisoquinoline derivative (R = Me) yielded a mixture of diastereoisomers, suggesting a more complex mechanism.

In a similar process, although not on cyclopropane, bromoisofenchol has been rearranged to campholenaldehyde, most probably by halide abstraction by silver ion (Scheme 3.29).[47] This reaction and related ones have been patented for the production of perfumes and essences.[48]

3.4 [3,3]-SIGMATROPIC REARRANGEMENTS

3.4.1 With Acyl as Migrating Groups

Rearrangements of propargyl esters with silver salts were first mentioned by Zakharova in the mid-1940s.[49] He described the conversion of 3-chloro-3-methylbut-1-yne into a mixture of acetates in which the allenic acetate, 1-acetoxy-3-methylbut-1,2-diene, was the major compound (Scheme 3.30). Although this product could arise from a silver assisted S_N2' reaction, it could also be produced from the substitution product through rearrangement, probably catalyzed by silver ions.

The first nonambiguous report of rearrangement of esters derived from propargyl alcohols promoted by silver salts was due to Saucy and coworkers at Hoffmann-La Roche.[50] Their observations on the transformation of acetate derived from

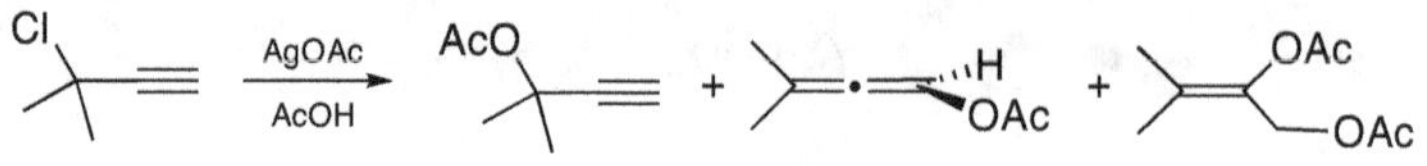

Scheme 3.30

dehydrolinalol clearly showed the superiority of silver salts over copper and gold salts and over zinc oxide, which had been used earlier for this rearrangement.[51]

Although already postulated in 1956 as [3,3]-sigmatropic rearrangement,[52] the mechanism, especially the stereochemical outcome of this reaction, was studied only 10 years later. In an elegant studies on C17-ethynyl steroids of biological relevance, Benn showed that treatment of 17α-ethynylandrost-5-ene-3β, 17β-diol with 5 mol% of silver perchlorate in dry acetone at reflux for 96 h cleanly afforded two isomers in 42% and 23% yields, respectively (Scheme 3.31). This poor selectivity was easily explained by equilibration experiments. Indeed, treatment of either pure acetoxyallene with 5 mol% of silver perchlorate in refluxing acetone for 24 h gave an equimolar mixture of both acetoxyallenes.[52]

Taking into account this lack of stereoselectivity and equilibration between allenic stereoisomers, Benn proposed a mechanism in which silver ion facilitates the acyloxy shift on formation of π complex (or a bridged ionic silver intermediate).[52] The thus-formed acetoxyallenes would remain coordinated to silver, allowing the formation of an organosilver allyl cation. Free rotation at this stage would provide both acetoxyallene isomers after reversible loss of silver (Scheme 3.32).

Further mechanistic work with optically active and labeled compounds confirmed that the rate-determining step in such rearrangement was the silver coordination to the alkyne moiety and revealed that the silver-catalyzed allene epimerization was 2–40 times faster than the propargyl ester rearrangement.[53]

Interestingly, these authors[53] and later others[54] showed the importance of the silver counterion in such rearrangements. While silver tetrafluoroborate or silver nitrate gave the [3,3]-sigmatropic shift product mainly or exclusively, silver trifluoroacetate yielded the dienyl acetate through isomerization from the allenic ester (Scheme 3.33).

It is worth noting that a silyl group at the propargyl position did not interfere with the silver-catalyzed rearrangement of a propargyl acetate, despite the potential for further evolution of the product (Scheme 3.34).[55]

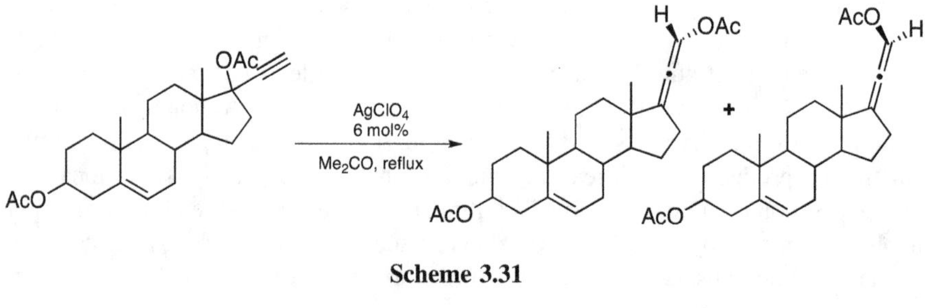

Scheme 3.31

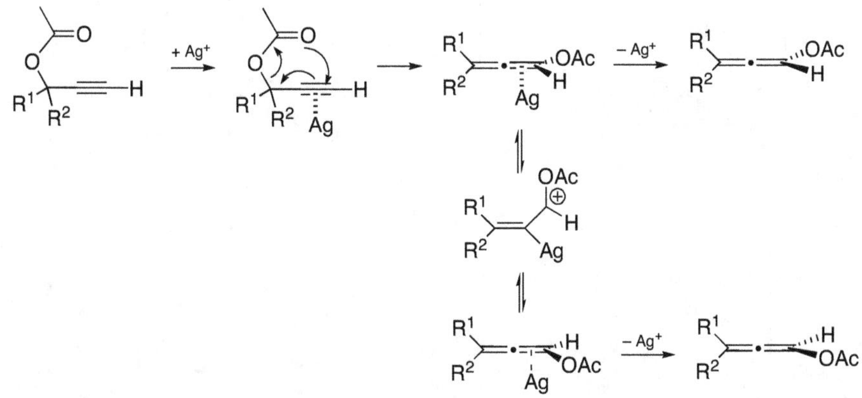

Scheme 3.32

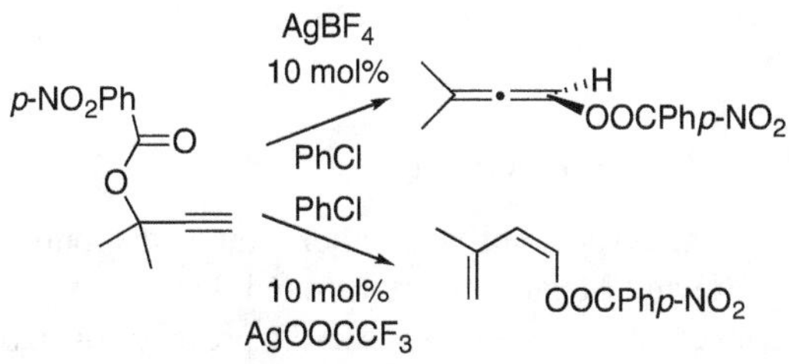

Scheme 3.33

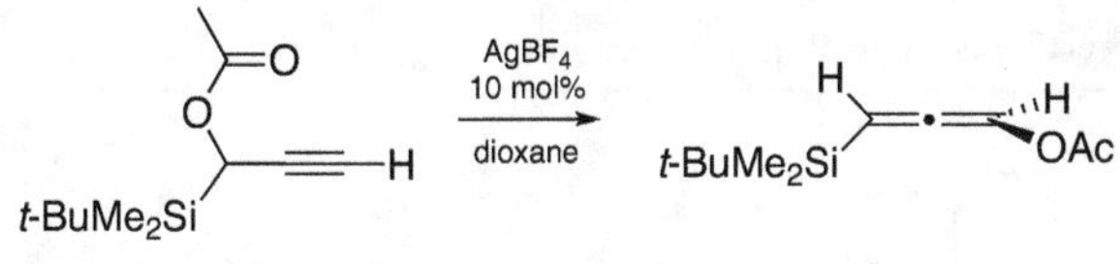

Scheme 3.34

[3,3]-Sigmatropic rearrangements catalyzed by silver salts have been used in total synthesis. The synthesis of three monoterpene compounds that comprise the pheromone of male boll weevil *Anthonomus grandis* has been achieved using the silver-carbonate catalyzed rearrangement of an acetoxycyclohexyl acetylene as the key step (Scheme 3.35).[56]

Ley et al. also applied this method to the synthesis of sesquiterpenes through a strategy involving a Diels–Alder reaction. Taking into account the effect of conditions and silver counterion on allene isomerization (see Scheme 33), they obtained the diene partner via isomerization of the acetoxyallene produced on treatment of a propargyl acetate with catalytic amount of silver hexafluoroantimonate (Scheme 3.36).[57]

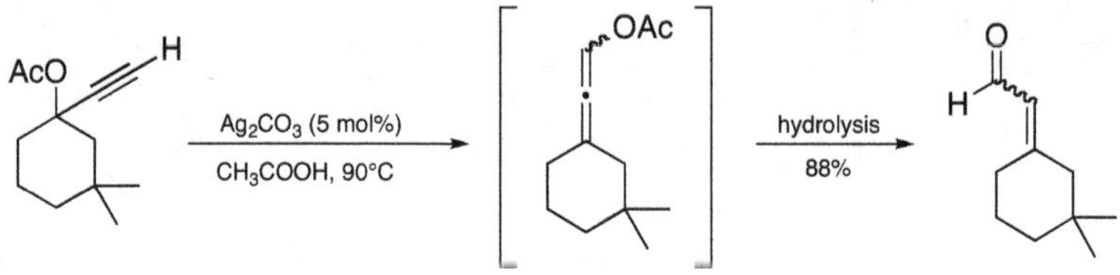

Scheme 3.35

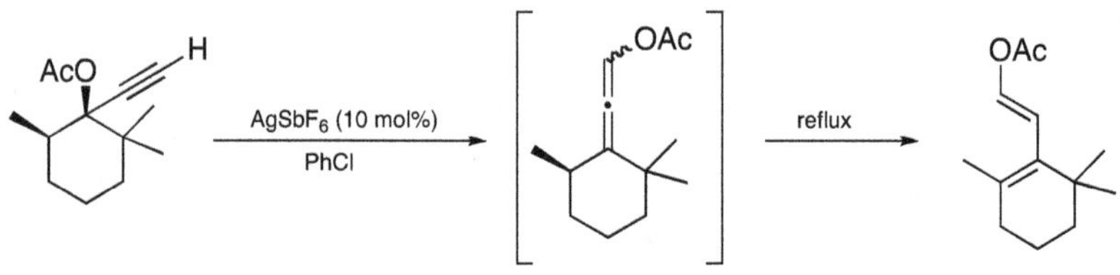

Scheme 3.36

3.4.2 With Vinyl as Migrating Groups

Reasoning that [3,3]-sigmatropic shifts of other than acyl groups could a priori be induced by silver, Goré and Malacria demonstrated that propargyl and homoallylic alcohols could be converted to hexadienones in the presence of silver salts through a oxy-Cope-type rearrangement (Table 3.2).[58] Silver nitrate and especially silver triflate proved to be the most effective promoters for this reaction. On the basis of

TABLE 3.2. Oxy-Cope-Type Rearrangements

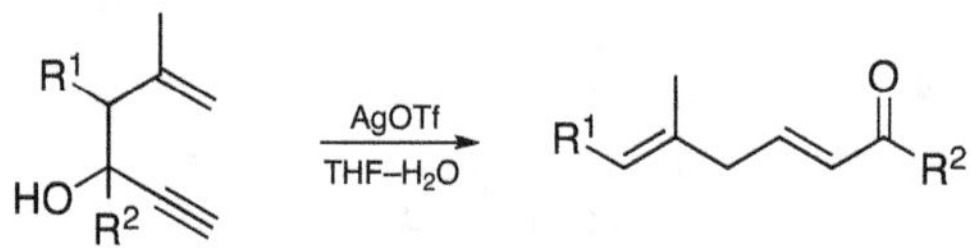

R^1	R^2	Temperature (°C)	Time (h)	Yield (%)
H	H	60	1	55
H	n-C_5H_{11}	40	7	73
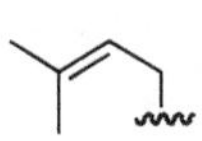	Me	40	20	57
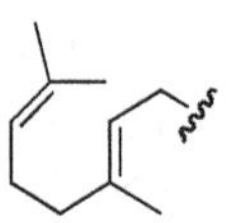	Me	60	30	55
-$(CH_2)_{10}$-		60	24	40

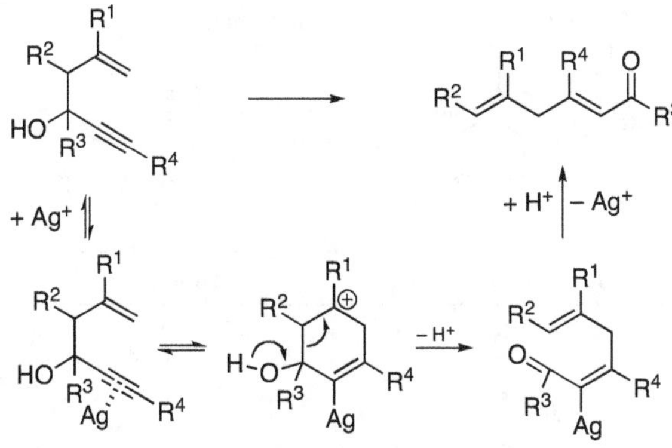

Scheme 3.37

mechanistic considerations, these authors devised a catalytic version of the reaction, using a combination of silver nitrate and potassium nitrate as catalysts.

In contrast to the preceding mechanisms proposed for [3,3]-sigmatropic shifts, the mechanism of the silver-catalyzed oxy-Cope rearrangement was proposed as a stepwise process (Scheme 3.37). As usual, the reaction would be initiated by silver coordination to the alkyne moiety. Nucleophilic attack of this complex by the double bond would then lead to a cyclic cationic vinylsilver intermediate. Fragmentation would then give the dienone.

This reaction exhibits strong analogy with the silver-promoted cationic aza-Cope rearrangement described by Overman et al. As a new route toward alkaloids, these authors showed that cyanomethylamines carrying unsaturated chain led to pyrrolidine or piperidine derivatives on treatment with silver salts depending on the substituent (Scheme 3.38).[59,60]

Further mechanistic investigations led to ambiguous conclusions.[61] Rearrangement of stereodefined aminoalcohols led to single product, although epimerization sometimes occurred. Aminoalcohols carrying an electron-deficient vinyl group could be rearranged, although in low yield, while aminoalcohols with electron-rich alkene readily reacted (Table 3.3). In both series, the silver counterion seemed to play an important role, favoring (or not) the rearrangement (Scheme 3.39). Some of these results suggested a classical cationic aza-Cope rearrangement, for which silver only

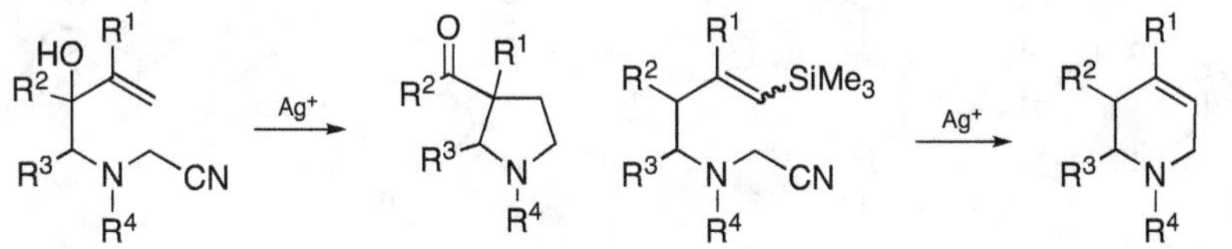

Scheme 3.38

TABLE 3.3. Silver-Mediated Aza-Cope Rearrangements

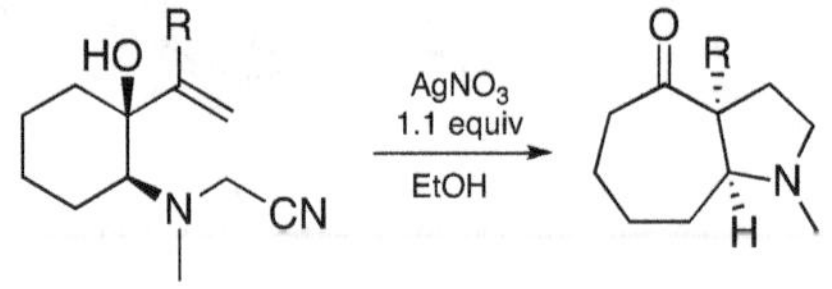

R	Temperature (°C)	Time (h)	Yield (%)
H	40	19	64[a]
Me	25	1	78
SPh	25	1	76
SO_2Ph	25	30	20
OEt	25	30	20[b]

[a] Performed in $CHCl_3$–pyridine to avoid epimerization.
[b] Performed in ether with $AgOOCCF_3$ (see Scheme 3.39).

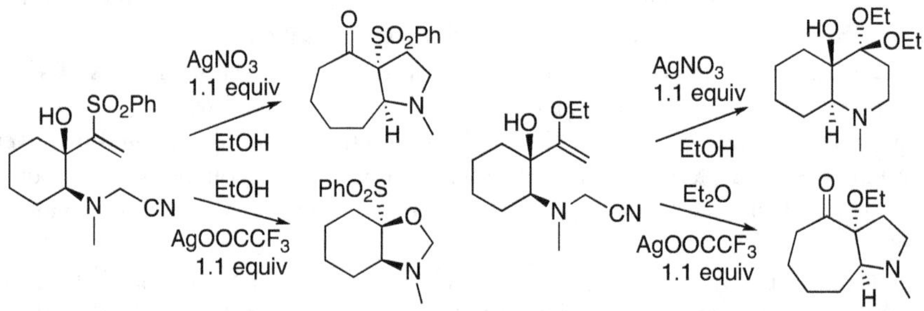

Scheme 3.39

initiated the cascade of events by generating a key iminium intermediate, while others support a more complex role of silver in such rearrangements.

Nevertheless, such rearrangements offer a concise route to some alkaloids. For example, the total synthesis of crinine was achieved with the silver-promoted formation of hydroindolone as the key step (Scheme 3.40).[59]

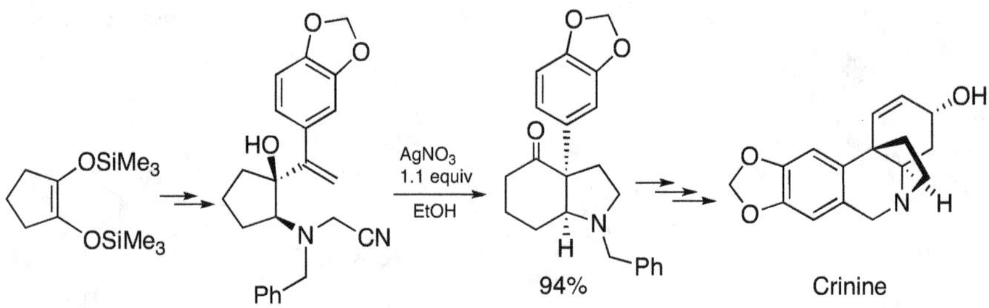

Scheme 3.40

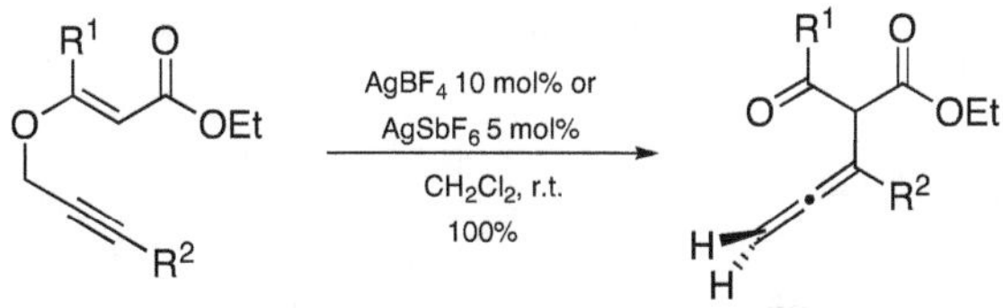

Scheme 3.41

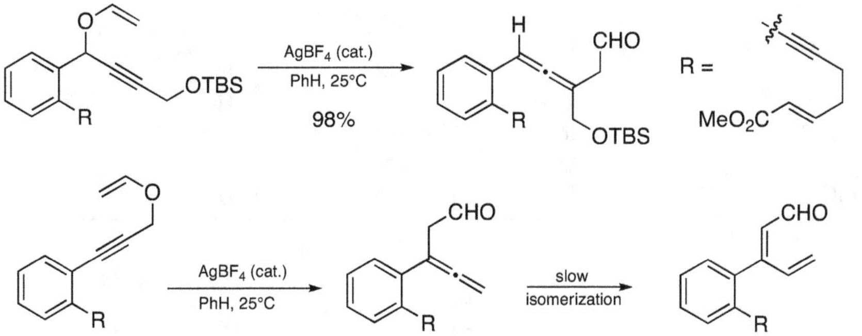

Scheme 3.42

In another related and well-known [3,3]-sigmatropic shift usually performed under thermal conditions, the propargyl–Claisen rearrangement,[62] silver salts were also able to catalyze the reaction. Silver tetrafluoroborate and hexafluoroantimonate proved to be the best catalysts for this reaction, leading quantitatively to allenic β-ketoesters when starting from propargyl ethers derived from β-ketoesters (Scheme 3.41).[63,64]

To improve the thermal tandem [3,3]-sigmatropic rearrangement–enyne allene cyclization, Grissom et al.[65] focused their attention on the transformation of 2-propynylvinyl ethers to allenes using silver salts as catalysts. The tetrafluoroborate proved to be the most effective, quantitatively yielding the expected allenyl aldehyde (Scheme 3.42). However, isomerization problems appeared in some cases (see Scheme 3.33).

3.4.3 With Migrating Groups Analogous to Acyl

The ease of silver-promoted [3,3]-sigmatropic rearrangements of propargyl acetates and other esters has led to the investigation of the sigmatropic shift of groups electronically analogous to acyloxy groups. As for the thermal versions, phosphates have been used as such analogs, and they compared favorably to acetates and benzoates (Table 3.4). Sulfonates were also used but seemed less reactive.[66]

In 2004, Gevorgyan et al. reported that a phosphatyloxy alkynone was also able to smoothly give a phosphatyloxy allenone under silver catalysis. The authors reported only a single example (Scheme 3.43).[67]

TABLE 3.4. Silver-Catalyzed [3,3]-Sigmatropic Rearrangements

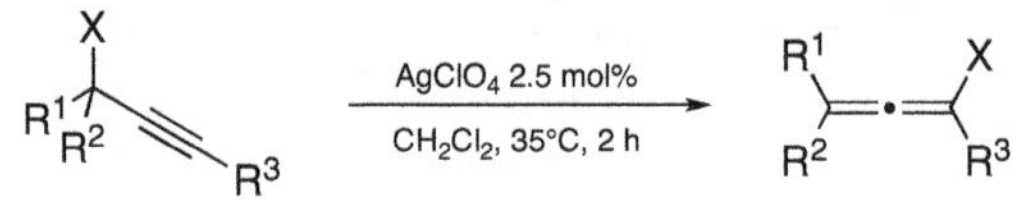

X	R^1	R^2	R^3	Yield (%)
CH_3CO_2	Me	Me	Me	46
—	Me	Me	H	68
—	t-C_4H_9	Me	H	63
$(C_2H_5O)_2P(O)O$	Me	Me	Me	62
—	Me	Me	H	54
—	t-C_4H_9	Me	H	47
p-NO_2PhCO_2	Me	Me	Me	46
—	Me	Me	H	37
CF_3CO_2	Me	Me	H	33

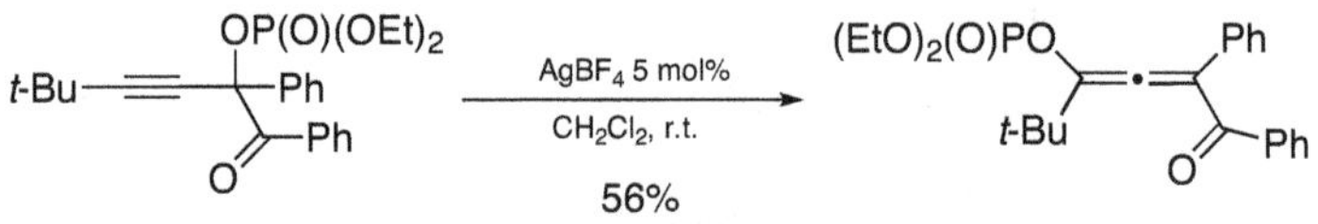

Scheme 3.43

3.4.4 [3,3]-Sigmatropic Rearrangement and Cyclization Cascades

In more sophisticated approaches, [3,3]-sigmatropic rearrangements have been incorporated in cascade reactions in which each step could be catalyzed by silver salts. Usually, the silver promoted [3,3]-sigmatropic shift is combined with a silver-catalyzed cyclization,[68] as outlined in the general scheme below for alkynyl compounds (Scheme 3.44).

Hiyama et al.[69] showed that monoacetylated 1,4-butynediols nicely afforded acetoxy allenols on silver-catalyzed rearrangement. These compounds were ideal substrates for silver-catalyzed cyclization, so the overall sequence could directly be performed in one pot, leading to substituted 2,5-dihydrofurans in high overall yields (Table 3.5). In this process, 5–10 mol% of silver perchlorate or tetrafluoroborate was used in refluxing benzene.

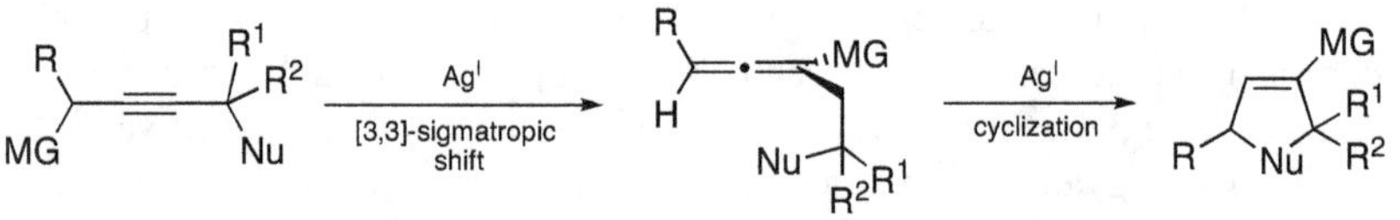

Scheme 3.44

TABLE 3.5. Silver-Catalyzed 2,5-Dihydrofurans Enolacetate Preparation

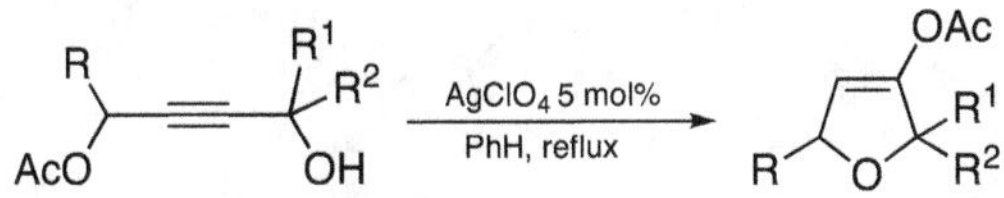

Entry	Starting Material	Products	Yield (%)
1	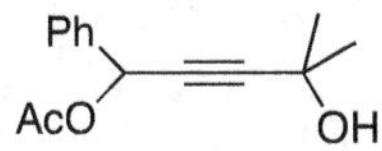	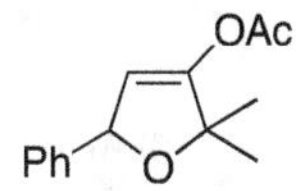	84
2	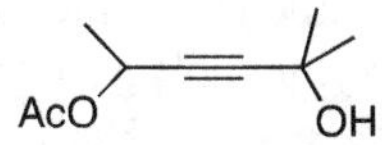	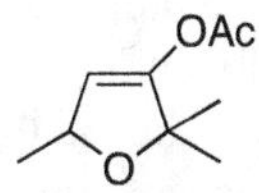	99
3	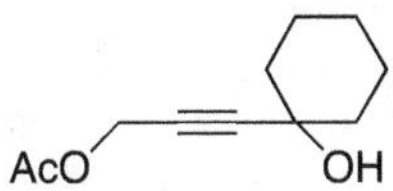	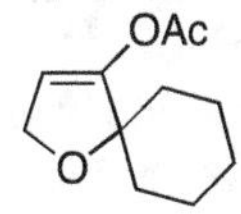	63
4	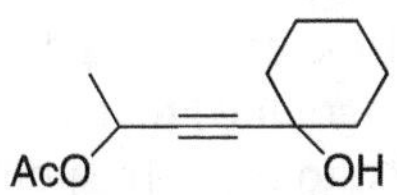	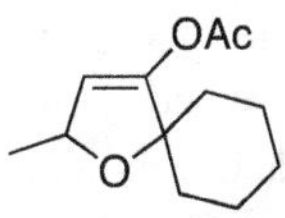	80
5	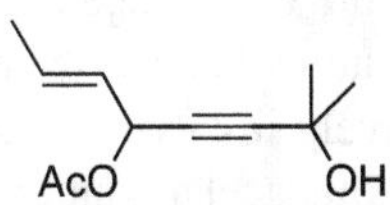	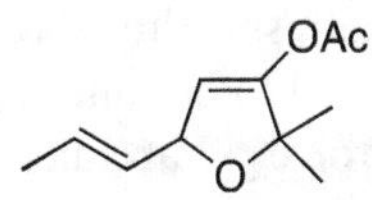	74
6	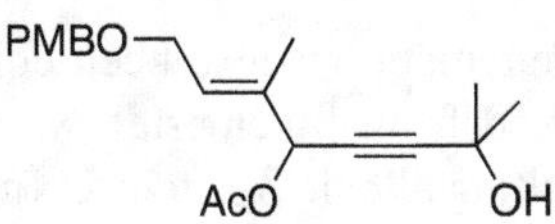	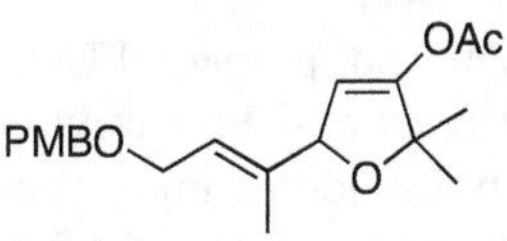	63
7	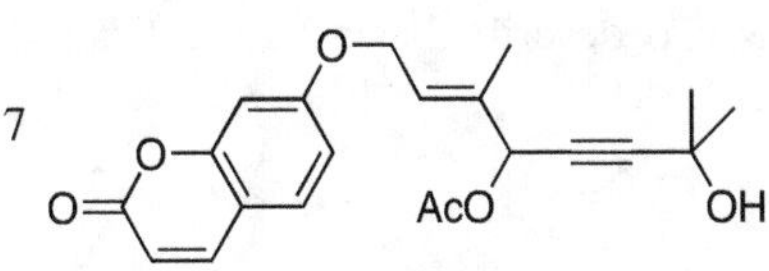	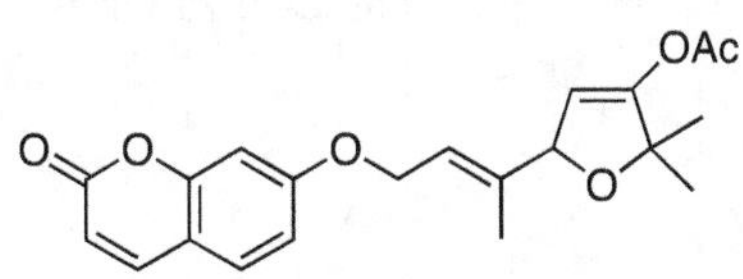	61

This rearrangement–cyclization cascade was applied to natural products synthesis. For example, DDQ oxidation of the appropriate dihydro-3(2*H*)-furanone enol acetates (Table 3.5, entries 1 and 7) afforded bullatenone and the antitumor agent geiparvarin in a very rapid route (Fig. 3.1).[69,70]

Toward the total synthesis of (−)-ascofuranone (Scheme 3.45), an antibiotic with hypolipidemic, antihypertensive, and antitumor properties, several improvements have been made. A pivaloyl ester was used instead of an acetyl ester.[71] More

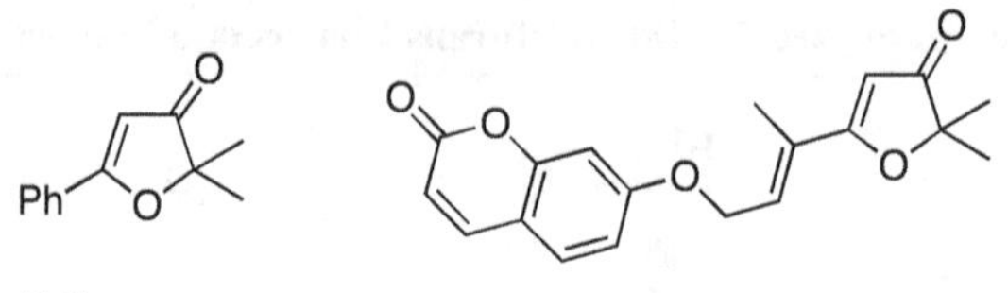

Figure 3.1. Synthetic applications of the rearrangement–cyclization cascade toward natural products.

importantly, chirality transfer was studied and showed to be quantitative, indicating that cyclization was more rapid than isomerization of the allene intermediate. Indeed, various enantiomerically enriched *tert*-butylcarbonyloxybutynols gave the expected 2,5-dihydropyrans with almost the same enantiomeric excess. The chirality was introduced via asymmetric reduction of the starting alkynones with (*S*)-Alpine Borane® (Aldrich-Sigma).[72–74]

Another cascade reaction was also nicely set up using alkynones instead of alkynols. Allenones were produced in situ through a [3,3]-sigmatropic rearrangement of propargyl acetates and led directly, under these conditions, to substituted furans (Table 3.6).[77] Interestingly, the key allenone could be isolated when the reaction was run at room temperature and with a phosphate as the migrating group (see above). In refluxing dichloroethane, this allenone readily cyclized to the corresponding furan. Moreover, both phosphate and sulfonate groups could be engaged in the same reaction, leading to tetrasubstituted furans from substituted allenones (Scheme 3.46).

This cascade proved to be quite complex, successively involving a [3,3]-sigmatropic shift of the propargyl acetate, 1,2 migration of this acetate, and cycloisomerization (Scheme 3.47).

The silver-catalyzed, propargyl Claisen rearrangement has also been combined with cyclization reactions. Propargyl ethers derived from β-ketoesters were rearranged by silver hexafluoroantimonate and the resulting allenic β-ketoesters cyclized on treatment with base, leading to the formation of 2*H*-pyrans in moderate to excellent yields (Scheme 3.48).[75] The cyclization process was described as a base-catalyzed

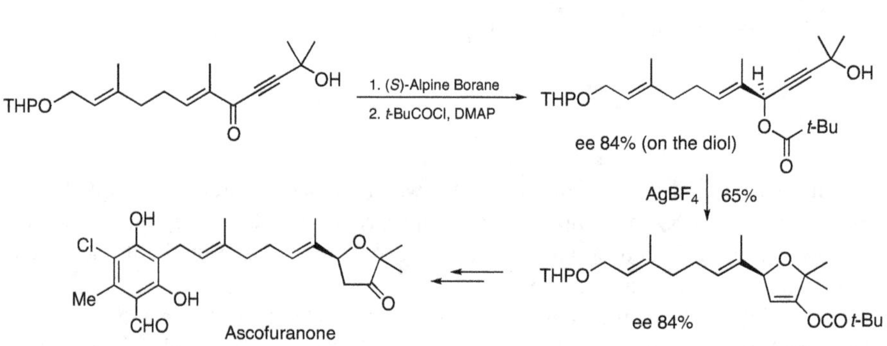

Scheme 3.45

TABLE 3.6. Silver-Catalyzed Preparation of Tetrasubstituted Furans

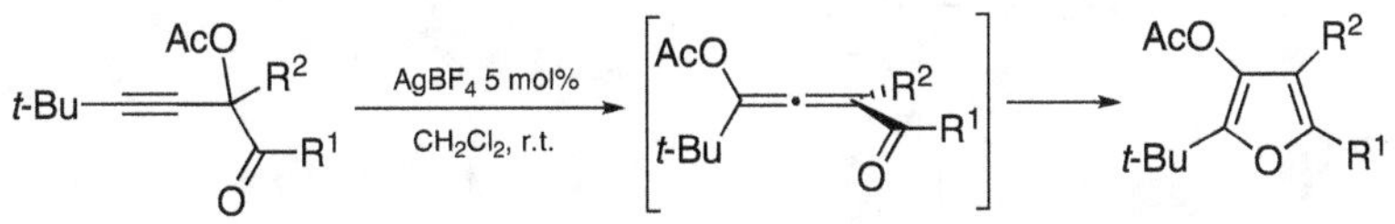

Entry	Substrate	Product	Time (min)	Yield (%)
1	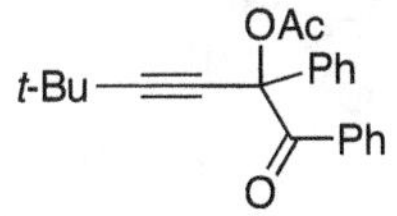	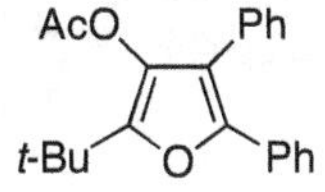	2	>99
2	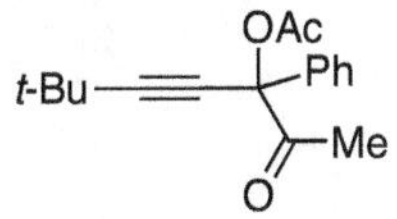	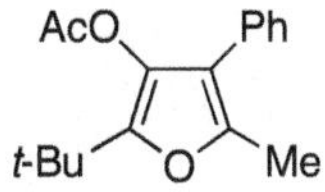	15	73
3	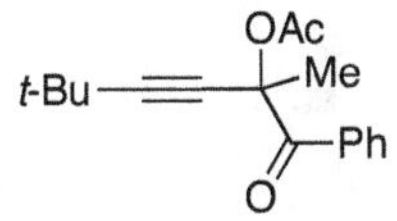	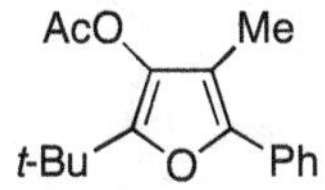	15	84
4	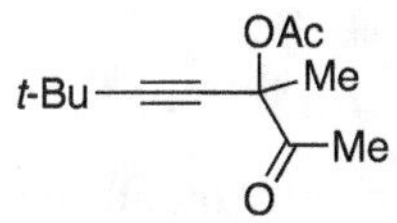	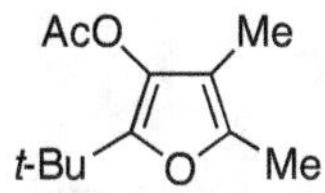	15	90
5	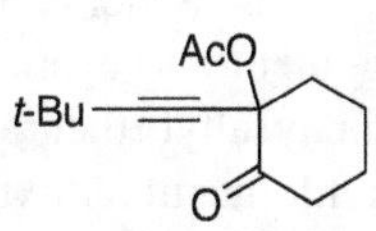	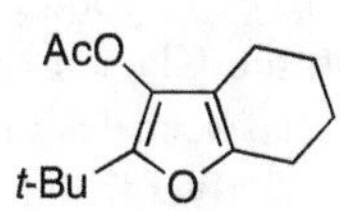	10	86

isomerization followed by a 6π-oxa electrocyclization, although silver-catalyzed cyclization of the allenone could occur.

Again on the basis of the silver-catalyzed propargyl Claisen rearrangement, an interesting one-pot process leading to highly substituted pyrroles was devised by Kirsch et al.[64] The allenic β-ketoester obtained through this rearrangement was directly treated with amines and then with a catalytic amount of a gold catalyst. The

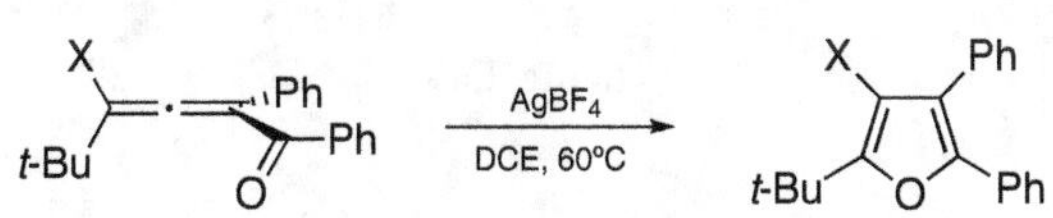

X = –OP(O)(OEt)$_2$, 77% yield

X = –OTs, 82% yield

Scheme 3.46

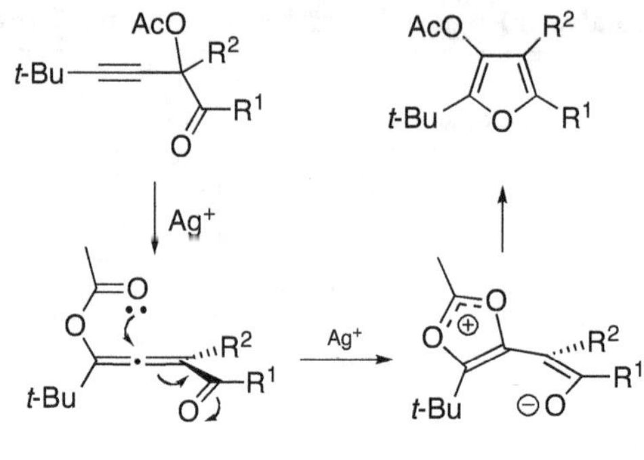

Scheme 3.47

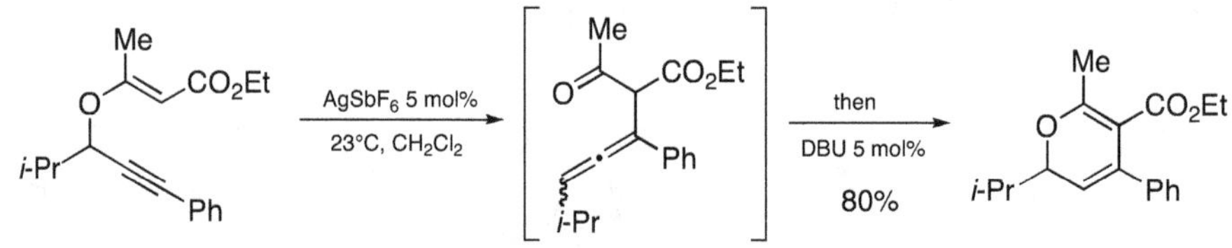

Scheme 3.48

intermediate enaminoester added to the external double bond of the allenic system in a 5-*exo-dig* cyclization, giving pyrroles in good to high yields (Scheme 3.49).

A cascade reaction relying on the Claisen rearrangement and cyclization, both catalyzed by copper or silver salts, has been described. Phenyl allyl ether rearranged in the presence of catalytic amounts of silver triflate or perchlorate in refluxing toluene, and the resulting 2-allylphenol cyclized to give the corresponding dihydrobenzofuran (Scheme 3.50).[76]

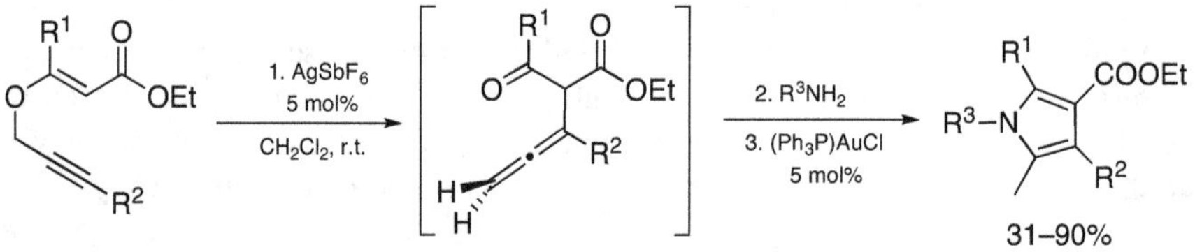

Scheme 3.49

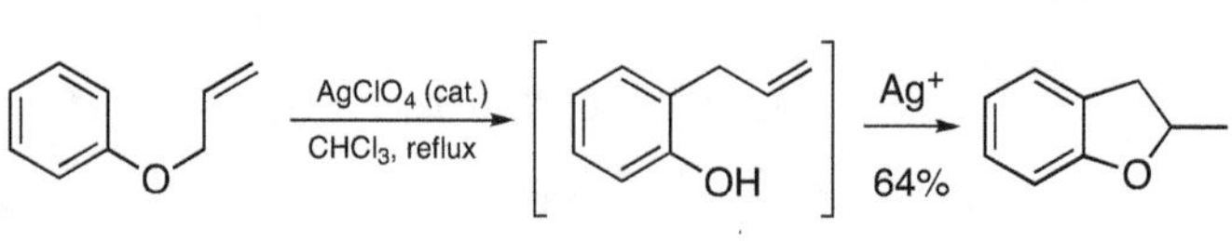

Scheme 3.50

3.5 [2,3]-SIGMATROPIC REARRANGEMENTS

Following their investigations on nitrene, carbene, and oxo transfer reactions catalyzed by fluorinated silver tris(pyrazoyl)borate (see Chapter 6 on nitrene chemistry), Lovely et al. looked for a combined carbene transfer and [2,3]-sigmatropic rearrangement. On the basis of these mechanistic considerations, these authors showed that diazoacetates, indeed, reacted with allyl halides in the presence of this silver catalyst to give α-halo-γ-unsaturated esters (Scheme 3.51).[77]

The substitution pattern in the product obtained through this sequence was in agreement with the proposed mechanism (Table 3.7). For example, crotyl chloride gave the 2-chloro-3-methyl hex-4-enoates (entries 3 and 4), and propargyl bromides gave the allenic products (e.g., entries 7 and 8). Yields were usually better with the more substituted *tert* -butyl diazoacetate.

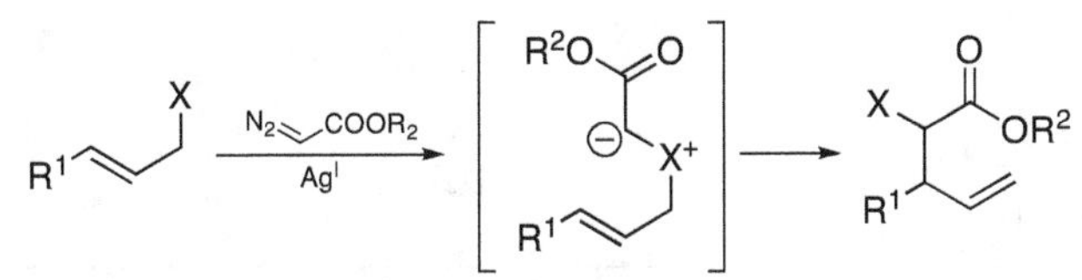

Scheme 3.51

TABLE 3.7. Silver-Catalyzed Preparation of α-Haloacetates

Entry	Halide	R^2	Product	Yield (%)
1	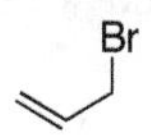	Et	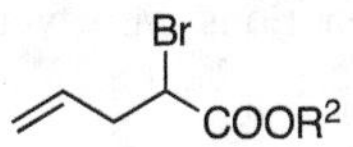	75
2		*t*-Bu		65
3	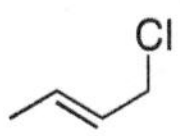	Et	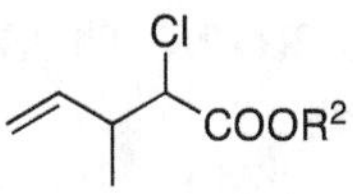	86[a]
4		*t*-Bu		96[a]
5	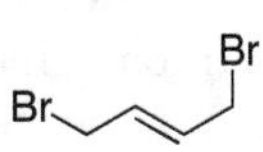	Et	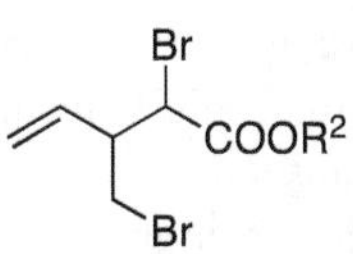	70[a]
6		*t*-Bu		80[a]
7	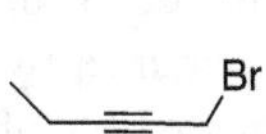	Et	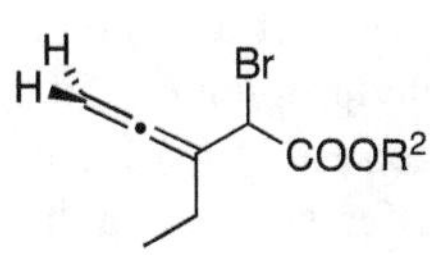	71
8		*t*-Bu		74

[a]Obtained as a 1 : 1 mixture of *syn* : *anti* isomers.

TABLE 3.8. Silver-Catalyzed Preparation of Allenic Sulfones

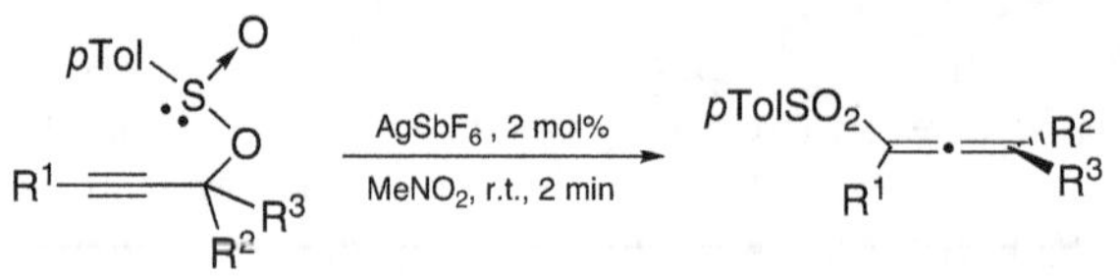

Entry	R^1	R^2		R^3	Yield (%)
1	H	H		Me	97
2	H	Me		Me	97
3	H	H		H	98
4	H	H		Et	96
5	Me	H		H	99
6	H	Me		Et	97
7	H	Et		Et	97
8	Ph	H		H	99
9	H	Ph		H	99
10	Et	H		Me	98
11	H		$-(CH_2)_4-$		98
12	H	H		Me	99

In a related reaction, Harmata and Huang showed that allenic sulfones were easily prepared from propargylic sulfinates (Table 3.8). Catalyst screening showed that silver nitrate was ineffective, whereas silver tetrafluoroborate and hexafluoroantimonate worked well in dichloromethane, chloroform, or nitromethane at room temperature. All reactions were very rapid, within 2 min, and were very effective in terms of yield.[78]

3.6 [1,2]-SIGMATROPIC REARRANGEMENTS

3.6.1 1,2-Aryl or Alkenyl Migration

Extending their work on α-bromoaryl ketones (see Scheme 3.5), Giordano et al. have reported a [1,2]-sigmatropic rearrangement assisted by silver ion.[79] Indeed, alkylacetals of primary and secondary α-halogenated aryl ketones furnished alkyl esters of α-aryl alkanoic acids in high yields using silver tetrafluoroborate in an alcoholic medium (Table 3.9).

More recently, Usui et al. studied the stereochemical aspects of this rearrangement.[80] (*S*)-2-Phenylpropionic acid was stereoselectively obtained by the $AgBF_4$-catalyzed rearrangement of (*S*)-2-chloropropiophenone dimethyl acetal with a complete chirality transfer, suggesting that the rearrangement proceeds with the backside phenyl group participation at the C–Cl bond, that is, with an Ag^+-aided, phenyl-assisted intramolecular S_N2 mechanism (Scheme 3.52).

TABLE 3.9. Silver-Mediated [1,2]-Sigmatropic Rearrangements

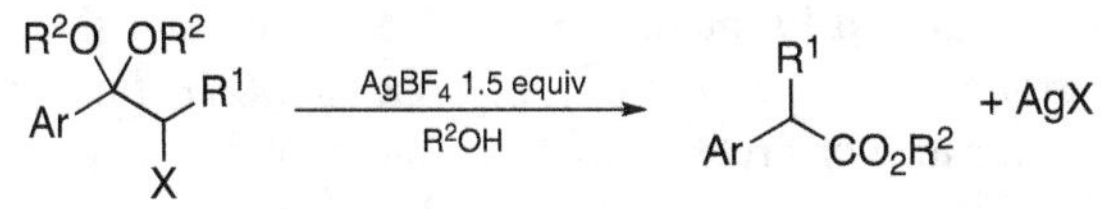

Aryl Group	R^1	R^2	X	Time (h)	Yield (%)
4-Methoxyphenyl	H	Me	Br	2	98
4-Methoxyphenyl	H	Me	I	0.25	87
4-Methoxyphenyl	H	Me	Cl	2	98
4-Methoxyphenyl	H	Et	Br	3	98
4-Methoxyphenyl	Me	Me	Br	1.5	98
2-Naphthyl	H	Me	Br	14	98

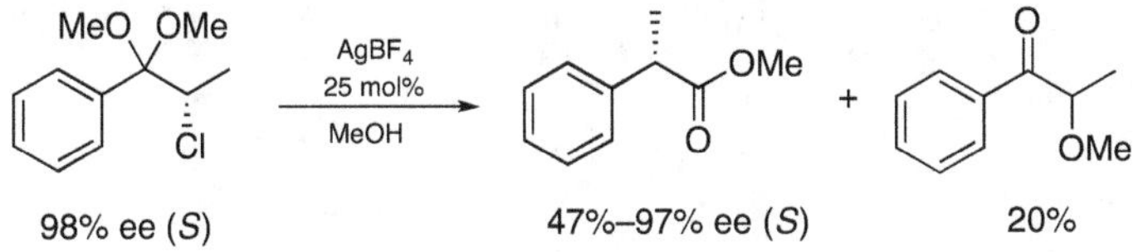

Scheme 3.52

Similarly, silver hexafluoroantimonate was also able to promote a fast [1,2]-sigmatropic rearrangement of *trans*-4-aryl- or *trans*-4-alkenyl-3-bromo-4,6-dimethyl-3,4-dihydro-2-pyrones in dichloromethane through substitution of the halide by migration of the aryl or alkenyl group to the 3 position. This rearrangement afforded the corresponding 3-substituted 2-pyrones in high yields (Table 3.10).[81] It is noteworthy that the debromination–migration process is probably concerted as supported by the fact that *cis* compounds did not take part in this reaction.

TABLE 3.10. Silver-Assisted Preparation of α-Pyrones

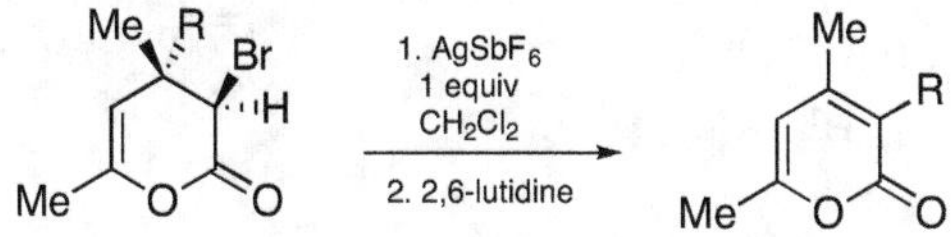

R	*cis* : *trans*	Time (min)	Yield (%)
p-Tolyl	97 : 3	15	100
p-Anisyl	85 : 15	15	80
Phenyl	80 : 20	15	47
1-Propenyl	97 : 3	15	84
(*Z*)-1-Hexenyl	97 : 3	15	93

3.6.2 1,2-Alkyl Migration

Kirsch et al.[82] have discovered that numerous metal salts, including silver salts, were able to catalyze the rearrangement of α-hydroxyalkynones to 3(2*H*)-furanones, introducing an original strategy for the construction of such compounds. Interestingly, spirocyclic furanones could be obtained starting from alkynyl cycloalkanones (Scheme 3.53).

Although not the best catalyst, $AgSbF_6$ led to the rearranged product in 50% yield. This cascade reaction probably started with the well-known cyclization of the ketone to the alkyne on silver coordination, giving a cyclic oxonium intermediate that rearranged to furanone via an alkyl 1,2-migration (Scheme 3.54).[82]

A similar reaction was described by Nakamura et al. as recently as 2009. Here again, silver salts were not the best catalysts. Nevertheless, silver triflate was able to rearrange 2-alkynyl tetralones to 1,2-disubstituted naphthalenes (Scheme 3.55).[83]

Scheme 3.53

3.6.3 1,2- or 1,5-Alkyl Migration

Gevorgyan et al. also observed alkyl migration while studying the cyclization of substituted allenones in the presence of various metal salts. While again not the best catalysts, silver hexafluorophosphate or triflate could be used in toluene or dichloromethane, giving substituted furans in high yields (Scheme 3.56).[84]

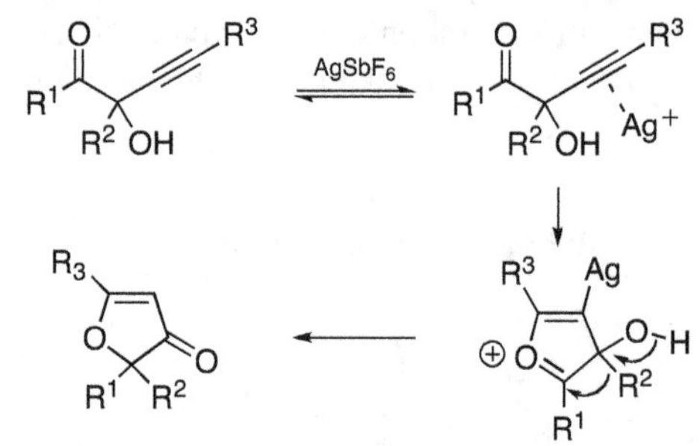

Scheme 3.54

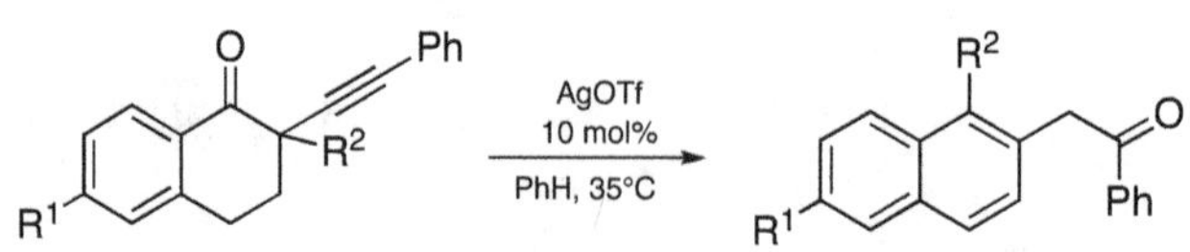

Scheme 3.55

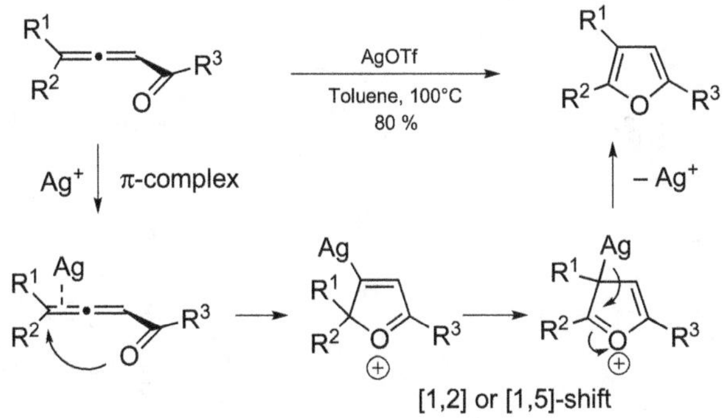

Scheme 3.56

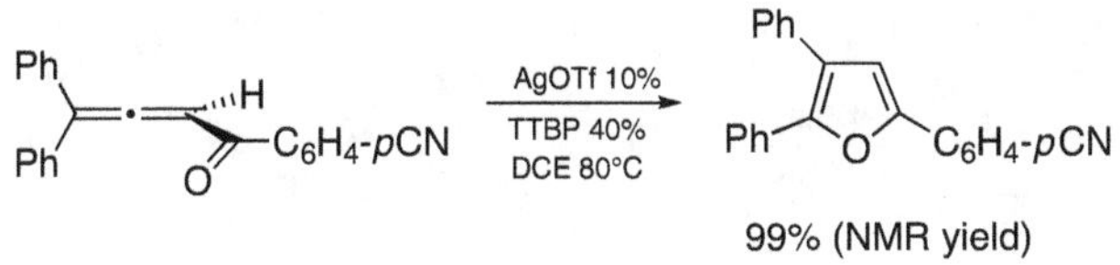

Scheme 3.57

From a mechanistic perspective, the reaction is similar to those described above. Silver-catalyzed cyclization of the ketone to the allene gave a cyclic oxonium intermediate. A [1,2]- or [1,5]-alkyl shift modified the sigma skeleton leading to an alkylsilver intermediate, which on elimination gave a trisubstituted furan.

In the presence of hindered base, such as 2,4,6-tris(*tert*-butyl)pyrimidine (TTBP), and in refluxing dichloroethane, this reaction became quantitative (Scheme 3.57).[85]

3.6.4 1,2 versus 3,3 Migrations

Although acyloxy, phosphatyloxy, and tosyl allenones were obtained from the corresponding propargyl alcohol derivatives via silver catalysis through an overall process that resembled [3,3]-sigmatropic rearrangement (see Sections 3.41 and 3.43), the mechanism was not fully proved and is still in question.

The same holds true for the cyclization of these allenones to trisubstituted furans (see Section 3.4.4). More recently, new insights were obtained using labeled substrates.[86] Two processes could explain the position of the migrating group on the furan ring: a [3,3]-sigmatropic rearrangement followed by a 1,2 shift or a direct 1,2 shift leading to a carbenoid intermediate (Scheme 3.58). The position of the labeled oxygen atom in the final furan product would confirm the pathway used: double inversion leading to the unlabeled oxygen atom linked to the carbon (pathway A) or simple inversion leading to the labeled oxygen atom link to the carbon (pathway B).

Treatment of labeled acyloxy alkynones with silver tetrafluoroborate in dichloromethane at room temperature exclusively gave a single furan, in which the labeled oxygen atom was connected to the ring carbon (Scheme 3.59). This observation was consistent with a 1,2-acyl migration but not with a [3,3]-sigmatropic rearrangement.

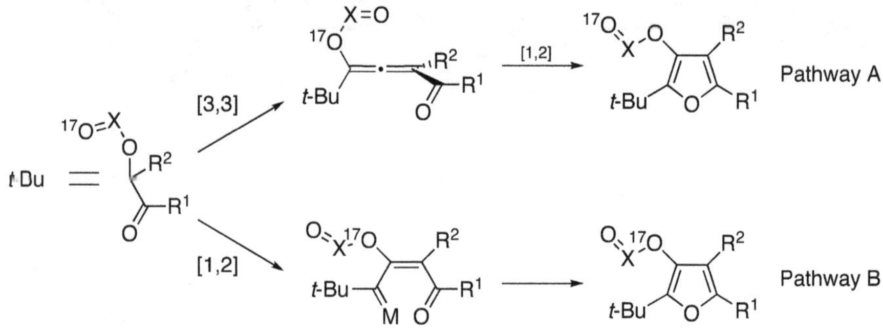

Scheme 3.58

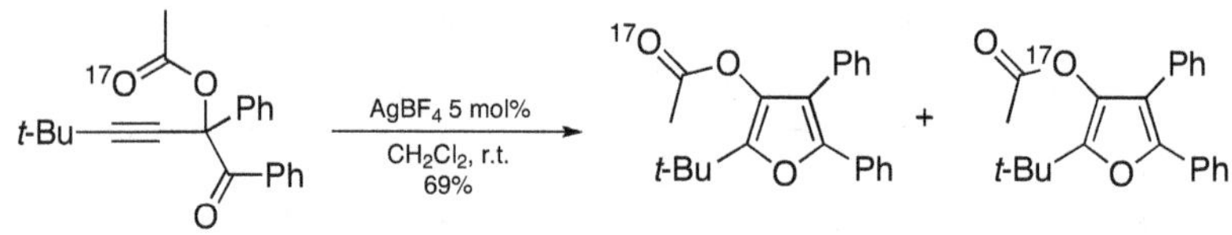

Scheme 3.59

Similarly, a single phosphatyloxy allene in which the labeled oxygen atom was not connected to the allene was obtained from a phosphatyloxy alkyone after treatment with silver tetrafluoroborate. This observation definitively ruled out a [3,3]-sigmatropic rearrangement and supported 1,2 migration. It is worth noting that during cyclization of the phosphatyloxy allenone, the labeled oxygen atom was again not connected to the furan ring. The result strongly suggested 1,2 migration with retention (Scheme 3.60).

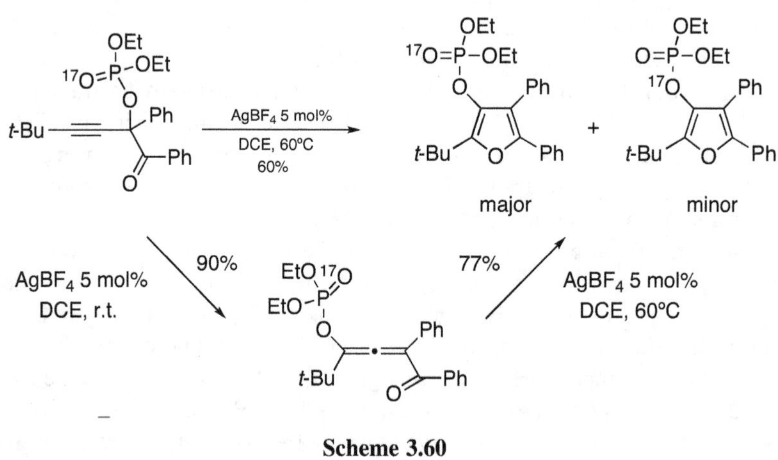

Scheme 3.60

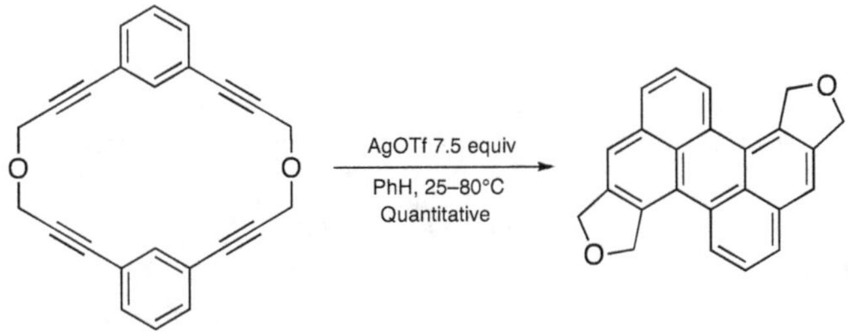

Scheme 3.61

These results clearly demonstrated that, depending on substrates, apparent [3,3]-sigmatropic shifts actually resulting from 1,2 shifts could occur. Further studies are thus clearly needed to better understand the mechanism of apparent sigmatropic rearrangements catalyzed by silver and other metals.

3.7 MISCELLANEOUS

Looking for new complexes, Yamaguchi et al. observed that a cyclic diaryltetrayne underwent rearrangement in the presence of 2.5 equiv of silver triflate to a difuroperylene, which was isolated in 30% yield. The reaction could be made nearly quantitative by addition of two other portions of silver triflate after 3 and 6 h while stirring (Scheme 3.61). The formation of perylene proceeds by a complex silver-mediated process involving *sp–sp* and *sp–sp*^{2} cyclization.[87]

3.8 CONCLUSION

Silver salts can be used as catalysts or promoters for a variety of reactions involving σ-bond migration and skeletal rearrangement. Bonds in strained systems directly interact with silver, leading to bond breaking and formation of a cation. These events lead to skeletal rearrangement through bond migration. In a similar way, halogenated cyclic compounds can also be rearranged on treatment with silver ions. The initiation of such rearrangements is often similar to the preceding, since strained cyclic intermediates are usually produced as the first step.

Silver-mediated rearrangement of propargyl esters is a process that has been known since the mid-1940s. Since then, this [3,3]-sigmatropic rearrangement has been extended to other sigmatropic rearrangements with a wide variety of participating groups and different applications.

This chemistry has been revamped more recently by its combination with cyclization reactions, most of them also promoted by silver salts.[69] More recent years have thus witnessed the development of silver-catalyzed cascade reactions leading to heterocycles under mild conditions.[68]

This broad range of reactions mediated by silver reveals the importance of silver salts as catalysts or promoters in organic chemistry and especially in synthesis.

REFERENCES

1. (a) Bagus, P. S.; Lee, Y. S.; Pitzer, K. S., *Chem. Phys. Lett.* **1975**, *33*, 408–411; (b) Pyykkö, P., *Angew. Chem., Int. Ed.* **2004**, *43*, 4412–4456; (c) Gorin, D. J.; Toste, F. D., *Nature* **2007**, *446*, 395–403.
2. Wolff, L., *Justus Liebigs Ann. Chem.* **1902**, *325*, 129–195.
3. Arndt, F.; Eistert, B., *Ber. Deutsch. Chem. Ges.* **1935**, *68*, 200–208.
4. Gill, G. B., in *Comprehensive Organic Synthesis*, Trost, B. M., Fleming, I., Pattenden, G., eds., Pergamon Press, New York, **1991**, Vol. 3, pp. 887–912.
5. Meier, H.; Zeller, K. P., *Angew. Chem. Int. Ed.* **1975**, *14*, 32–43.
6. Koch, K.; Podlech, J., *Synth. Commun.* **2005**, *35*, 2789–2794.
7. Sudrik, S. G.; Maddanimath, T.; Chaki, N. K.; Chavan, S. P.; Chavan, S. P.; Sonawane, H. R.; Vijayamohanan, K., *Org. Lett.* **2003**, *5*, 2355–2358.
8. Sudrik, S. G.; Chaki, N. K.; Chavan, V. B.; Chavan, S. P.; Chavan, S. P.; Sonawane, H. R.; Vijayamohanan, K., *Chem. Eur. J.* **2006**, *12*, 859–864.
9. Baldwin, J. B.; Herchen, S. R.; Singh, P. D., *Biochem. J.* **1980**, *186*, 881–887.
10. Plucinska, K.; Liberek, B., *Tetrahedron* **1987**, *43*, 3509–3517.
11. Tilekar, J. N.; Patil, N. T.; Dhavale, D. D., *Synthesis* **2000**, 395–398.
12. Giordano, C.; Castaldi, G.; Casagrande, F.; Abis, L.; Donegani, I. G., *Tetrahedron Lett.* **1982**, *23*, 1385–1386.
13. De Kimpe, N.; Stevens, C., *Tetrahedron* **1990**, *46*, 6753–6770.
14. Gassman, P. G.; Fox, B., *J. Am. Chem. Soc.* **1967**, *89*, 338–342.
15. Gassman, P. G.; Cryberg, R. L., *J. Am. Chem. Soc.* **1968**, *90*, 1355–1356.
16. Davies, J. W.; Malpass, J. R.; Walker, M. P., *J. Chem. Soc. Chem. Commun.* **1985**, 685–686.
17. Durrant, M. L.; Malpass, J. R.; Walker, M. P., *J. Chem. Soc. Chem. Commun.* **1985**, 687–689.
18. Rautenstrauch, V., *Chem. Commun.* **1969**, 1122–1123.
19. Durrant, M. L.; Malpass, J. R., *Tetrahedron* **1995**, *51*, 7063–7076.
20. Davies, J. W.; Durrant, M. L.; Naylor, A.; Malpass, J. R., *Tetrahedron* **1995**, *51*, 8655–8664.
21. Krow, G. R.; Lin, G.; Yu, F.; Sonnet, P. E., *Org. Lett.* **2003**, *5*, 2739–2741.
22. Paquette, L. A.; Wilson, S. E.; Henzel, R. P., *J. Am. Chem. Soc.* **1972**, *94*, 7771–7779.
23. Paquette, L. A.; Henzel, R. P.; Wilson, S. E., *J. Am. Chem. Soc.* **1972**, *94*, 7780–7788.
24. Paquette, L. A., *Angew. Chem., Int. Ed.* **1972**, *11*, 328–329.
25. Masamune, S.; Sakai, M., *J. Am. Chem. Soc.* **1971**, *93*, 4610–4611.
26. Masamune, S.; Sakai, M.; Westberg, H. H.; Yamaguchi, H., *J. Am. Chem. Soc.* **1971**, *93*, 4611–4613.
27. Paquette, L. A.; Blount, J. F., *J. Am. Chem. Soc.* **1980**, *102*, 644–650.
28. Christl, M.; Brunn, E.; Roth, W. R.; Lennartz, H. W., *Tetrahedron* **1989**, *45*, 2905–2915.
29. Taylor, R. T.; Paquette, L. A., *J. Org. Chem.* **1978**, *43*, 242–250.
30. Billups, W. E.; McCord, D. J.; Maughon, B. R., *Tetrahedron Lett.* **1994**, *35*, 4493–4496.
31. Halton, B.; Dixon, G. M.; Forman, G. S., *Arkivoc* **2006**, *xii*, 38–45.
32. Leftin, J. H.; Gil-Av, E., *Tetrahedron Lett.* **1972**, *13*, 3367–3370.

33. Peelen, F. C.; Landheer, I. J.; De Wolf, W. H.; Bickelhaupt, F., *Rec. Trav. Chim. Pays-Bas* **1986**, *105*, 326–331.
34. Padwa, A.; Blacklock, T. J.; Loza, R., *J. Org. Chem.* **1982**, *47*, 3712–3721.
35. Eaton, P. E.; Cassar, L.; Halpern, J., *J. Am. Chem. Soc.* **1970**, *92*, 6366–6368.
36. Paquette, L. A.; Ward, J. S., *Tetrahedron Lett.* **1972**, *13*, 4909–4912.
37. Paquette, L. A.; Ward, J. S.; Boggs, R. A.; Farnham, W. B., *J. Am. Chem. Soc.* **1975**, *97*, 1101–1112.
38. Turnblom, E. W.; Katz, T. J., *J. Am. Chem. Soc.* **1973**, *95*, 4292–4311.
39. Mehta, G.; Ravikrishna, C.; Ravikumar, K., *J. Chem. Soc. Chem. Commun.* **1994**, 2321–2322.
40. Dauben, W. G.; Buzzolini, M. G.; Schallhorn, C. H.; Whalen, D. L., *Tetrahedron Lett.* **1970**, *10*, 787–790.
41. Reese, C. B.; Shaw, A., *J. Chem. Soc. Perkin Trans. 1* **1975**, 2422–2435.
42. Bach, R. D.; Holubka, J. W.; Willis, C. L., *J. Am. Chem. Soc.* **1982**, *104*, 3980–3987.
43. Bach, R. D.; Mazur, U.; Hamama, I.; Lauderback, S. K., *Tetrahedron* **1972**, *28*, 1955–1963.
44. Ito, S.; Ziffer, H.; Bax, A., *J. Org. Chem.* **1986**, *51*, 1130–1133.
45. Danheiser, R. L.; Morin, J. M.; Yu, M.; Basak, A., *Tetrahedron Lett.* **1981**, *22*, 4205–4208.
46. Ludivine, J.-G.; Pauvert, M.; Collet, S.; Guingant, A.; Evain, M., *Tetrahedron* **2007**, *63*, 11250–11259.
47. Anhalt, K.; Sprung, I.; Schulze, K., *Monatsh. Chem.* **2003**, *134*, 1593–1606.
48. Schulze, K.; Uhlig, H., German (East) Patent (**1988**), GEXXA8 DD254382 (CAN 111:23747).
49. Zakharova, A. I., *Zh. Obsch. Khim.* **1945**, *15*, 429–437.
50. Saucy, G.; Marbet, R.; Lindlar, H.; Isler, O., *Helv. Chim. Acta* **1959**, *42*, 1945–1955.
51. Landor, P. D.; Landor, S. R., *J. Chem. Soc.* **1956**, 1015–1019.
52. Benn, W. R., *J. Org. Chem.* **1968**, *33*, 3113–3118.
53. Schlossarcyk, H.; Sieber, W.; Hesse, M.; Hansen, H.-J.; Schmid, H., *Helv. Chim. Acta* **1973**, *56*, 875–944.
54. Cookson, R. C.; Cramp, M. C.; Parsons, P. J., *J. Chem. Soc. Chem. Commun.* **1980**, 197–198.
55. Sakaguchi, K.; Okada, T.; Shinada, T.; Ohfune, Y., *Tetrahedron Lett.* **2008**, *49*, 25–28.
56. Pelletier, S. W.; Mody, N. V., *J. Org. Chem.* **1976**, *41*, 1069–1071.
57. Hollinshead, D. M.; Howell, S. C.; Ley, S. V.; Mahon, M.; Ratcliffe, N. M., *J. Chem. Soc. Perkin Trans. 1* **1983**, 1579–1589.
58. Bluthe, N.; Goré, J.; Malacria, M., *Tetrahedron* **1986**, *42*, 1333–1344.
59. Overman, L. E.; Mendelson, L. T.; Jacobsen, E. J., *J. Am. Chem. Soc.* **1983**, *105*, 6629–6637.
60. Flann, C.; Malone, T. C.; Overman, L. E., *J. Am. Chem. Soc.* **1987**, *109*, 6097–6107.
61. Jacobsen, E. J.; Levin, J.; Overman, L. E., *J. Am. Chem. Soc.* **1988**, *110*, 4329–4336.
62. Overman, L. E., *Angew. Chem., Int. Ed.,* **1984**, *23*, 579–586.
63. Suhre, M. R.; Reif, M.; Kirsch, S. F., *Org Lett.* **2005**, *7*, 3925–3927.
64. Binder, J. T.; Kirsch, S. F., *Org Lett.* **2006**, *8*, 2151–2153.

65. Grissom, J. W.; Klingberg, D.; Huang, D.; Slattery, B. J., *J. Org. Chem.* **1997**, *62*, 603–626.
66. Oelberg, D. G.; Schiavelli, M. D., *J. Org. Chem.* **1977**, *42*, 1804–1807.
67. Sromek, A. W.; Kel'in, A. V.; Gevorgyan, V., *Angew. Chem. Int. Ed.* **2004**, *43*, 2280–2282.
68. (a) Weibel, J.-M.; A. Blanc, A.; Pale, P., *Chem. Rev.* **2008**, *108*, 3149–3173; (b) Alvarez-Corral, M.; Munoz-Dorado, M.; Rodriguez-Garcia, I., *Chem. Rev.* **2008**, *108*, 3174–3198.
69. Saimoto, H.; Hiyama, T.; Nozaki, H., *J. Am. Chem. Soc.* **1981**, *103*, 4975–4977.
70. Saimoto, H.; Hiyama, T.; Nozaki, H., *Bull. Chem. Soc. Jpn.* **1983**, *56*, 3078–3087.
71. Saimoto, H.; Kusano, Y.; Hiyama, T., *Tetrahedron Lett.* **1986**, *27*, 1607–1610.
72. Shigemasa, Y.; Yasui, M.; Ohrai, S.-I.; Sasaki, M.; Sashiwa, H.; Saimoto, H., *J. Org. Chem.* **1991**, *56*, 910–913.
73. Saimoto, H.; Ohrai, S.-I.; Sashiwa, H.; Shigemasa, Y.; Hiyama, T., *Bull. Chem. Soc. Jpn.* **1995**, *68*, 2727–2734.
74. Saimoto, H.; Masaru, Y.; Ohrai, S.-I.; Oikawa, H.; Yokoyama, K.; Shigemasa, Y., *Bull. Chem. Soc. Jpn.* **1999**, *72*, 279–284.
75. Menz, H.; Kirsch, S. F., *Org. Lett.* **2006**, *8*, 4795–4797.
76. Ito, Y.; Kato, R.; Hamashima, K.; Kataoka, Y.; Oe, Y.; Ohta, T.; Furukawa, I., *J. Organomet. Chem.* **2007**, *692*, 691–697.
77. Krishnamoorthy, P.; Browning, G. R.; Singh, S.; Siuvappa, R.; Lovely, C. J.; Dias, H. V. R., *Chem. Commun.* **2007**, 731–733.
78. Harmata, M.; Huang, C., *Adv. Synth. Catal.* **2008**, *350*, 972–974.
79. Giordano, C.; Castaldi, G.; Casagrande, F.; Belli, A., *J. Chem. Soc. Perkin Trans. 1* **1982**, 2575–2581.
80. Usui, S.; Matsumoto, T.; Ohkubo, K., *Tetrahedron Lett.* **1998**, *39*, 9755–9758.
81. Kume, T.; Iwasaki, H.; Yamamoto, Y.; Akiba, K., *Tetrahedron Lett.* **1988**, *29*, 3825–3828.
82. Kirsch, S. F.; Binder, J. T.; Liébert, C.; Menz, H., *Angew. Chem. Int. Ed.* **2006**, *45*, 5878–5880.
83. Chan, C. S.; Araki, T.; Nakamura, I.; Terada, M., *Tetrahedron Lett.* **2009**, *50*, 216–218.
84. Dudnik, A. S.; Gevorgyan, V., *Angew. Chem., Int. Ed.* **2007**, *46*, 5195–5197.
85. Dudnik, A. S.; Sromek, A. W.; Rubina, M.; Kim, J. T.; Kel'in, A. V.; Gevorgyan, V., *J. Am. Chem. Soc.* **2008**, *130*, 1440–1452.
86. Schwier, T.; Sromek, A. W.; Yap, D. M. L.; Chernyak, D.; Gevorgyan, V., *J. Am. Chem. Soc.* **2007**, *129*, 9868–9878.
87. Yamaguchi, Y.; Kobayashi, S.; Wakamiya, T.; Matsubara, Y.; Yoshida, Z.-I., *J. Am. Chem. Soc.* **2000**, *122*, 7404–7405.

4

SILVER(I)-MEDIATED ELECTROCYCLIC PROCESSES

TINA N. GRANT AND FREDERICK G. WEST
Department of Chemistry, University of Alberta, Edmonton, Canada

4.1 INTRODUCTION

The halophilic character of the silver(I) cation is traditionally exploited to assist in removal of halides from metal complexes in order to open a coordination site and, in effect, initiate catalyst reactivity. Analogously, silver(I) salts can be used to facilitate the removal of halides from organic molecules to reveal reaction pathways that involve cationic intermediates. In the simplest scenarios, silver(I) has been used to

Silver in Organic Chemistry Edited by Michael Harmata

assist in the abstraction of halides during solvolysis or substitution reactions.[1] A more intriguing role for silver(I) in organic chemistry is its ability to trigger electrocyclic ring opening of the halocyclopropane moiety, resulting in a reactive cationic species. These cationic processes can be standalone reactions that result in simple, ring-opened products, or they can be used to initiate domino reactions that generate more complex and, in some instances, polycyclic products.[2]

This chapter focuses on the use of silver(I) salts as Lewis acids in the ring-opening reactions of halocyclopropanes. Both the classical and most recent examples of this type of chemistry are examined with emphasis on the use of this simple transformation to build up complex intermediates that are potentially useful in the construction of natural products.

4.1.1 Ring-Opening Reactions of Halocyclopropanes

Halocyclopropanes are readily available substrates that can be utilized in a variety of different transformations, including reduction to furnish monohalocyclopropanes and/or simple cyclopropanes, substitution reactions, and elimination reactions to provide cyclopropene compounds. In all of these cases, the integrity of the three-membered ring is preserved as a result of careful manipulation of the halogen substituents. Although these reactions can be useful for an organic chemist, the most intriguing aspect of halocyclopropane chemistry is the ability of this functionality to undergo facile cationic ring opening under thermal conditions to provide an allyl cation **I**, which can further undergo nucleophilic trapping or elimination to provide olefinic products (**2** and **3**, Fig. 4.1). The 2π-electron ring opening process obeys the conservation of orbital symmetry rules described by Woodward and Hoffmann,[3] and thereby proceeds in a disrotatory fashion with concomitant loss of a halide anion. The direction of disrotation is strongly influenced by stereoelectronic factors.[4] Orbitals that are involved in the breaking of the C–C σ bond must rotate in a direction that assists halide departure by overlapping with the σ* orbital of the dissociating halogen (Fig. 4.2). This stereoelectronic effect implies that chemoselective removal of one of the halogens would result in the preferential formation of one allyl cation over the other. The directionality of ring opening has a distinct impact on ring expansion reactions involving dibromobicyclo[*n*-1,0]alkanes (Fig. 4.3).[5] When medium-size rings are being formed ($n = 2$–4), the disrotation occurs selectively with loss of the *endo*-bromide. Rotation in this direction affords the *E*-isomer, which

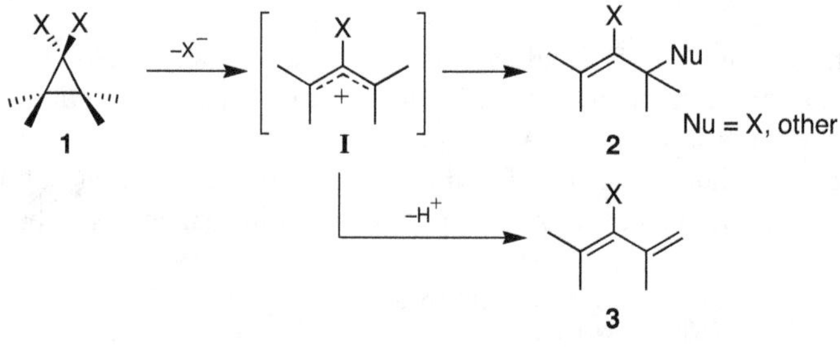

Figure 4.1. Cationic ring opening of *gem*-dihalocyclopropanes.

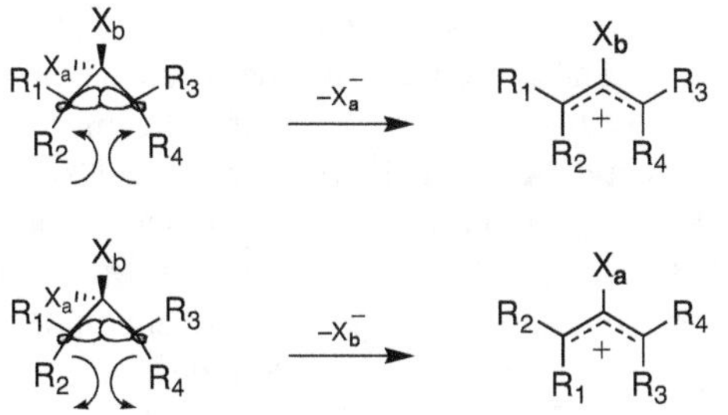

Figure 4.2. Stereoelectronic factors affecting cationic ring opening.

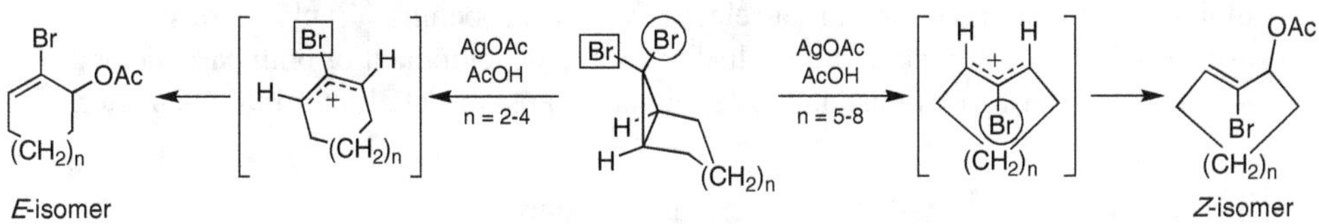

Figure 4.3. The direction of disrotation in ring expansion reactions.

minimizes ring strain during product formation. However, when larger rings are being generated under the same conditions ($n = 5$–8), the *Z*-isomer is preferentially formed, with removal of the more accessible *exo*-bromide. Formation of the *Z*-isomer is actually a favorable process if the developing ring strain permits, since untoward $A_{1,3}$ steric interactions that would arise during ring opening in the other direction are avoided.

Porter et al. attempted to manipulate the directionality of ring opening when proposing a general synthesis of polyunsaturated fatty acid hydroperoxides.[6] In this work, the vinylcyclopropyl bromide **4** was synthesized and subjected to silver trifluoroacetate and hydrogen peroxide in diethyl ether at room temperature (Scheme 4.1). It was anticipated that disrotatory ring opening would preferentially occur in the direction that assists with loss of the bromide to afford the cationic species **II**. The authors maintained that trapping of this cation with hydrogen peroxide would occur faster than isomerization, and loss of stereochemistry and, in fact, the reaction

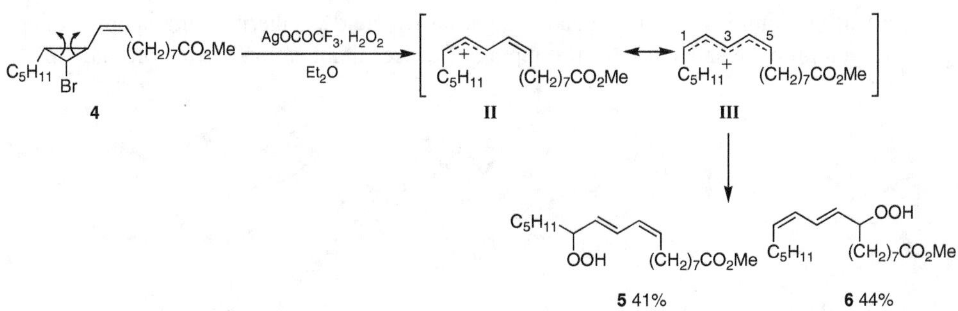

Scheme 4.1. Porter's approach to polyunsaturated fatty acid hydroperoxides.

did furnish the predicted hydroperoxides **5** and **6** as the major products in 41% and 44% yields, respectively. These products are the result of trapping the anticipated cationic intermediate **III** at the C1 and C5 positions. The minor compounds resulting from reaction of this vinylcyclopropyl bromide were also examined and were determined to arise from either product isomerization or ring opening in the less favored direction of disrotation. This "alternate" pathway of ring opening likely occurs with some assistance from the alkenyl substituent on the cyclopropane, since the π bond is situated *trans* to the departing bromide.

Porter's work illustrates the impact of substituent effects on the ring opening of halocyclopropanes. Although the influence of the alkenyl substituent was weak in this particular example, it has been shown that substitution on the halocyclopropane moiety can either impair or promote electrocyclic ring opening.[7] Exploitation of these aspects of ring opening could be valuable for the development of both cascade and asymmetric transformations, and is discussed in greater detail later in this chapter.

4.1.2 Silver(I)-Assisted Ring-Opening Reactions

While cationic ring opening of halocyclopropanes can be induced under strictly thermal conditions, it is most often performed in the presence of a Lewis acid.[8] The Lewis acids commonly used in these reactions are silver(I) salts due the inherent halophilicity of the silver(I) cation. The silver(I)-mediated reactions can be carried out at lower temperatures due to activation of the departing halide by coordination to silver(I) in what has been described as a highly concerted "push–pull" mechanism.[1a] Under these conditions the halide-substituted carbon atom bears slightly increased positive character, which enables the cationic ring opening to proceed under mild conditions (Fig. 4.4).

The presence of silver(I) is additionally used to sequester halide anions liberated in the ring-opening process, effectively removing free halide species from the reaction mixture. This sequestration prolongs the lifetime of the haloallyl cation, which can then be intercepted by another nucleophile or undergo rearrangement reactions. In this respect, the thermal and silver(I)-mediated ring opening processes can be used to afford different products from the same starting material.

An illustration of this divergent reactivity can be observed in Banwell's use of *gem*-dibromocyclopropane **7** to construct an intermediate used in an approach to the Amaryllidaceae alkaloids (Scheme 4.2).[9] When **7** was heated in the absence of Lewis acid, allyl bromide **8** was produced in quantitative yield via thermal ring opening of the cyclopropane and subsequent trapping of the resultant cation by free bromide in

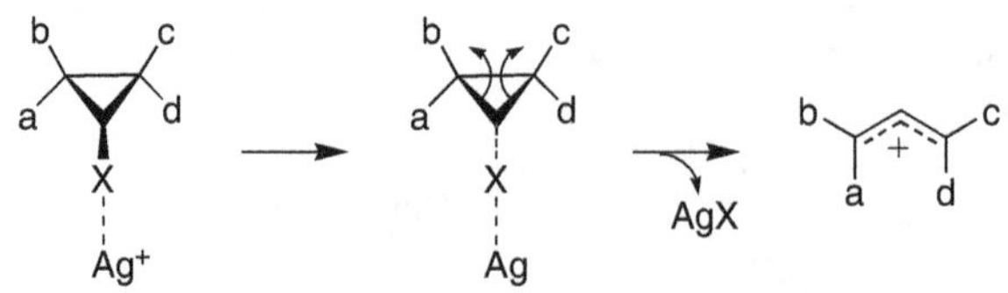

Figure 4.4. Silver(I) assistance in the ring opening of halocyclopropanes.

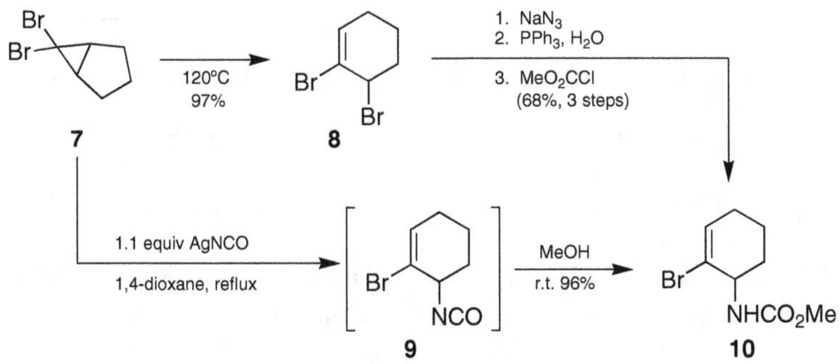

Scheme 4.2. Thermal and silver(I)-assisted ring opening.

the reaction mixture. The allyl bromide was further functionalized (68% over three steps) to the desired carbamate **10**. Banwell was able to dramatically improve the synthesis of the carbamate by developing a one-pot reaction, which involved treatment of *gem*-dibromocyclopropane **7** with silver isocyanate, followed by the addition of methanol.

Under these conditions, silver-assisted electrocyclic ring opening provided the haloallyl cation, which was subsequently trapped by isocyanate anion. Interception of the cationic species with isocyanate was successful since bromide was removed from the reaction mixture as a precipitate (AgBr). Finally, treatment of intermediate **9** with methanol furnished the desired carbamate in 96% yield. This example demonstrates the usefulness of the silver(I)-mediated process. Removal of free halide from the reaction mixture affords a long-lived cationic species that can be captured by a different nucleophile, such as solvent, the silver(I) counteranion, or an intramolecular nucleophile. This reactivity has been exploited in many different ways throughout the years and is examined in greater detail later in this chapter.

4.2 NUCLEOPHILIC TRAPPING OF CATIONIC INTERMEDIATES

The construction of complex intermediates from simple and readily available starting materials has been accomplished using the electrocyclic ring-opening reaction of halocyclopropanes. This is typically achieved through interception of the cationic haloallyl intermediate by solvent, the silver(I) counteranion, or some alternate tethered heteroatom or carbon-based nucleophile. Examples of these processes are described below.

4.2.1 Solvolysis Reactions

The earliest report of the solvolysis of a halocyclopropane to an allyl alcohol was presented by Roberts and Chambers in 1951.[10] These early examples did not utilize silver(I) to assist in halide departure. Instead, the halocyclopropane was simply dissolved in solvent and allowed to react at elevated temperatures. For example, in

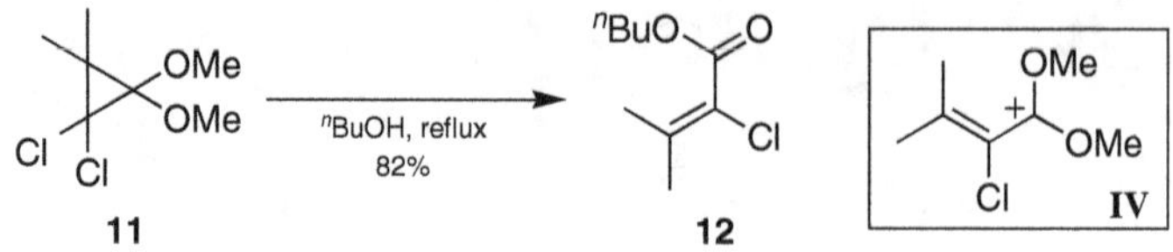

Scheme 4.3. Unassisted solvolysis of halocyclopropanes.

1959 McElvain and Weyna provided evidence for the formation of α-chloroacrylic esters as a result of subjecting *gem*-dichlorocyclopropanes, such as **11**, to refluxing *n*-butyl alcohol (Scheme 4.3).[11] This reaction was believed to proceed through trapping of the stabilized haloallyl cation (**IV**) by *n*-butanol to afford an orthoester that readily collapsed to the observed product. When the solvent participates as a nucleophile in the capture of a cationic species, it may not be necessary for the involvement of silver(I) as a halide sequestering agent since there is a large excess of the nucleophile present; however, silver(I) salts are typically used in order to lower the reaction temperatures and effect a milder solvolysis of halocyclopropanes.[12]

Unfortunately, in some instances participation by solvent can interfere with a desired reaction pathway and afford unwanted side products. In 2004, Banwell and Sydnes[13] disclosed their efforts toward the total synthesis of plant-derived phenanthroquinolizidine alkaloids such as julandine.[14] Their initial strategy involved the intermolecular trapping of an allylic cation with 2-piperidinemethanol **14** (Scheme 4.4). The desired *gem*-dibromocyclopropane precursor **13** was derived from dibromocarbene addition to the respective olefin (**12**) in 78% yield using standard phase transfer conditions developed by Makosza and Wawrzyniewicz.[15] When ring opening of **13** was induced using a variety of silver(I) salts, the desired intermolecular trapping product **15** was not observed. Instead, the cationic species was trapped by either solvent (CF_3CH_2OH, or $(CH_3)_2CHOH$) or the silver(I) counteranion.

In an effort to overcome this issue, the authors decided to approach the synthesis of **15** in a different manner, inducing solvolysis of the *gem*-dibromocyclopropane with silver acetate in the presence of acetic acid to furnish an allylic acetate. The acetate

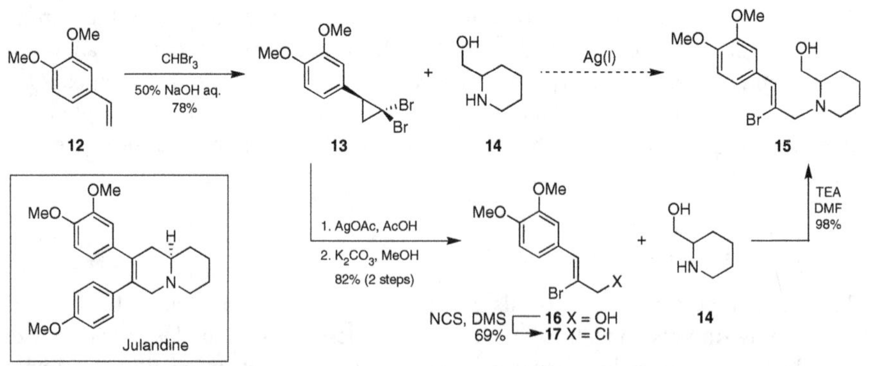

Scheme 4.4. Steps toward the synthesis of phenanthroquinolizidine alkaloids.

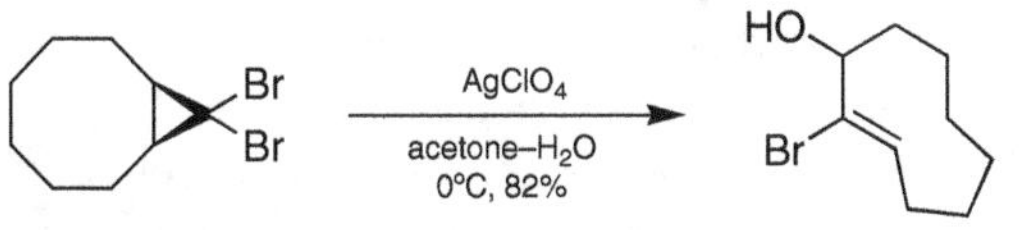

Scheme 4.5. Silver(I)-promoted solvolysis reaction with ring expansion.

was then smoothly converted to the allylic alcohol **16**, with a yield of 82% over the two steps. Subsequently, **16** was treated with *N*-chlorosuccinimide in dimethyl sulfide to afford allyl chloride **17**, which was then coupled with 2-piperidinemethanol to provide the desired product **15** in 98% yield. Although the planned intermolecular trapping process was unsuccessful, solvolysis of *gem*-dibromocyclopropane **13** provided a useful synthetic intermediate to allow eventual assembly of the desired vinyl bromide.

In many instances, however, solvolysis of a halocyclopropane is deliberately accomplished in order to install an essential vinyl halide or hindered olefin. In 2000, Murphy and coworkers performed a silver ion-mediated ring expansion of *gem*-dibromocyclopropane **18** in wet acetone to afford the allylic alcohol **19** in 82% yield (Scheme 4.5).[16] Under these conditions the desired cyclononene product was obtained as an inseparable mixture of *E*- and *Z*-isomers (7 : 93). Interestingly, two sets of peaks observed in the ^{1}H NMR spectrum indicated that the *Z*-isomer existed as two separate conformers at room temperature. This intermediate was subsequently used in Murphy's approach to the radical-based preparation of tricyclic indoles.

One particular advantage to using silver(I) to promote the electrocyclic ring opening of halocyclopropanes is that deliberate selection of the counteranion can result in products derived from cation interception by the silver salt. An example of this mode of trapping was illustrated earlier in the conversion of a halocyclopropane to an allylic carbamate via trapping with silver isocyanate (Scheme 4.2).[9,17] One obvious drawback to this mode of trapping is the necessity for a large excess of the silver salt in order to achieve efficient trapping. Nonetheless, an excellent example of the utility of this reaction can be observed in Banwell's approach to the ABC ring system observed in paclitaxel (Scheme 4.6).[18] In this route, the complex intermediate

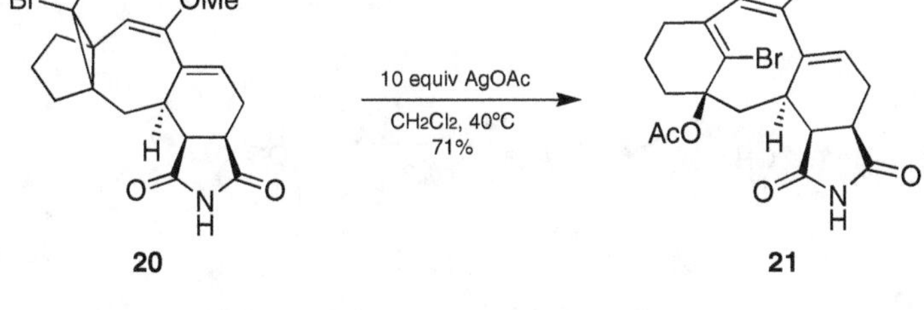

Scheme 4.6. An approach to paclitaxel.

20 was assembled in five short steps from 5-methoxyindane. The *gem*-dibromocyclopropane was then subjected to silver acetate in dichloromethane at 40°C to generate the product **21**. Under these conditions the silver counteranion was involved in regioselective capture of the haloallylic cation derived from ring opening of the halocyclopropane, resulting in the installation of necessary oxygenation and a very strained bridgehead olefin.

4.2.2 Intramolecular Trapping with Heteronucleophiles

Halocyclopropane compounds that contain an internal nucleophile have proved to be very successful substrates for cascade reactions. On silver(I)-mediated ring opening of the halocyclopropane moiety, the tethered nucleophile can intercept the cationic haloallyl intermediate affording cyclic and/or polycyclic products. Typically these reactions utilize at least one equivalent of the silver(I) salt to assist in ring opening and completely sequester free halide ions from the reaction mixture, which effectively eliminates intermolecular trapping of the cationic species by halide. Early examples of intramolecular trapping in this manner were performed using the silver(I)-mediated ring opening of *gem*-dibromocyclopropanes tethered to a carboxylic acid or alcohol moiety (Scheme 4.7).[19] In this scenario, electrocyclic ring opening afforded a haloallyl cation that was readily trapped by the pendent nucleophile to furnish lactone and cyclic ether products in good to excellent yields. In a similar fashion, *gem*-dihalocyclopropanecarboxylic acids **26** can be used as substrates for intramolecular trapping of the haloallyl cation (Scheme 4.8).[20] In this study, the substrates were transformed to the corresponding butenolide products in excellent yields, provided the right combination of silver salt and solvent were utilized. When methanol, isopropyl alcohol, or 2,2,2-trifluoroethanol (TFE), was used as solvent in these reactions, solvolysis products were always observed to some minor extent. When **26b** was treated with silver trifluoroacetate in hexafluoroisopropanol (HFP), it underwent ring opening to afford the expected butenolide **28** in 82% yield, along

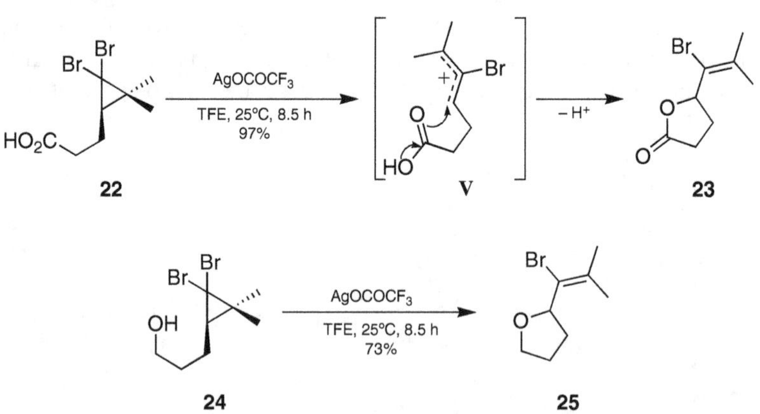

22 **V** **23** **24** **25**

Scheme 4.7. Synthesis of lactones and cyclic ethers from halocyclopropanes.

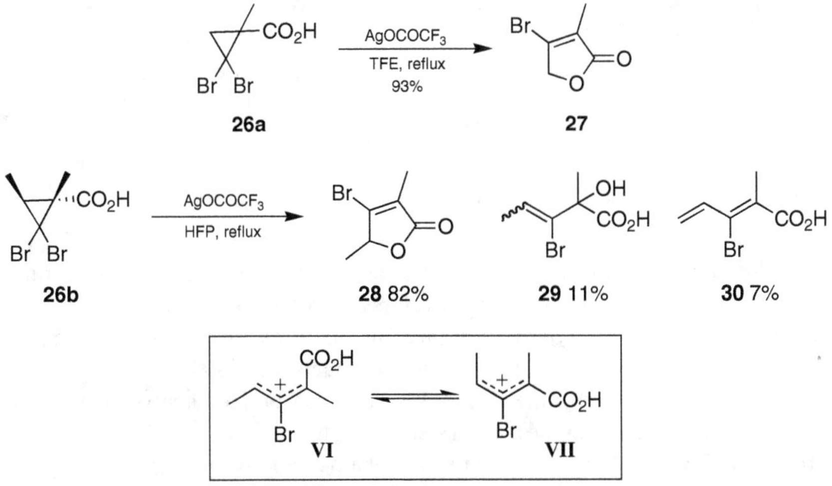

Scheme 4.8. Butenolide formation from *gem*-dihalocyclopropanecarboxylic acids.

with the hydroxy acid **29** and diene **30**. The formation of these minor products can be attributed to intermediacy of two cationic intermediates (**VI** or **VII**), only one of which (**VI**) can proceed immediately to the butenolide product.

Cha and coworkers have used the silver(I)-mediated ring opening of a *gem*-dichlorocyclopropane to construct the interesting bicyclo[6.4.0]dodecanone derivative **32** (Scheme 4.9).[21] The halocyclopropane compound was synthesized from the corresponding enol ether, which was constructed using an ingenious oxy-Cope rearrangement of *cis*-1,2-dialkenylcyclopropanols. Silyloxy substitution on a halocyclopropane moiety typically results in the formation of an unsaturated ketone on silver(I)-promoted ring opening; however, the additional oxygen present in Cha's substrate **31** was poised to participate in the trapping of the intermediate haloallyl cation and afford the ketal **32** in 70% yield. This result implies that the silyloxy substituted allyl cation has a substantial lifetime, allowing trapping processes to occur readily in place of immediate termination to the carbonyl moiety.

Nitrogen-based nucleophiles have been thoroughly examined in the context of trapping the cationic intermediate generated from silver(I)-mediated halocyclopro-

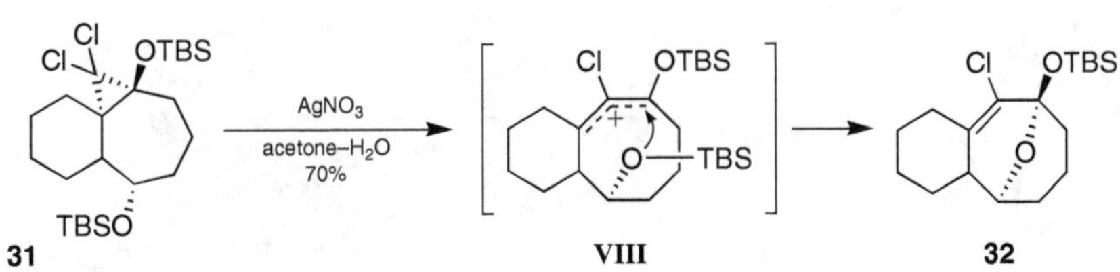

Scheme 4.9. Intramolecular ketal formation from ring opening of a halocyclopropane.

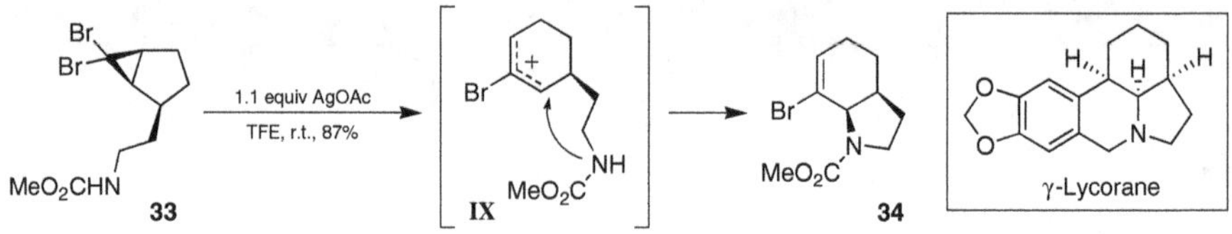

Scheme 4.10. Key step in the synthesis of γ-lycorane.

pane ring opening. Banwell has been the most prominent figure in this area of halocyclopropane chemistry and has developed some very elegant ways to utilize intramolecular trapping to construct highly functionalized intermediates toward the synthesis of a number of alkaloid natural products. In 1994, Banwell and Wu presented a racemic total synthesis of γ-lycorane that utilized a silver(I)-promoted electrocyclic ring opening as the pivotal step for the formation of an advanced bicyclic intermediate (Scheme 4.10).[22] In this reaction, halocyclopropane **33** was subjected to silver acetate in 2,2,2-trifluoroethanol. These conditions induced ring opening to the 2-bromoallyl cation **IX** and subsequent interception by the nitrogen of a tethered carbamate moiety to afford the bicyclic product **34** in 87% yield.

A similar strategy was used in the ensuing syntheses of the crinine-type alkaloids maritinamine and *epi*-maritinamine.[23] The key step in the synthesis of these natural products is depicted in Scheme 4.11, in which the *gem*-dichlorocyclopropane substrate **35** was subjected to 6 equiv of silver tetrafluoroborate in tetrahydrofuran to afford the cyclized product **36** in moderate yields after reinstallation of the carbamate protecting group. Interestingly, the authors noted that the two epimeric forms of the halocyclopropane compound undergo the cascade cyclization sequence at different rates, with one isomer reacting completely within 8 h and the other requiring prolonged reaction times. The resulting bicyclic product was carried on to complete the synthesis of both maritinamine and its C3 epimer.

The intramolecular nitrogen-trapping protocol used by Banwell has also been successfully exploited in the assembly of spirocyclic frameworks relating to the aromatic erythrina alkaloids.[24] However, when it was applied to the synthesis of nonaromatic spirocycles, as found in histrionicotoxin, the flexible alkyl tether proved to be problematic.[25] In this study, *gem*-dichlorocyclopropane substrate **37** was initially subjected to silver(I) salts under a variety of conditions only to provide solvolysis and elimination products without any indication of trapping by the pendent

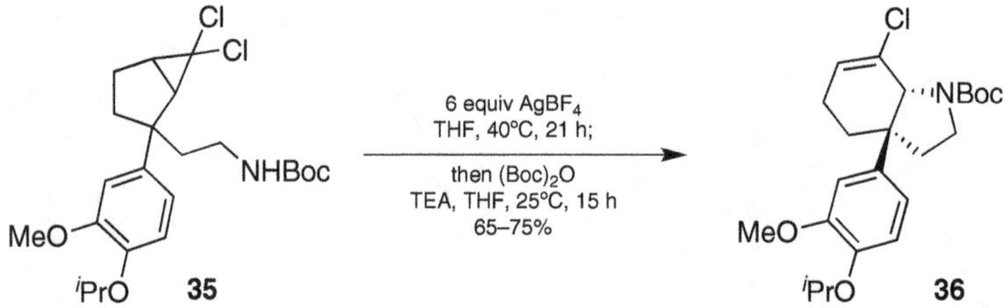

Scheme 4.11. Intramolecular nitrogen trapping en route to maritinamine.

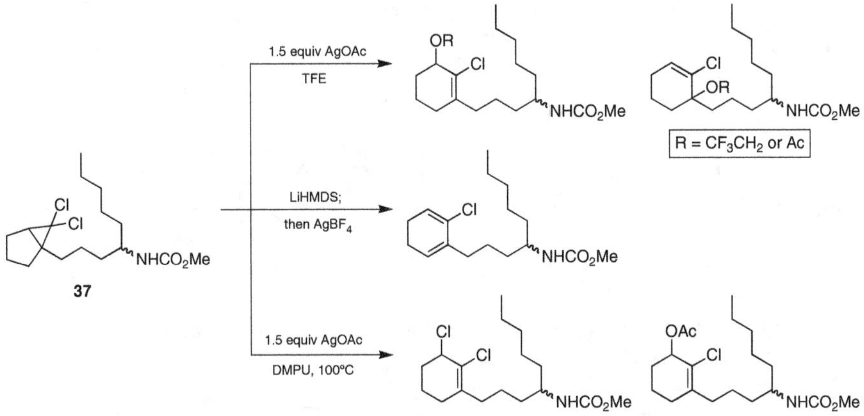

Scheme 4.12. Initial attempts at spirocyclization.

nitrogen-based nucleophile (Scheme 4.12). Subsequently, a *cis*-olefin was incorporated into the tether between the halocyclopropane and carbamate moieties in an effort to introduce an element of rigidity to the substrate (Scheme 4.13). This strategy was successful in inducing spirocyclization to provide a sample of the desired product **39**, albeit in poor yield, along with a small amount of polyene **40**.

The successful utilization of nitrogen-based nucleophiles in the previously described cascade reactions has allowed for the synthesis of complex polycyclic structures from simple and readily available starting materials. The fact that carbamates can participate as nucleophiles has provided the opportunity for development of diastereoselective ring closures onto the halocyclopropane-derived allyl cation.

4.2.3 Diastereoselective Reactions

In 2000, Banwell and coworkers reported their initial attempts toward diastereoselective ring closure onto the cation that results from silver(I)-promoted halocyclopropane ring opening. In this work, a concise approach to the synthesis of bridged seven- and eight-membered nitrogen heterocycles was described in which a chiral auxiliary was used to impart stereoselectivity in the pivotal electrocyclic ring-opening/nitrogen-trapping step (Scheme 4.14).[26]

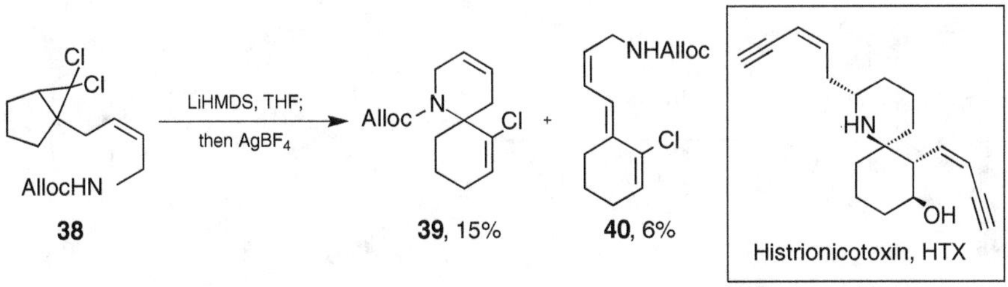

Scheme 4.13. Assembly of 1-azaspiro[5.5]undecane framework.

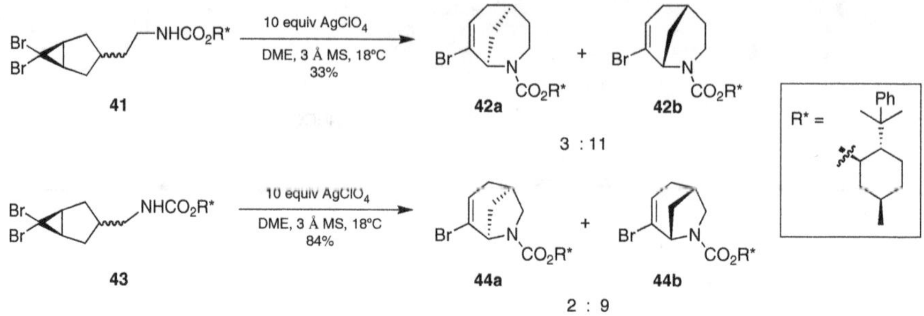

Scheme 4.14. Chiral auxiliary approach to diastereoselective cyclizations.

Previously, chiral carbamates were used by Banwell to afford facile separation of diastereomers at appropriate stages in the synthesis of various alkaloid natural products.[9,27,28] In those instances, the chiral carbamates were prepared by silver (I)-mediated ring opening of a halocyclopropane, followed by trapping with cyanate and subsequent treatment with (−)-menthol, a method analogous to that outlined earlier in this chapter (Scheme 4.2). However, in this example the chiral auxiliary is used for direct control of subsequent ring closure, affording 2-azabicyclo[3.3.1]non-7-enes **42** and 6-azabicyclo[3.2.1]oct-3-enes **44** in modest diastereoselectivities. In this work, the best results were obtained when the 8-phenylmenthyl-based auxiliary was used, suggesting that π stacking plays a role and may be further exploited to improve the selectivity of this cascade process.

More recently, Banwell has investigated the possible influence of remote stereogenic centers on the intramolecular capture of halocyclopropane-derived allyl cations (Scheme 4.15).[29] During the enantioselective total synthesis of *ent*-erythramine and 3-*epi*-erythramine, *gem*-dichlorocyclopropane **45** was prepared in nine steps from known starting materials and then subjected to silver tetrafluoroborate after deprotonation with lithium hexamethyldisilazide. Subsequent deprotection afforded a separable diastereomeric mixture of spirocyclic products **46a** and **46b** in 26% and 30% yields, respectively. This result implies that the remote stereogenic center is ineffective at controlling the stereoselectivity of the silver(I)-mediated ring opening/nitrogen-trapping reaction of **45**, perhaps due to the lack of steric bulk provided by the methoxy group as well as its distance from the developing spirocenter.

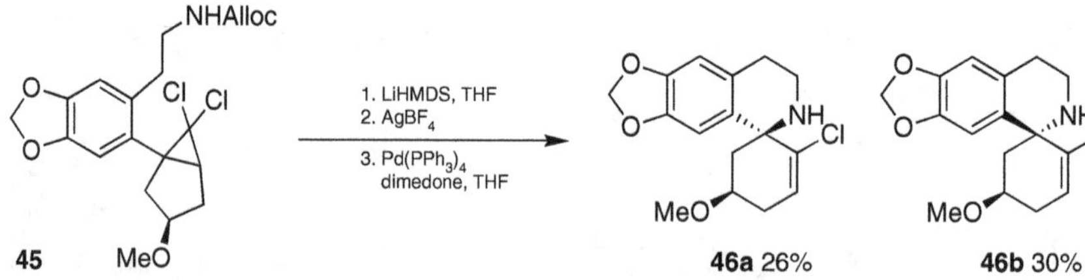

Scheme 4.15. An attempt to remotely control the stereoselectivity of ring closure.

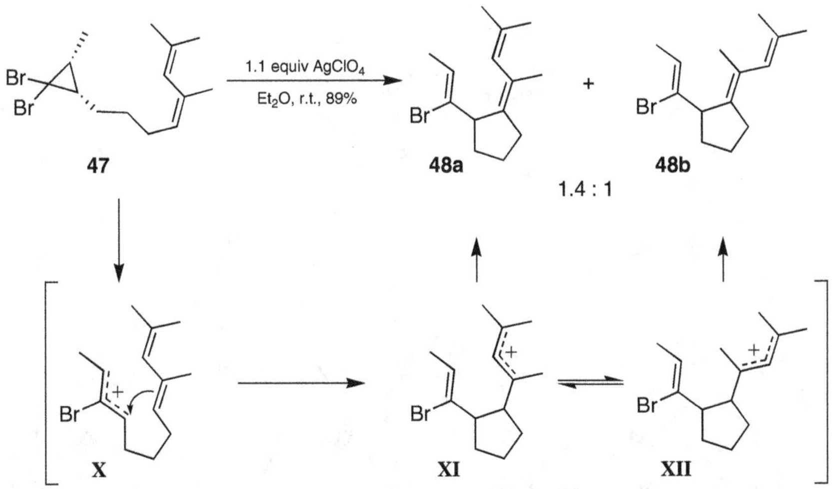

Scheme 4.16. Cationic cyclization of halocyclopropane dienes.

4.2.4 Carbon–Carbon Bond Formation

The silver(I)-mediated ring opening of halocyclopropanes has been used to construct complex frameworks through the inter- and intramolecular trapping of cationic intermediates with heteronucleophiles. An obvious extension of this work is the involvement of carbon-based nucleophiles to form new carbon–carbon bonds. In 1996, Kostikov and coworkers reported the participation of aromatic solvents in the capture of halocyclopropane-derived allyl cations even in the absence of silver(I).[30] However, this early example of intermolecular attack by a carbon nucleophile is one of very few such reports. In the same year, Gassman et al. reported cationic cyclizations of *gem*-dibromocyclopropanes tethered to remote diene moieties (Scheme 4.16).[31]

They found that treatment of diene **47** with 1.1 equiv of silver perchlorate in diethyl ether afforded the cyclized trienes **48a** and **48b** in a ratio of 1.4 : 1 and 89% yield. The mechanism of this reaction was believed to involve initial ring opening of the *gem*-dibromocyclopropane followed by trapping of the haloallyl cation **X** to furnish another cationic allyl species. The observed scrambling of alkene geometry in the products can be explained by facile rotation of the carbon–carbon bond that connects the allyl cation to the cyclopentane in **XI** and **XII**. These authors also investigated the effect of substitution on the terminal olefin by removing the *gem*-dimethyl groups. This substrate **49** provided the expected triene in 50% yield, but also resulted in the formation of bicyclic product **51** in 13% yield. This second product is most likely constructed from two successive cation–olefin cyclizations (Scheme 4.17), followed by oxidation to the aromatic compound.

In 2004, Banwell and Sydnes reported an unintentional but intriguing result of intermolecular "pseudodimer" formation as a result of carbon–carbon bond formation during their efforts toward the synthesis of phenanthroquinolizidine alkaloids (Scheme 4.18).[13]

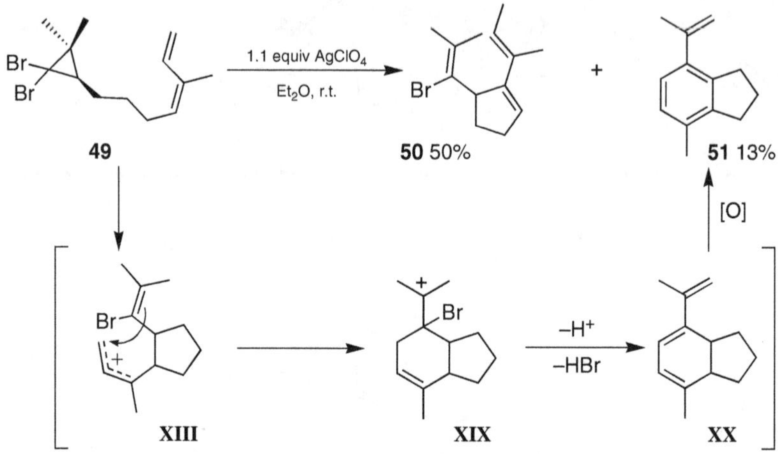

Scheme 4.17. Two cation–olefin cyclization events to afford an aromatic indane.

On treatment with silver tetrafluoroborate in acetonitrile, the *gem*-dibromocyclopropane **13** underwent electrocyclic ring opening and subsequent capture by another molecule of **13** in an S_EAr process. This "dimerization" reaction occurred despite the presence of an able, nitrogen-based nucleophile, piperidine **14**, in large excess. Despite efforts to effect nitrogen trapping of the cationic intermediates, no solution to this problem could be found, which led to the design of a new approach to the synthesis of the target compounds.

Carbon-based nucleophiles have now been thoroughly explored in the design of both inter- and intramolecular trapping reactions of haloallyl cations. Banwell and coworkers have presented examples of both pyrrole and indole participation in these processes, resulting in the construction of important alkaloid scaffolds.[32] Banwell's most recent contribution to this area was described in his work toward the quick assembly of hapalindole and fischerindole structures (Scheme 4.19).[33] Indoles **53a** and **53b** were able to capture the cationic intermediate generated from treatment of *gem*-dibromocyclopropane **54** with silver tetrafluoroborate in tetrahydrofuran, providing the tricyclic products **55a** and **55b** in 67–77% yields. The trapping products could be carried on to afford tetracycle **56** (or its *cis* ring-fused isomer) in four steps.

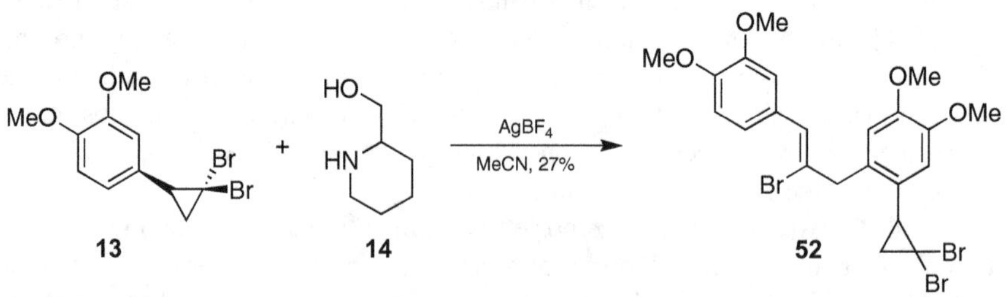

Scheme 4.18. Unexpected carbon–carbon bond formation.

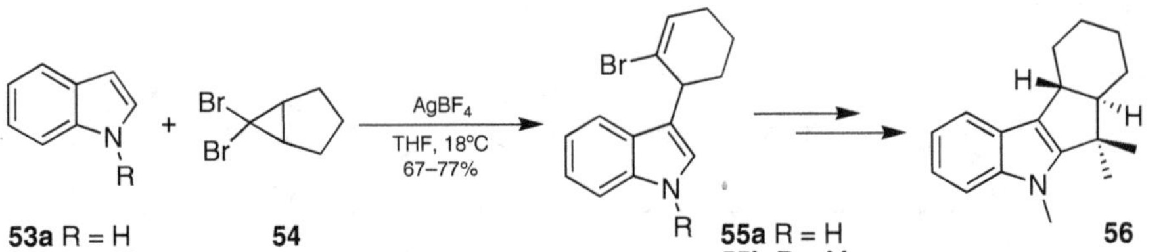

Scheme 4.19. Intermolecular trapping of the allyl cation by indole nucleophiles.

This strategy allowed for the rapid construction of complex, polycyclic skeletons in good yields from simple starting materials.

In the same publication, an enantioselective process was attempted wherein commercially available (2*R*,3*R*)-butane-2,3-diol was used to generate the chiral cyclopentene **57**, which was cyclopropanated to afford *gem*-dibromocyclopropane **58** (Scheme 4.20). Unfortunately, when this substrate was subjected to the reaction conditions outlined above, product **59** was obtained as a 1 : 1 mixture of diastereomers. This result implies that selectivity in these trapping processes is unaffected by the presence of a chiral auxiliary on the remote carbon of the cyclopentane framework.

The silver(I)-mediated electrocyclic ring opening of halocyclopropanes has been used to induce extensive skeletal rearrangements in *gem*-dibromospiropentanes, providing rapid construction of naphthalenes and/or indenes (Scheme 4.21).[34] A variety of Lewis acids, Brønsted acids, and solvent effects were carefully examined before optimal conditions were identified. It was found that subjection of spirocycle **60** to silver acetate in trifluoroacetic acid afforded rearrangement products **61** and **62** in moderate to good yields. The proposed mechanism of the reaction is illustrated in Scheme 4.21.

Silver(I)-promoted ring opening of the *gem*-dibromocyclopropane would result in the formation of haloallyl cation **XXI**, which is the common intermediate in the synthesis of both naphthalene and indene products. It was observed that naphthalene compounds were preferentially formed when electron-withdrawing groups were placed on the aromatic ring. In these substrates, haloallyl cation **XXI** would be destabilized and therefore readily undergo another ring-opening process to afford cation **XXII**. This new cation would then be trapped in a Friedel–Crafts reaction to provide the observed naphthalene products **61**. In instances where the aromatic ring was substituted with electron-donating groups, the indene products **62** were

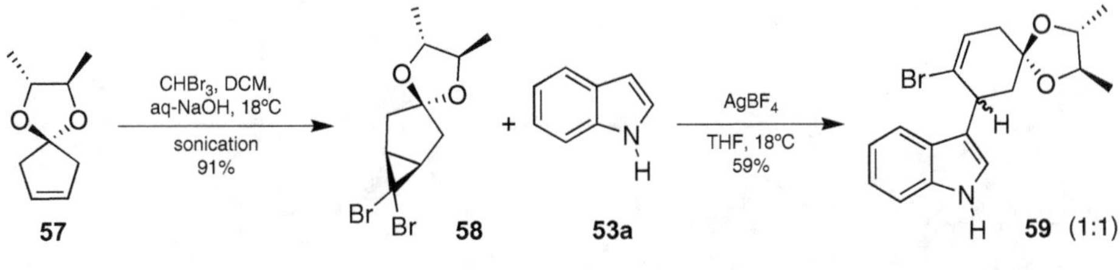

Scheme 4.20. Chiral auxiliary approach to diastereoselective trapping processes.

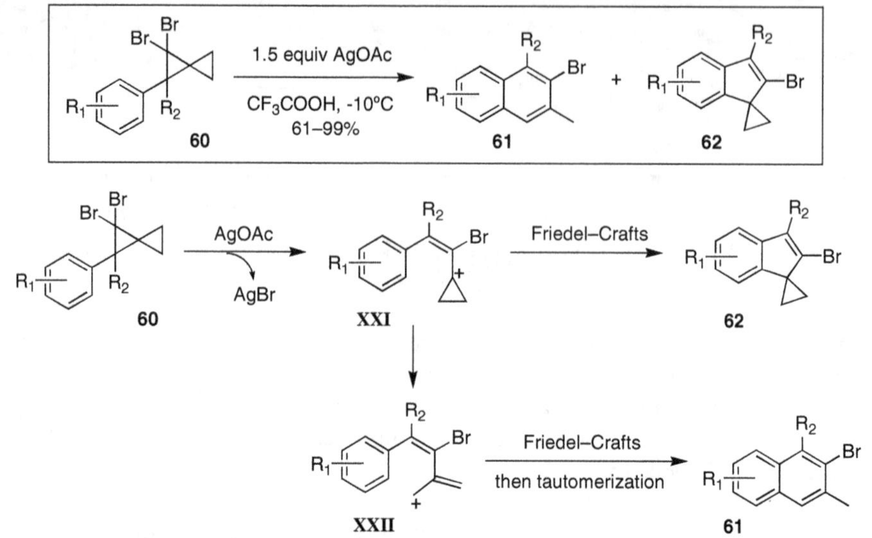

Scheme 4.21. Skeletal rearrangement of *gem*-dibromospiropentanes.

preferentially formed. This can be explained in a similar fashion, since these substrates would afford long-lived, stabilized haloallyl cations **XXI** that could be immediately captured in a Friedel–Crafts reaction instead of undergoing further ring opening processes.

4.3 THE SILVER(I)-PROMOTED NAZAROV REACTION

In 2006, the unique silver(I)-mediated cationic ring opening of *gem*-dihalocyclopropanes was examined as an interesting entry into the design of a novel Nazarov cyclization.[35] The Nazarov reaction is an electrocyclization process that typically transforms a cross-conjugated dienone, **63**, into a cyclopentenone product, **64**, by means of conrotatory ring closure (Fig. 4.5). Since its initial discovery in 1941,[36] the Nazarov reaction has been keenly investigated and important advances have been made that expand both the scope and utility of this process. Halocyclopropane substrates were investigated as novel substrates in the Nazarov reaction since it was well documented that cationic intermediates derived from ring opening were

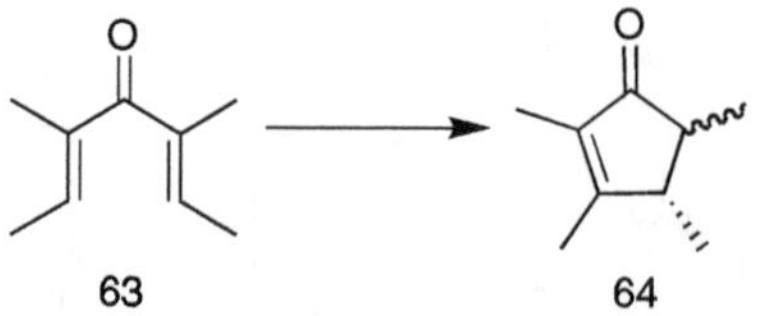

Figure 4.5. Nazarov cyclization.

accessible using very mild reaction conditions. It was believed that appropriately substituted *gem*-dihalocyclopropanes could be used to access a pentadienyl cationic species capable of undergoing Nazarov cyclization.

4.3.1 Development and Initial Findings

In an effort to mimic Nazarov cyclization with nontraditional substrates, the design of a sequential 2π-electron disrotatory ring opening and 4π electrocyclization was undertaken (Fig. 4.6). 1,1-Dihalo-2-(silyloxy)-2-vinylcyclopropanes **65** were determined to be optimal substrates for this investigation, since computational experiments[7] have shown that alkenyl substitution on *gem*-dihalocyclopropanes accelerates the ring-opening process relative to hydrogen. Oxygen substitution at the same position on the cyclopropane moiety was also shown to assist in the disrotatory ring opening, which would promote the efficient generation of haloallyl cation **XXIII**. Because of the presence of the vinyl substituent, this cation might also be viewed as a pentadienyl species **XXIV**, which is analogous to the cationic intermediate observed during conventional Nazarov cyclizations. It was believed that the pentadienyl cation would undergo conrotatory electrocyclization to furnish a 2-silyloxycyclopentenyl cation (**XXV**) that would experience traditional elimination and subsequent hydrolysis to provide α-halocyclopentenone products **66** and/or **67**. This type of sequential ring opening and electrocyclization involving *gem*-dihalocyclopropanes has not previously been studied, although conceptually similar protocols have been reported.[37,38]

Initial experiments to investigate the use of *gem*-dihalocyclopropanes in the Nazarov reaction were performed on methyl vinyl ketone, a simple and readily available substrate. Methyl vinyl ketone was smoothly converted to 2-triisopropylsilyloxydiene **69** in low yield after immediate purification on an alumina column (Scheme 4.22). Preparations of both *gem*-dichlorocyclopropane compound **70** and the *gem*-dibromocyclopropane compound **71** were investigated. While dichlorocyclopropanation proceeded smoothly, the analogous dibromocyclopropanation reaction afforded a mixture of compounds, including the desired product **71**, which could not be separated. This observation prompted selection of *gem*-dichlorocyclopropane compounds for use in development of the new Nazarov methodology.

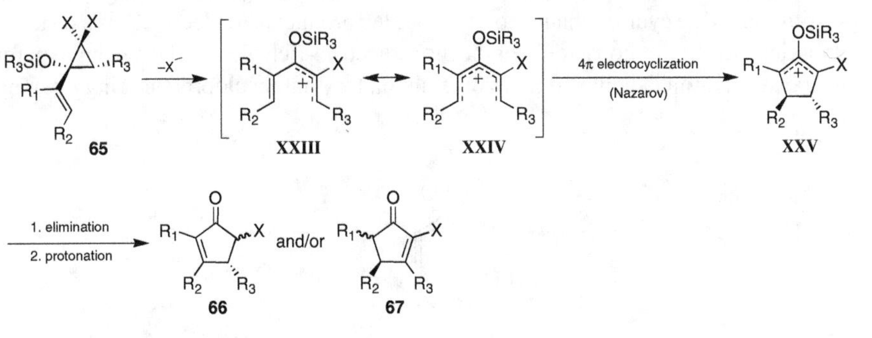

Figure 4.6. Concept behind *gem*-dihalocyclopropanes in the Nazarov reaction.

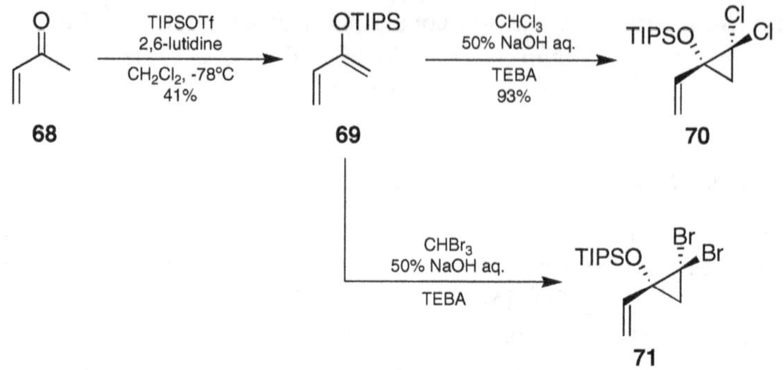

Scheme 4.22. The synthesis of 1,1-dihalo-2-(silyloxy)-2-vinylcyclopropanes.

The choice of silyl substituent was based on the durability of a triisopropylsilyl (TIPS) group relative to other, less bulky silyl groups. Other silyl groups, such as *tert*-butyldimethylsilyl and triethylsilyl, were also investigated during optimization of the reaction conditions, but it became obvious that a more robust silyl substituent was needed to prevent premature termination of the reaction by desilylation to afford 2-chlorodienones such as **73** (Fig. 4.7).

Substrate **70** was exposed to a variety of reaction conditions in a qualitative investigation used to assess the practicality of the proposed ring opening and 4π-electrocyclization sequence, in which it was found that treatment with 1.5 equiv of silver tetrafluoroborate in dichloromethane afforded the α-chlorocyclopentenone **74** after prolonged stirring at room temperature (Fig. 4.8). Further investigation showed that this mode of reactivity was, in fact, general, and could be applied to a number of readily accessible *gem*-dichlorocyclopropane substrates to afford the corresponding α-chlorocyclopentenone products in moderate to good yields. It was later observed that both yields and reaction times for these sequences could be improved by treatment of the starting materials with 1 equiv of silver tetrafluoroborate in refluxing acetonitrile (yields depicted in parentheses in Fig. 4.8).[39]

The majority of *gem*-dichlorocyclopropane substrates examined in this study provided the desired α-chlorocyclopentenones as a result of sequential electrocyclic ring opening and Nazarov cyclization. In general, those substrates lacking additional substitution on the cyclopropane moiety provided products **75** selectively as a result of regioselective elimination to deliver the more electron-rich olefin. The mechanism for this transformation is believed to involve disrotatory halocyclopropane ring opening

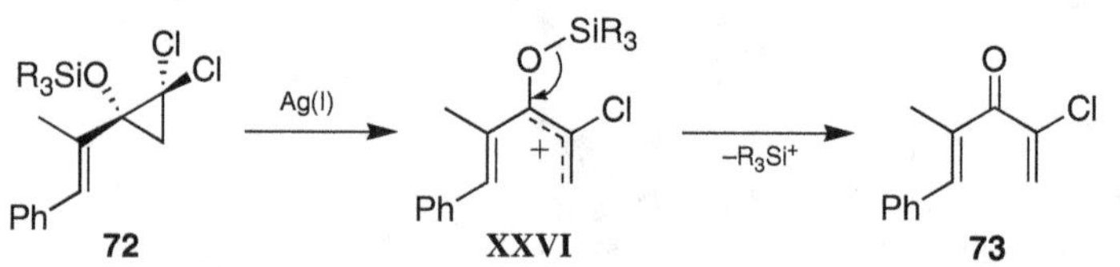

Figure 4.7. Termination of the reaction via desilylation of the haloallyl cation.

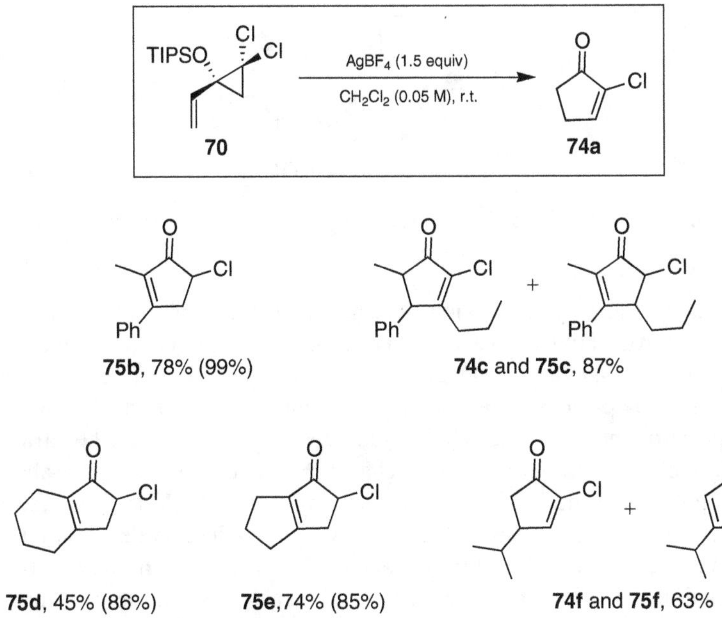

Figure 4.8. The halocyclopropane Nazarov reaction.

and in situ generation of a pentadienyl cation capable of undergoing Nazarov cyclization, as outlined previously. This proposal is supported by the observation silver(I) salts do not promote cyclization of unactivated α-chlorodienones under the standard reaction conditions (Scheme 4.23).

4.3.2 Interrupted Nazarov Reactions

To further expand the scope of this new silver(I)-mediated reaction sequence, *interrupted* Nazarov cyclizations were explored using the halocyclopropane chemistry, an investigation that was prompted by the discovery of an intriguing result. It was found that treatment of the phenethyl-substituted compound **76** with 1.5 equiv of silver tetrafluoroborate in dichloromethane provided benzohydrindenone **77** as the sole product, with no apparent formation of the simple α-chlorocyclopentenone (Scheme 4.24). This prompted an examination of appropriately substituted *gem*-dichlorocyclopropane substrates in analogous interrupted Nazarov processes to ascertain the scope of this new cascade reaction sequence.

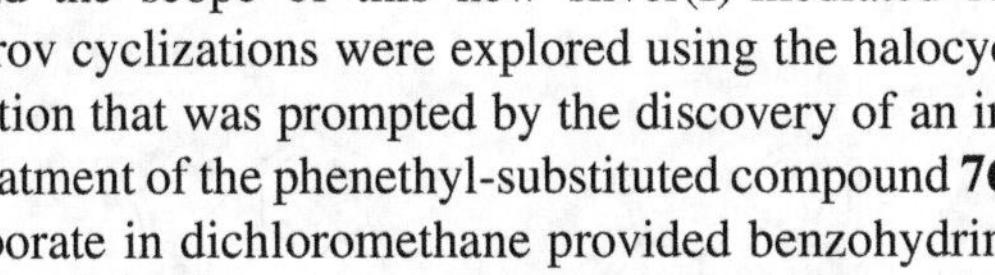

Scheme 4.23. Treatment of chlorodienones with $AgBF_4$.

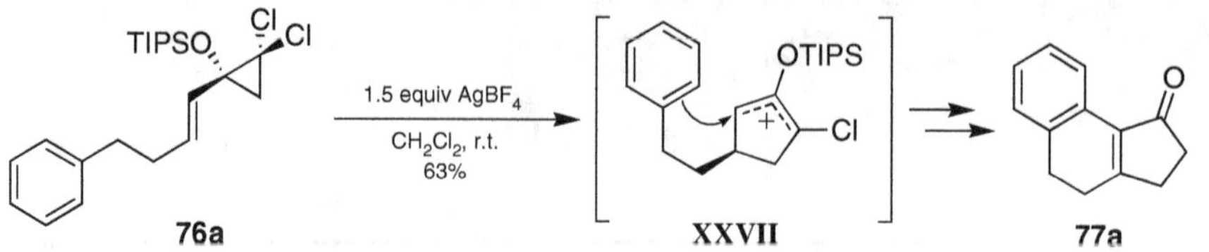

Scheme 4.24. An interrupted Nazarov reaction using *gem*-dichlorocyclopropanes.

Although arene trapping has previously been demonstrated in traditional Nazarov reactions,[40] the participation of an unactivated aromatic ring was unprecedented. This result implied that 2-silyloxycyclopentenyl cations (**XXVII**) derived from ring opening of the halocyclopropane moiety were much more reactive than were traditional Nazarov intermediates. This increased electrophilicity may be attributed to the presence of an electron-withdrawing chlorine substituent as well as the absence of stabilizing alkyl substitution on the cyclopentenyl cation **XXVII**. These cationic intermediates appear to be capable of reacting with even electron-deficient nucleophiles, which presents the possibility for new modes of trapping in the interrupted Nazarov reaction. To assess the generality of this reaction sequence, a number of substrates with one or more pendent arene moieties were synthesized and subjected to the optimized silver(I) reaction conditions (Fig. 4.9).[41]

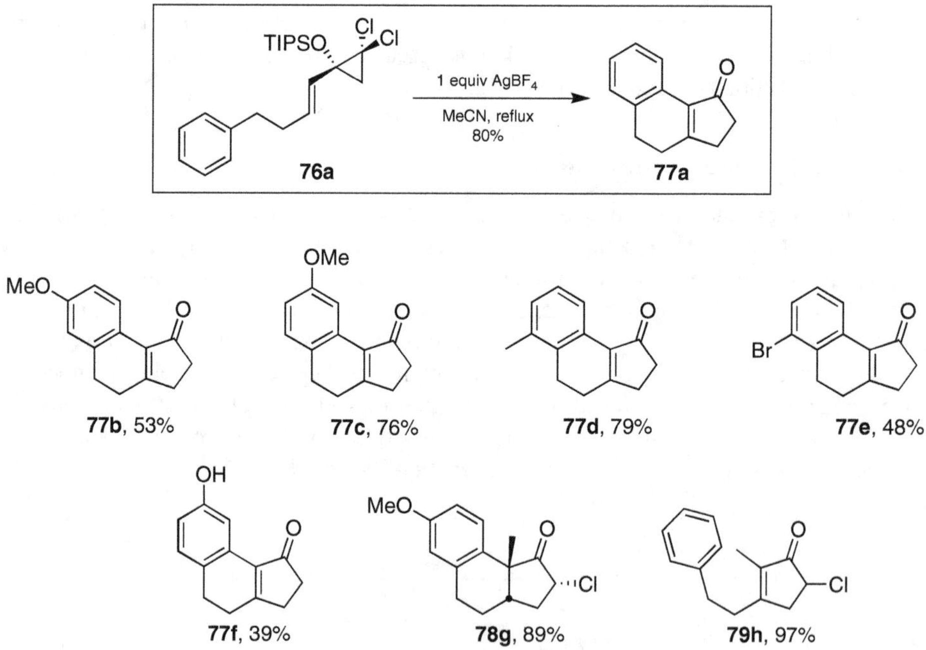

Figure 4.9. Silver(I)-mediated interrupted Nazarov reactions.

In most cases, the desired products were isolated in moderate yields and as a single regioisomer due to electrophilic attack at the least hindered position of the aromatic ring (i.e., *para* to the methoxy substituent in **77b**). Also, all of the tricyclic products (**77**) had experienced dehydrochlorination and possessed a ring fusion alkene that was conjugated to the aromatic ring and ketone functionalities. Remarkably, a deactivated aromatic ring was capable of trapping the reactive 2-silyloxycyclopentenyl cation to furnish a bromosubstituted benzohydrindenone product (**77e**). This type of electron-deficient product would not be accessible from the conventional interrupted Nazarov reaction of divinyl ketone substrates, which demonstrates the value of this complementary trapping process. Another interesting observation was made when a silyl-protected phenol was subjected to silver tetrafluoroborate in refluxing acetonitrile, prompting deprotection of the labile phenolic protecting group to provide benzohydrindenone **77f** in 39% after an extended reaction time.

When the effect of α substitution on the vinyl moiety was examined, the importance of an electron-rich aromatic trap was demonstrated. None of the desired trapping product was observed with a deactivated phenyl group (**79h**) due to stabilization of the 2-silyloxycyclopentenyl cation by alkyl substitution; however, when the 3-methoxyphenyl moiety was used in place of a nonsubstituted phenyl, the chlorocyclopentanone **78g** was isolated in very good yield. The *cis* relationship between the hydrogen and methyl groups at the ring fusion, as well as the relative stereochemistry of the chlorine moiety, were confirmed by single-crystal X-ray crystallography. These results provided important insight into the mechanism of this cascade process (Fig. 4.10). It was proposed that the reaction proceeds through the typical silver(I)-assisted disrotatory ring opening of the *gem*-dichlorocyclopro-

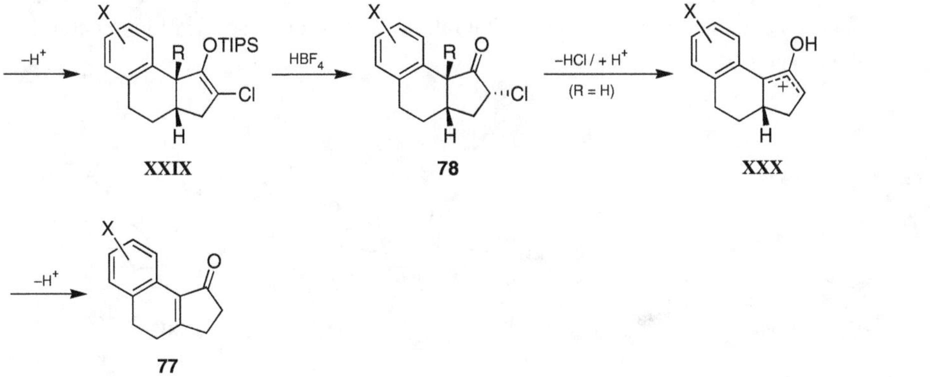

Figure 4.10. Proposed mechanism for arene trapping.

pane moiety, followed by 4π electrocyclization (Nazarov cyclization) to provide the 2-silyloxycyclopentenyl cation **XXVII**. This highly reactive cationic species would then be captured by a pendent arene nucleophile to afford the tricyclic intermediate **XXIX** on rearomatization. Protodesilylation of the silyl enol ether from the least-hindered, convex face of **XXIX** would then result in the formation of chlorocyclopentanones **78**. In the absence of alkyl substitution at the bridgehead position, compounds **78** could experience dehydrochlorination to provide the benzohydrindenone products **77**. Since elimination of HCl to generate an olefin is not observed when "R" is a methyl group, it was proposed that double-bond formation proceeds through a second oxyallyl cationic intermediate **XXX**,[42] followed by conventional Nazarov-type elimination and protonation to furnish the final products.

In an effort to provide experimental evidence for the formation of a second oxyallyl cationic intermediate, a *gem*-dichlorocyclopropane substrate was envisaged with two internal nucleophiles: one to participate in the initial interrupted Nazarov reaction, and the other to capture the second cationic species. Surprisingly, when the carefully designed substrate **80** was subjected to the optimized reaction conditions, an alternate mode of trapping occurred to generate the intriguing bridged bicyclic product **81** (Scheme 4.25).

This reactivity proved to be a general process, providing the unique products in moderate yields following cyclopropanation and immediate treatment with silver tetrafluoroborate. These structures revealed that a cascade sequence was proceeding stereoselectively in every case to furnish a single product as the result of conrotatory 4π electrocyclization, electrophilic aromatic substitution at the least hindered position on the arene moiety (*para* to the MeO) in favor of six-membered ring formation, and desilylation with protonation from the *exo* face of the bicyclic product. Dehydrochlorination to form a second cationic intermediate did not occur in this case, due to structural restrictions imposed by the bridged architecture of **81**.

The cascade sequences presented herein demonstrate unprecedented modes of reactivity in Nazarov chemistry that are initiated by the silver(I)-promoted ring opening of halocyclopropanes. The ease with which the *gem*-dichlorocyclopropanes can be prepared, the relatively mild reaction conditions, and the efficiency of these processes make these substrates attractive intermediates for an application in natural product synthesis.

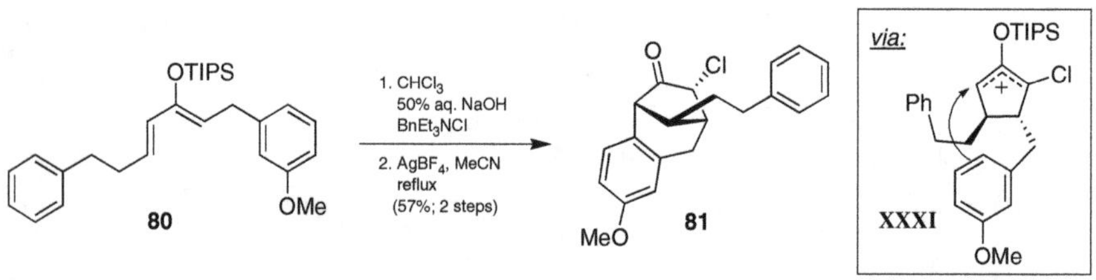

Scheme 4.25. The formation of bridged bicyclic products.

4.4 CONCLUDING REMARKS

Silver(I) has been used extensively in the development of halocyclopropane chemistry. The electrocylic ring opening of substituted halocyclopropanes can be used to afford many useful synthetic intermediates, both simple and complex, by introduction of a heteroatom- or carbon-based nucleophile to capture the cationic intermediates. Silver(I) can assist in the removal of the halogen from these substrates, inducing ring opening at low temperatures and under mild reaction conditions compared to the unassisted ring-opening processes. In these instances, silver(I) is also helpful in the sequestration of free halide in reaction mixtures, providing long-lived cationic species that can proceed along alternate trapping or rearrangement pathways. This chemistry has also been used in the more recent design of a new approach to the Nazarov reaction, wherein the silver(I)-promoted electrocyclic ring opening of 1,1-dihalo-2-(silyloxy)-2-vinylcyclopropanes was used to initiate Nazarov cyclization, leading to the formation of simple cyclopentenones. This sequence was further utilized in the development of novel trapping pathways that can be used to generate benzohydrindenone and bridged architectures in a single transformation. From this discussion, it is clear that the silver(I)-mediated electrocylic ring opening of halocyclopropanes has great potential for future applications in organic synthesis.

REFERENCES

1. (a) Pasto, D. J.; Garves, K., *J. Org. Chem.* **1967**, *32*, 778–781; (b) Masuike, T.; Furukawa, N.; Oae, S., *Bull. Chem. Soc. Jpn.* **1971**, *44*, 448–450; (c) Cohen, T.; Solash, J., *Tetrahedron Lett.* **1973**, *27*, 2513–2516; (d) Cookson, P. G.; Davies, A. G.; Roberts, B. P., *J. Chem. Soc. Chem. Commun.* **1976**, 1022–1023.
2. For reviews on this topic, see (a) Kostikov, R. R.; Molchanov, A. P.; Hopf, H., Small ring compounds in organic synthesis, *Top. Curr. Chem.* **1990**, *155*, 41–80; (b) Banwell, M. G.; Reum, M. E., in *Advances in Strain in Organic Chemistry*, Halton, B., ed., JAI Press, Greenwich, CT, **1991**, Vol. 1, pp. 19–64; (c) Fedorynski, M., *Chem. Rev.* **2003**, *103*, 1099–1132; (d) Halton, B.; Harvey, J., *Synlett* **2006**, *13*, 1975–2000.
3. Woodward, R. B.; Hoffmann, R., *Conservation of Orbital Symmetry*, Academic Press, New York, **1970**.
4. (a) de Puy, C. H.; Schnack, L. G.; Hausser, J. W.; Wiedemann, W., *J. Am. Chem. Soc.* **1965**, *87*, 4006–4006; (b) Loozen, H. J. J.; Robben, W. M. M.; Richter, T. L.; Buck, H. M., *J. Org. Chem.* **1976**, *41*, 384–385;(c) Loozen, H. J. J.; de Haan, J. W.; Buck, H. M., *J. Org. Chem.* **1977**, *42*, 418–422.
5. Ito, S.; Ziffer, H.; Baz, A., *J. Org. Chem.* **1986**, *51*, 1130–1133.
6. Porter, N. A.; Ziegler, C. B.; Khouri, F. F.; Roberts, D. H., *J. Org. Chem.* **1985**, *50*, 2252–2258.
7. Faza, O. N.; López, C. S.; Álvarez, R.; de Lera, Á. R., *J. Org. Chem.* **2004**, *69*, 9002–9010.
8. Marvell, E. N., *Thermal Electrocyclic Reactions*, Academic Press, New York, **1980**.
9. Banwell, M. G.; Cowden, C. J., *Aust. J. Chem.* **1994**, *47*, 2235–2254.
10. Roberts, J. D.; Chambers, V. C., *J. Am. Chem. Soc.* **1951**, *73*, 5034–5040.

11. McElvain, S. M.; Weyna, P. L., *J. Am. Chem. Soc.* **1981**, *10*, 2579–2588.
12. (a) Sandler, S. R., *J. Org. Chem.* **1967**, *32*, 3876–3881;(b) Parham, W. E.; Yong, K. S., *J. Org. Chem.* **1968**, *33*, 3947–3948;(c) Parham, W. E.; Yong, K. S., *J. Org. Chem.* **1970**, *35*, 683–685.
13. Banwell, M. G.; Sydnes, M. O., *Aust. J. Chem.* **2004**, *57*, 537–548.
14. Hart, N. K.; Johns, S. R.; Lamberton, J. A., *Aust. J. Chem.* **1968**, *21*, 2579–2581.
15. Makosza, M.; Wawrzyniewicz, M., *Tetrahedron Lett.* **1969**, *10*, 4659–4662.
16. Murphy, J. A.; Scott, K. A.; Sinclair, R. S.; Martin, C. G.; Kennedy, A. R.; Lewis, N., *J. Chem. Soc. Perkin Trans. 1* **2000**, 2395–2408.
17. Banwell, M. G.; Cowden, C. J.; Gable, R. W., *J. Chem. Soc. Perkin Trans. 1* **1994**, 3515–3517.
18. Banwell, M. G.; Gable, R. W.; Peters, S. C.; Phyland, J. R., *J. Chem. Soc. Chem. Commun.* **1995**, 1395–1396.
19. Danheiser, R. L.; Morin, J. M. Jr.; Yu, M.; Basak, A., *Tetrahedron Lett.* **1981**, *22*, 4205–4208.
20. Sydnes, L. K.; Mungaroo, R.; Aanesen, B. A., *Acta Chem. Scand.* **1998**, *52*, 1386–1391.
21. Lee, J.; Kim, H.; Cha, J. K., *J. Am. Chem. Soc.* **1995**, *117*, 9919–9920.
22. Banwell, M. G.; Wu, A. W., *J. Chem. Soc. Perkin Trans. 1* **1994**, 2671–2672.
23. Banwell, M. G.; Harvey, J. E.; Jolliffe, K. A., *J. Chem. Soc. Perkin Trans. 1* **2001**, 2002–2005.
24. Stanislawski, P. C.; Willis, A. C.; Banwell, M. G., *Org. Lett.* **2006**, *8*, 2143–2146.
25. Banwell, M. G.; Vogt, F.; Wu, A. W., *Aust. J. Chem.* **2006**, *59*, 415–425.
26. Banwell, M.; Edwards, A.; Harvey, J.; Hockless, D.; Willis, A., *J. Chem. Soc. Perkin Trans. 1* **2000**, 2175–2178.
27. Banwell, M. G.; Cowden, C. J.; Mackay, M. F., *J. Chem. Soc. Chem. Commun.* **1994**, 61–62.
28. Banwell, M. G.; Harvey, J. E.; Hockless, D. C. R., *J. Org. Chem.* **2000**, *65*, 4241–4250.
29. Stanislawski, P. C.; Willis, A. C.; Banwell, M. G., *Chem. Asian J.* **2007**, *2*, 1127–1136.
30. Kostikov, R. R.; Varakin, G. S.; Molchanov, A. P.; Ogloblin, K. A., *Zh. Org. Khim.* **1996**, *32*, 367.
31. Gassman, P. G.; Tan, L.; Hoye, T. R., *Tetrahedron Lett.* **1996**, *37*, 439–442.
32. Banwell, M. G.; Beck, D. A. S.; Stanislawski, P. C.; Sydnes, M. O.; Taylor, R. M., *Curr. Org. Chem.* **2005**, *9*, 1589–1600.
33. Banwell, M. G.; Ma, X.; Taylor, R. M.; Willis, A. C., *Org. Lett.* **2006**, *8*, 4959–4961.
34. Wu, L.; Shi, M., *Tetrahedron Lett.* **2009**, *50*, 1636–1638.
35. Grant, T. N.; West, F. G., *J. Am. Chem. Soc.* **2006**, *128*, 9348–9349.
36. Nazarov, I. N.; Zaretskaya, I. I., *Izvest. Akad. Nauk. SSSR Ser. Khim.* **1941**, 211–224.
37. (a) Paquette, L. A.; Hamme, A. T.; Ku, L. H.; Doyon, J.; Kreuzholz, R., *J. Am. Chem. Soc.* **1997**, *119*, 1242–1253; (b) Paquette, L. A.; Liu, Z.; Ramsey, C.; Gallucci, J. C., *J. Org. Chem.* **2005**, *70*, 8154–8161.
38. (a) Gee, S. K.; Danheiser, R. L., *J. Org. Chem.* **1984**, *49*, 1674–1678;(b) Danheiser, R. L.; Brisbois, R. G.; Kowalczyk, J. J.; Miller, R. F., *J. Am. Chem. Soc.* **1990**, *112*, 3093–3100.

39. Grant, T. N., *Modified Nazarov Reactions and Ring Expansion Chemistry: Useful Methodologies for the Construction of Carbocyclic and Heterocyclic Compounds,* Ph. D. dissertation, Univ. Alberta, Edmonton, AB, **2008**.

40. Browder, C. C.; Marmsäter, F. P.; West, F. G., *Org. Lett.* **2001**, *3*, 3033–3035.

41. Grant, T. N.; West, F. G., *Org. Lett.* **2007**, *9*, 3789–3792.

42. The formation of cyclic oxyallyl cations from dehydrochlorination of chlorocyclopentanones is precedented in the literature; see Harmata, M.; Elomari, S.; Barnes, C. L., *J. Am. Chem. Soc.* **2004**, *82*, 375–385.

5

SILVER-CATALYZED CYCLOISOMERIZATION REACTIONS

PHILIPPE BELMONT

Laboratory for Organometallic Chemistry, Heterocycles and Biological Targets, Unit 176: Conception, Synthesis and Targeting of Biomolecules, Institut Curie and CNRS, Paris, France

5.1 INTRODUCTION

The impact of coinage metals (copper, silver, and gold) in organic synthesis is becoming important, especially with the worldwide impact of green chemistry, since their use as catalysts is part of the 12 "principles of green chemistry."[1] Moreover, silver salts are special since silver cations are being considered and evaluated as alternative biocides in water distribution systems, showing the great usefulness and nontoxicity (to humans) of monovalent silver cation.[2]

Although silver chemistry has a long history in organic chemistry, it has typically been used in stoichiometric amounts and is developed mostly for anion metathesis

Silver in Organic Chemistry Edited by Michael Harmata

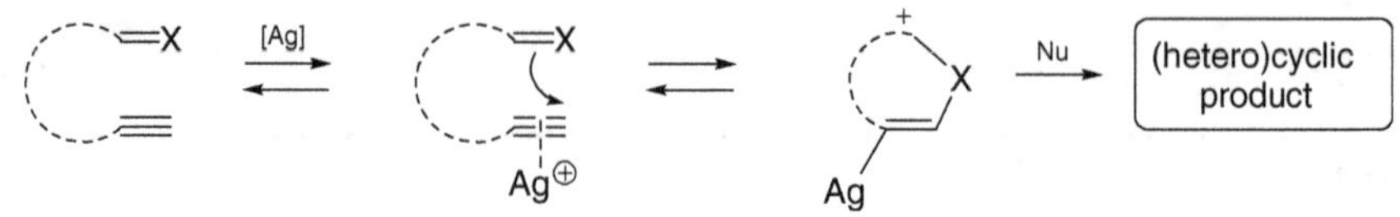

Scheme 5.1. Cycloisomerization diversity in silver chemistry.

(anion exchange, halogen scavengers) and oxidation reactions. More recent reviews pinpoint the current (r)evolution in silver chemistry.[3–10] In fact, with the new gold rush,[11–24] silver appeared as a partner to access more reactive cationic species. (Note that silver has also been used with other metals for this purpose: Pd, Cu, Rh, Ru, Pt). More recently catalysis with silver salts has become widespread, and this can be also explained by the σ- and π-Lewis acidity properties of silver(I) complexes,[25] leading to a variety of chemical transformations. Therefore, exploring new catalytic reactions with silver complexes is of great interest, and this work has been nicely presented in several reviews.[3–10,26]

The chemical reactions possible with silver catalysis are multiple and cover cycloadditions, cycloisomerizations, allylations, aldol reactions, and even C–H bond activation. Also, asymmetric versions are known, even though they still need to be improved.[3–10]

Our interest in this chapter is in silver-catalyzed cycloisomerization reactions. Therefore, we shall present different silver-catalyzed cycloisomerization reactions as a function of the nucleophilic and electrophilic moiety. Cycloisomerization reactions including the classical ene–yne cycloisomerization (with X = CHR, Scheme 5.1), and the related heterocyclization reactions with heteroatoms embedded in unsaturated systems (X = NR, O; Scheme 5.1) belong to the same reaction family. In addition, the alkynyl part can be exchanged for an allene unit. Internal or external nucleophiles (Nu) can then stabilize, through cascade reactions, the positive charge created.[24]

Heterocyclization reactions with saturated moieties (alcohols, amines, thiols, etc.) or acids on unsaturated counterparts (alkenes, allenes, alkynes, etc.) are not covered in this chapter since they are addition, and not isomerization, reactions. Silver is also widely used as an activating agent for producing highly reactive metallic cations (anion metathesis), which, in turn, may catalyze cycloisomerization reactions. This aspect is covered only when the silver control experiments give substantial positive results.

A selection of the literature is also necessary in order to give the reader an overview of silver chemistry in the field of cycloisomerization reactions. Therefore, this chapter is not intended to be an exhaustive review of the literature, since more recent specialized reviews can be accessed for that purpose.[3–10,26]

5.2 CYCLOISOMERIZATION OF C=O ONTO C=C=C

Following preliminary observations on allenyl alcohols from Goré[27–31] and Balme[32] along with Claesson and Olsnon,[33] Marshall et al. demonstrated in a seminal publication[34] that allenyl ketones (R^3 = alkyl, Scheme 5.2) or allenyl aldehydes

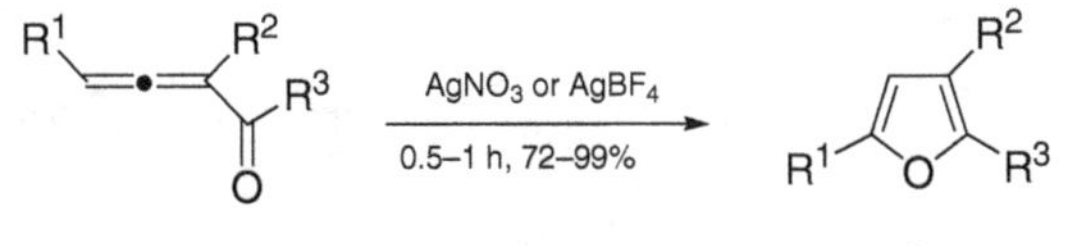

Scheme 5.2. Marshall's work.[34–38]

($R^3 = H$) **1** could selectively undergo a cycloisomerization reaction to produce various furan rings **2** under mild conditions with rhodium(I) or silver(I) catalysts.

This transformation was further studied and the catalyst load could be decreased to 0.2 equiv.[35–38] A mechanism was proposed through deuterium incorporation experiments, and the conclusion was that there was no 1,2 shift of the deuterium present in the starting material (**A**, Scheme 5.3) since exclusive formation of furans **D** deuterated on position 3 could be explained by the presence of an external source of deuterium (such as D_2O). Therefore, it is believed that after silver(I) coordination to the allenyl system (**A**, Scheme 5.3), the attack by the carbonyl oxygen may lead to an oxo cation intermediate **B**. Finally, proton lost would generate silver furan **C** that would lead to furan **D** after silver release (Scheme 5.3).[39]

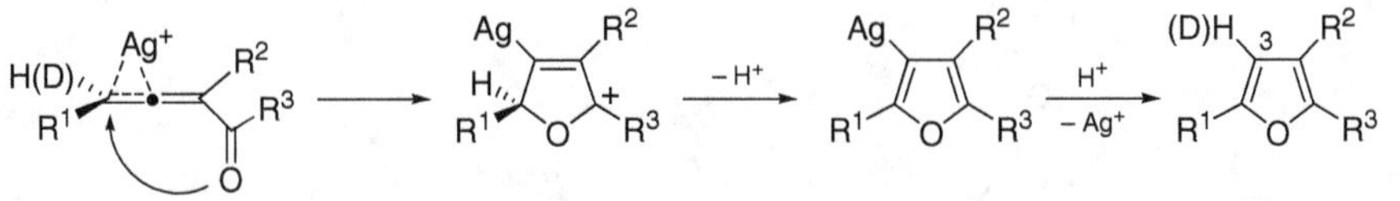

Scheme 5.3. Marshall's mechanism.[39]

This methodology has been nicely used in the total synthesis of several natural products such as racemic kallolide B (**3**), (Scheme 5.4),[37,38] a member of the pseudopterolide diterpene family.

Thereafter, Hashmi et al. showed that this transformation (Scheme 5.2) was more efficiently catalyzed with palladium(II)[40] and gold(I)[41] catalysts.

Moreover, following the cycloisomerization reaction, a tandem dimerization reaction is also possible on the same substrates under Pd^{II}, Ag^{I}, and Au^{III} catalysis, leading to different substituted furans (**4** or **6**) depending on the nature of the catalyst used (Scheme 5.5). Indeed, from compound **3** (Scheme 5.5), palladium(II) catalysis led to a 59% yield of **4**, whereas silver(I) and gold(III) catalysis led to furans **5** and **6**.[41]

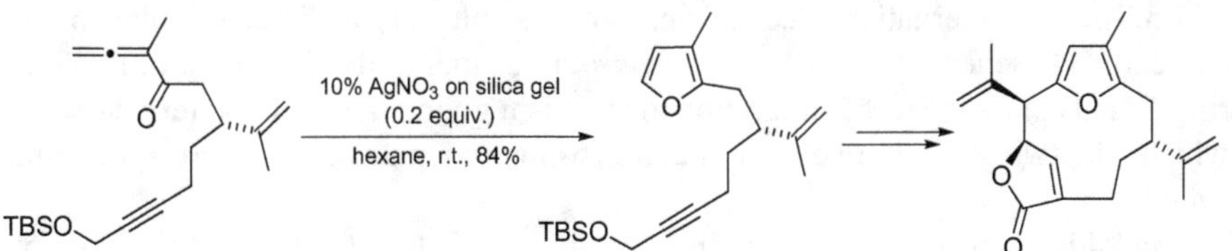

Scheme 5.4. Total synthesis of racemic kallolide B.[37,38]

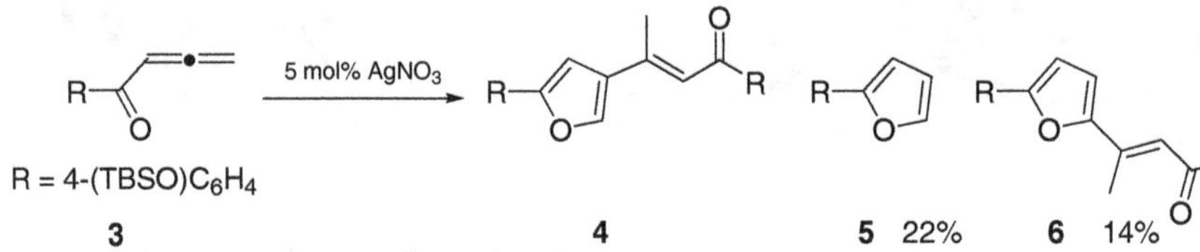

Scheme 5.5. Cycloisomerization/dimerization reaction.[41]

Hashmi's team[41] noticed that the cycloisomerization/dimerization reactions leading to **4** or **6** required, under the same reaction conditions (1 mol% catalyst), over a week for $AgNO_3$, about an hour with $PdCl_2(MeCN)_2$ and only about a minute with $AuCl_3$.

Gevorgyan et al. in several landmark publications[42,43] reported the silver-catalyzed ($AgBF_4$) 1,2 migration of various functional groups such as acyloxy, phosphatyloxy, and sulfonyloxy, positioned on an *sp*^{2} carbon from an allenyl unit. The migration was accompanied by a cycloisomerization reaction leading to the efficient synthesis of functionalized furans. Therefore, from acyloxyallene **E** (Scheme 5.6), the 1,2 migration might generate a dioxolenium **F**, and cycloisomerization would lead to furan **G**.

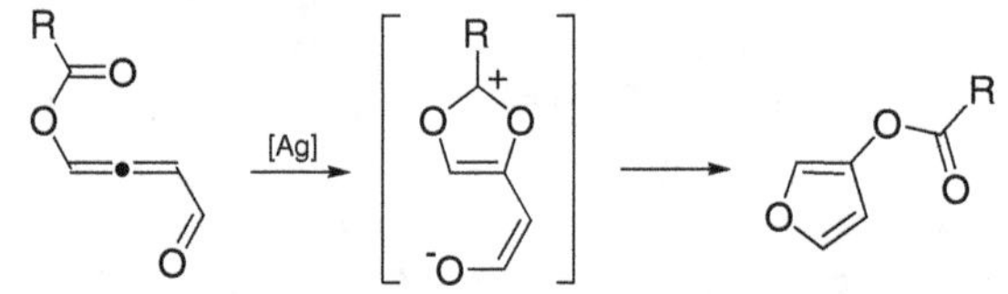

Scheme 5.6. Proposed 1,2-migration pathway for furan synthesis.[42]

So, from acyloxy, phosphatyloxy, and sulfonyloxy alkynyl derivatives **7–9** (Scheme 5.7), on $AgBF_4$ catalysis, allenyl compounds **10–12** are formed, leading to substituted furans **13–15**.

Note that alkynyl derivative **9** (Scheme 5.7) could not be isolated since during its synthesis, it directly formed allenyl substrate **12**. Also, allene intermediate **11** could be isolated (in 56% yield) when the reaction was conducted at room temperature.[42]

Mechanistic studies provided through ^{17}O labeling[43] of **H** (Scheme 5.8) showed that the nature of the migrating group was important. The 1,2-migration mechanism proposed earlier was confirmed for the migration of the acyloxy group (Schemes 5.6 and 5.7). Two alternative mechanisms are possible: (1) a direct trapping of the carbenoid (**I**, Scheme 5.8) by the ketone group leading to the formation of the furan ring **J**; and (2) a second 1,2-migration mechanism giving the allenyl intermediate **K**, which via two competitive oxirenium/dioxolenium pathways, affords the furan ring **J**.[43]

In addition to this latter work, a silver-catalyzed 1,2-alkyl/aryl shift is also possible with allenyl ketones (**16**), giving, as before, highly substituted furans (**17**, Scheme 5.9).[44] Not surprisingly, Gevorgyan et al. observed that the migratory aptitude

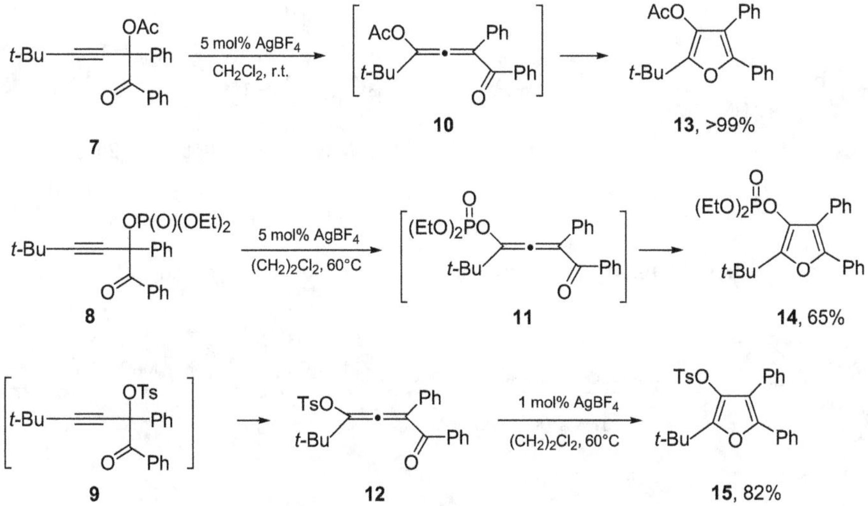

Scheme 5.7. Access to functionalized furans.[42,43]

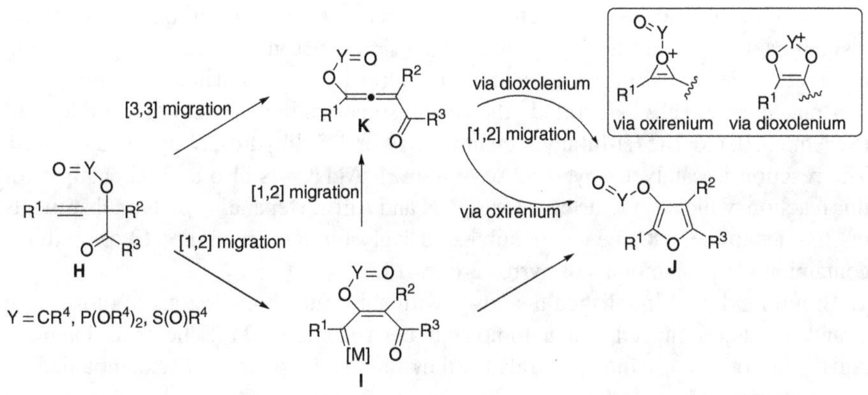

Scheme 5.8. Mechanistic pathways based on ^{17}O labeling.[43]

of a phenyl versus a methyl group was greater than 100 to 1, which is in agreement with their known migratory aptitudes.

π-Philic (Ag^I, Cu^I, Au^I) and oxophilic (Lewis acid catalysis; e.g., $InCl_3$) activation are also both possible, and this reaction has been thoroughly studied.[45] Note that the

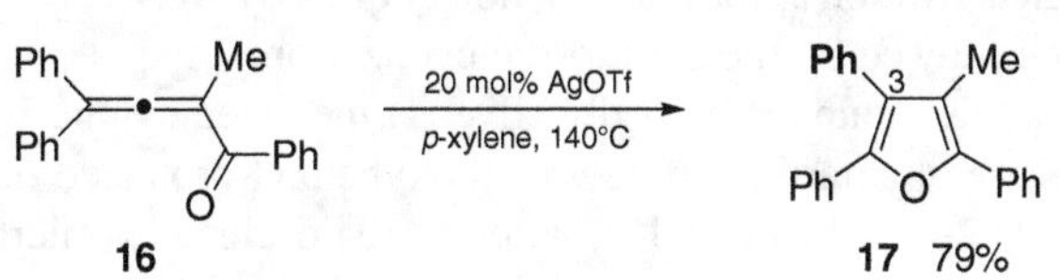

Scheme 5.9. [1,2]-phenyl migration.[44]

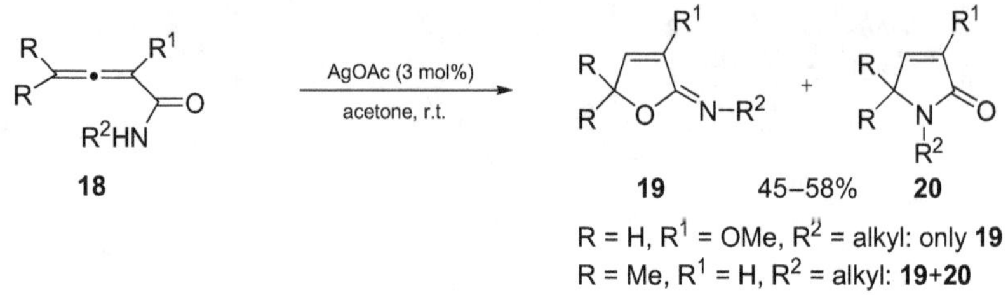

Scheme 5.10. Cycloisomerization versus simple heterocyclization.[50]

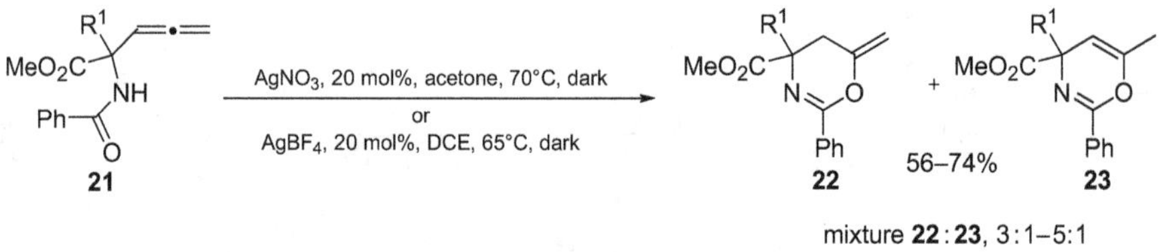

Scheme 5.11. Access to oxazines.[51]

use of a proton scavenger such as 2,4,6-tris-*tert*-butylpyrimidine (TTBP) has a significant impact on reactivity, which is a sign of Brønsted acid catalysis during metal-catalyzed transformations, and this has already been noted in the literature.[46–49]

Brandsma et al.[50] reported the cycloisomerization of 2,3-dienamides **18** (Scheme 5.10) to 2-(5*H*)-furanylidenamines **19** or 1,5-dihydro-2*H*-pyrrol-2-ones **20**. The reaction is catalyzed by AgOAc (3 mol%). $AgNO_3$ is also a good catalyst for this reaction, which is not the case for AgCN and AgBr. Depending on the substituents on 2,3-dienamides **18**, one can obtain exclusively furanylidenamines **19** or mixtures containing various amounts of pyrrol-2-ones **20**.

Brummond et al. published a study[51] with substrates homologous to those from Brandsma et al.[50] Indeed, using amino acid–derived allenes **21** (Scheme 5.11), under $AgBF_4$ or $AgNO_3$ (20 mol%) catalysis, they obtained oxazines **22** accompanied by minor isomers **23**.

5.3 CYCLOISOMERIZATION OF C=O ONTO C≡C

Agrofoglio et al. reported[52] the synthesis of furo[2,3-*d*]pyrimidines **25** via $AgNO_3$-catalyzed 5-*endo-dig* cyclization (Scheme 5.12). Furo[2,3-*d*]pyrimidines **25** were obtained through activation of the alkynyl moiety of **24** by $AgNO_3$ (0.2 equiv), giving intermediate **L**, followed by the 5-*endo-dig* cyclization.

Van der Eycken's group developed a silver(I)-mediated synthesis of substituted furo[2,3-*b*]pyrazines.[53] Starting from *p*-methoxybenzyl-protected 3,5-dichloropyrazine-2(1*H*)-ones **26** (Scheme 5.13), after a regioselective microwave-assisted Sonogashira reaction with various terminal alkynes, the cycloisomerization reaction could occur using AgOTf (2 mol%) with trifluoroacetic acid (TFA, 5 equiv) to yield

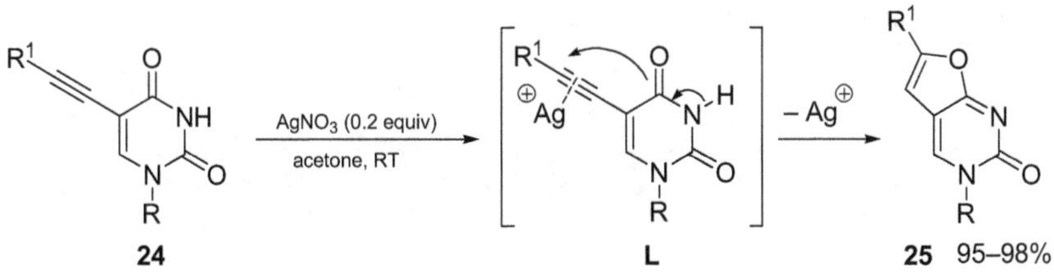

Scheme 5.12. Access to the furo[2,3-*d*]pyrimidines.[52]

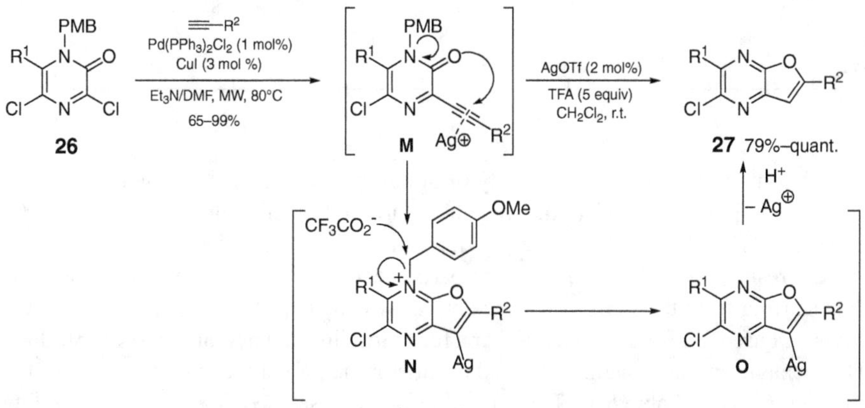

Scheme 5.13. Furo[2,3-*b*]pyrazine synthesis.[53]

2-chlorofuro[2,3-*b*]pyrazines **27**. Control experiments showed that both AgOTf and TFA were needed for high yields and a low catalyst load.[53] The PMB deprotection was proposed to occur right after the cyclization as a result of the formation of a cationic intermediate **N**, which reacts with the TFA anion (**M–N–O**, Scheme 5.13). Finally, the silver-linked furopyrazine intermediate **O** regenerates the catalyst on protonolysis and produces furopyrazines **27** in the range of 79% to quantitative (quant.) yield.

Porco's team reported the formation of benzopyrylium salts **29** (Scheme 5.14) from the action of silver and gold catalysts on *o*-alkynylbenzaldehydes **28**.[54] Although this reaction proceeds much faster with gold than with silver, the benzopyrylium formation is of interest and will be discussed further.

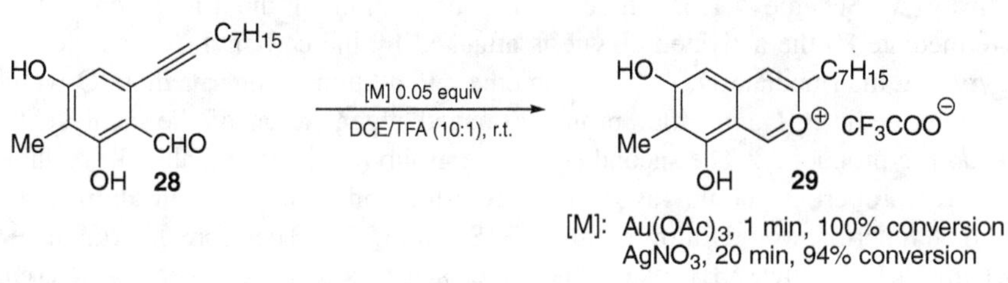

Scheme 5.14. Benzopyrylium synthesis.[54]

TABLE 5.1. Silver Diversity for a Versatile Reaction[55]

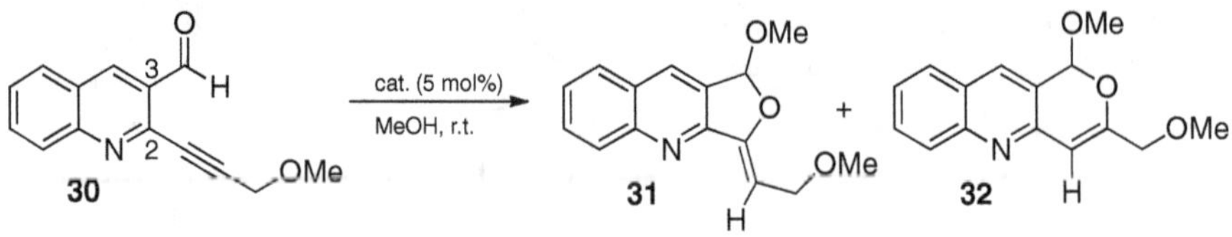

[Ag] Complex	Reaction Time	5-*exo*-dig product **31** (Conversion %)	6-*endo-dig* product **32** (Conversion %)	pK_a
AgOTf	4 h	—	100	<0
AgOAc	12 h	48	52	4.8
Ag_2O	10 min	100	—	15.7

Belmont et al. worked on a variety of quinoline derivatives bearing a substituted alkynyl group on position 2 of the ring and a carbonyl function on position 3 (e.g., **30**, Table 5.1). They accessed diverse furoquinoline and pyranoquinoline cores (e.g., **31** and **32**, respectively, Table 5.1).[55] One particularly interesting aspect of this work is that, starting from the same quinoline derivatives and depending on the nature of the silver salt used, one could access either a furoquinoline or a pyranoquinoline product. This behavior was rationalized from the value of the pK_a of the conjugate acid of the silver counterion (Table 5.1). Three families of silver salts could be found depending on the results of the cyclization reaction. Counterions whose conjugate acids had a negative pK_a value led to a 6-*endo-dig* product (e.g., AgOTf). Those with a pK_a value over 10 led to a 5-*exo-dig* product (e.g., Ag_2O). Finally, those with a pK_a range between 2 and 5 afforded mixtures of products (e.g., AgOAc, Table 5.1).

This observation is not related to traces of base or acid from the silver salts used since control experiments ruled out this possibility. It was known from the literature that the 5-*exo-dig* versus 6-*endo-dig* cyclization mode could depend on the nature of the carbonyl group,[56,57] of the alkyne substituent,[58,59] and of the nature[60,61] and oxidation state[62] of the metallic source used. Also, work from Yamamoto[25] demonstrated the importance of both σ- and π-Lewis acidity properties of silver(I) complexes. Therefore, depending on the silver salt used, two mechanistic pathways were proposed (pathways A and B, Scheme 5.15).

The reactivity of the first group (e.g., AgOTf, Table 5.1) could be explained by pathway A (Scheme 5.15), where after coordination of the triple bond of **30** (intermediate **P**) the activated alkyne is attacked by the carbonyl oxygen yielding a pyrylium intermediate **Q**.[63,64] Then, methanol addition on intermediate **Q** would give intermediate **R** and, after protonation and regeneration of the catalyst, the 6-*endo-dig* product **32**. The second silver salt group (e.g., Ag_2O, Table 5.1), whose oxidative properties should favor direct coordination to the oxygen atom of the carbonyl function, would follow pathway B (Scheme 5.15). Therefore, intermediate **S** (Scheme 5.15, pathway B), where the silver salt acts as a Lewis acid,[25,65] could give acetal **T**, which on cyclization would give complex **U**. Expulsion of Ag^+ would

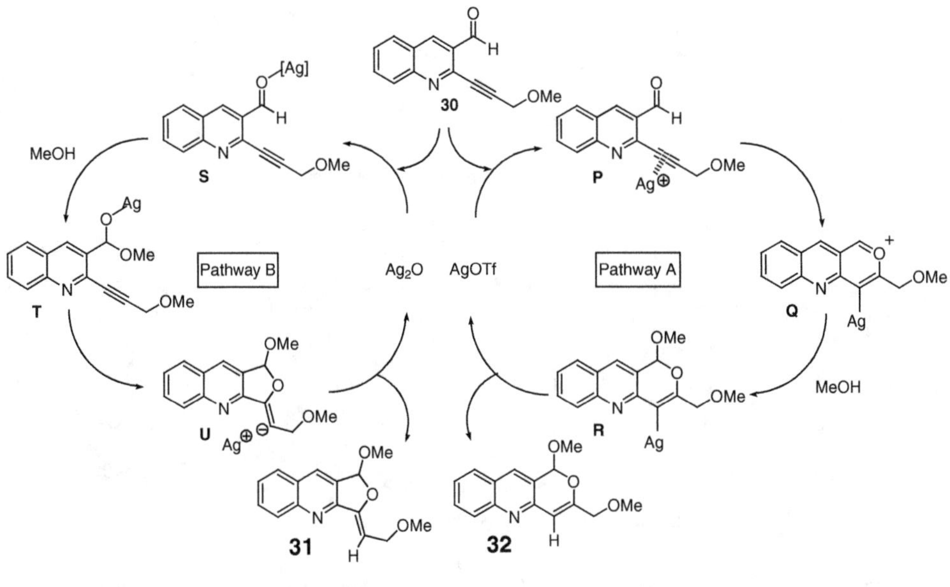

Scheme 5.15. Proposed reaction pathways.[55]

yield the 5-*exo-dig* product **31**. The third silver salt group (e.g., AgOAc, Table 5.1) could engage in both pathways, leading to poor regioselectivity.

Wu's team published additional results, related to Porco's[54] and Belmont's[55] previous work on 2-alkynylbenzaldehyde substrates using AgOTf catalysis (see Scheme 5.14 and Table 5.1), but in their case diethyl phosphite [$HPO(OEt)_2$] was used as the nucleophile instead of methanol.[66]

Padwa et al.[67] have used 3-oxo-*N*-propargylamides and esters to build furopyrrolidinones and furodihydrofuranones via a silver(I) alkyne activation (Schemes 5.16 and 5.17).

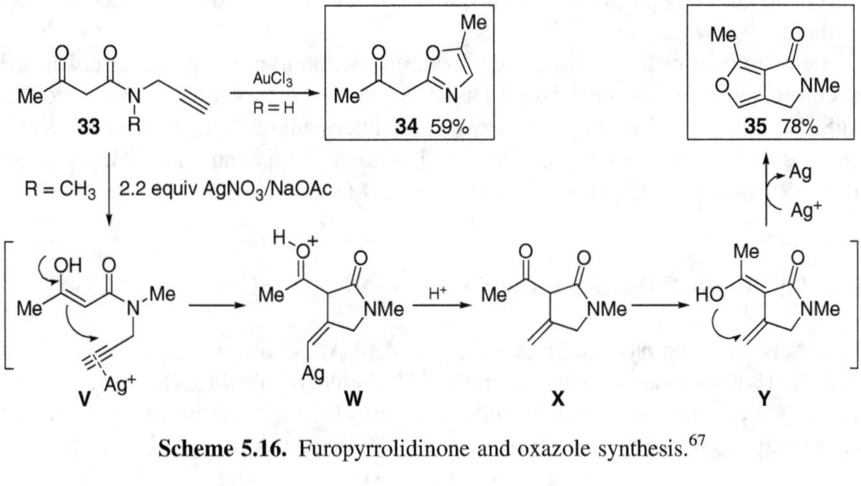

Scheme 5.16. Furopyrrolidinone and oxazole synthesis.[67]

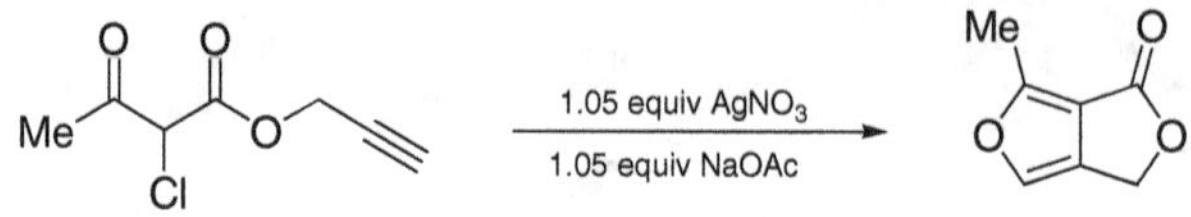

36 **37** 82%

Scheme 5.17. Furodihydrofuranone synthesis.[67]

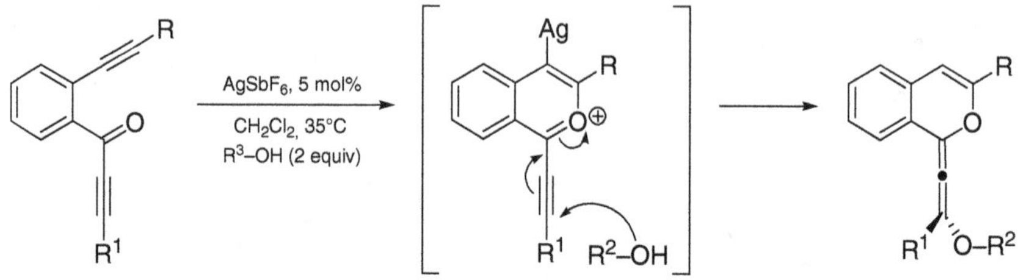

38 **Z** **39** 65–99%

Scheme 5.18. Allenylisochromene synthesis.[68]

The authors used silver salts since gold salts catalyzed the reaction with R=H (giving oxazole **34**, Scheme 5.16) but not with R=Me. Moreover, only traces of the desired furopyrrolidinone were formed with the use of a cationic gold species activated with silver additives. Therefore, silver traces were thought to be the active reagent. Indeed, on activation of compound **33** mediated by $AgNO_3$ in the presence of sodium acetate (Scheme 5.16), the enol moiety **V** can then accomplish a nucleophilic attack to produce the pyrrolidinone **W** and after protonolysis give compound **X**. Pyrrolidinone **Y** (the enol version of **X**) can, in turn, be subject to an oxidative cyclization to yield the furopyrrolidinone **35**. Two equivalents of silver salts are needed for the activation step and the oxidative cyclization.

However, using a chlorinated version of a related ester compound **36** (Scheme 5.17) required only one silver equivalent to prepare pyrrolofuranone **37**, since chlorine elimination directly provided the final product without the need of a silver-mediated oxidation.

Yamamoto described a cascade cyclization reaction to prepare allene-substituted isochromenes **39** (Scheme 5.18).[68] Diynones **38**, when submitted to the action of $AgSbF_6$ (5 mol%), formed a benzopyrylium intermediate **Z** (identified by NMR spectroscopy), which could undergo a 1,4-Michael-type addition with alcohol nucleophiles to produce allenylisochromenes **39** (Scheme 5.18).

5.4 CYCLOISOMERIZATION OF C=N ONTO C=C=C

This transformation has been less widely studied. We chose two publications from Gallagher's team to show a few examples.[69,70] Allenic oximes **40** (Scheme 5.19) were treated with $AgBF_4$ to produce cyclic nitrones **41**. Since these nitrones were not sufficiently stable, they were trapped with 1,3-dipolarophiles to give bicyclic or

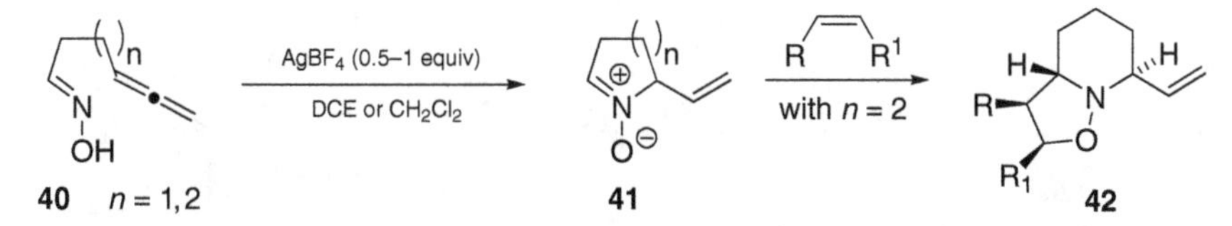

Scheme 5.19. Cyclic nitrones and their 1,3-dipolar adducts.[70]

tricyclic adducts **42** (Scheme 5.19). Therefore, *trans*-2,6-disubstituted piperidines and *trans*-2,5-pyrrolidines could be obtained using this sequence.[69,70]

This methodology has been used to produce various structures, including hexahydroazepines[71] and pyrrolizidine alkaloids.[69,72]

5.5 CYCLOISOMERIZATION OF C=N ONTO C≡C

Larock's group reported[73] the synthesis of isoquinolines **44** (Scheme 5.20) by the electrophilic cyclization of iminoalkynes **43**. This reaction has been extensively studied with various electrophiles such as I_2, ICl, ArSCl, and ArSeCl and also under palladium catalysis. Among the electrophiles tested, $AgNO_3$ produced interesting results since several heterocycles **44** (Scheme 5.20) could be prepared on the activation of iminoalkynes **43** with 5 mol% of this reagent (leading to intermediate **AA**, Scheme 5.20). The *tert*-butyl group is lost from intermediate **BB** during the reaction as an isobutene to afford product **44** (Scheme 5.20). Note that this reaction is also possible with pyridyliminoalkynes **43**.

Porco et al.[74] accessed pyrroloisoquinoline heterocycles **46** by employing a domino process. Initially, a metal-catalyzed cycloisomerization of alkynyl *N*-benzylidene glycinates **45** gave azomethine ylide **CC** (Scheme 5.21), which underwent

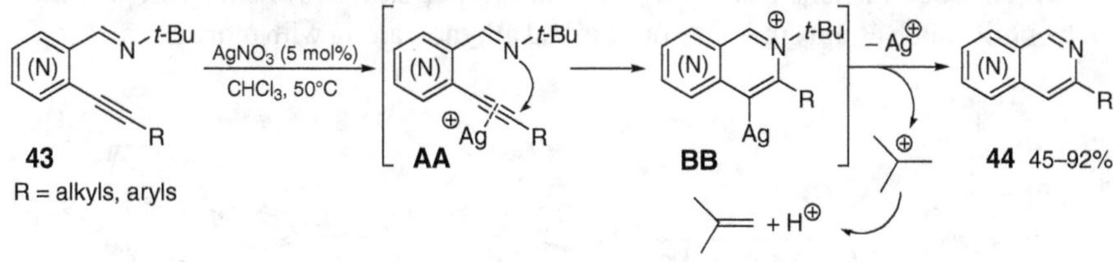

Scheme 5.20. Isoquinoline and aza derivatives.[73]

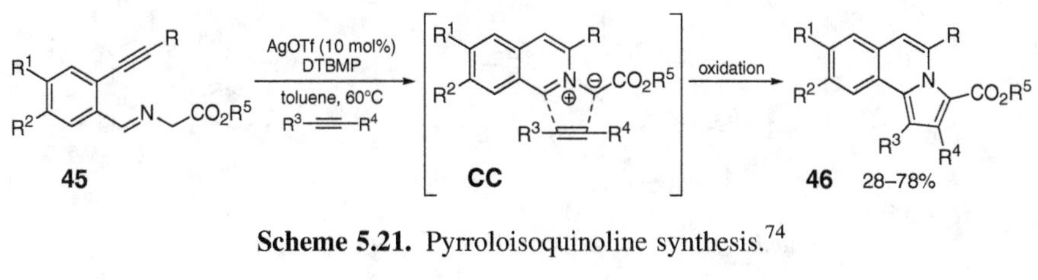

Scheme 5.21. Pyrroloisoquinoline synthesis.[74]

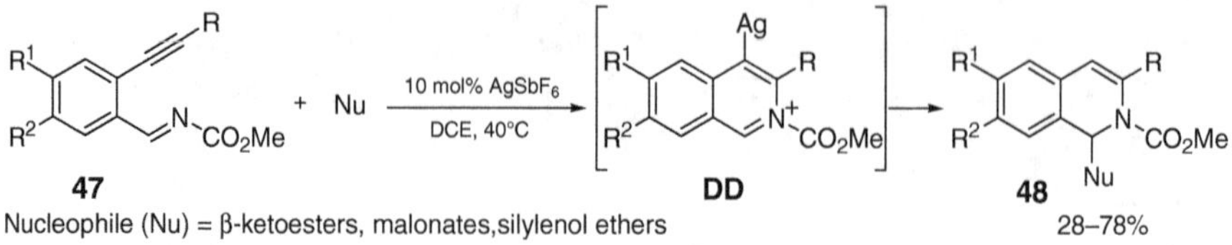

Scheme 5.22. Access to the dihydroisoquinoline core.[75]

a dipolar cycloaddition with an activated alkyne such as dimethylacetylene dicarboxylate (DMAD), leading to **46** after air oxidation.

Both AgOTf and $AgSbF_6$ (10 mol%) were the best catalysts for this transformation, and the reaction was conducted in the presence of an hindered base, 2,6-di-*tert*-butyl-4-methylpyridine (DTBMP). The azomethine ylide **CC** intermediate has been identified in a test reaction in the absence of DMAD, via mass spectrometry. The scope and limitations have been thoroughly studied for reaching pyrroloisoquinoline scaffold, which is a substructure of the lamellarin alkaloid family.

The same research group developed a tandem cycloisomerization–addition reaction on *o*-alkynylacylimines **47** (Scheme 5.22) to generate the dihydroisoquinoline core **48**.[75] This type of reaction had been reported by Asao et al. and will be discussed shortly.[76]

The best catalyst for this transformation was $AgSbF_6$ (10 mol%), and β-ketoesters, malonates, and silyl enol ethers have been used for the nucleophilic addition on the pyridinium intermediate **DD**. The dihydroisoquinolines **48** have been further used in several reactions in order to assemble the framework of various alkaloids. One example is given in the formation of dihydroisoquinoline **49**, bearing a pendent α,β-unsaturated ketone. Compound **49** can rearrange to the tetracycle **50** (related to the core structure of karachine, Scheme 5.23), using TMSOTf, via a tandem Michael addition–Mannich reaction process (intermediates **EE** and **FF**).

As mentioned earlier, Asao et al.[76] have developed similar systems with particular nucleophiles such as nitromethane or terminal alkynes, along with more classical ones

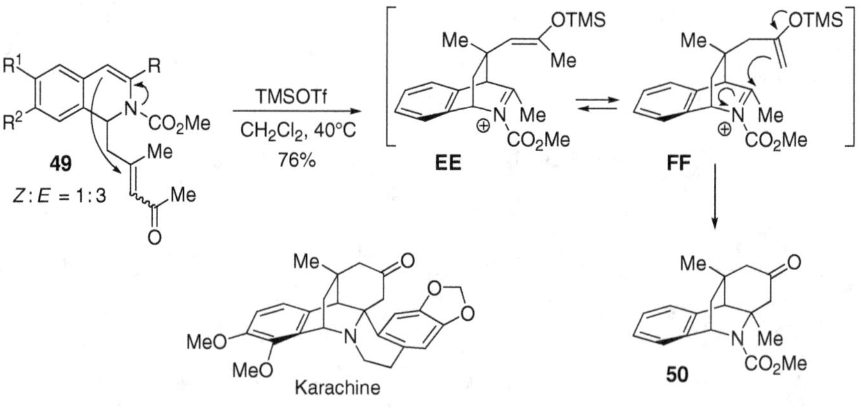

Scheme 5.23. Toward karachine alkaloid structure.[75]

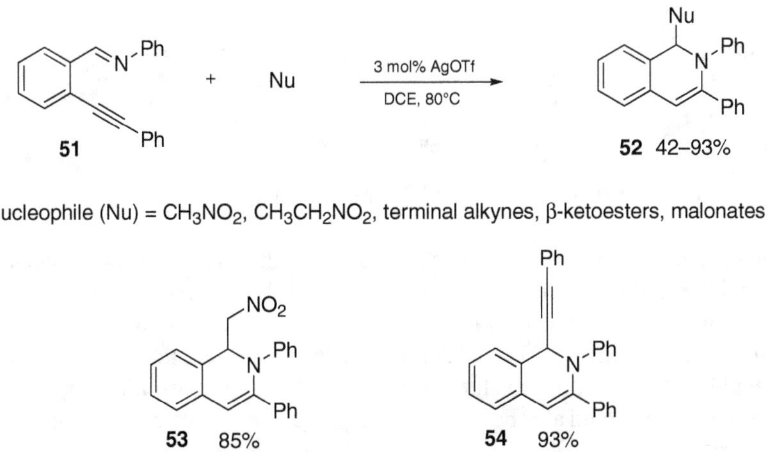

Scheme 5.24. Dihydroisoquinolines synthesis.[76]

(β-ketoesters, malonates). *o*-Alkynylaryl aldimines **51** (Scheme 5.24) can thus produce 1,2-dihydroisoquinolines **52** via AgOTf (3 mol%) catalysis. It is worth noting that nucleophiles such as nitromethane and terminal alkynes give interesting products such as **53** and **54**, respectively (Scheme 5.24).

The mechanism proceeds through activation of the alkynyl moiety (**GG**, Scheme 5.25), leading to the isoquinolinium intermediate **HH**. After addition of

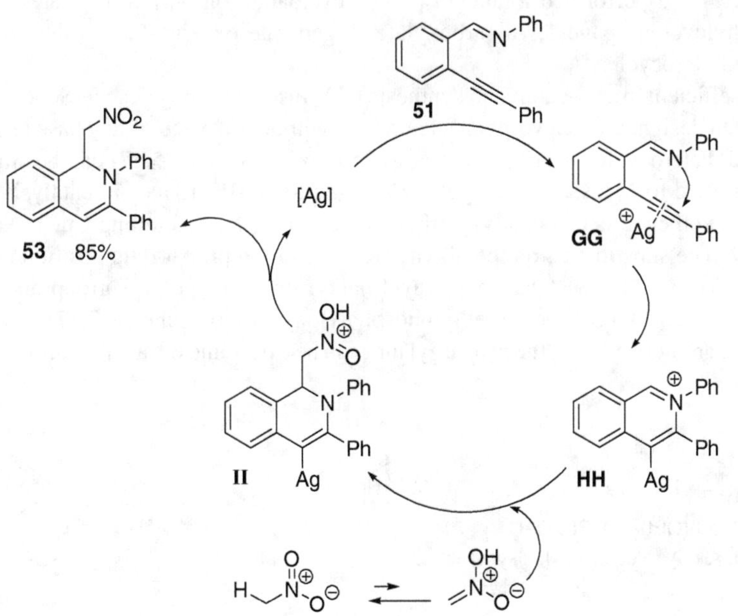

Scheme 5.25. Proposed mechanism.[76]

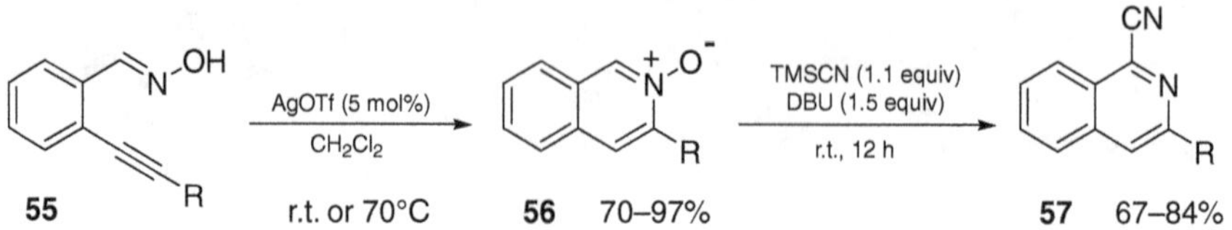

55 **56** 70–97% **57** 67–84%

Scheme 5.26. Synthesis of isoquinoline-*N*-oxides.[80]

nitromethane, and protonolysis of intermediate **II**, the silver catalyst is regenerated along with dihydroisoquinoline **53** (Scheme 5.25).[76]

Wu's group reported studies related to previous results from the groups of Porco and Asao (see Schemes 5.22 and 5.24, respectively). Indeed, the iminium formed in situ could be the substrate for further addition of diethyl phosphite[77] or enamine[78] nucleophiles, and could also be reduced with sodium borohydride,[79] yielding various 1,2-dihydroisoquinolines.

Shin's group[80] also used AgOTf (5 mol%) to catalyze the formation of isoquinoline-*N*-oxide structures **56** (Scheme 5.26) from cycloisomerization of 2-alkynylbenzaldoximes **55**. This synthesis is of interest since these structures are usually made by oxidation of the parent nitrogen heterocycles.

Other silver salts can efficiently catalyze this reaction at room temperature. These include $AgSbF_6$ and $AgNTf_2$ and also some gold salts such as Au(IMes)OTf. Some of the isoquinoline-*N*-oxide derivatives **56** (Scheme 5.26) can be further converted in 1-cyanoisoquinolines **57** by reacting with TMSCN and DBU as a base.

Wu's group worked on the same transformation (**55–56**, Scheme 5.26) with AgOTf (5 mol%) and performed a tandem [3 + 3] cycloaddition on intermediates **56** with dimethylcyclopropane-1,1-dicarboxylate to generate oxazine-fused dihydroisoquinoline heterocycles.[81]

An efficient, metal-catalyzed synthesis of *N*-fused heterocycles **59** (Scheme 5.27) has been designed by Gevorgyan's team,[82,83] without the need of any base or ligand. Indeed, heterocycles bearing a propargyl group **58** (Scheme 5.27) can be smoothly transformed to *N*-fused heterocycles **59** owing to $AgBF_4$ (3 mol%) catalysis.

Some other salts can catalyze efficiently this reaction, including CuCl, AuI, and $AgPF_6$. The substituents on the alkyne (R^2) and at the propargylic position (R^1) are rather diverse, since R^2 can be a (cyclo)alkyl or a (hetero)aryl group and R^1 an acetyloxy, an O-TBS, or a diethylphosphatyloxy group (Scheme 5.27). Since the aromatic moiety bearing the propargyl unit can be a pyridine **60**, a quinoxaline **61**, or a

n = 0, 1
A = N or S
R^1 = OP(O)(OEt)$_2$, OTBS, OAC
R^2 = H, Alk, Ar, Het, alkenyl, alkynyl

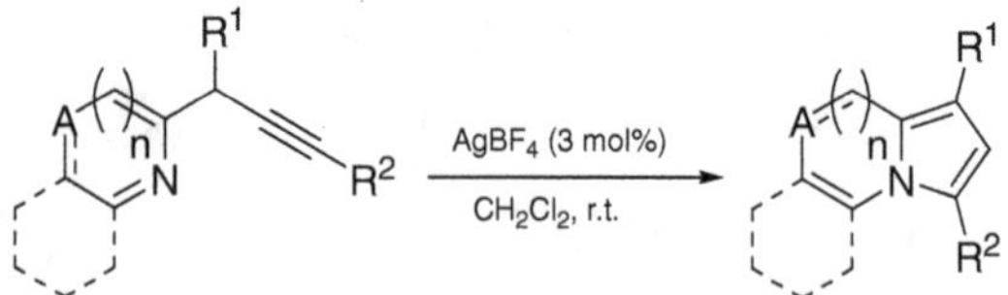

58 **59** 52–95%

Scheme 5.27. Fused *N*-containing heterocycles.[82,83]

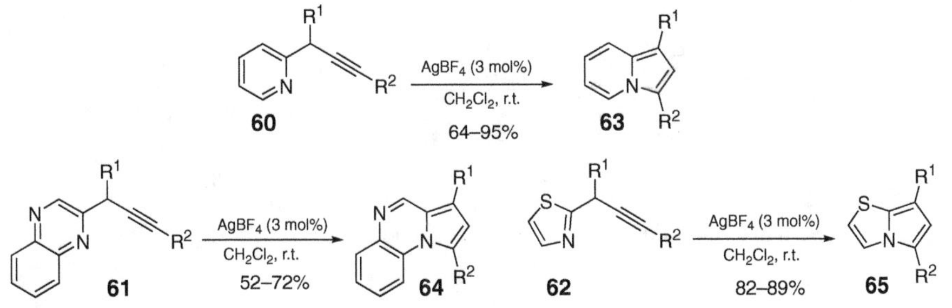

Scheme 5.28. Fused-ring diversity.[82,83]

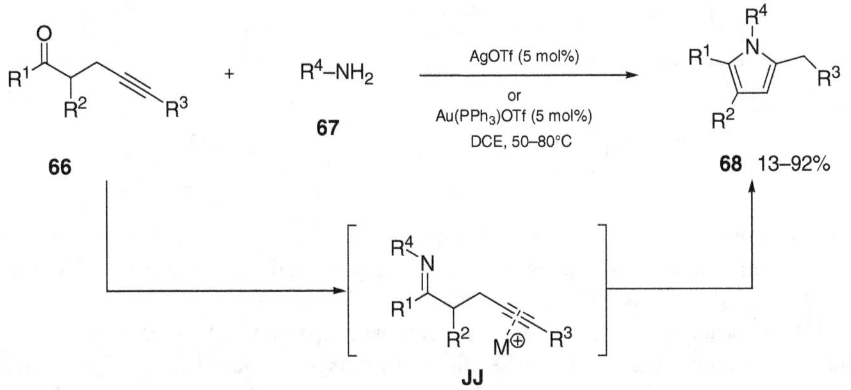

Scheme 5.29. Access to the pyrrole scaffold.[84]

thiazole **62**, after the cycloisomerization one can obtain indolizines **63**, pyrroloquinoxalines **64**, and pyrrolothiazoles **65**, respectively (Scheme 5.28).

Dake's group[84] published an interesting report in which AgOTf and cationic gold (I) complexes were compared for their use in the synthesis of the pyrrole scaffold. From β-alkynyl ketones **66** (Scheme 5.29) and various primary amines **67**, the imine intermediates **JJ** were formed in situ and the intramolecular cyclization produced various pyrroles **68**. Both catalysts AgOTf (5 mol%) or Au(PPh_3)OTf (5 mol%) were efficient, but the reaction proceeded more rapidly with silver catalysis.

5.6 ENE–YNE CYCLOISOMERIZATION: C=C ONTO C≡C

Few examples of ene–yne cycloisomerization reactions are seen in the literature. The first results for ene-yne cycloisomerizations were with systems bearing an heteroatom (amine or oxygen) next to the alkene counterpart (forming an enamine or an enol ether). Indeed, Dake's group reported the cyclization of enesulfonamides on alkynes (**69–70**, Scheme 5.30) under catalysis by platinum and silver salts.[85] Catalysis using AgOTf (1–4 mol%) was particularly efficient with systems such as **69** (Scheme 5.30)

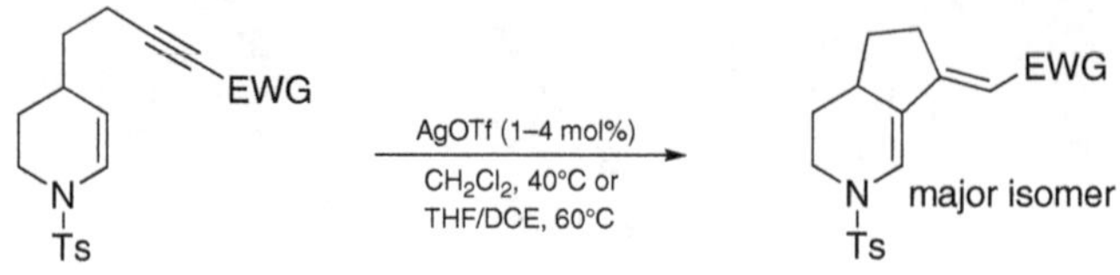

69 **70** 31–99%

Scheme 5.30. Ene-sulfonamides cyclization on activated alkynes.[85]

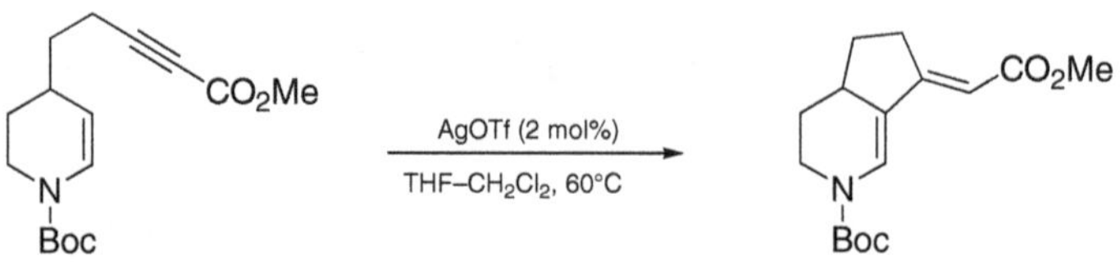

71 **72** 91%

Scheme 5.31. Access to Boc-protected product.[85]

bearing an electron-withdrawing group on the alkynyl moiety (e.g., ketones, esters, amides, nitriles). Usually the use of silver catalysts required lower catalyst loading and gave higher yields compared with platinum(II) catalysis.

Also, in order to obtain a nitrogen protecting group that would be easier to remove than the *p*-toluenesulfonyl group (Ts), the *tert*-butoxycarbamate ($BocNR_2$) derivative **71** (Scheme 5.31) was also engaged in the same reaction to efficiently yield the desired dienyl product **72**.

The diene products have been used in the Diels–Alder reaction with success. For example, the reaction of eneyne **73** with α-methylacrolein gave the tricyclic compound **74** in a nice 93% yield over two steps (Scheme 5.32).

Tanaka et al. published a route to 2-pyridones from the cycloisomerization of *N*-alkenyl alkynylamides.[86] This reaction did not work with *N*-alkenyl alkynylamide **75** (Scheme 5.33) using $AuClPPh_3$, even in refluxing dichloroethane (DCE). However, the cationic complex $AuPPh_3BF_4$ ($AuClPPh_3$ + $AgBF_4$) gave a high yield of the desired pyridinone **76** at room temperature.

The silver salt $AgBF_4$ was the only catalyst tested along with $AuPPh_3BF_4$ that was efficient at room temperature.

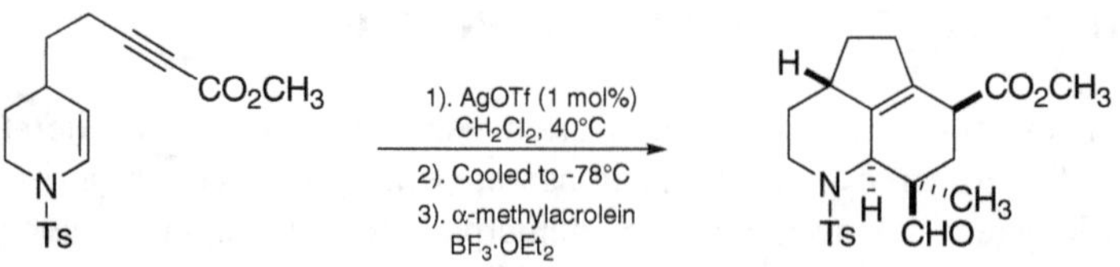

73 **74** 93%

Scheme 5.32. Tandem cycloisomerization–cycloaddition reaction.[85]

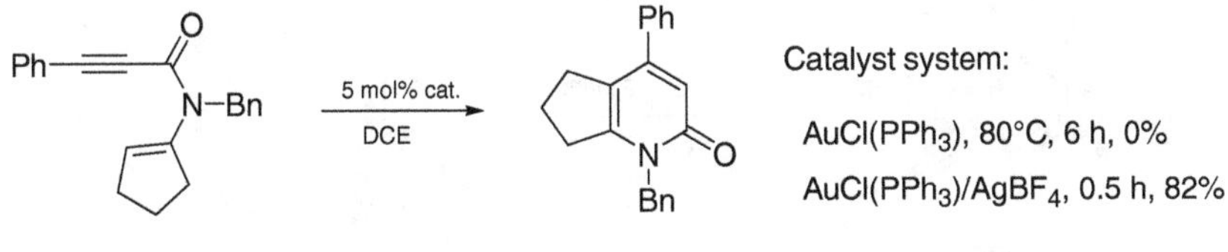

75 76

Scheme 5.33. Pyridone synthesis.[86]

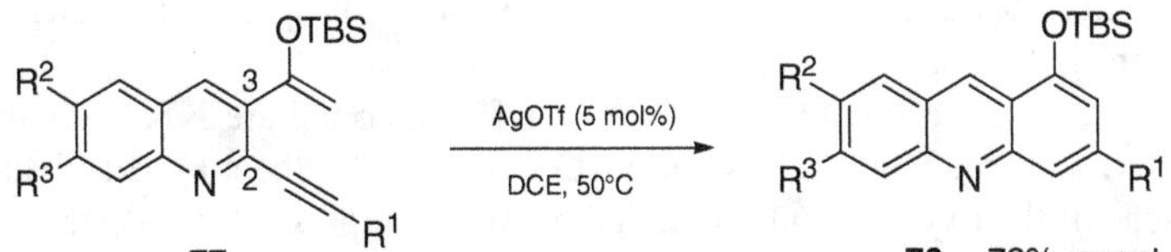

77 78 72%–quant.

Scheme 5.34. Polysubstituted acridine synthesis.[87]

Belmont's group[87] reported a cycloisomerization reaction on quinolines **77** (Scheme 5.34) bearing a silyl enol ether group on position 3 and an alkynyl group on position 2, leading to acridine derivatives **78**.

Although this reaction worked nicely with an activated gold species ($AuPPh_3SbF_6$), they studied this reaction under silver catalysis with a broad range of complexes. Of all the silver salts tested, only those having a negative pK_a for the conjugate acid of silver counterion were efficient in the reaction: $AgSbF_6$, $AgPF_6$, AgOTf, and $AgNO_3$. All the other silver salts tested (Ag_2SO_4, AgF, AgOAc, Ag_2O, AgO) could not catalyze the reaction. Complexation of the silver salt to the alkynyl

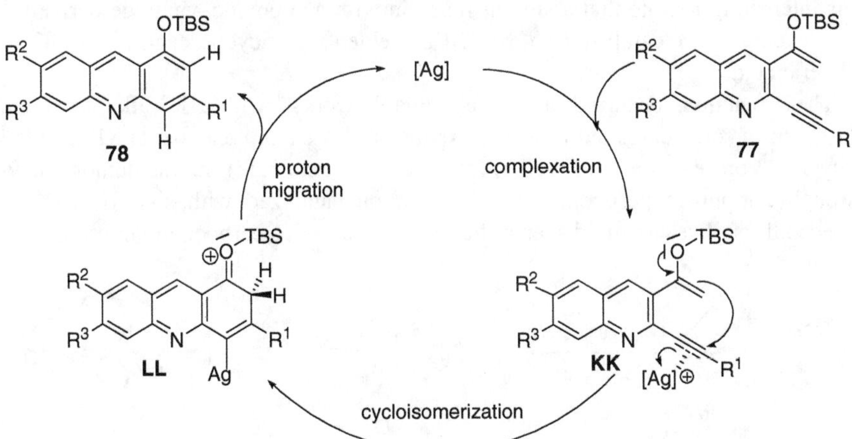

Scheme 5.35. Proposed mechanism pathway.[87]

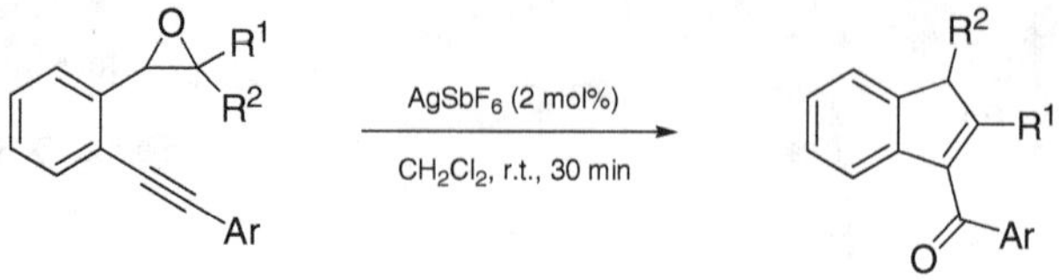

70 **00** 71–09%

Scheme 5.36. Synthesis of indenyl ketones.[89]

moiety seems compulsory, and this can explain why some other silver salts are not effective, since they are best known for their oxidative properties and therefore are more oxophilic than π-acidic.[25,65]

Therefore, after the complexation step on **77**, intermediate **KK** (Scheme 5.35) could undergo an ene–yne cycloisomerization assisted by the alkyne activation and the presence of the oxygen atom to yield the silver complex intermediate **LL**. Then, proton migration and regeneration of the catalyst would give product **78**. A large variety of acridine derivatives were formed with R^1 group such as alkyls, (hetero) aryls, and even the ferrocenyl group.[87] It is remarkable that Echavarren's group[88] succeeded, using a silver catalysis, in reacting 1,6-ene-ynes bearing an allyl stannane to yield vinyl-stannanes along with tin-free compounds.

5.7 OTHER TRANSFORMATIONS

Liu et al. obtained diverse carbocyclic and heterocyclic frameworks by employing a silver(I)-catalyzed cycloisomerization of epoxyalkynes.[89] With $AgSbF_6$ (2 mol%), epoxides **79** were transformed in 3-1*H*-indenyl ketones **80** in high yields (Scheme 5.36).

The reaction has been exemplified with a variety of aryl-substituted alkynes, but it is interesting to note that a similar reaction with an epoxide–ynamide derivative **81** (Scheme 5.37), catalyzed with $AgBF_4$, yielded the tricyclic central core **82** of gibberic acid.

These reactions occurred, as in Gevorgyan's work,[44] with a 1,2-alkyl/aromatic shift, and in some cases led to a ring expansion; for example, epoxide **81** gave the tricycle **82** on reaction with $AgBF_4$ (5 mol%) (Scheme 5.37). It was demonstrated through trapping experiments that the reaction catalyzed with silver(I) did not proceed through a carbenoid species but instead through a carbocation intermediate.

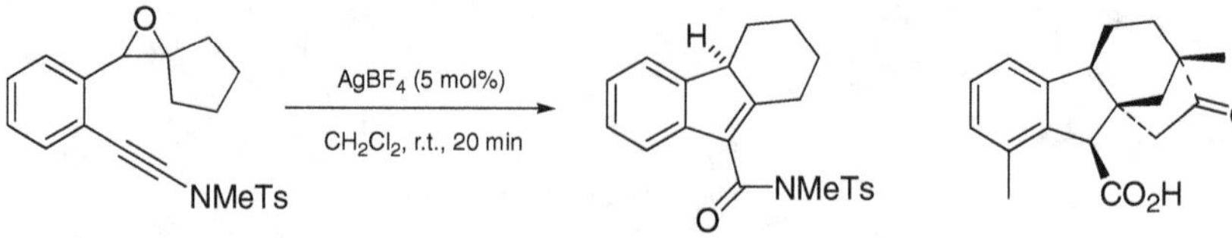

81 **82** 57% Gibberellic acid

Scheme 5.37. Access to the central core of gibberic acid.[89]

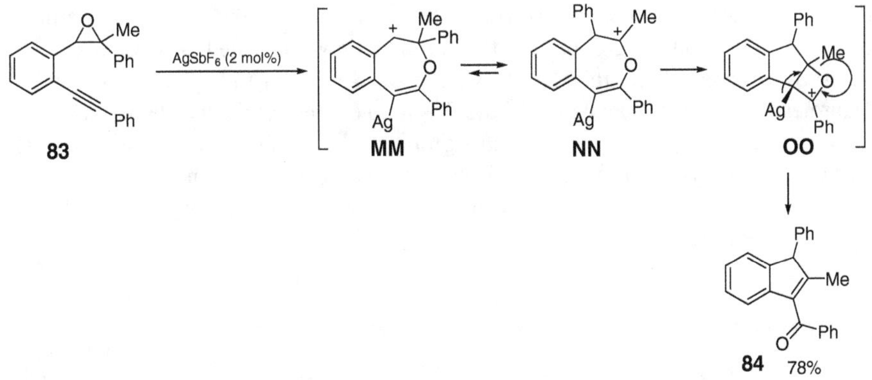

Scheme 5.38. Proposed mechanism pathway.[89]

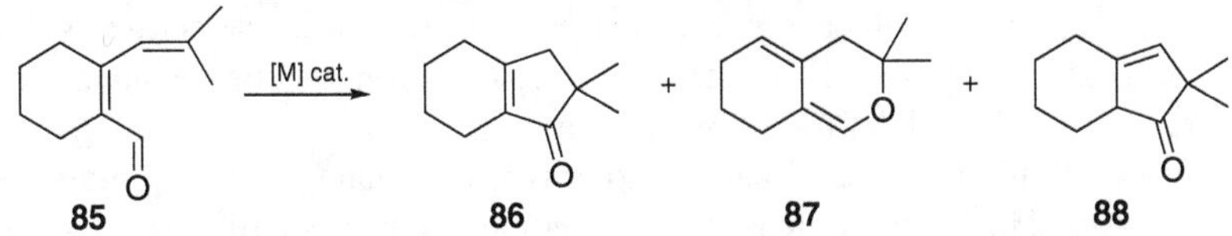

Scheme 5.39. Chemical diversity under silver and gold catalyses.[90]

Therefore, from epoxide **83** (Scheme 5.38), a 1,2-phenyl shift in intermediate **MM** would provide oxonium derivative **NN**, which would furnish, after rearrangement of the oxetane intermediate **OO**, the desired ketone **84**. It is interesting to note that on the same substrates cationic gold derivatives (activated with silver salts) did not lead to the same final compounds, showing the unique reactivity of silver salts.

Liu's group[90] also reported the metal-catalyzed formation of 2-cyclopentenone **86** from *cis*-2,4-dien-1-al **85** (Scheme 5.39). It is interesting to note that with the same starting material, $AuClPPh_3$ and AuCl catalyzed the formation of pyrano derivative **87** (32% and 63%, respectively), whereas the cationic gold derivatives $AuPPh_3SbF_6$ ($AuClPPh_3$ + $AgSbF_6$) gave only traces of 3-cyclopentenone **88** (15%). Therefore, silver salts were tested alone and, indeed, AgOTf efficiently catalyzed the formation of 2-cyclopentenone **86** (91%, Scheme 5.39).

Wu et al.[91] published a tandem addition–cyclization reaction using 2-alkynylbenzenamines **89** (Scheme 5.40). The reaction proceeded under AgOTf catalysis

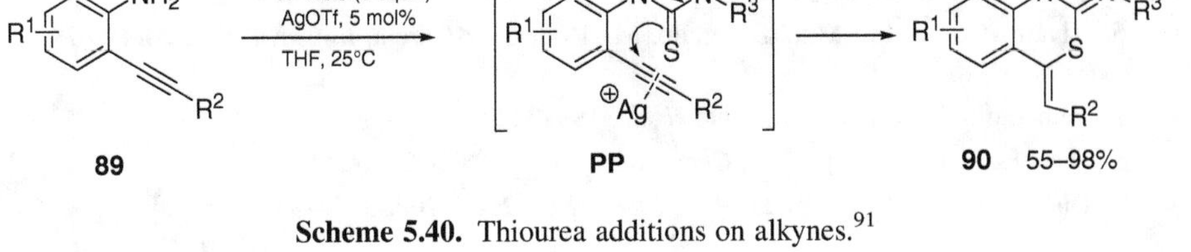

Scheme 5.40. Thiourea additions on alkynes.[91]

(from 1 to 5 mol%) via thiourea intermediates **PP**. Because of the high electrophilicity of the isothiocyanate group, the tandem reactions provided a family of 2,4-dihydro-1*H*-benzo[*d*][1,3]thiazines **90** (Scheme 5.40) in fair to high yields (55–98%) regardless of the electronic nature of the R^3 group on the isothiocyanate partner.

Very interesting work from Kirsch's group was not reported here since silver(I) was used to catalyze a propargyl–Claisen rearrangement on ene–yne systems leading to allenyl ketones and not the following cycloisomerization step where gold(I)[92] or DBU[93] were used.

5.8 CONCLUSION

This overview of silver-catalyzed cycloisomerization reactions shows the great flexibility of the method and its usefulness in reaching the core of numerous heterocycles.

On comparison with other metallic species, we are able to detect areas where efforts are still needed, such as the cycloisomerization of simple nonactivated ene–yne systems, as well as C–H bond activation.

About two-thirds of the applications for silver in organometallic homogeneous catalysis reported in the literature involve activation of other metallic sources through anion metathesis to generate cationic species. Nevertheless, we can clearly see that silver has gained attention more recently not only for halogen scavenging or oxidation purposes but also for its own reactivity, which is sometimes very distinct from that of other related metals such as copper and gold. Further research is needed to carefully establish the reactivity of the metallic center and therefore clearly eliminate the possibility of simple Brønsted acid catalysis due to catalyst instability.

We may wonder whether, following today's "gold rush" in homogeneous catalysis, silver will be the next in line!

REFERENCES

1. Anastas, P.; Warner, J., *Green Chemistry: Theory and Practice*, Oxford Univ. Press; New York, **1998**.
2. Wu, M. Y.; Suryanarayanan, K.; van Ooij, W. J.; Oerther, D. B., *Water Sci. Technol.* **2007**, *55*, 413–419.
3. Lipshutz, B. H.; Yamamoto, Y., *Chem. Rev.* **2008**, *108*, 2793–2795.
4. Naodovic, M.; Yamamoto, H., *Chem. Rev.* **2008**, *108*, 3132–3148.
5. Weibel, J. -M.; Blanc, A.; Pale, P., *Chem. Rev.* **2008**, *108*, 3149–3173.
6. Alvarez-Corral, M.; Munoz-Dorado, M.; Rodriguez-Garcia, I., *Chem. Rev.* **2008**, *108*, 3174–3198.
7. Yamamoto, Y., *Chem. Rev.* **2008**, *108*, 3199–3222.
8. Dias, H. V. R.; Lovely, C. J., *Chem. Rev.* **2008**, *108*, 3223–3238.
9. Diaz-Requejo, M. M.; Perez, P. J., *Chem. Rev.* **2008**, *108*, 3379–3394.

10. Patil, N. T.; Yamamoto, Y., *Chem. Rev.* **2008**, *108*, 3395–3442.
11. Hashmi, A. S. K., *Chem. Rev.* **2007**, *107*, 3180–3211.
12. Hashmi, A. S. K., *Angew. Chem. Int. Ed.* **2005**, *44*, 6990–6993.
13. Hashmi, A. S. K., *Gold Bull.* **2004**, *37*, 51–65.
14. Li, Z.; Brouwer, C.; He, C., *Chem. Rev.* **2008**, *108*, 3239–3265.
15. Arcadi, A., *Chem. Rev.* **2008**, *108*, 3266–3325.
16. Jimenez-Nunez, E.; Echavarren, A. M., *Chem. Rev.* **2008**, *108*, 3326–3350.
17. Gorin, D. J.; Sherry, B. D.; Toste, F. D., *Chem. Rev.* **2008**, *108*, 3351–3378.
18. Muzart, J., *Tetrahedron* **2008**, *64*, 5815–5849.
19. Jimenez-Nunez, E.; Echavarren, A. M., *Chem. Commun.* **2007**, 333–346.
20. Zhang, L.; Sun, J.; Kozmin, S. A., *Adv. Synth. Catal.* **2006**, *348*, 2271–2296.
21. Asao, N., *Synlett* **2006**, 1645–1656.
22. Dyker, G., *Angew. Chem. Int. Ed.* **2000**, *39*, 4237–4239.
23. Hashmi, A. S. K.; Rudolph, M., *Chem. Soc. Rev.* **2008**, *37*, 1766–1775.
24. Kirsch, S. F., *Synthesis* **2008**, 3183–3204.
25. Yamamoto, Y., *J. Org. Chem.* **2007**, *72*, 7817–7831.
26. Halbes-Letinois, U.; Weibel, J. -M.; Pale, P., *Chem. Soc. Rev.* **2007**, *36*, 759–769.
27. Chilot, J. J.; Doutheau, A.; Gore, J., *Tetrahedron Lett.* **1982**, *23*, 4693–4696.
28. Arseniyadis, S.; Gore, J., *Tetrahedron Lett.* **1983**, *24*, 3997–4000.
29. Chilot, J. J.; Doutheau, A.; Gore, J.; Saroli, A., *Tetrahedron Lett.* **1986**, *27*, 849–852.
30. Bluthe, N.; Gore, J.; Malacria, M., *Tetrahedron* **1986**, *42*, 1333–1344.
31. Bernard, D.; Doutheau, A.; Gore, J., *Tetrahedron* **1987**, *43*, 2721–2732.
32. Balme, G., Ph.D. thesis, Univ. Claude Bernard, Lyon, **1979**.
33. Olsson, L. I.; Claesson, A., *Synthesis* **1979**, 743–745.
34. Marshall, J. A.; Robinson, E. D., *J. Org. Chem.* **1990**, *55*, 3450–3451.
35. Marshall, J. A.; Wang, X. J., *J. Org. Chem.* **1991**, *56*, 4913–4918.
36. Marshall, J. A.; Wang, X. J., *J. Org. Chem.* **1992**, *57*, 3387–3396.
37. Marshall, J. A.; Wallace, E. M.; Coan, P. S., *J. Org. Chem.* **1995**, *60*, 796–797.
38. Marshall, J. A.; Sehon, C. A., *J. Org. Chem.* **1995**, *60*, 5966–5968.
39. Marshall, J. A.; Bartley, G. S., *J. Org. Chem.* **1994**, *59*, 7169–7171.
40. Hashmi, A. S. K.; Rupert, T. L.; Knoefel, T.; Bats, J. W., *J. Org. Chem.* **1997**, *62*, 7295–7304.
41. Hashmi, A. S. K.; Schwarz, L.; Choi, J. -H.; Frost, T. M., *Angew. Chem. Int. Ed.* **2000**, *39*, 2285–2288.
42. Sromek, A. W.; Kel'in, A. V.; Gevorgyan, V., *Angew. Chem. Int. Ed.* **2004**, *43*, 2280–2282.
43. Schwier, T.; Sromek, A. W.; Yap, D. M. L.; Chernyak, D.; Gevorgyan, V., *J. Am. Chem. Soc.* **2007**, *129*, 9868–9878.
44. Dudnik, A. S.; Gevorgyan, V., *Angew. Chem. Int. Ed.* **2007**, *46*, 5195–5197.
45. Dudnik, A. S.; Sromek, A. W.; Rubina, M.; Kim, J. T.; Kel'in, A. V.; Gevorgyan, V., *J. Am. Chem. Soc.* **2008**, *130*, 1440–1452.
46. Hashmi, A. S. K., *Catal. Today* **2007**, *122*, 211–214.

47. Li, Z.; Zhang, J.; Brouwer, C.; Yang, C. -G.; Reich, N. W.; He, C., *Org. Lett.* **2006**, *8*, 4175–4178.
48. Rosenfeld, D. C.; Shekhar, S.; Takemiya, A.; Utsunomiya, M.; Hartwig, J. F., *Org. Lett.* **2006**, *8*, 4179–4182.
49. Rhee, J. U.; Krische, M. J., *Org. Lett.*, **2005**, *7*, 2493–2495.
50. Nedolya, N. A.; Schlyakhtina, N. I.; Zinov'eva, V. P.; Albanov, A. I.; Brandsma, L., *Tetrahedron Lett.* **2002**, *43*, 1569–1571.
51. Mitasev, B.; Brummond, K. M., *Synlett* **2006**, 3100–3104.
52. Aucagne, V.; Amblard, F.; Agrofoglio, L. A., *Synlett* **2004**, 2406–2408.
53. Ermolat'ev, D. S.; Mehta, V. P.; Van der Eycken, E. V., *Synlett* **2007**, 3117–3122.
54. Zhu, J.; Germain, A. R.; Porco, J. A. JJr., *Angew. Chem. Int. Ed.* **2004**, *43*, 1239–1243.
55. Godet, T.; Vaxelaire, C.; Michel, C.; Milet, A.; Belmont, P., *Chem. Eur. J.* **2007**, *13*, 5632–5641.
56. Barluenga, J.; Vazquez-Villa, H.; Ballesteros, A.; Gonzalez, J. M., *J. Am. Chem. Soc.* **2003**, *125*, 9028–9029.
57. Yue, D.; Della Ca, N.; Larock, R. C., *Org. Lett.* **2004**, *6*, 1581–1584.
58. Jong, T. T.; Leu, S. J., *J. Chem. Soc. Perkin Trans. 1* **1990**, 423–424.
59. Genin, E.; Toullec, P. Y.; Antoniotti, S.; Brancour, C.; Genet, J. -P.; Michelet, V., *J. Am. Chem. Soc.* **2006**, *128*, 3112–3113.
60. Gulias, M.; Rodriguez, J. R.; Castedo, L.; Mascarenas, J. L., *Org. Lett.* **2003**, *5*, 1975–1977.
61. Alonso, F.; Beletskaya, I. P.; Yus, M., *Chem. Rev.* **2004**, *104*, 3079–3159.
62. Patil, N. T.; Yamamoto, Y., *J. Org. Chem.* **2004**, *69*, 5139–5142.
63. Asao, N.; Aikawa, H.; Yamamoto, Y., *J. Am. Chem. Soc.* **2004**, *126*, 7458–7459.
64. Asao, N.; Takahashi, K.; Lee, S.; Kasahara, T.; Yamamoto, Y., *J. Am. Chem. Soc.* **2002**, *124*, 12650–12651.
65. Asao, N.; Nogami, T.; Takahashi, K.; Yamamoto, Y., *J. Am. Chem. Soc.* **2002**, *124*, 764–765.
66. Yu, X.; Ding, Q.; Wang, W.; Wu, J., *Tetrahedron Lett.* **2008**, *49*, 4390–4393.
67. Verniest, G.; Padwa, A., *Org. Lett.* **2008**, *10*, 4379–4382.
68. Patil, N. T.; Pahadi, N. K.; Yamamoto, Y., *J. Org. Chem.* **2005**, *70*, 10096–10098.
69. Lathbury, D. C.; Shaw, R. W.; Bates, P. A.; Hursthouse, M. B.; Gallagher, T., *J. Chem. Soc. Perkin Trans. 1* **1989**, 2415–2424.
70. Lathbury, D.; Gallagher, T., *Tetrahedron Lett.* **1985**, *26*, 6249–6252.
71. Shaw, R.; Lathbury, D.; Anderson, M.; Gallagher, T., *J. Chem. Soc. Perkin Trans. 1*, **1991**, 659–660.
72. Lathbury, D.; Gallagher, T., *J. Chem. Soc. Chem. Commun.* **1986**, 1017–1018.
73. Huang, Q.; Hunter, J. A.; Larock, R. C., *J. Org. Chem.* **2002**, *67*, 3437–3444.
74. Su, S.; Porco, J. A., Jr., *J. Am. Chem. Soc.* **2007**, *129*, 7744–7745.
75. Su, S.; Porco, J. A., Jr., *Org. Lett.* **2007**, *9*, 4983–4986.
76. Asao, N.; Yudha, S. S.; Nogami, T.; Yamamoto, Y., *Angew. Chem. Int. Ed.* **2005**, *44*, 5526–5528.
77. Sun, W.; Ding, Q.; Sun, X.; Fan, R.; Wu, J., *J. Comb. Chem.* **2007**, *9*, 690–694.

78. Ding, Q.; Wu, J., *Org. Lett.* **2007**, *9*, 4959–4962.
79. Ding, Q.; Yu, X.; Wu, J., *Tetrahedron Lett.* **2008**, *49*, 2752–2755.
80. Yeom, H. -S.; Kim, S.; Shin, S., *Synlett* **2008**, 924–928.
81. Ding, Q.; Wang, Z.; Wu, J., *Tetrahedron Lett.* **2009**, *50*, 198–200.
82. Seregin, I. V.; Schammel, A. W.; Gevorgyan, V., *Org. Lett.* **2007**, *9*, 3433–3436.
83. Seregin, I. V.; Schammel, A. W.; Gevorgyan, V., *Tetrahedron* **2008**, *64*, 6876–6883.
84. Harrison, T. J.; Kozak, J. A.; Corbella-Pane, M.; Dake, G. R., *J. Org. Chem.* **2006**, *71*, 4525–4529.
85. Harrison, T. J.; Dake, G. R., *Org. Lett.* **2004**, *6*, 5023–5026.
86. Imase, H.; Noguchi, K.; Hirano, M.; Tanaka, K., *Org. Lett.* **2008**, *10*, 3563–3566.
87. Godet, T.; Belmont, P., *Synlett* **2008**, 2513–2517.
88. Porcel, S.; Echavarren, A. M., *Angew. Chem., Int. Ed.* **2007**, *46,* 2672–2676.
89. Lin, G. -Y.; Li, C. -W.; Hung, S. -H.; Liu, R. -S., *Org. Lett.* **2008**, *10*, 5059–5062.
90. Lo, C. -Y.; Lin, C. -C.; Cheng, H. -M.; Liu, R. -S., *Org. Lett.* **2006**, *8*, 3153–3156.
91. Ding, Q.; Wu, J., *J. Comb. Chem.* **2008**, *10*, 541–545.
92. Binder, J. T.; Kirsch, S. F., *Org. Lett.* **2006**, *8*, 2151–2153.
93. Menz, H.; Kirsch, S. F., *Org. Lett.* **2006**, *8*, 4795–4797.

6

SILVER-CATALYZED NITRENE TRANSFER REACTIONS

ZIGANG LI, DAVID A. CAPRETTO, AND CHUAN HE
Department of Chemistry, University of Chicago, Chicago, Illinois

6.1 INTRODUCTION

Nitrogen is a prevalent element in naturally occurring molecules, and chemical formation of new carbon–nitrogen bonds has broad applications in both industry and academic research. Despite more recent advances in carbon–nitrogen bond formation, new facile methods are still sought after. Great advances have been made since the late 1990s, particularly in the development of the efficient hydroamination of

Silver in Organic Chemistry Edited by Michael Harmata

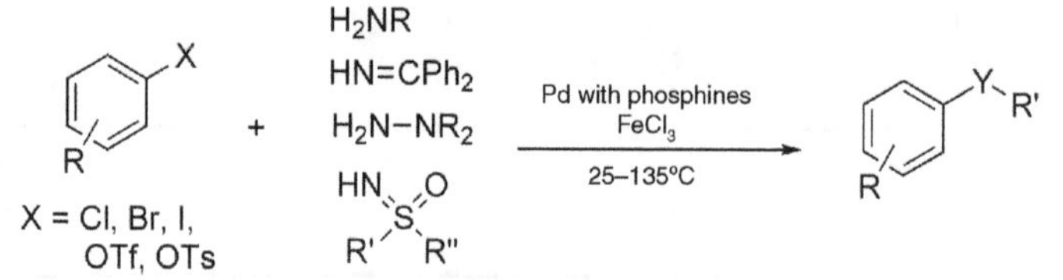

Scheme 6.1. Buchwald–Hartwig coupling chemistry for C–N bond formation.

unsaturated carbon–carbon bonds and coupling reactions between amines (even ammonia) and haloarenes (Scheme 6.1).[1]

The nitrene transfer reaction is another promising method for installing new C–N bonds.[2] The most commonly used nitrene precursors are a class of molecules called iminoiodanes (such as PhI=NNs, where Ns = 4-nitrobenzenesulfonyl), which are typically stabilized with electron-withdrawing groups. Iminoiodanes are relatively stable and can be easily synthesized and stored for reasonable periods of time at lower temperatures; reactions using iminoiodanes are safe in general.[3] Other commonly used nitrene precursors include organic azides and chloramines such as TsN · NaCl. However, the preparation of azides and anhydrous chloramines carries a risk of explosion.[4] More reactive nitrene precursors, such as TsN · NaBr and PhBr=NNs, are also being increasingly used.[5]

The aziridination of olefins, which forms a three-membered nitrogen heterocycle, is one important nitrene transfer reaction. Aziridination shows an advantage over the more classic olefin hydroamination reaction in some syntheses because the three-membered ring that is formed can be further modified. More recently, intramolecular amidation and intermolecular amination of C–H bonds into new C–N bonds has been developed with various metal catalysts. When compared with conventional substitution or nucleophilic addition routes, the direct formation of C–N bonds from C–H bonds reduces the number of synthetic steps and improves overall efficiency.[2] After early work on iron, manganese, and copper,[6] Muller, Dauban, Dodd, Du Bois, and others developed different dirhodium carboxylate catalyst systems that catalyze C–N bond formation starting from nitrene precursors,[7] while Che studied a ruthenium porphyrin catalyst system extensively.[8] The rhodium and ruthenium systems are

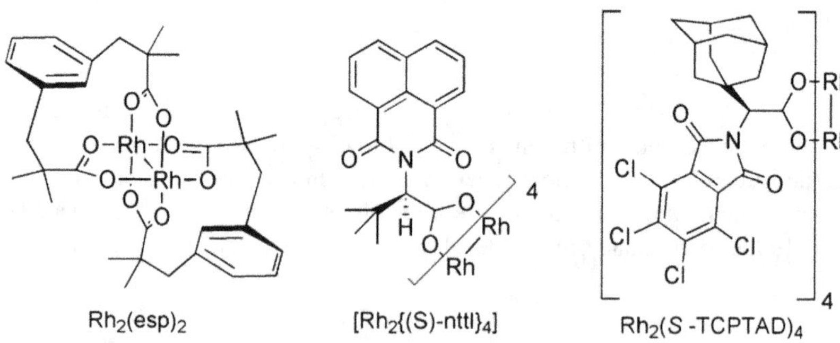

Figure 6.1. Rhodium–carboxylate catalysts for C–N bond formation.

currently the most intensely studied, while copper,[9] nickel,[10] iron,[11] cobalt,[12] and gold catalysts[13] also show nitrene transfer activity (Fig. 6.1).

In addition to rhodium and ruthenium, silver catalysts have also been investigated, and this chapter discusses these silver-based nitrene transfer reactions.[2a] In discussing the work chronologically, we hope that the readers can get an idea of the evolution of thinking in the research process.

6.2 AZIRIDINATION

6.2.1 Chloramine-T as Nitrene Precursor

As stated in the introduction, chloramine-T (where T denotes three crystalline water molecules) is a commonly used nitrene precursor, which is commercially available and costs less than do most other nitrene sources. The benefit of a silver salt in nitrene transfer reactions with chloramine-T is surprisingly simple. Because silver chloride is insoluble in most solvents, substoichiometric amounts of silver salts (like silver nitrate) can be used to remove the chloride from chloramine to facilitate the release of a free nitrene radical, which can aziridinate olefins. Since the amount of silver is near stoichiometric, it should not be called silver-based catalysis, although turnover numbers (TONs) higher than 1 have been observed in some cases.

In 2001, Rai first reported such a transformation.[14] The work specifically noted that the reaction favors aprotic solvents such as THF. This statement was challenged by Komatsu shortly thereafter, who found no reactivity in THF but high conversions in CH_2Cl_2 or benzene for the same reaction.[15] Solvent grade or quality may play an important role; less likely but still possible, the grade or quality of silver nitrate used may also be a factor.

In Komatsu's report, a stoichiometric amount of $AgNO_3$ and chloramine-T produced 43–92% yields of the aziridine products from different olefin substrates. An advantage for this reaction is that the chloramine-T could be directly used without any dehydration. An important mechanistic study using 1,6-dienes was also carried out, which yielded bicyclic pyrrolidines. The formation of bicyclic pyrrolidines together with the reaction sensitivity toward oxygen, a good radical inhibitor, support a free nitrene radical mechanism (Scheme 6.2). The formation of AgCl precipitate and subsequent generation of a nitrene radical intermediate is well accepted; however, whether the metal salt plays any other role is not clear.

6.2.2 Iminoiodanes as Nitrene Precursors

Silver has three synthetically useful oxidation states: Ag(I), Ag(II), and Ag(III).[16] Free Ag(III) is very unstable, and most currently known Ag(III) complexes are stablized with electron-donating and/or sterically demanding ligands.[17] It is known that Ag(I) can be oxidized to Ag(II) with strong oxidants such as persulfates. Nitrogen-based ligands such as pyridines are commonly used to stabilize high-valence metal ions.[18] In 2003, He and coworkers utilized a pyridine-supported silver catalyst and reported the first silver-catalyzed aziridination of olefins.[19]

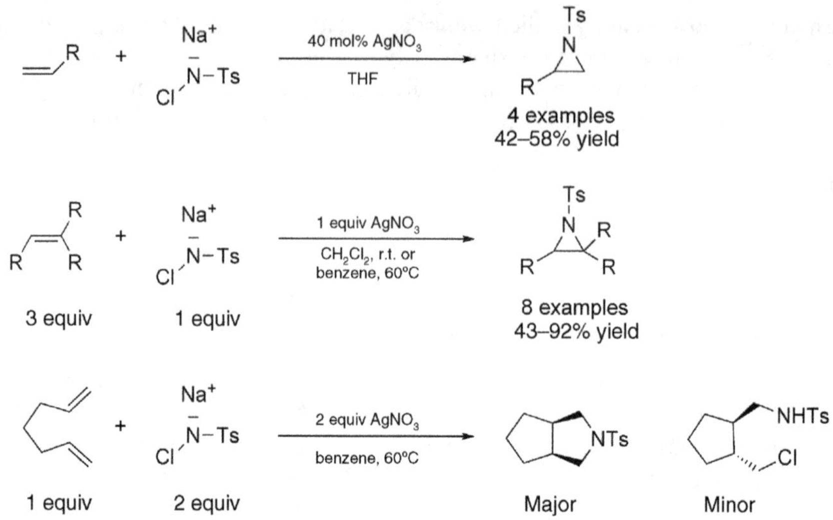

Scheme 6.2. Olefin aziridination using chloramine-T as a nitrene source.

With the tridentate 4,4′,4″-tri-*tert*-butyl-2,2′:6′,2″-terpyridine (^{t}Bu$_3$tpy) ligand, a unique disilver complex crystallized with high yield in acetonitrile. NMR studies showed that the complex forms within 5 min when a silver salt and ligand are mixed together in CD_3CN at ambient temperature. The catalyst structure was fully characterized to show a unique dinuclear silver(I) [Ag$_2$(^{t}Bu$_3$tpy)$_2$(NO$_3$)](NO$_3$) (Fig. 6.2). The Ag–Ag distance is 2.842(2) Å, which indicates a fairly strong silver(I)–silver(I) interaction. Both silvers are five-coordinate if the silver-silver interaction is included.

The aziridination catalysis was carried out in situ with mixed ligand and silver salt, which gives results similar to those obtained using the pure [Ag$_2$(^{t}Bu$_3$tpy)$_2$(NO$_3$)] (NO$_3$) crystal. The ^{t}Bu$_3$tpy ligand shows superior reactivity over other pyridine

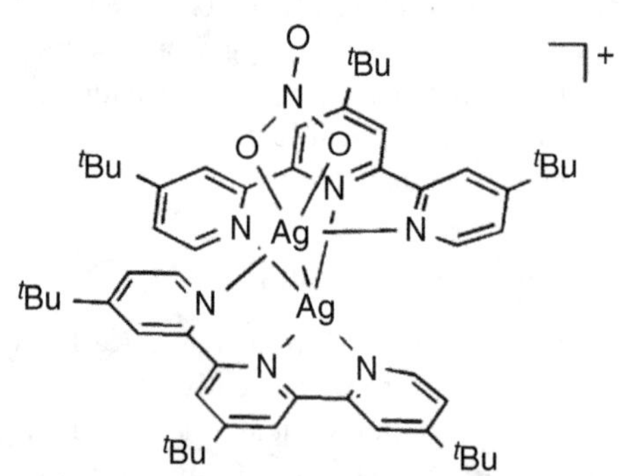

[Ag$_2$(^{t}Bu$_3$tpy)$_3$(NO$_3$)](NO$_3$)

Figure 6.2. Structure of disilver nitrene transfer catalyst.

ligands, while the choice of silver salt does not play an important role as long as the salt is soluble in the solvent used. Interestingly, this disilver(I) complex has some structural features similar to those of di-Rh(II) compounds, which were previously used to catalyze a similar reaction. Whether the structural similarity implies the same reaction mechanism with rhodium or other systems is presently unclear. One aspect of this research hampering the elucidation of mechanism in all these systems is the insolubility of iminoiodanes in common organic solvents.

The aziridination works for both aromatic and aliphatic olefins, including less active linear terminal olefins. Most reactions proceed in good yield at room temperature. The use of *cis*-stilbene at 0°C gives predominately *cis* aziridine product in about 90 : 10 *cis* : *trans* ratio (Table 6.1). The conservation of *cis* structure suggests that a discrete silver nitrene intermediate is involved in the reaction path. Because of the unique disilver structure and unlikely formation of a silver(III) species, the authors suspect that a bridged nitrene intermediate between the two silver atoms may be responsible for this transformation in which each silver atom donates one electron to the nitrenoid. However, further research is necessary to prove this hypothesis and a fast radical reaction mechanism cannot be eliminated on the basis of current evidence.

TABLE 6.1. Substrate Scope for Olefin Aziridination by Silver–Terpyridine Catalyst

$$\text{Olefin} + \text{PhI=NTs} \xrightarrow[\text{CH}_3\text{CN, r.t.}]{\text{5 mol\% } [\text{Ag}_2(^t\text{Bu}_3\text{tpy})_3(\text{NO}_3)](\text{NO}_3)}$$

Olefin	Aziridine	Isolated Yield (%)
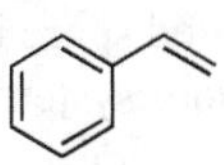	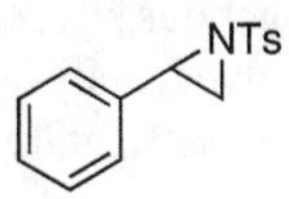	91
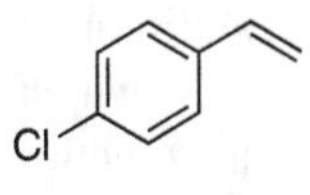	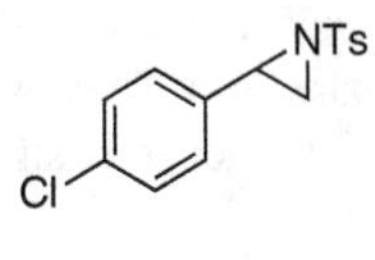	74
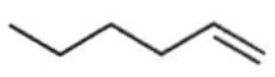	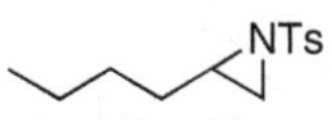	71
	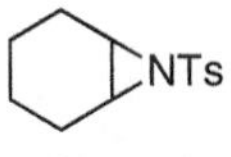	81
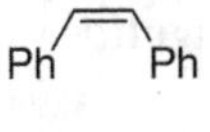	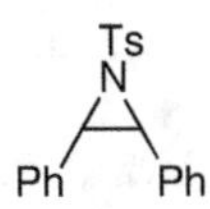	86[a]

[a] Reaction performed at 0°C.

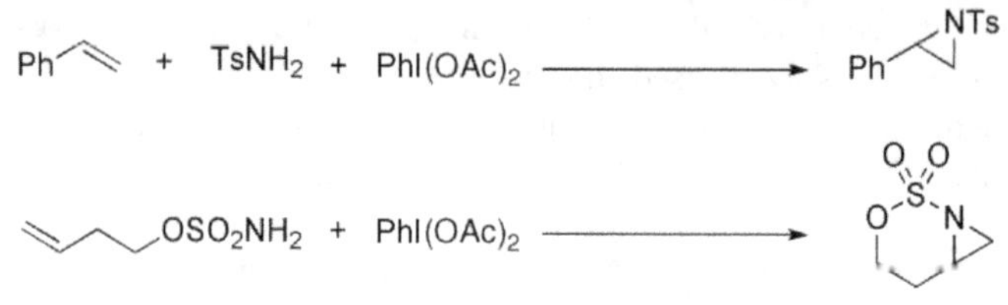

Scheme 6.3. Silver–terpyridine catalyst can catalyze olefin aziridination using an in situ prepared nitrene.

The reaction was further investigated with electrospray ionization–mass spectrometry (ESI-MS). Two different iminoiodanes were used, PhI$=$NTs and PhI$=$NSO$_2$Ph, and the results suggest an intermediate containing a disilver core plus a nitrene moiety ($=$NTs or $=$NSO$_2$Ph). The masses corresponded to these species, and the 14 D weight difference between the two nitrene sources further proved the involvement of the nitrene moeity. This catalyst can also mediate olefin epoxidation, although at very low efficiencies. Although the synthesis of PhI$=$NTs is relatively easy, it is more attractive to directly use the sulfonamide precursor and some inexpensive oxidant to generate the nitrene source in situ. Rhodium- or copper-based catalysts are able to perform this one-pot reaction, and silver was able to as well. [Ag$_2$(^{t}Bu$_3$tpy)$_2$(NO$_3$)](NO$_3$) catalyzed the direct aziridination of styrenes using TsNH$_2$ and PhI(OAc)$_2$. Unfortunately, aliphatic olefins could only be converted using separately prepared iminoiodanes (Scheme 6.3).

6.2.3 Heterogeneous Silver Catalysis

This chapter focuses mostly on homogenous silver catalysis, so only limited space is spent on discussion of heterogeneous silver catalysis. However, heterogenous systems are important. Heterogeneous silver-based catalysts are well known in industrial ethylene epoxidation and methanol/formaldehyde oxidation.[20] Friend et al. showed that the silver(110) single-crystal surface can mediate the oxidative addition of ammonia to styrene to form phenylaziridines. They used oxygen as the oxidant, which shows the promising potential of heterogeneous silver catalysts for olefin functionalization (Scheme 6.4).[21]

6.3 SULFIDE AND SULFOXIDE IMINATION

Sulfilimines and sulfoximines are useful synthetic building blocks for chiral ligands and pseudopeptides. Compared to other preparation methods, the direct imination of corresponding sulfides and sulfoxides is a more convenient and straightforward synthetic route.

Different metal complexes have shown the ability to catalyze these imination reactions, such as rhodium, copper, and iron.[22] In 2005, Bolm found that the disilver(I) complex described in Section 6.2.2 catalyzes the imination of sulfides and sulfoxides

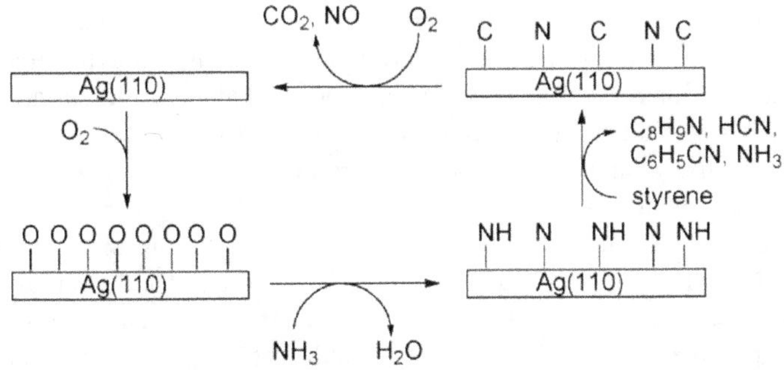

Scheme 6.4. Heterogenous silver can facilitate C–N bond formation.

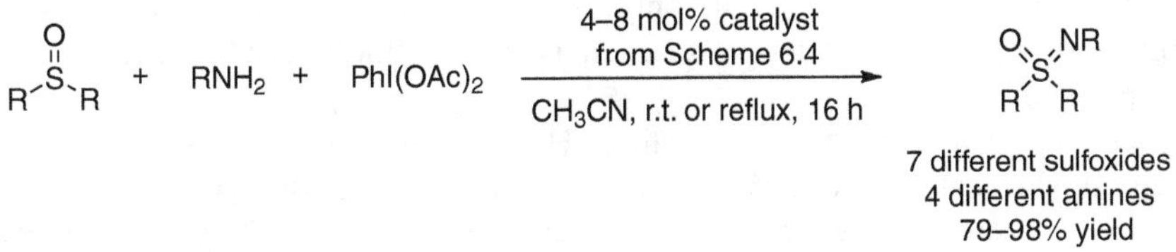

Scheme 6.5. Sulfoxide imination using silver–terpyridine catalyst.

with good to excellent yields.[23] Simple sulfonamides can be used directly with commercially available oxidants such as $PhI(OAc)_2$ in acetonitrile at room temperature. Sulfoxides can give sulfoximine products, and sulfides can afford sulfilimines, while double iminations were not detected (Scheme 6.5). In addition, the imination of an enantioenriched sulfoxide yielded a sulfoximine (after deprotection) with the same enantiomeric excess value. This finding also suggests a conserved nitrene transfer process and that the disilver(I) **1** system possibly could be utilized in other catalytic nitrene transfer chemistry.

6.4 AMIDATION

When compared to aziridination or sulfoxide imination, efficient amidation of C–H bonds is a more formidable task since it involves the activation of relatively inert C–H bonds. With an increase in the difficulty of the reaction comes an increase in the value of products, as many are derived from cheap simple hydrocarbons. In addition to making more valuable materials in fewer steps, a better mechanistic understanding of nitrene transfer can also be gleaned, leading to the development of efficient catalysts.

6.4.1 Intramolecular Amidation

Intramolecular amidation provides a useful way to directly install a new C–N bond at a designated position and is a powerful method for introducing new synthons.

Generally, hydrocarbons tethered with sulfamate esters or carbamates are suitable substrates and afford five- or six-membered ring products through intramolecular C–H amidation. For reasons not yet understood, carbamates and sulfamate esters prefer to form five- and six-membered rings, respectively. In general, this important and useful selectivity can avoid formation of product mixtures, a major drawback for similar organic transformations.

After successful application of the silver catalyst shown in olefin aziridination (Section 6.1.1), He and coworkers showed that intramolecular amidation was possible with both hydrocarbon-tethered carbamates and sulfamate esters.[24] They found that only the $^{t}Bu_3$tpy silver complex could catalyze efficient intramolecular amidation, while other pyridine ligands gave either dramatically lower yields or complicated product mixtures. In an interesting control study, both copper and gold were also tested in this reaction. Both the copper and gold $^{t}Bu_3$tpy complexes can mediate olefin aziridination, but only silver can catalyze intramolecular C–H amidation, indicating that the silver catalyst forms a more reactive metal nitrene intermediate.

Both tertiary and secondary aliphatic C–H bonds can undergo this transformation to give products in good to excellent yields. In some cases, 4-^{t}Bu-pyridine was added to both increase the conversion and afford a cleaner reaction. Notably, no mixed five- and six-membered ring products were observed for any substrates (Table 6.2). To gain mechanistic perspective, the amidation reaction was performed with a carbamate substrate derived from (*S*)-2-methyl-1-butanol. The resulting product completely retained its (*S*) configuration, which suggests a conserved nitrene insertion process. However, a short-lived radical process cannot be ruled out. Unfortunately, the $^{t}Bu_3$tpy–silver complex could not catalyze intermolecular C–H amination. This lack of activity led to the search for and discovery of a more reactive silver catalyst that can perform intermolecular C–H amination.

6.4.2 Intermolecular Amination with Phenanthroline Ligands

Intermolecular amination is highly sought after and is the most straightforward way to install a new nitrogen-based functional group. However, because of the difficulty of this reaction, published explorations into this field were almost nonexistent in the past. More recently, the field has grown quickly as several different groups independently reported their breakthroughs on the topic. Several examples of intermolecular C–H amination of simple hydrocarbons with copper, rhodium, gold, and silver systems have been reported.[7–9,13,25]

The He group continued to search for more active silver-based nitrene transfer catalysts. Considering the reactivity, metal nitrene intermediates are generally electrophilic and less donating ligands may enhance the electrophilicity of the intermediate. With this in mind, various pyridine-type ligand systems were tested. In agreement with their hypothesis, the fused-pyridine bidentate ligand bathophenanthroline [4,7-diphenyl-1,10-phenanthroline (BP)] showed the best reactivity. Regular phenanthroline showed good reactivity, but solubility was poor, making the diphenyl version more acceptable. This catalyst allowed intramolecular amidation to

TABLE 6.2. Intramolecular C–H Amidation by Silver-Terpyridine Catalyst

$$\text{Substrate} + \text{PhI(OAc)}_2 \xrightarrow[\text{CH}_3\text{CN, 82°C}]{\text{5 mol\% } [\text{Ag}_2(^t\text{Bu}_3\text{tpy})_3(\text{NO}_3)](\text{NO}_3)}$$

Substrate	Product	Isolated Yield (%)
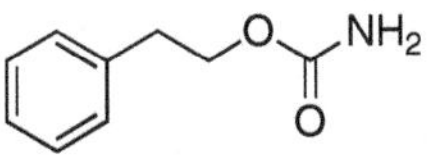	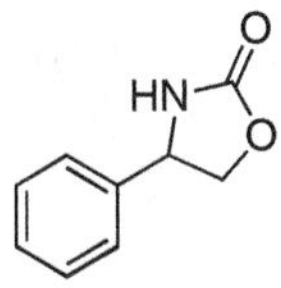	81
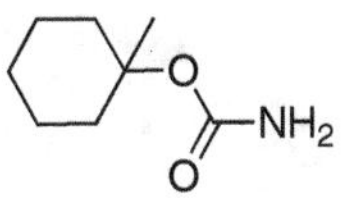	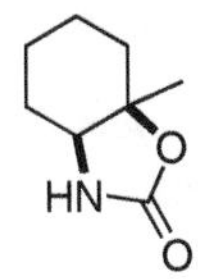	85
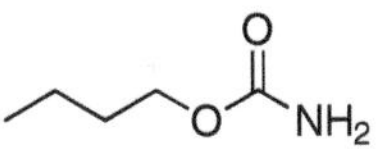	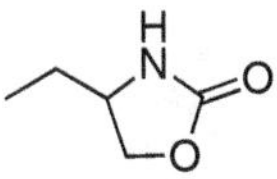	48
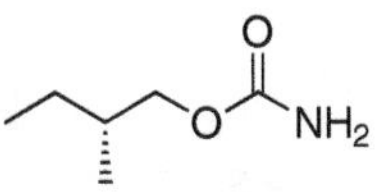	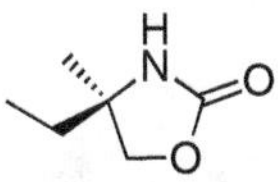	73

proceed at 50°C and improved yields significantly when compared to the terpyridine system using the same substrates (see Section 6.4.1).

More importantly, this silver system catalyzes the intermolecular amination of hydrocarbons, as shown in Table 6.3. In addition to aminating weaker benzylic C–H bonds, stronger aliphatic C–H bonds such as those in cyclohexane were also reactive. Although yields with more inert hydrocarbons were modest with the bathophenanthroline system, the discovery of the first silver-catalyzed intermolecular amination opens opportunities for further developments. This reaction favored tertiary cyclic sp^3 C–H bonds over secondary cyclic sp^3 C–H bonds, and showed limited success with simple linear alkanes. No conversion was observed with any aromatic C–H bonds. The compound $NsNH_2$ was tested as the nitrene precursor with different oxidants. The use of $PhI(OAc)_2$ as oxidant gave the expected amination product with a lower yield, while persulfate and peroxides showed no reactivity.

Similar to that of the ^{t}Bu$_3$tpy–silver catalyst, the crystal structure of the bathophenanthroline–silver complex has a disilver(I) core and two BP ligands stacked over each other, similar to the corresponding palladium(II)–BP complex.

TABLE 6.3. Silver–Bathophenanthroline Catalysis of a Mild Intermolecular C-H Amination

$$\text{Substrate} + \text{PhI=NNs} \xrightarrow[\text{CH}_2\text{Cl}_2,\ 50^\circ\text{C},\ 4\ \text{Å MS}]{5\ \text{mol\%}\ \text{Ag}_2(\text{BP})_2\text{OTf}}$$

Substrate	Product	Isolated Yield (%)
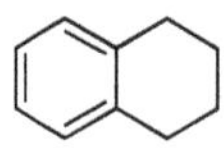	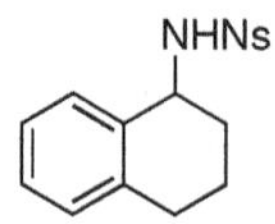	70
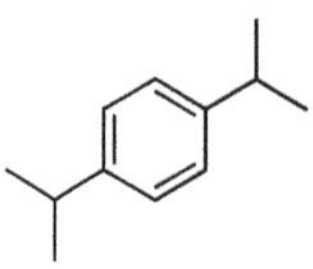	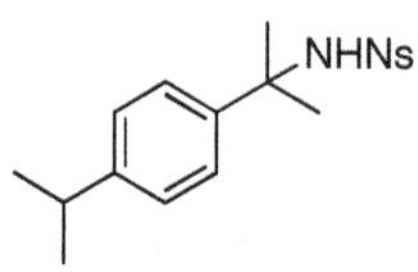	57
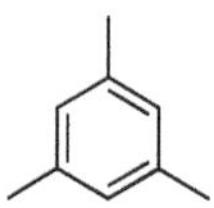	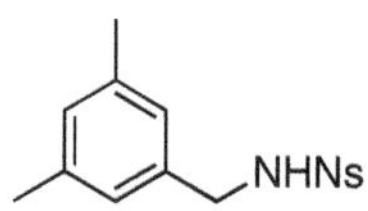	25
	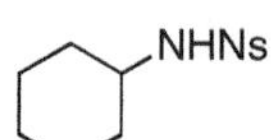	40
	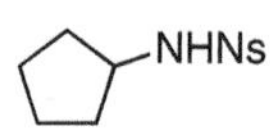	35
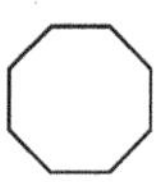	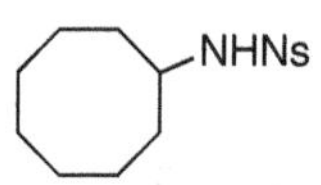	33
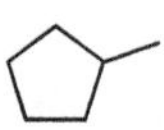	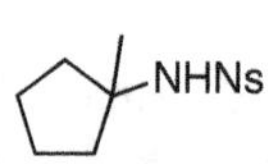	39

The silver–silver distance is 3.386 Å, indicating a weak silver–silver interaction (Fig. 6.3). Observing the disilver structure for the second time, the authors studied the importance of the disilver structure on reactivity. By testing different phenanthrolines, they found that 2,9-substituted phenanthrolines did not catalyze the amination reaction. This result was supported by structural evidence—the silver-2,9-dimethyl-phenanthroline complex did not form disilver structure; two ligands coordinated one silver in a tetrahedral geometry, blocking the approach of any other molecule to the silver center. Even one bulky mesityl group in the 2 position inhibits the reactivity completely (Scheme 6.6).

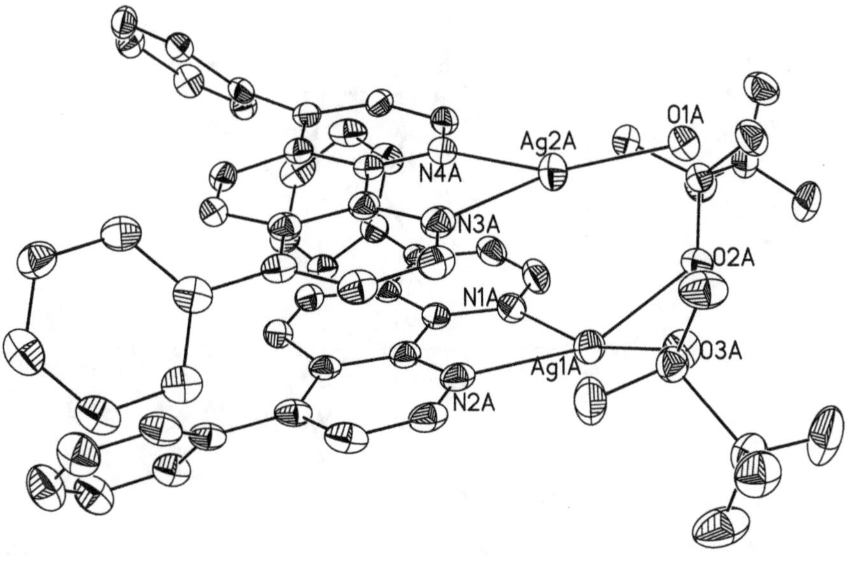

Figure 6.3. X-ray crystallographic structure of silver–bathophenanthroline catalyst system: Ag1A is coordinated with two triflate molecules, while Ag2A is coordinated with a water molecule.

The development of new phenanthroline ligands with electron-withdrawing modifications on the 4 and 7 positions may be a way to enhance the reactivity further.

6.4.3 Intermolecular Amination Based on Pyrazolylborate Ligands

Shortly after the discovery by He and coworkers, Díaz-Requejo and Pérez et al. reported an intriguing silver–pyrazolylborate system that catalyzes the amination of aliphatic hydrocarbons.[26] Earlier, they had developed a similar copper–pyrazolylborate system that could also catalyze the C–H amination reaction. However, a different ligand was required with the silver catalyst system.

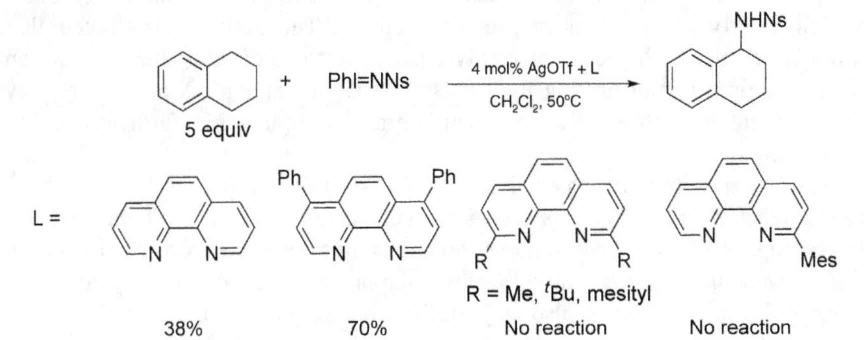

Scheme 6.6. Ligand effects on intermolecular C–H amination catalysis.

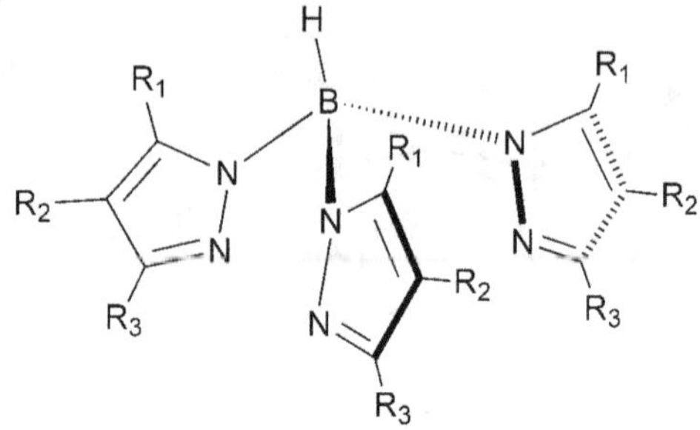

Tp^{type}	R_1	R_2	R_3
Tp^{Br3}	Br	Br	Br
Tp^{Ms}	H	H	Mesityl
Tp^{*}	Me	H	Me
$Tp^{*,Br}$	Me	Br	Me

Figure 6.4. Pyrazolylborate ligands used for silver-based intermolecular C–H amination.

They chose four different pyrazolylborate ligands to test ligand effects on the catalyst, as shown in Figure 6.4, and found Tp^{Br3} to be the ligand with the most electron-withdrawing groups and Tp^{Ms} the one with the most steric hindrance, while $Tp^{*,Br}$ and Tp^{*} were intermediate in electronics and sterics. In the silver case, the $Tp^{*,Br}$ ligand generally gave the highest yield, while Tp^{Br3} was the least efficient. Interestingly, they found that when simple linear alkanes were used, the amination products from primary sp^3 C–H bond and secondary sp^3 C–H bond activation were detected. The activation of a primary sp^3 C–H bonds is unique and rarely detected with other catalysts. The results are shown in Table 6.4. The AgT_p^{Ms} complex gave the highest ratio of primary sp^3 C–H bond activation, possibly due to the sterically demanding mesityl group, which should generate the smallest catalytic pocket and thus prefer smaller, primary CH_3 groups.

The conversion of simple hydrocarbons into corresponding amine products is significantly higher than in previous reports. The authors also noted that NMR experiments showed that catalyst peaks reappeared after the completion of the reaction, demonstrating the robustness of the catalyst. A high TON may be achieved with this silver–pyrazolylborate system with further reaction optimization.

When 2 equiv (relative to catalyst) of 2,6-di-*tert*-butylhydroxytoluene (BHT, a radical reaction inhibitor) was added to the reaction, the yield of the reaction decreased by half. A radical trap experiment using cyclohexane was carried out in CCl_4 as well, and together with 40% amination product, 12% of chlorocyclohexane was also detected. This result also suggests that a radical pathway is operative. When a 1 : 2.3 mixture of *cis*- and *trans*-2 pentene was reacted with PhI=NTs (Scheme 6.7), a 1 : 1.5 *cis* : *trans* aziridine product was isolated, which suggests the mechanism may

TABLE 6.4. Substrate Scope for Silver–Pyrazolylborate-Catalyzed Aliphatic C–H Amination

$$\text{R–H} + \text{PhI=NTs} \xrightarrow[\text{no solvent, 80°C, 16 h}]{\text{5 mol\% AgTp}^{*,\text{Br}}} \text{R–NHTs}$$

Substrate	Product(s)	Isolated Yield (%)[a]
	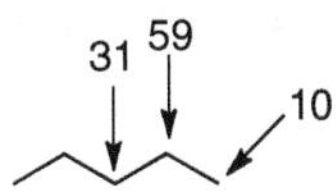	65
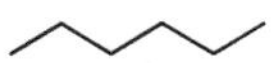	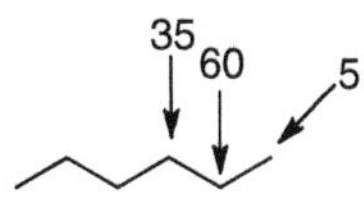	70
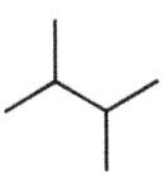	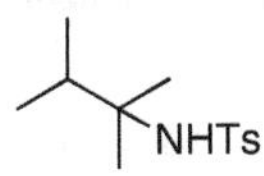	75
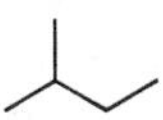	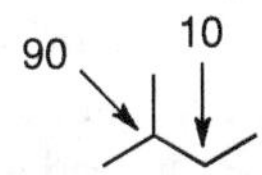	80
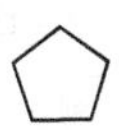	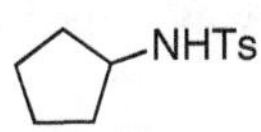	80
	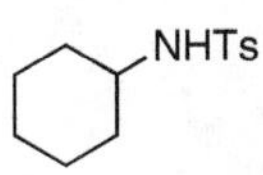	90

[a] Yield indicates combined yield. Arrows refer to percent product at the denoted position.

be part stepwise radical process and part discrete nitrene transfer. Generally, *trans*-aziridines are more stable and these *cis* : *trans* ratios were cited as given from the original report.

This observation fits with results previously described by Che et al. with the porphyrin ruthenium nitrenoid. Even when isolated ruthenium nitrenoid was used,

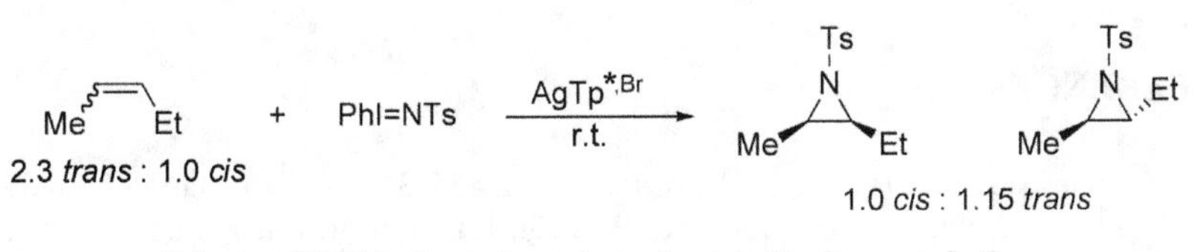

Scheme 6.7. Mechanistic probes of catalytic nitrene transfer.

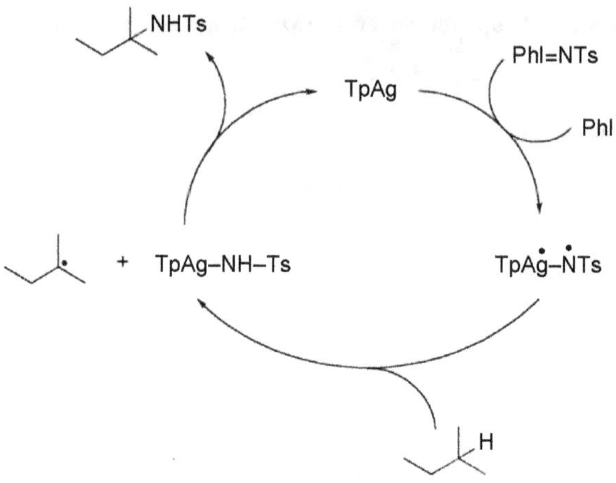

Scheme 6.8. Proposed mechanism for silver–pyrazolylborate-catalyzed nitrene transfer.

trans-aziridine products were still detected from *cis*-olefin substrates, and sometimes as the predominant product. Current results on silver-catalyzed nitrene transfer reactions, indicate that silver probably can interact with iminoiodanes to generate a silver nitrene precursor. This precursor can lead to reactions via either a concerted metal nitrene or a stepwise radical pathway, depending on the substrate and reaction conditions (Scheme 6.8).

6.5 CONCLUSION

Since the original report on aziridination in 2003, silver-catalyzed nitrene transfer reactions have been developed into both intramolecular C–H amidation and intermolecular C–H amination. Currently, three promising ligand groups have been successful, each showing unique reactivity. All are worth further investigation, but new ligand groups will undoubtedly be discovered. Current knowledge indicates nitrogen-based or sterically demanding ligands to be most promising. How to convert these systems into synthetically meaningful enantioselective systems is a challenging task. With the relatively low price of silver salts, the stability of silver complexes, and high yields observed with current systems, the future for silver-catalyzed nitrene transfer is bright.

REFERENCES

1. For reviews, see (a) Hartwig, J. F., *Nature* **2008**, *455*, 314–322; (b) Hartwig, J. F., *Acc. Chem. Res.* **2008**, *41*, 1534–1544; (c) Hartwig, J. F., *Angew. Chem. Int. Ed.* **1998**, *37*, 2046–2067; (d) Surry, D. S.; Buchwald, S. L., *Angew. Chem. Int. Ed.* **2008**, *47*, 6338–6361;

(e) Wolfe, J. P.; Wagaw, S.; Marcoux, J.-F.; Buchwald, S. L., *Acc. Chem. Res.* **1998**, *31*, 805–818; (f) Correa, A.; Bolm, C., *Adv. Synth. Catal.* **2008**, *350*, 391–394.

2. For reviews, see (a) Li, Z.; He, C., *Eur. J. Org. Chem.* **2006**, 4313–4322; (b) Watson, I. D. G.; Yu, L.; Yudin, A. K., *Acc. Chem. Res.* **2006**, *39*, 194–206; (c) Du Bois, J., *Chemtracts* **2005**, *18*, 1–13; (d) Dauban, P.; Dodd, R. H., *Synlett* **2003**, 1571–1586; (e) Mueller, P.; Fruit, C., *Chem. Rev.* **2003**, *103*, 2905–2919; (f) Hilt, G., *Angew. Chem. Int. Ed.* **2002**, *41*, 3586–3588.
3. (a) Evans, D. A.; Faul, M. M.; Bilodeau, M. T., *J. Org. Chem.* **1991**, *56*, 6744–6746; (b) Heuss, B. D.; Mayer, M. F.; Dennis, S.; Hossain, M. M., *Inorg. Chim. Acta* **2003**, *342*, 301–304; (c) Taylor, S.; Gullick, J.; McMorn, P.; Bethell, D.; Bulman, P.; Philip, C.; Hancock, F. E.; King, F.; Hutchings, G. J., *Top. Catal.* **2003**, *24*, 43–50.
4. (a) Ando, T.; Kano, D.; Minakata, S.; Ryu, I.; Komatsu, M., *Tetrahedron* **1998**, *54*, 13485–13494; (b) Vyas, R.; Chanda, B. M.; Bedekar, A. V., *Tetrahedron Lett.* **1998**, *39*, 4715–4716; (c) Ando, T.; Minakata, S.; Ryu, I.; Komatsu, M., *Tetrahedron Lett.* **1998**, *39*, 309–312.
5. (a) Vyas, R.; Gao, G.-Y.; Harden, J. D.; Zhang, P. X., *Org. Lett.* **2004**, *6*, 1907; (b) Vyas, R.; Chanda, B. M.; Bedekar, A. V., *Tetrahedron Lett.* **1998**, *39*, 4715–4716; (c) Vyas, R.; Gao, G.-Y.; Harden, J. D.; Zhang, P. X., *Org. Lett.* **2004**, *6*, 1907–1910; (d) Ochiai, M.; Kaneaki, T.; Tada, N.; Miyamoto, K.; Chuman, H.; Shiro, M.; Hayashi, S.; Nakanishi, W., *J. Am. Chem. Soc.* **2007**, *129*, 12938–12939.
6. (a) Breslow, R.; Gellman, S. H., *J. Chem. Soc. Chem. Commun.* **1982**, *24*, 1400–1401; (b) Breslow, R.; Gellman, S. H., *J. Am. Chem. Soc.* **1983**, *105*, 6728–6729; (c) Li, Z.; Quan, R. W.; Jacobsen, E. N., *J. Am. Chem. Soc.* **1995**, *117*, 5889–5890.
7. (a) Guthikonda, K.; Du Bois, J., *J. Am. Chem. Soc.* **2002**, *124*, 13672–13673; (b) Dauban, P.; Saniere, L.; Tarrade, A.; Dodd, R. H., *J. Am. Chem. Soc.* **2001**, *123*, 7707–7708; (c) Espino, C. G.; Du Bois, J., *J. Am. Chem. Soc.* **2002**, *124*, 12950–12951; (d) Espino, C. G.; Fiori, K. W.; Kim, M. J.; Du Bois, J., *J. Am. Chem. Soc.* **2004**, *126*, 15378–15379; (e) Liang, C.; Collet, F.; Robert-Peillard, F.; Müller, P.; Dodd, R. H.; Dauban, P., *J. Am. Chem. Soc.* **2008**, *130*, 343–344; (f) Fiori, K. W.; Du Bois, J., *J. Am. Chem. Soc.* **2007**, *129*, 562–568.
8. (a) Liang, J. -L.; Yuan, S. -X.; Huang, J. -S.; Yu, W. -Y.; Che, C. -M., *Angew. Chem. Int. Ed.* **2002**, *41*, 3465–3468; (b) Liang, J. -L.; Huang, J. -S.; Yu, X. Q.; Zhu, N.; Che, C. -M., *Chem. Eur. J.* **2002**, *8*, 1563–1572.
9. (a) Díaz-Requejo, M. M.; Belderraíén, T. R.; Nicasio, M. C.; Trofimenko, S.; Peréz, P. J., *J. Am. Chem. Soc.* **2003**, *125*, 12078–12079; (b) Fructos, M. R.; Trofimenko, S.; Díaz-Requejo, M. M.; Pérez, P. J., *J. Am. Chem. Soc.* **2006**, *128*, 11784–11791.
10. Mindiola, D. J.; Hillhouse, G. L., *Chem. Commun.* **2002**, *17*, 1840–1841.
11. Mahy, J. P.; Battioni, P.; Mansuy, D., *J. Am. Chem. Soc.* **1986**, *108*, 1079–1080.
12. Gao, G. -Y.; Harden, J. D.; Zhang, X. P., *Org. Lett.* **2005**, *7*, 3191–3193.
13. (a) Li, Z.; Ding, X.; He, C., *J. Org. Chem.* **2006**, *71*, 5876–5880; (b) Li, Z.; Capretto, D. A.; Rahaman, R. O.; He, C., *J. Am. Chem. Soc.* **2007**, *129*, 12058–12059.
14. Kumar, K. A.; Rai, L. K. M.; Umesha, K. B., *Tetrahedron* **2001**, *57*, 6993–6996.
15. Minakaa, S.; Kano, D.; Fukuoka, R.; Oderaotoshi, Y.; Komatsu, M., *Heterocycles* **2003**, *60*, 289–298.
16. Greenwood, N. N.; Earnshaw, A.; Chemistry of the Elements Elsevier Oxford **1997**.
17. Furuta, H.; Maeda, H.; Osuka, A., *J. Am. Chem. Soc.* **2000**, *122*, 803–804.

18. For reviews, see (a) Ishii, T.; Aizawa, N.; Kanehama, R.; Yamashita, M.; Sugiura, K. -I.; Miyasaka, H., *Coord. Chem. Rev.* **2002**, *226*, 113–124; (b) Srochinski, D.; Dziegiec, Y.; Grzejdziak, A., *Russ. J. Coord. Chem.* **1997**, *23*, 447–460; (c) Levason, W.; Spicer, M. D., *Coord. Chem. Rev.* **1987**, *76*, 45–120.
19. Cui, Y.; He, C., *J. Am. Chem. Soc.* **2003**, *125*, 16202–16203.
20. Anonymous, *Chem. Eng. News* **2006**, *84*, 59–62.
21. Liu, X.; Madix, R. J.; Friend, C. M., *J. Am. Chem. Soc.* **2006**, *128*, 14266–14267.
22. Mancheno, O. G.; Bolm, C., *Chem. Eur. J.* **2007**, *13*, 6674–6681.
23. Cho, G. Y.; Bolm, C., *Org. Lett.* **2005**, *7*, 4983–4985.
24. Cui, Y.; He, C., *Angew. Chem. Int. Ed.* **2004**, *43*, 4210–4212.
25. Li, Z.; Capretto, D. A.; Rahaman, R. H.; He, C., *Angew. Chem. Int. Ed.* **2007**, *46*, 5184–5186.
26. Gomez-Emeterio, B. P.; Urbano, J.; Diaz-Requejo, M. M.; Perez, P. J., *Organometallics* **2008**, *27*, 4126–4130.

7

SILVER-CATALYZED SILYLENE TRANSFER

Tom G. Driver

Department of Chemistry, University of Illinois, Chicago, Illinois

7.1 INTRODUCTION

Silver salts are well established to promote atom transfer reactions.[1–5] In 1946 Bachmann and coworkers reported that silver oxide facilitated the Wolff rearrangement[6] of α-diazocarbonyl compounds.[7] After this initial report, several other silver(I) reagents (including $AgNO_3$ and AgO_2CPh)[8,9] were identified to provide higher yields,

Silver in Organic Chemistry Edited by Michael Harmata

greater reproducibility, and lower temperatures for formation of the rearrangement product.[10] In contrast to the decades of research interest in silver-mediated carbene transfer reactions, the use of silver salts to promote reactions involving silylenes remained undiscovered until 2002.[11]

Ever since silylenes were proposed as reactive intermediates in 1964, substantial research effort has been directed toward understanding and harnessing their reactivity.[12] These divalent silicon reactive intermediates can be generated thermally or photochemically from a number of different sources and participate in a range of atom transfer reactions, including insertion and addition reactions. In addition to interest surrounding the chemistry involving free silylenes, the involvement of transition metal silylenoids in important industrial processes continues to inspire research efforts into their synthesis and reactivity.[13] The development of synthetic processes that exploit the reactivity of silylene and metal silylenoid reactive intermediates resulted from these research programs.

This chapter examines the synthesis and reactivity of transition metal silylenoids and silylmetal complexes to provide context for the focus of this chapter: silver-mediated silylene transfer reactions.

7.2 REACTIVITY AND ATTRIBUTES OF METAL SILYLENOIDS AND SILYLMETAL COMPLEXES

7.2.1 Synthesis of Transition Metal Complexes of Silylenes[13,14]

The proposal of metal silylene complexes as reactive intermediates in the mechanism of industrial processes such as hydrosilylation[15] and dehydrogenative coupling of silanes to polysilanes[16] stimulated interest in the synthesis of these subvalent silicon compounds.[13,14] For example, certain hydrosilylation catalysts such as $(Ph_3P)_2Pt(C_2H_4)$ and $(Ph_3P)_3RhCl$ were found to be silane-specific; while Et_2SiH_2 was an acceptable substrate, Et_3SiH was not. This specificity was attributed to a metal silylenoid intermediate [$Et_2Si{=}M$].[17] Similarly, the polymerization of $PhSiH_3$ by Cp_2TiMe_2 was proposed to occur via a titanium silylenoid intermediate [$Cp_2Ti{=}SiHR$] that formed whenever another silyl group was added to the growing polymer.[16]

Isolable early transition metal silylenoids could be synthesized by photolysis of silylmetal complexes.[18–30] Photolysis of iron disilyl complex **1** triggered a 1,2-silyl shift to produce a complex **2** as a 2 : 1 mixture of diastereomers (Scheme 7.1).[19] Ogino and coworkers interpreted the nearly planar relationship of the atoms at each silicon and the short Fe–Si bonds (2.207 and 2.222 Å) as evidence of bis(silylene) character. They concluded that complex **2** is best described as the combination of the two mesomers **3** and **4**. Light-induced silicon–silicon bond cleavage was also used for the synthesis of tungsten silylene complexes;[31–33] the exposure of tungsten complex **5** to ultraviolet radiation leads to the loss of CO and formation of tungsten silylene complex **7**.[33] When the starting tungsten complex bore a cyclopentadiene ligand or when pentamethyldisilane was used, a base-stabilized (THF) tungsten complex was formed.[32]

Late transition metal silylenoid complexes were first synthesized by exploiting the lability of silicon–triflate bonds (Scheme 7.2).[34–37] Tilley and coworkers reported that

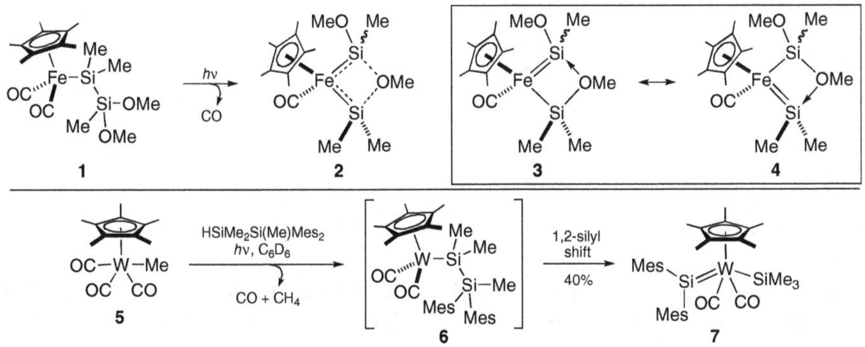

Scheme 7.1. Light-induced 1,2-silyl shift to synthesize early transition metal silylenoid complexes.

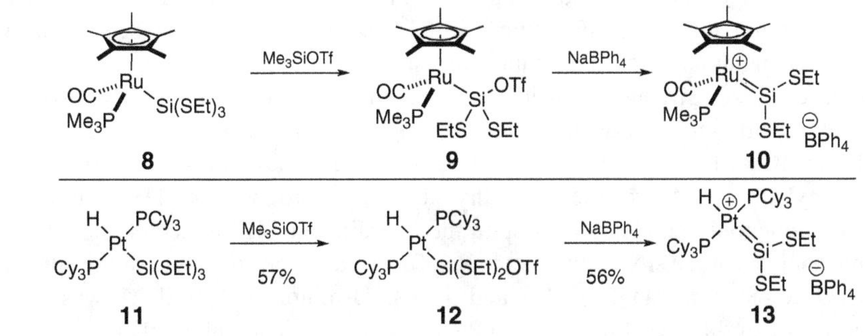

Scheme 7.2. Synthesis of late transition metal silylenoid complexes.

ruthenium silylenoid could be accessed in two steps from the silyl ruthenium cyclopentadienyl complex **8**. Ethyl thiolate abstraction by trimethylsilyl triflate generated **9**, which formed the silylenoid complex (**10**) on exposure to sodium tetraphenylborate.[38] A similar strategy was employed for the synthesis of platinum silylenoid **13**.[38] Downfield ^{29}Si {^{1}H} NMR resonances (δ 251 and 309 ppm for **10** and **13**, respectively) characterized these metal silylenoid complexes. In the presence of a stabilizing base (such as Et_2O),[39] an upfield shift to δ 89 ppm in the ^{29}Si {^{1}H} NMR was observed for a platinum silylenoid.

Tilley and coworkers reported a more direct procedure for the synthesis of late transition metal silylenoid complexes by reaction with silanes [equation (7.1)].[36,37] Exposing iridium complex **14** to dimesitylsilane afforded iridium silylenoid **15**.[36] In addition to dimesitylsilane, other silanes could be used, including diphenyl-, diethyl-, or dimethylsilane. Primary silanes such as mesitylsilane and 2,4,6-triisopropylphenylsilane were also tolerated as substrates.[37]

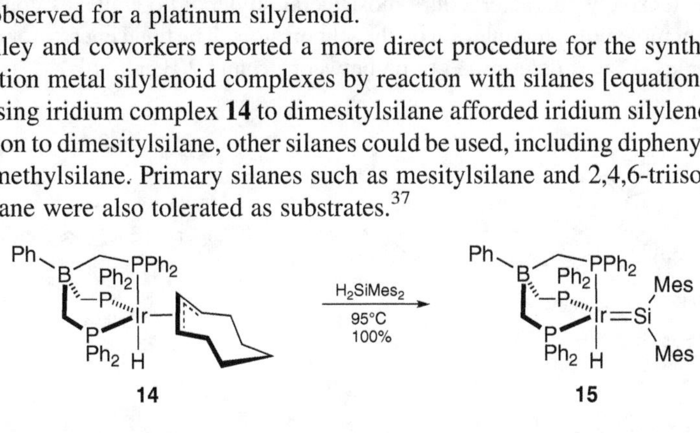

(7.1)

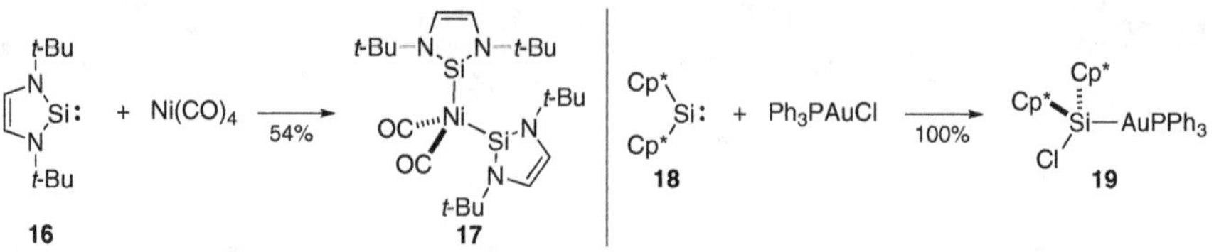

Scheme 7.3. Generation of transition metal silylenoid complexes from stable silylenes.

The coordination of stable silylenes to metal complexes was also reported to produce transition metal silylenoids.[40,41] Exposure of unsaturated silylene **16** to $Ni(CO)_4$ resulted in the formation of the disilylene-substituted tetrahedral nickel complex **17** (Scheme 7.3).[42,43] Similarly, mixing Ph_3PAuCl with decamethylsilicocene **18** produced the silylgold complex **19**.[44] The $^{29}Si\ \{^1H\}$ NMR spectrum of **19** (δ 78 ppm) revealed its silylenoid character. In addition to nickel and gold, other metals, including tungsten,[41] platinum,[45] iron,[40] and ruthenium,[46] have been utilized to form silylmetal complexes of stable silylenes.

Metal-catalyzed silylene extrusion from hexamethylsilirane **20** was also used to generate transition metal silylenoids [equation (7.2)].[47] In the presence of a metalloporphyrin, $[Os(TTP)]_2$ (where TTP = the dianion of *meso*-tetra-*p*-tolylporphyrin), hexamethylsilirane **20** extruded dimethylsilylene to produce the THF-stabilized osmium silylenoid complex **21**. Despite the coordinated THF molecule, crystallographic analysis of the silylosmium complex **21** revealed the sp^2 nature of the silicon atom (Os–C–Si bond angles 121° and 117°). Osmium silylenoid **21** was also synthesized from the reaction of $[Os(TPP)]^{2-}$ and dimethyldichlorosilane.[47]

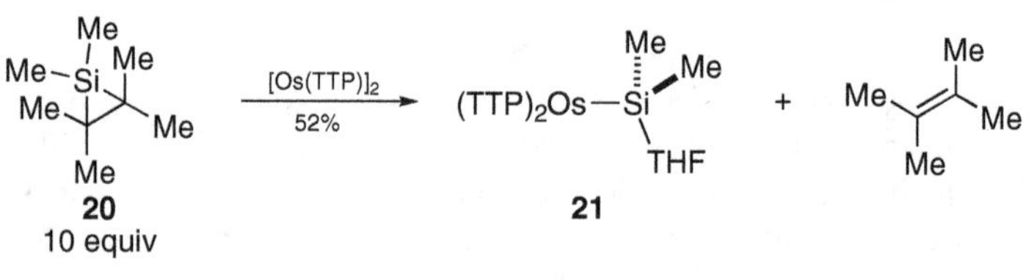

(7.2)

The structure and bonding of metal silylene complexes varied from those of their carbon analogs. While Fischer-type metal carbene complexes without solvent adducts have been extensively characterized,[48] most metal silylenoid complexes contain a bound solvent molecule or counterion on the silicon atom. The bond energy for donor silicon complex **22** was determined to be between 15 and 20 kcal/mol:[49,50]

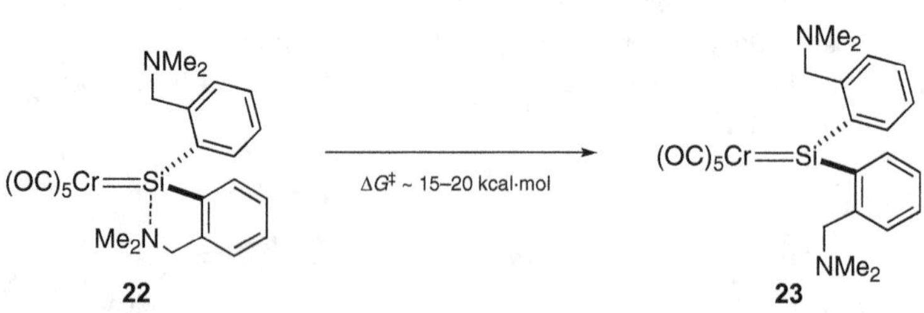

(7.3)

The bonding picture of a metal with a divalent silicon atom emerged through computational studies to explain the need for the incorporation of a Lewis base at silicon.[51,52] Similar to the bonding scheme of a metal carbene, the Fischer-type silylene $(OC)_5Mo{=}SiH_2$ contains an *sp* σ donor and *p* π acceptor. Silicon differs from carbon in that π bonding is less efficient. This less efficient π bonding leads to a high polarity of the metal silicon double bond and a nearly barrier-free rotation.[51]

7.2.2 Reactivity of Transition Metal Silylenoids

The reactivity of early transition metal silylenoid complexes is still emerging. An example of the chemistry that these complexes can participate in is the sila-Wittig reaction [equation (7.4)].[53] In this transformation, a metathesis occurred between the chromium silylenoid **24** and the dimethyl carbonate to afford a new Schrock carbene, **25**, and the trimerized product **26**. This methodology allowed access a new carbene complex that eluded previous synthetic efforts.

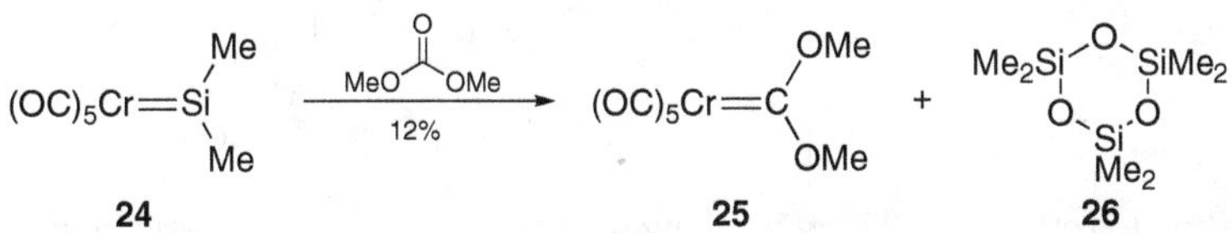

(7.4)

The reactivity of platinum silylenoid **27** was explored with traditional silylene trapping reagents. While the silylenoid did not react with triethylsilane or 2,3-dimethyl-1,3-butadiene, phenylacetylene was a viable substrate, providing the metallocyclohexadiene **29** (Scheme 7.4).[54] The formation of platinum complex **29** was hypothesized to occur via platinum cyclobutene intermediate **28**, which formed on insertion of the acetylene into the platinum–silicon bond. A second molecule of phenylacetylene was then inserted into the remaining platinum–silicon bond to provide the observed product.

7.2.3 Transition Metal Silylenoid Complex–Catalyzed Hydrosilation Reactions

Cationic ruthenium silylenoid complexes function as efficient hydrosilation catalysts. Tilley and Glaser reported that 1 mol% of **30** catalyzed the hydrosilation of

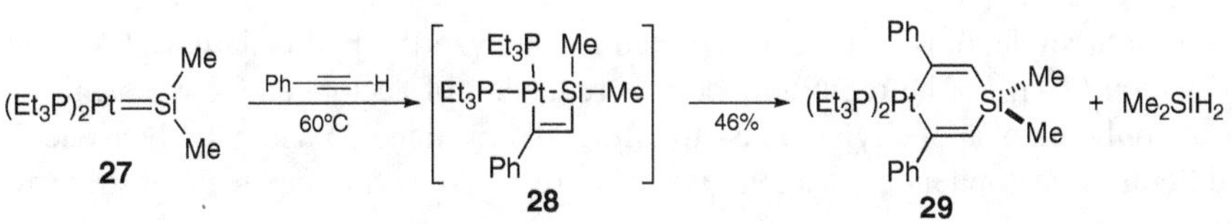

Scheme 7.4. Reactivity of platinum dimethylsilylenoid toward phenylacetylene.

1-methylcyclohexene with phenylsilane [equation (7.5)].[55] This process occurred in a diastereo- and regioselective fashion to produce silane **31** as a single isomer. Other petroleum adducts, including 1-hexene, ethene, and styrene, were efficiently hydrosilated. In addition to phenylsilane, *n*-hexylsilane could be used without reduction of catalytic efficiency. While this reaction tolerated mono-, di-, or trisubstituted olefins, it was limited to primary silanes; no reaction was observed with diethyl- or diphenylsilane, and the hydrosilation product (e.g., **31**) was not converted to the tertiary silane.

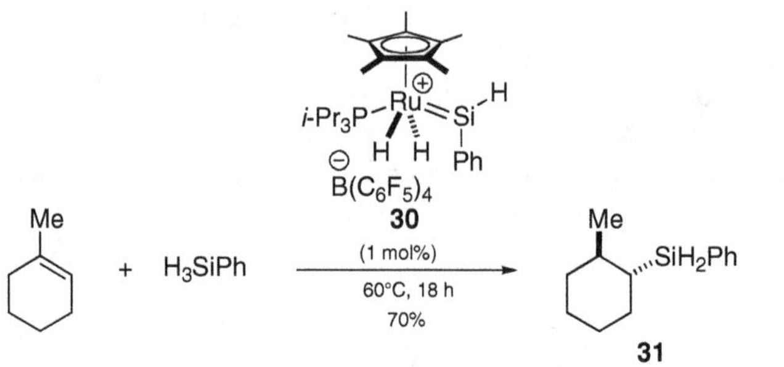
(7.5)

In addition to ruthenium, Tilley and coworkers also reported that cationic iridium silylenoid complexes were efficient olefin hydrosilation catalysts [reaction (7.6)].[56] This silylene complex catalyzes the hydrosilation of unhindered mono- or disubstituted olefins with primary silanes to produce secondary silanes with *anti* Markovnikov selectivity. Iridium catalyst **32** exhibited reactivity patterns similar to those of ruthenium **30**; only primary silanes were allowed as substrates. In contrast to **30**, cationic iridium **32** catalyzed the redistribution of silanes. Exposing phenylsilane to 5 mol% of **32** in the absence of olefin produced diphenylsilane, phenylsilane, and silane.

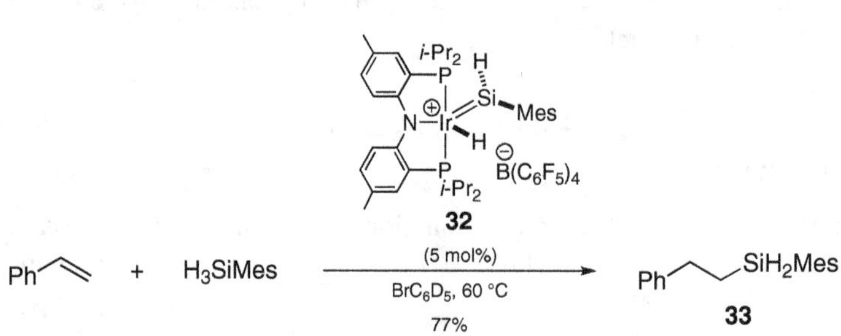
(7.6)

Iridium silylenoids were also reported to catalyze the hydrosilation of ketones [equation (7.7)].[57] The room temperature reduction of acetophenone was achieved using only 5 mol% of silylenoid **34** to afford alkoxysilane **35** and ethyl benzene. In addition to acetophenone, acetone was also reported as a competent substrate for hydrosilation.

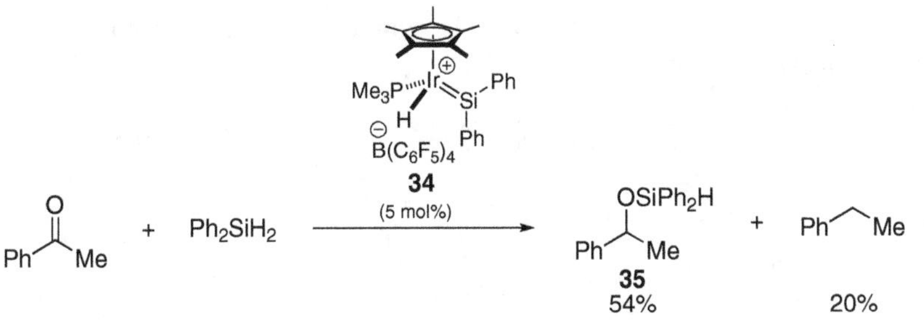

(7.7)

7.2.4 Transition Metal Silylenoid–Catalyzed Atom Transfer Reactions

While trisubstituted silanes are inactive reagents for the above-mentioned hydrosilation processes, under appropriate conditions they can be sources of silylene. Disilanes, in particular, form silylenoids on exposure to platinum complexes and carbonyl compounds (Scheme 7.5).[58] Hayashi and Okamoto reported that exposure of pentamethyldisilane ($HSiMe_2SiMe_3$) to 3 mol% of (cod)$PtCl_2$ and β,β-dimethyl-substituted phenylenone **36** produced a 10 : 1 mixture of oxasilacyclohexene **38** and oxasilacyclopentene **39**. While the process works well for β,β-disubstituted phenylenones, replacement of one of the β-methyl groups with a hydrogen atom changes the product distribution to favor formation of the oxasilacyclopentene (**42** or **43**). Higher yields of oxasilacyclopentene were obtained when pentaphenyldisilane **41** was used as the source of silylene.[59]

Tetrasubstituted silanes are also sources of silylene. Suginome and coworkers reported that palladium catalyzed the transfer of dimethylsilylene, formed from silylborane **44**, to alkynes [equation (7.8)].[60] Exposure of silylborane **48** and alkyne **49** to substoichiometric amounts of palladium and P(t-Bu)$_2$(2-biphenyl) afforded 2,4-disubstituted silole **50**. This process tolerates a variety of functionality including silyl ether-, dimethylamino-, and trifluoromethyl groups. In addition to aliphatic terminal acetylenes, arylacetylenes were also competent substrates. For

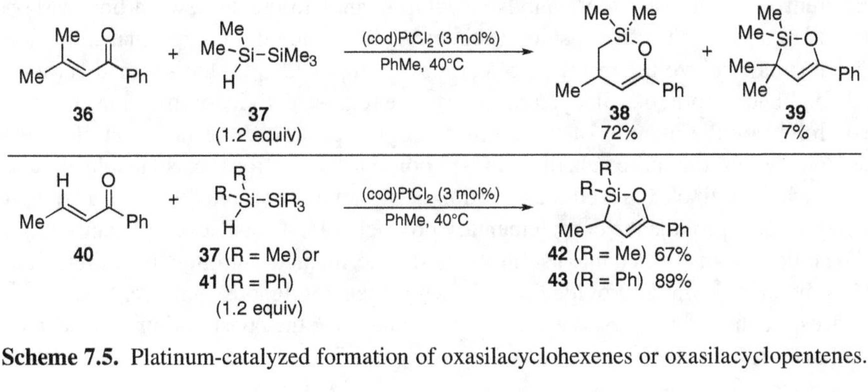

Scheme 7.5. Platinum-catalyzed formation of oxasilacyclohexenes or oxasilacyclopentenes.

arylacetylenes, triphenylphosphine could be employed as the phosphine ligand without attenuation of the regioselectivity.

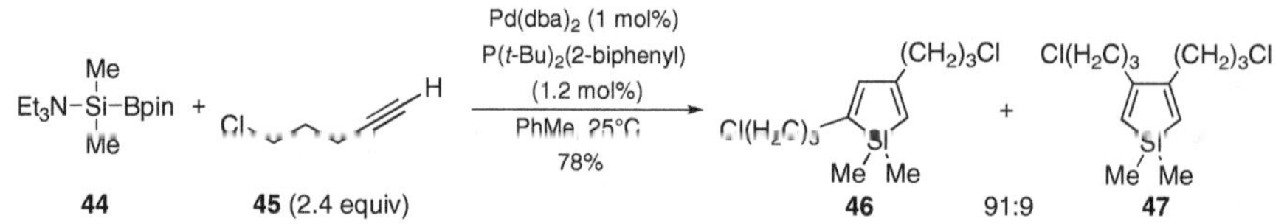

(7.8)

Palladium(0) complexes also catalyze the transfer of di-*tert*-butylsilylene from *cis*-dimethylsilacyclopropane **48** to alkynes (Scheme 7.6).[61] In the presence of 5 mol % of $(Ph_3P)_2PdCl_2$, the formation of dimethylsilacyclopropene **51** could be achieved at 110°C. In the absence of the palladium catalyst, silylene transfer from silacyclopropane **48** to 3-hexyne occurred over 3 days at 130°C. While a reasonable mechanism was proposed involving palladacycle **49** as an intermediate, an alternative mechanism could involve palladium silylenoid intermediate **50**.

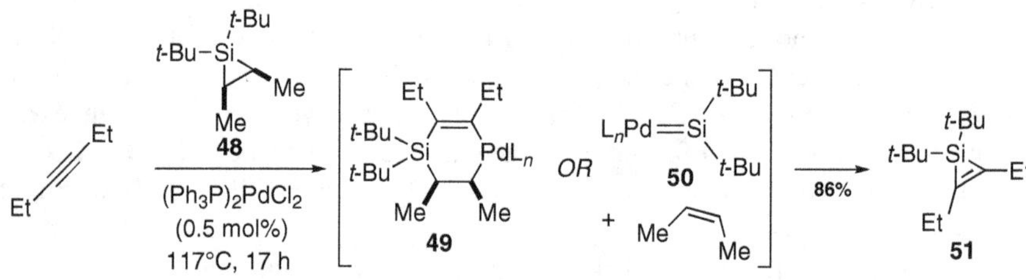

Scheme 7.6. Palladium-catalyzed di-*tert*-butylsilylene transfer to an internal alkyne.

7.3 SILACYCLOPROPANES AS IMPORTANT SYNTHETIC INTERMEDIATES

To develop new methods for organic synthesis, Woerpel and coworkers exploited the inherent reactivity of di-*tert*-butylsilacyclopropanes to create new carbon–carbon bonds in a stereoselective fashion (Scheme 7.7).[62] They discovered that transition metal salts catalyze the insertion of carbonyl compounds into the strained carbon–silicon bond to form oxasilacyclopentanes. The regioselectivity of insertion could be controlled by the identity of the catalyst. Copper promoted the insertion of crotonaldehyde into the more substituted C–Si bond of **52** to afford oxasilacyclopentane **53**,[63] whereas zinc catalyzed the insertion of butyraldehyde into the less substituted bond of **52** to provide the complementary product, **54**.[64] Oxasilacyclopentanes (e.g., **55**) could be transformed into useful synthetic intermediates through oxidation of the C–Si bond,[65,66] which provided diol **56** with three contiguous stereocenters.

Despite these advances, the conditions required for the construction of silacyclopropanes limited their synthetic utility (Scheme 7.8). These heterocycles can be

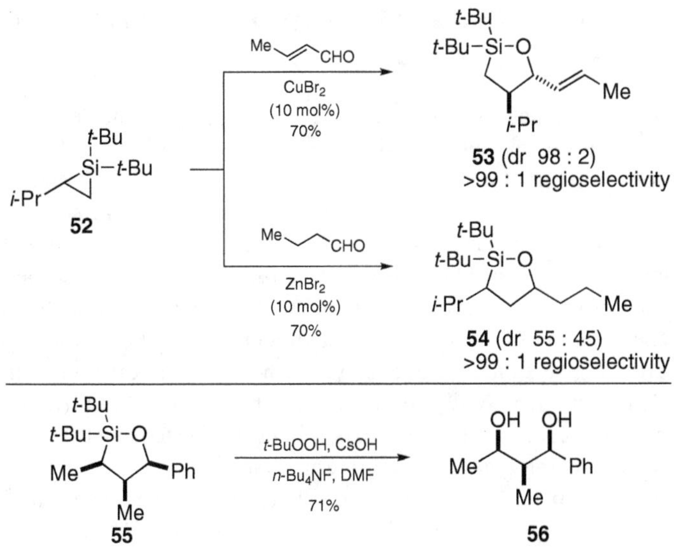

Scheme 7.7. Synthetic utility of silacyclopropanes.

prepared by reaction of an olefin with a silylene or a silylenoid species. Strongly reducing conditions [Li(0) and t-Bu_2SiCl_2] are needed to access the requisite lithium silylenoid (**57**).[67] Alternatively, free silylene **60** could be generated thermally or photochemically from cyclic silanes, such as **58** or **59**.[12j,68–73] Each of these methods, however, limited the functionality that could be introduced on the silacyclopropane. To overcome these limitations, Woerpel and coworkers set out to develop conditions that generated a silylenoid species at low temperatures and exhibited broad functional group tolerance.

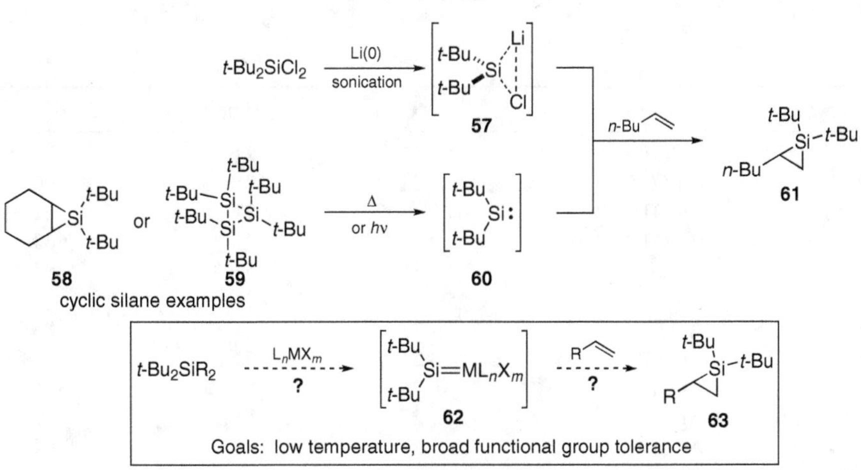

Scheme 7.8. Synthesis of silacyclopropanes.

7.4 SILVER-MEDIATED TRANSFER OF DI-*tert*-BUTYLSILYLENE TO OLEFINS

Exploration of the reactivity of cyclohexene silacyclopropane led Woerpel and coworkers to discover that the inclusion of metal salts enabled silylene transfer to monosubstituted olefins at reduced temperatures (Table 7.1).[11,74] A dramatic reduction in the temperature of transfer was observed when cyclohexene silacyclopropane was exposed to copper, silver, or gold salts. Silver salts were particularly effective at decomposing **58** (entries 6–11). The use of substoichiometric quantities of silver triflate enabled *n*-hexene silacyclopropane **61** to be formed quantitatively at −27°C (entry 6). The identity of the counterion did affect the reactivity of the silver salt. In general, better conversions were observed when noncoordinating anions were employed. While the reactivity differences could be attributed to the solubility of the silver salt in toluene, spectroscopic experiments suggested that the anion played a larger role in stabilizing the silylenoid intermediate.

Under these optimized conditions, di-*tert*-butylsilylene could be transferred to a range of acyclic and cyclic olefins (Schemes 7.9 and 7.10).[11,74] The method was not sensitive to the steric nature of the R substituent; nearly quantitative silylene transfer to olefins bearing *n*-butyl, isopropyl, or *tert*-butyl groups was observed. Vinylsilanes were also tolerated as substrates. Olefins containing silyl ether, benzyl ether, and pivolate substituents were all effective traps of di-*tert*-butylsilylene.

Silver-catalyzed silylene transfer to disubstituted olefins was also possible (Scheme 7.10).[11,74] The transformation was stereospecific; *cis*-2-butene afforded *cis*-dimethylsilacyclopropane **66a** and *trans*-2-butene generated *trans*-dimethylsila-

TABLE 7.1. Effect of a Metal Salt on the Temperature of Di-*tert*-Butylsilylene Transfer

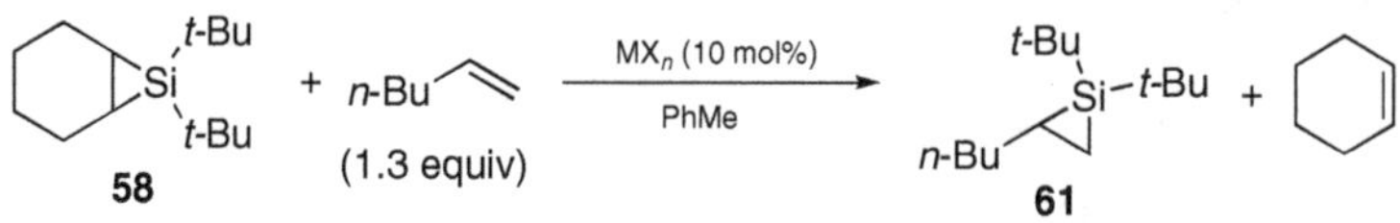

Entry	MX_n	T (°C)	Time	Yield (%)[a]
1	$Zn(OTf)_2$	55	12 h	60
2	$Cu(OTf)_2$	25	17 min	96
3	$Cu(OTf)_2$	0	2 h	62
4	$(CuOTf)_2{\cdot}PhH$	0	2 h	62
5	$CeCl_3$	25	15 h	87
6	AgOTf	−27	2 h	90
7	AgO_2CCF_3	−27	2 h	86
8	$AgSbF_6$	−27	5 h	>95
9	AgOBz	0–25	8 h	>95
10	$AgBF_4$	25	45 min	99
11	Ag_3PO_4	25	16 h	94

[a] As determined using H^1 NMR spectroscopy.

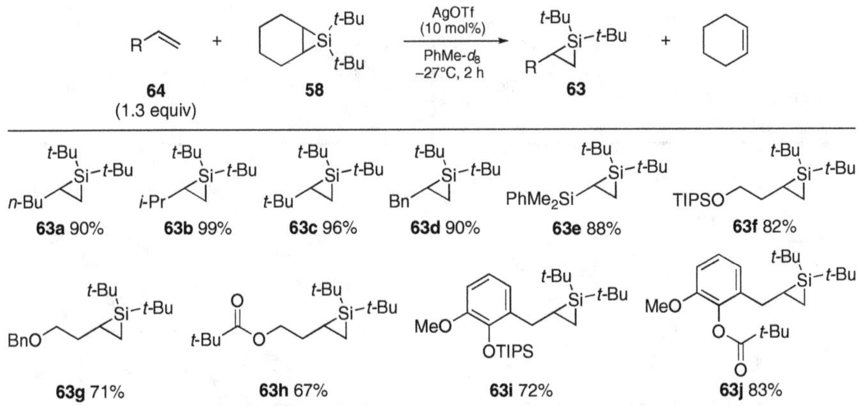

Scheme 7.9. Silver-catalyzed di-*tert*-butylsilylene transfer to monosubstituted olefins.

cyclopropane **66b** as single isomers. Di-*tert*-butylsilylene transfer was also diastereoselective. Exposure of 2-substituted cyclopentene **65d** or α-pinene to reaction conditions afforded **66d** and **66e** as single diastereomers. Diastereoselective silylene transfer was not limited to cyclic olefins; silacyclopropane **66f** was formed from **65f** as a 9 : 1 mixture of diastereomers.

One drawback of this methodology was the sensitivity of the product silacyclopropane to the reaction conditions, which prevented isolation of the product. Woerpel and coworkers were able to overcome this debilitating limitation by changing the anion of silver catalyst to trifluoroacetate and reducing the catalyst load to <1 mol% (Scheme 7.11).[74] The combination of these two modifications enabled the isolation of the silacyclopropane product after careful Schlenk filtration. Isolation of the silacyclopropane allowed for either zinc-[64,75] or copper-mediated functionalization of the silacyclopropane to produce 1,2- or 1,3-disubstituted oxasilacyclopentanes.[63]

Alternatively, the silacyclopropane could be functionalized in situ through the addition of 30 mol% of zinc bromide (or zinc iodide) and a carbonyl compound to the reaction mixture, which produced oxasilacyclopentane **70** (Scheme 7.12).[11,74]

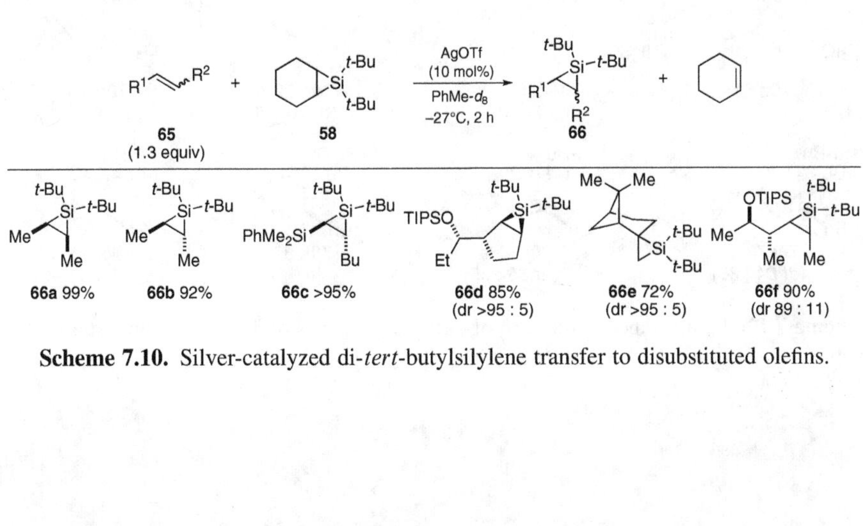

Scheme 7.10. Silver-catalyzed di-*tert*-butylsilylene transfer to disubstituted olefins.

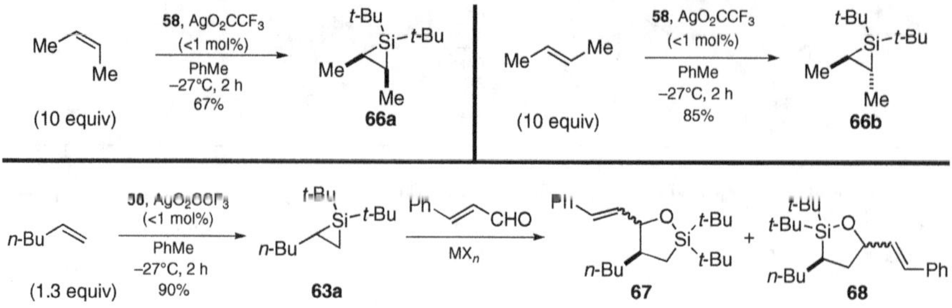

MX_n	Yield, (%)	67 : 68
CuI	85%	96 : 4
$ZnBr_2$	92%	1 : 99

Scheme 7.11. Formation of either 1,2- or 1,3-disubstituted oxasilacyclopentanes.

This functionalization was limited to the formation of 1,3-substituted oxasilacyclopentanes, as exposure of the in situ–generated silacyclopropane to substoichiometric amounts of copper salts did not produce the complementary 1,2-disubstituted oxasilacyclopentanes.

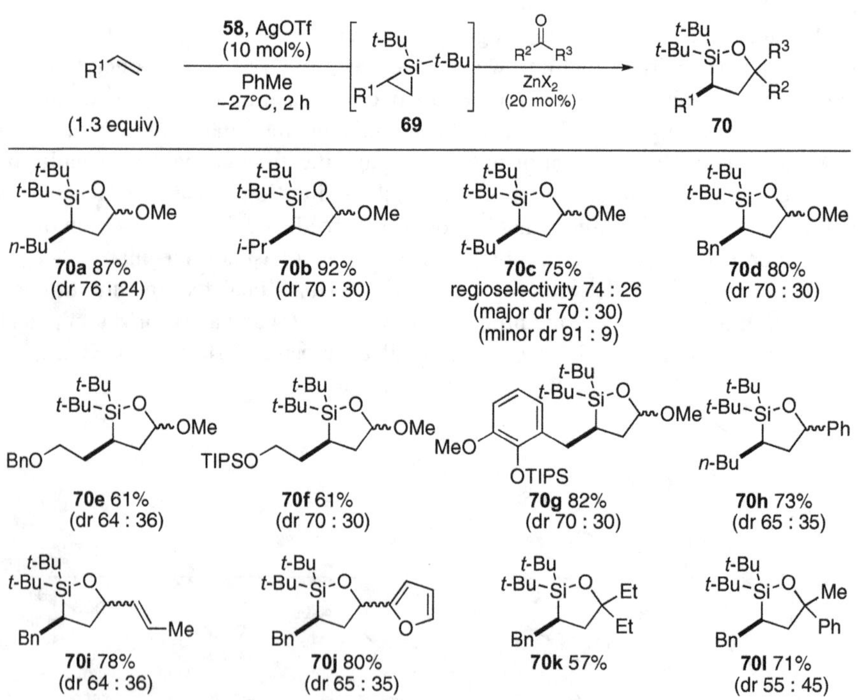

Scheme 7.12. In situ functionalization of silacyclopropanes derived from monosubstituted olefins.

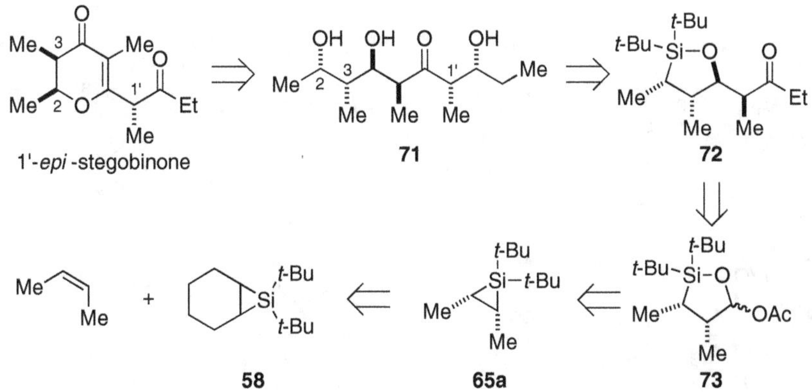

Scheme 7.13. Retrosynthesis of 1′-*epi*-stegobinone.

The synthetic utility of silacyclopropanes was illustrated in the total synthesis of 1′-*epi*-stegobinone (Scheme 7.13).[76] Woerpel and coworkers envisioned that every stereocenter in this pheromone[77] could be derived from the stereochemistry of the starting *cis*-dimethylsilacyclopropane **65a**. They believed that 1′-*epi*-stegobinone could be synthesized from polypropionate **71**. This fragment could be produced from an *anti,syn*-selective aldol reaction of ethyl ketone **72**, which is available from the functionalization of oxasilacyclopentane **73**. This oxasilacyclopentane is readily available from silacyclopropane **65a**, produced from di-*tert*-butylsilylene transfer from **58** to *cis*-2-butene. Implementation of the Woerpel methodology at the outset of the total synthesis would illustrate the power of the methodology to move large amounts of material through the silicon atom transfer step.

Gram quantities of oxasilacyclopentane **73** were prepared from *cis*-2-butene through silver-catalyzed di-*tert*-butylsilylene transfer from cyclohexene silacyclopropane **58** followed by copper-mediated formamide insertion (Scheme 7.14). After acetolysis, the resulting oxasilacyclopentane (**73**) participated in a nucleophilic substitution reaction with silyl enol ether **74** to produce ethyl ketone **72** diastereoselectively.[62,78–80] After protection of the ketone, the hindered C–Si bond was oxidized using conditions developed by Woerpel and coworkers to afford **75**.[66,81,82] The resulting polypropionate fragment was elaborated in 10 steps to 1′-*epi*-stegobinone. This first report of using silacyclopropanes as synthetic intermediates in total synthesis clearly demonstrates their potential as polyproprionate synthons.

Insight into the mechanism of silver-catalyzed silylene transfer from cyclohexene silacyclopropane to an olefin was obtained using bistriphenylphosphine silver triflate as a catalyst.[83] Woerpel and coworkers chose to employ ancillary ligands on silver to address the both the poor solubility of silver triflate as well as its propensity to decompose to afford a silver(0) mirror or precipitate. The addition of triphenylphosphine, however, attenuated the reactivity of the silver catalyst. For example, the reaction temperature needed to be raised from −27°C to 10°C to obtain a moderate rate of reaction (Scheme 7.15).

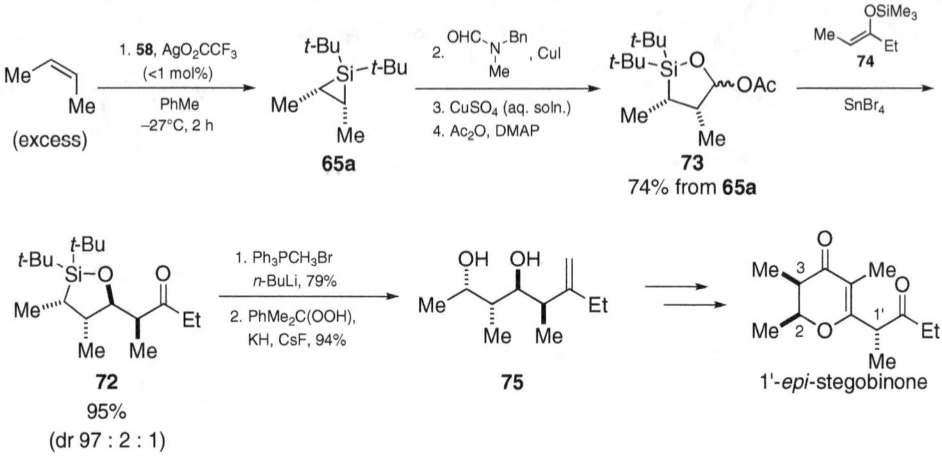

Scheme 7.14. Total synthesis of 1′-*epi*-stegobinone.

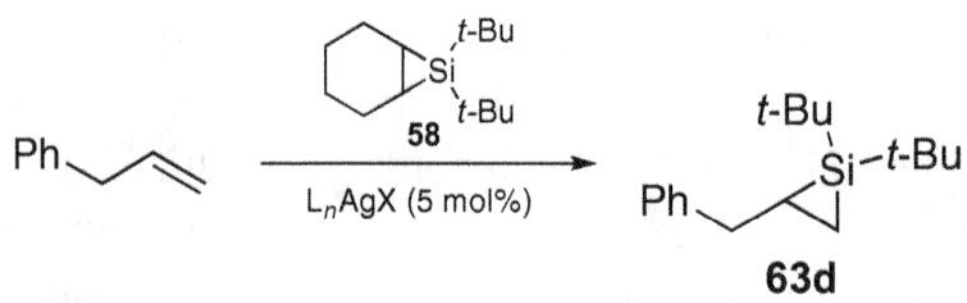

L_nAgX	T (°C)	$t_{1/2}$ (approx.)
AgOTf	−27	24 min
$(Ph_3P)_2AgOTf$	10	60 min

Scheme 7.15. Effect of phosphine ligand on rate of silylene transfer.

The relative reactivities of various silacyclopropanes were compared to gain insight into the reversibility of the reaction (Scheme 7.16).[83] Woerpel and coworkers observed that the reactivity of the silacyclopropane toward the silver catalyst depended on the ring size, and that cyclohexene silacyclopropane was the most reactive. Notably, benzyl-substituted silacyclopropane **63d** did not react when exposed to the silver complex. This lack of reactivity was interpreted as evidence in support of the irreversibility of silylene transfer to monosubstituted olefins.

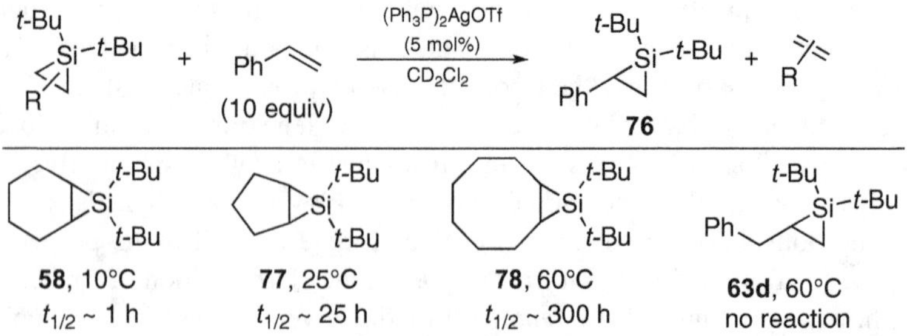

Scheme 7.16. Relative reactivities of silacyclopropanes.

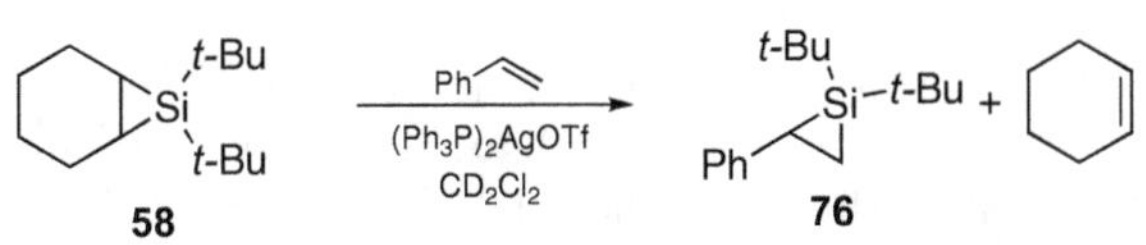

$\Delta H^‡ = 30(1)$ kcal/mol $\Delta S^‡ = 31(7)$ eu

Component	Kinetic order
58	1
styrene	<1
$(Ph_3P)_2AgOTf$	Saturation
Ph_3P	<1
Cyclohexene	<1

Scheme 7.17. Kinetic studies on silver-catalyzed silylene transfer.

To elucidate the mechanism of silver-catalyzed silylene transfer, kinetic studies were performed by Woerpel and coworkers (Scheme 7.17).[83] The reaction of cyclohexene silacyclopropane **58** and styrene in the presence of 5 mol% of $(Ph_3P)_2AgOTf$ was followed using ^1H NMR spectroscopy. The kinetic order in cyclohexene silacyclopropane **58** was determined to be 1. In contrast to the rate acceleration observed with increasing the concentration of **58**, inhibition of the rate of the reaction was observed when styrene, cyclohexene, or triphenylphosphine concentrations were increased. Saturation kinetic behavior in catalyst concentration was observed. Activation parameters were determined to be $\Delta H^‡ = 30(1)$ kcal/mol and $\Delta S^‡ = 31(7)$ eu (entropy units). Similar activation parameters were observed in toluene-d_8, which suggests that the mechanism is not dependent on the solvent.

A catalytically active silylsilver intermediate was observed using low-temperature ^{29}Si NMR and IR spectroscopy (Scheme 7.18).[83] Exposure of cyclohexene silacyclopropane **58** to the silver complex produced cyclohexene as well as a new species, which exhibited two doublets at 97 ppm ($J_{AgSi} = 260$ and 225 Hz) in the $^{29}Si\ \{^1H\}$ NMR spectrum. The $^{29}Si\ \{^1H\}$ NMR spectra of this species are consistent with a Lewis base–stabilized metal silylenoid. Tilley and coworkers have reported that $(Et_3P)_2Pt(H)$–$Si(St\text{-}Bu)_2(OTf)$ appears at 52 ppm,[39] and Jützi et al. observed

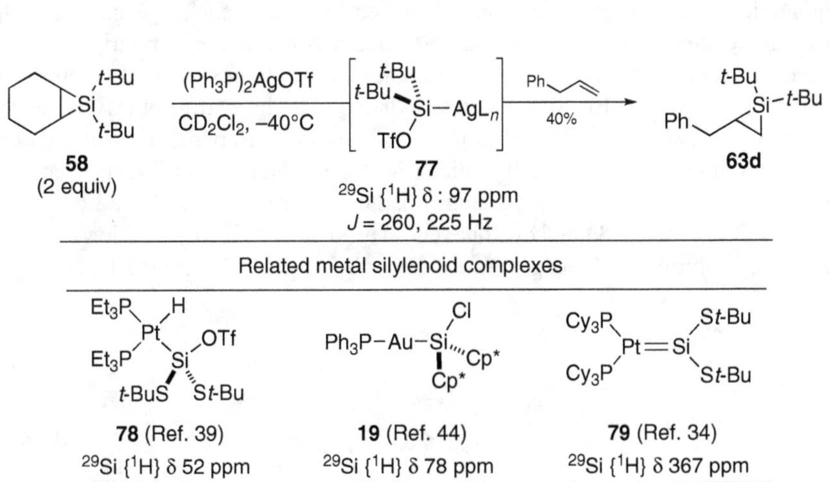

Scheme 7.18. Spectroscopic attributes of silylsilver intermediate **77**.

$Ph_3PAu–SiCp_2Cl$ at 78 ppm.[44] In contrast, neutral silylenoids appear much further downfield: Tilley et al. reported that $(Cy_2P)_2Pt{=}SiMes_2$ appeared at 367 ppm.[34] Infrared spectroscopy was also used to gain insight into the structure of the intermediate as ion-paired triflate salts exhibit stretches at 1270 and 1043 cm^{-1},[84] whereas covalent t-$BuMe_2SiOTf$ produces a spectrum with peaks at 1212 and 970 cm^{-1}. Low-temperature IR spectroscopy of the intermediate in CH_2Cl_2 revealed strong stretches at 1204 and 972 cm^{-1}, which is consistent with a monodentate binding of the triflate anion to the silicon atom. On the basis of these observations, the structure of the silver silylenoid intermediate was assigned to be silylsilver **77**.

The electronic nature of silylsilver intermediate was interrogated through intermolecular competition experiments between substituted styrenes and the silylsilver intermediate (**77**).[83] The product ratios from these experiments correlated well with the Hammett equation to provide a ρ value of -0.62 using σ_p constants (Scheme 7.19). Woerpel and coworkers interpreted this ρ value to suggest that this silylsilver species is electrophilic. Smaller ρ values were obtained when the temperature of the intermolecular competition reactions was reduced [$\rho = -0.71$ (8°C) and -0.79 (-8°C)]. From these experiments, the isokinetic temperature was estimated to be 129°C, which meant that the product-determining step of silver-catalyzed silylene transfer was under enthalpic control. In contrast, related intermolecular competition reactions under metal-free thermal conditions indicated the product-determining step of free silylene transfer to be under entropic control. The combination of the observed catalytically active silylsilver intermediate and the Hammett correlation data led Woerpel and colleagues to conclude that the silver functions to both decompose the sacrificial cyclohexene silacyclopropane as well as transfer the di-*tert*-butylsilylene to the olefin substrate.

Woerpel and coworkers analyzed the experimental data and constructed a catalytic cycle to describe silver-catalyzed di-*tert*-butylsilylene transfer from cyclohexene silacyclopropane to monosubstituted olefins (Scheme 7.20).[83] Dissociation of the oligomeric silver phosphine complex followed by ligand exchange generates the postulated catalytically active species, **81**, which contains only one triphenylphosphine and two styrene ligands. Coordination of cyclohexene silacyclopropane followed extrusion of t-Bu_2Si, generates silylsilver **83**. This extrusion was postulated to occur via transition state **82**. This transition state forms on coordination of triflate to the electrophilic silicon atom of **58**, which triggers the heterolysis of the strained C–Si bond in a transmetallation-like mechanism. This heterolysis is followed by β-silyl elimination to produce **83** and cyclohexene. Turnover-limiting cyclization occurs to form silacyclopropane **76** and generate the coordinatively unsaturated silver catalyst.

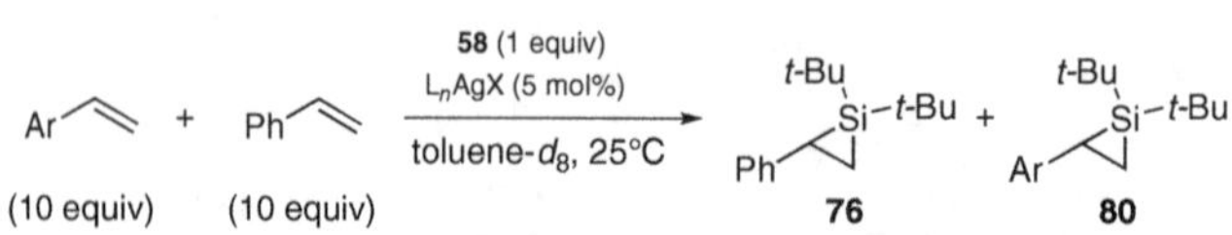

L_nAgX	ρ value
AgOTf	−0.59(2)
$(Ph_3P)_2AgOTf$	−0.62(2)

Scheme 7.19. Hammett correlation of intermolecular competition experiments.

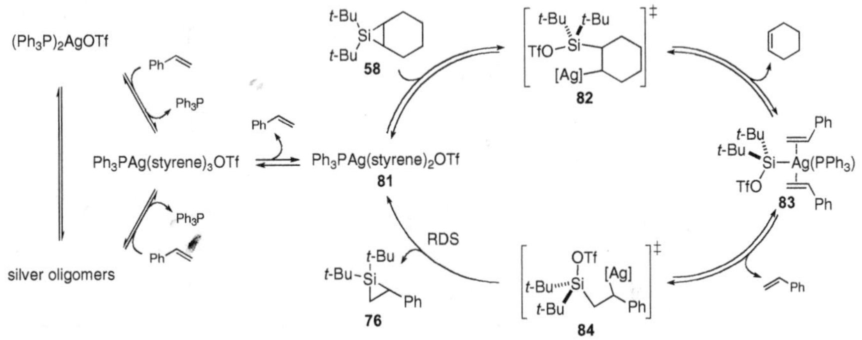

Scheme 7.20. Potential catalytic cycle for silver-mediated di-*tert*-butylsilylene transfer from cyclohexene silacyclopropane **58** to styrene.

An unexpected product containing the incorporation of two molecules of silylene was observed in the reaction of homoallylic ethers at elevated temperatures (Scheme 7.21).[85] At reduced temperatures, the reaction of homoallylic ethers with cyclohexene silacyclopropane **58** produced the expected silacyclopropane (Scheme 7.9, cf. **63g** and **63h**). Increasing the temperature, however, resulted in a rearrangement to afford allylsilane **86**. This transformation proved general and a range of silylmethyl allylic silanes could be accessed from mono- or geminal disubstituted homoallylic ethers. The resulting allylic silane was demonstrated to be a competent nucleophile through reaction with *N*-chlorosulfonyl isocyanate.

The mechanism of this transformation was probed using regiospecifically labeled substrates and silacyclopropanes (Scheme 7.22).[85] No crossover was observed in the reaction of **87** and **88**. Further, incorporation of deuterium into the silane and α position revealed that reorganization of the substrate had occurred. Monosubstituted silacyclopropanes were established as potential reactive intermediates by readily reacting in the presence of cyclohexene silacyclopropane **58** to give the silylmethylsilane **93**. Control experiments established that these monosubstituted silacyclopropanes did not extrude di-*tert*-butylsilylene under reaction conditions.

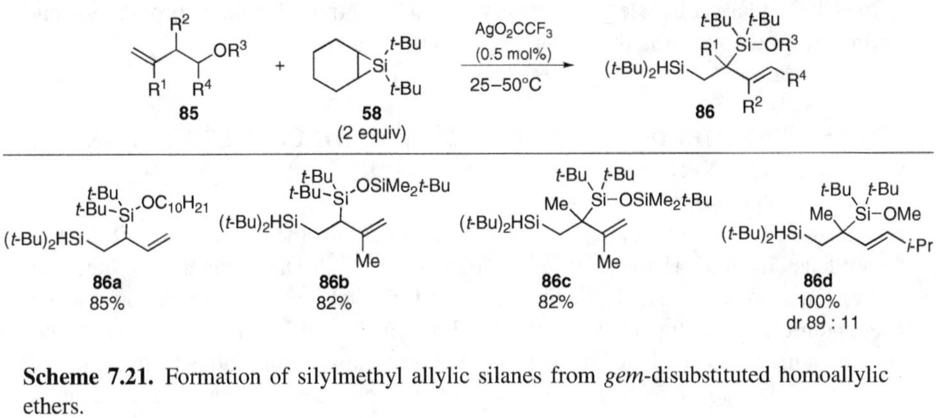

Scheme 7.21. Formation of silylmethyl allylic silanes from *gem*-disubstituted homoallylic ethers.

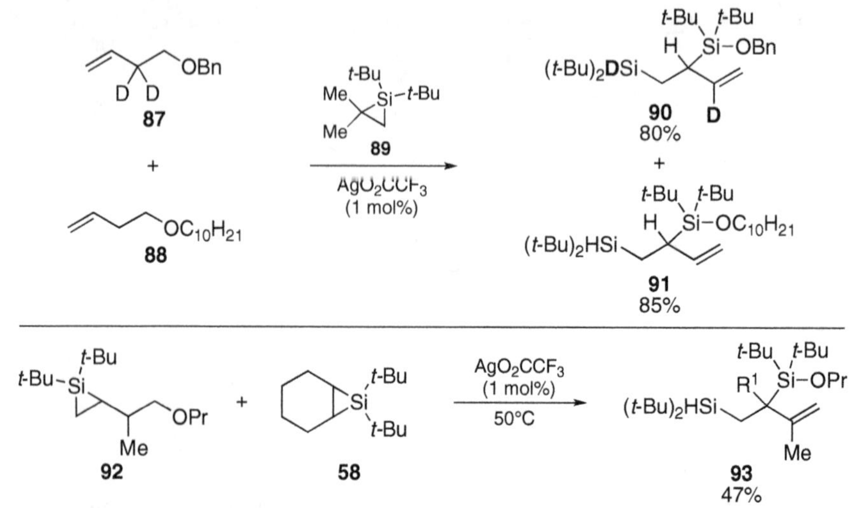

Scheme 7.22. Mechanistic experiments.

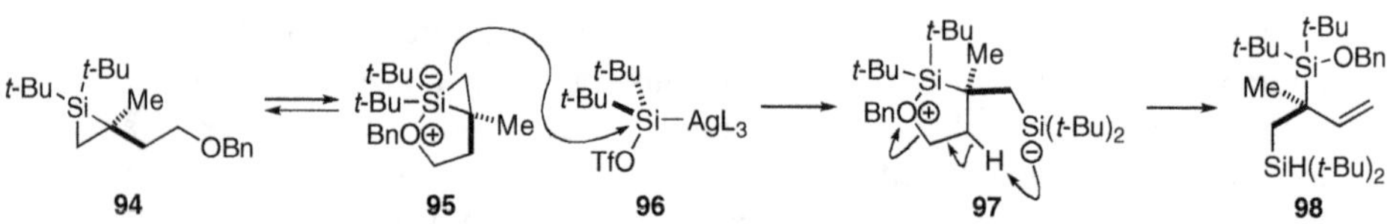

Scheme 7.23. Potential mechanism for formation of silylmethyl allylic silanes from *gem*-disubstituted homoallylic ethers.

From these observations, Woerpel and Cleary proposed a mechanism to account for allylic silane formation (Scheme 7.23).[85] Silacyclopropane **94** is formed from cyclohexene silacyclopropane **58** through silylene transfer. Coordination of the Lewis basic benzyl ether to the electrophilic silicon atom[86–88] generates pentacoordinate siliconate **95** and increases the nucleophilicity of the apical Si-C bond.[89] Electrophilic attack by silylsilver triflate **96** forms silyl anion **97**. Intramolecular deprotonation and elimination then affords the silylmethyl allylic silane.

7.5 SILVER-MEDIATED TRANSFER OF DI-*tert*-BUTYLSILYLENE TO ACETYLENES

Transition metal complexes were known to facilitate the addition of silylene to acetylenes from a variety of different sources.[60,61,90,91] These conditions, however, often required heating, and the initially formed silacyclopropene often incorporated a second molecule of the acetylene to afford a silole.[92,93] With their discovery of low-temperature silver-mediated di-*tert*-butylsilylene transfer conditions from cyclohexene silacyclopropane **58** to olefins, Woerpel and coworkers set out to investigate the

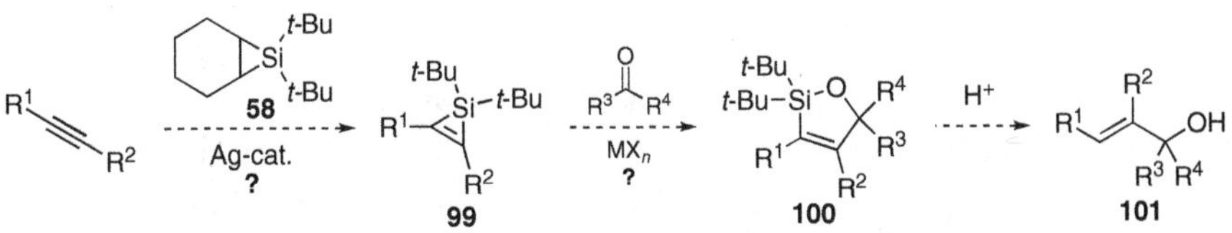

Scheme 7.24. Di-*tert*-butylsilylene transfer to alkynes.

possibility of transferring di-*tert*-butylsilylene to mono- and disubstituted acetylenes at low temperatures (Scheme 7.24). If conditions were found to prevent insertion of a second molecule of alkyne, then the resulting silacyclopropenes might be functionalized in situ with a ketone or an aldehyde to produce oxasilacyclopentenes.[94] Protonolysis of the C–Si bond would produce allylic alcohol **101**. Formally, the resulting method would represent a three-component reductive coupling reaction to produce an allylic alcohol from an alkyne and a carbonyl compound.

Woerpel and Clark identified silver phosphate as the optimal catalyst to promote di-*tert*-butylsilylene transfer from cyclohexene silacyclopropane to a variety of substituted alkynes (Scheme 7.25).[95] While this silver salt exhibited attenuated reactivity as compared to silver triflate or silver trifluoroacetate, it exhibited greater functional group tolerance. Both di- and monosubstituted silacyclopropenes were easily accessed. Terminal alkynes are traditionally difficult substrates for silylene transfer and typically insert a second molecule of the starting acetylene.[61,90–93] Consequently, the discovery of silver-mediated silylene transfer represents a significant advance as it enables further manipulation of monosubstituted silacyclopropenes. For enyne substrates, silylene transfer the alkynyl group was solely observed. The chemoselectivity of the formation of **99f** was attributed to ring strain as theoretical calculations suggest that silacyclopropenes are less strained than silacyclopropanes.[96,97]

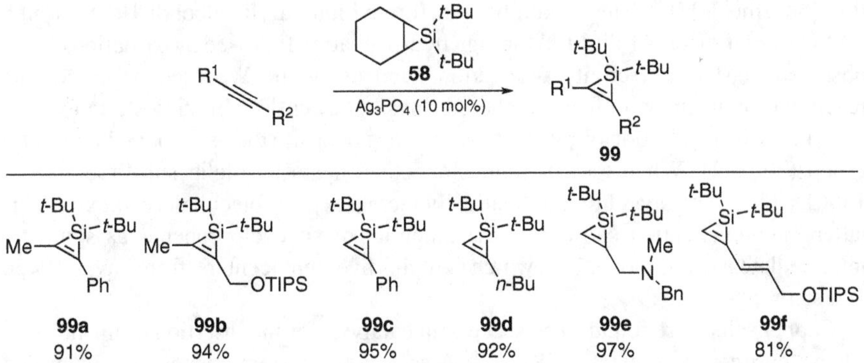

Scheme 7.25. Silver-catalyzed di-*tert*-butylsilylene transfer to di- or mono-substituted alkynes.

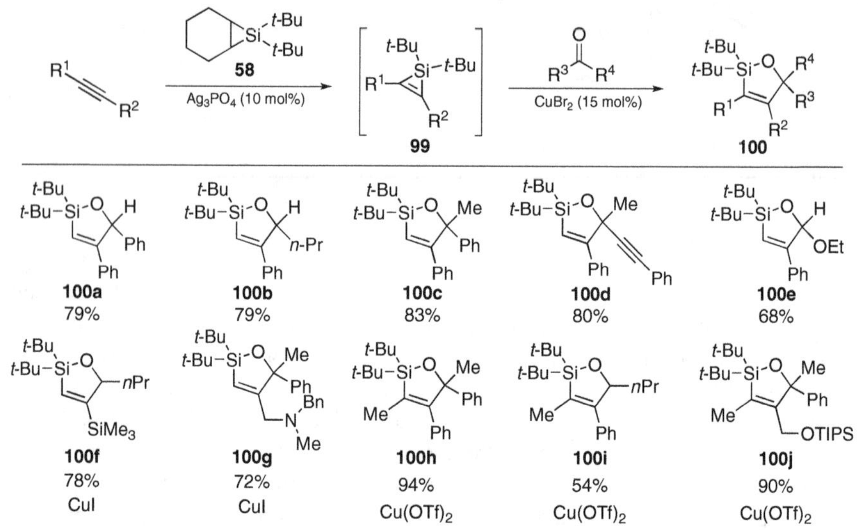

Scheme 7.26. Regioselective in situ functionalization of silacyclopropene.

The resulting silacyclopropenes were particularly reactive and challenging to isolate. To overcome this limitation, Clark and Woerpel developed a procedure for in situ functionalization of **99** (Scheme 7.26).[95] Copper(II) bromide was used to promote the regioselective insertion of aldehydes and ketones into the more substituted C–Si bond of **99** to afford oxasilacyclopentenes **100** in one flask from the acetylene starting material. Copper-catalyzed ring expansion proved to be quite general, tolerating a range of carbonyl compounds and functional groups. Regioselective oxasilacyclopentane formation (**100g–100i**) was even observed for disubstituted silacyclopropenes (**99g–99i**). The origin of selectivity was attributed to steric interactions between the silacyclopropene substituent and the *tert*-butyl groups, which arise in the transition state when insertion occurs into the less substituted C–Si bond.[63]

The resulting oxasilacyclopentenes (**100**) were amenable to further functionalization (Scheme 7.27).[95] They could be transformed into allylic alcohol **101h** through protodesilylation or 1,3-diol **103** through hydrogenation followed by oxidation.[65] The observed diastereoselectivity was rationalized using the Woerpel model for the reaction of reagents with five-membered rings.[98] Fewer destabilizing steric interactions occur from addition of hydrogen to the face opposite the pseudoaxial *tert*-butyl group (**TS-102**). When the oxasilacyclopentene bears an olefin substituent (e.g., **100k**), a [4 + 2] cycloaddition dramatically increases the molecular complexity. This latter functionalization showcases the potential of silacyclopropenes as synthetic intermediates by producing **105**, which contains five stereocenters, from enyne **104** in two steps.

Copper salts were found to be successful catalysts for the insertion of nitriles into silacyclopropenes (Scheme 7.28).[99,100] A screen of different copper salts identified $(CuOTf)_2{\cdot}PhMe$ as the optimal catalyst. This reaction occurs readily at room temperature for monosubstituted silacyclopropenes to afford azasilacyclopentadienes

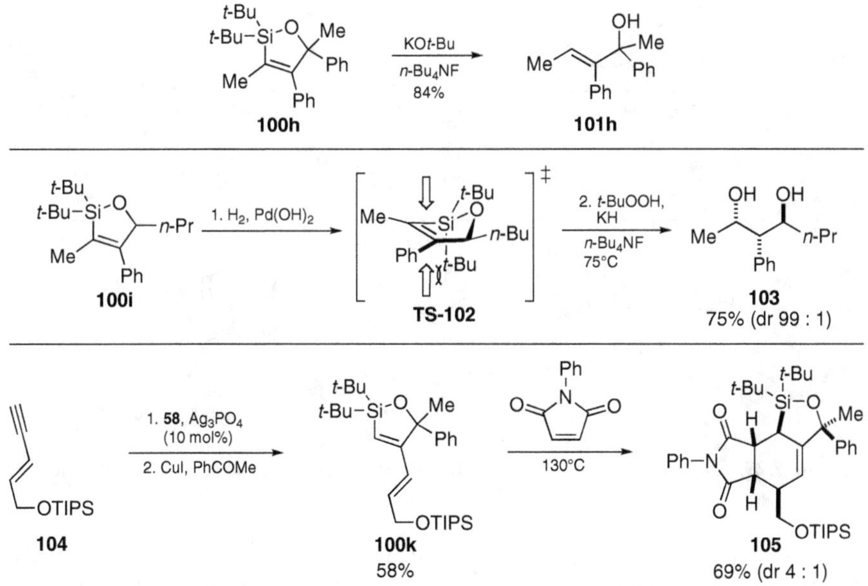

Scheme 7.27. Functionalization of silacyclopentenes to access important synthetic intermediates.

106 as single enamine tautomers. Only when sterically more congested nitriles (e.g., *i*-PrCN) were employed was the imine tautomer (**106f**) obtained. Higher temperatures were required for insertions into disubstituted silacyclopropenes (**99g–99j**). These

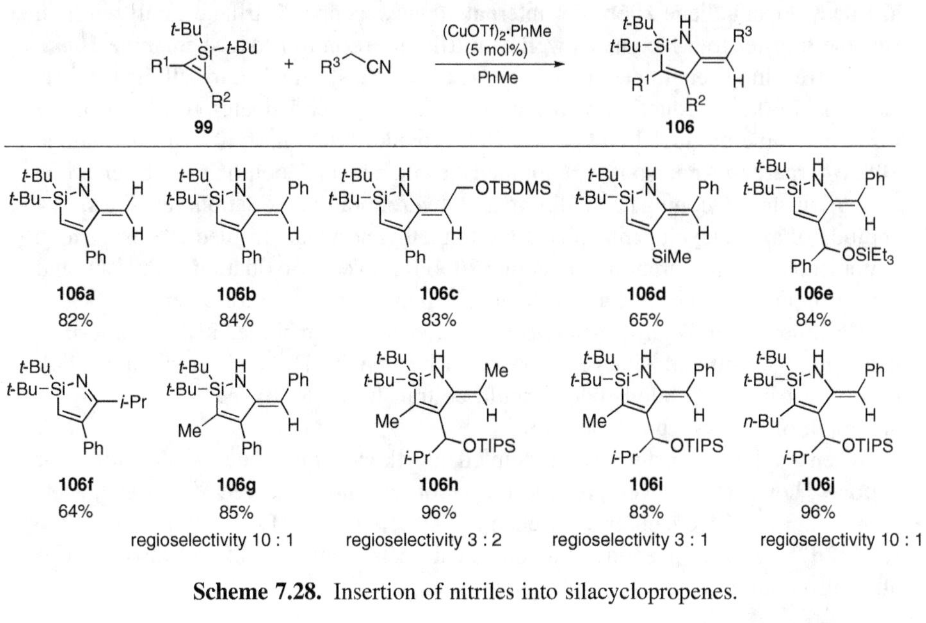

Scheme 7.28. Insertion of nitriles into silacyclopropenes.

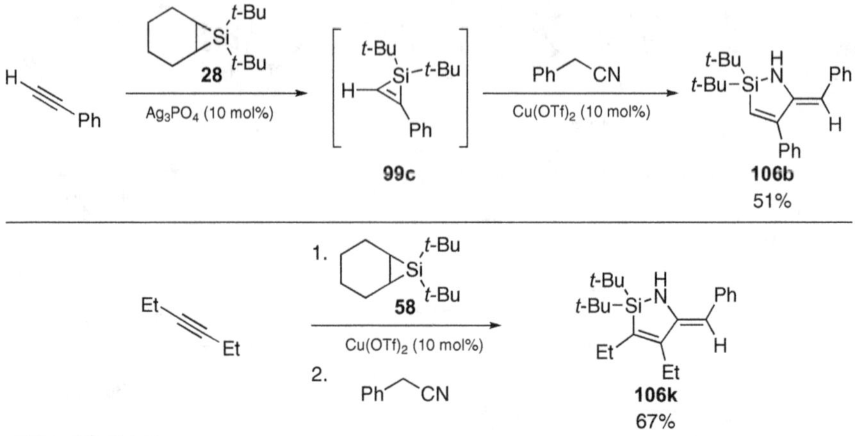

Scheme 7.29. Two-step, one-flask transformation of alkynes into azasilacyclopentadienes.

reactions were not quite as regioselective, although good selectivity was observed for **99g** and **99j**. It is apparent from the selectivity observed for the formation of **106g–106j** that both electronic and steric factors influence which C–Si bond the nitrile is inserted into.

Isolation of the air-sensitive silacyclopropene was avoided by development of a two-step, one-flask procedure, which transformed alkynes into the desired azasilacyclopentadienes (Scheme 7.29).[99] For terminal alkynes, silver phosphate was employed for di-*tert*-butylsilylene transfer and copper(I) triflate was used to promote nitrile insertion. These conditions successfully transformed phenylacetylene into azasilacyclopentadiene **106b**. For internal alkynes, copper(I) triflate catalyzed both silylene transfer to 3-hexyne as well as nitrile insertion to produce enamine **106k**.

The resulting enamines could be readily transformed into allylic amines (Scheme 7.30).[99] Reduction of the imine and methyl carbamate protection of the resulting amine produced vinylsilane **107**; protodesilylation of **107** afforded amine **108**. Woerpel and Anderson also exploited the inherent nucleophilicity of the enamine to participate in conjugate addition and hydroboration/oxidation reactions; 1,4 addition of azasilacyclopentadiene **106a** to methyl acrylate afforded **109** as a single tautomer, and hydroboration of enamine **106b** followed by oxidation with H_2O_2 and NaOH produced allylic 1,2-aminoalcohol **110** as a single stereoisomer.

The mildness of the di-*tert*-butylsilylene transfer was exploited in the synthesis of heteroatom-substituted silacyclopropenes (Scheme 7.31).[101] Woerpel and Clark observed that di-*tert*-butylsilylene could be transferred to alkynes bearing alkoxy, amino, thio, or boryl functional groups.

When silylene transfer was attempted to alkynes substituted with halides or sulfones, however, silacyclopropene formation was not observed.[101] Instead, acetylenic silanes **113** were observed (Scheme 7.32). Treatment of silacyclopropene **112c** (or **112d**) with acetophenone and substoichiometric amounts of CuI also induced alkyne formation.

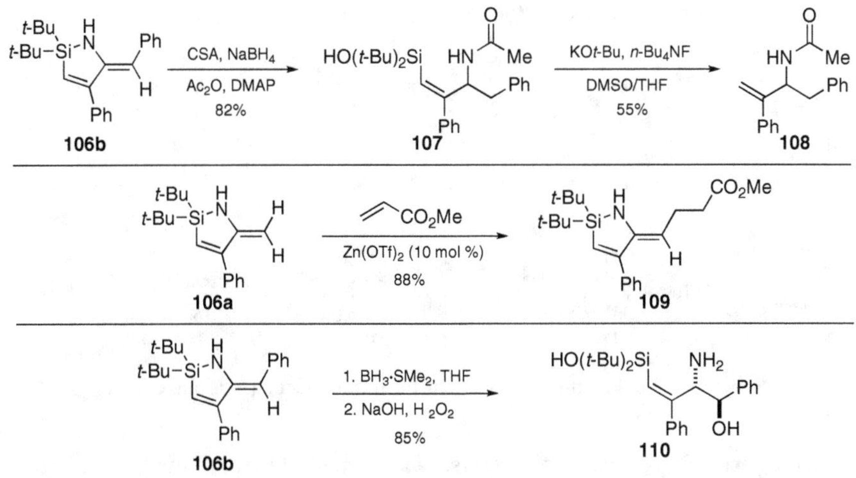

Scheme 7.30. Formation of allylic amines from azasilacyclopentadienes.

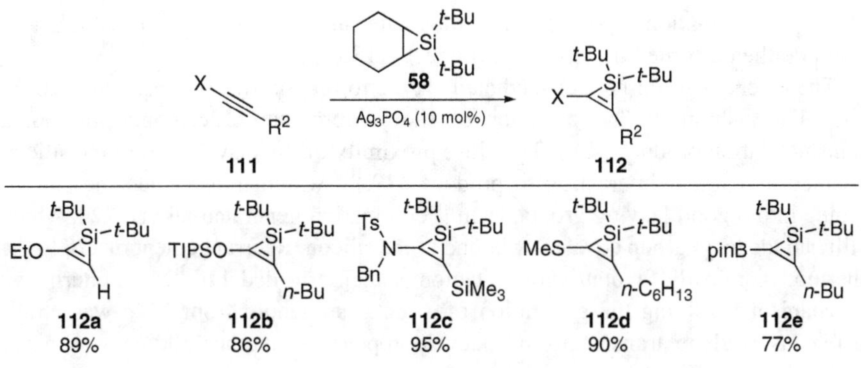

Scheme 7.31. Heteroatom-substituted silacyclopropene synthesis.

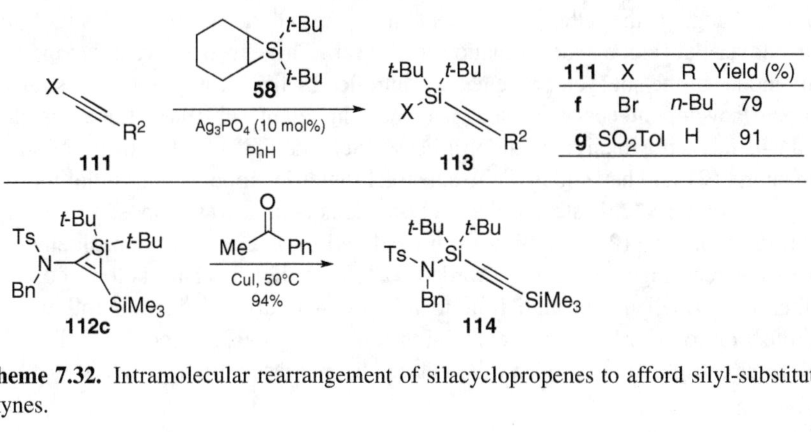

111	X	R	Yield (%)
f	Br	*n*-Bu	79
g	SO_2Tol	H	91

Scheme 7.32. Intramolecular rearrangement of silacyclopropenes to afford silyl-substituted alkynes.

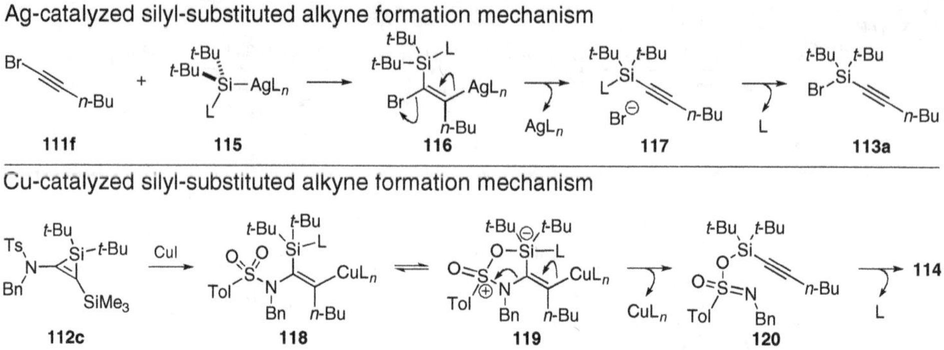

Scheme 7.33. Potential mechanism for intramolecular rearrangement.

On the basis of crossover experiments, Woerpel and Clark attributed the formation of alkynes to an intramolecular rearrangement mechanism (Scheme 7.33).[101] For silver-promoted alkyne formation, insertion of the acetylene into the Ag–Si bond produced **116**. Then β elimination of bromide occurred more rapidly than silacyclopropene formation to generate silyl-substituted alkyne **117**. Substitution of L with bromide then afforded the observed product (**113a**).

The mechanism for copper-mediated alkyne formation was proposed to occur in a similar fashion.[101] On formation of silacyclopropene **112c**, copper-promoted transmetalation produced **118**. The close proximity of the tosyl group to the silicon enabled a covalent interaction to produce **119**,[102] which transformed the sulfonamide into a good leaving group. Then β elimination generated alkyne **120**, which still had the tosyl group covalently bound to the silicon. Rearrangement of **120** to the thermodynamically favored sulfone tautomer then afforded **114**.[103] An alternative mechanism involving the generation of a metal silylenoid from **112c** was eliminated, as silylene transfer from silacyclopropene to another alkyne was never observed.

In contrast to aforementioned silacyclopropenes, copper-catalyzed ring expansion was achieved with silyloxy- or boryl-substituted silacyclopropenes (Scheme 7.34).[101,104] Simply changing the ether substituent from an alkyl group to a silyl group enabled oxasilacyclopentene formation from **112b**. These results reveal that the reactivity modes of the silacyclopropene are controlled by the identity of the substituent.

Oxasilacyclopentenes were shown to be competent substrates for a scandium triflate–catalyzed Mukaiyama aldol process (Scheme 7.35).[104] Exposure of silacyclopentene **121** and benzaldehyde to substoichiometric amounts of scandium triflate produced ketone **122** diastereoselectively.[105] This ketone was proposed to form by addition of enolate **123**, resulting from desilylation of **121**,[106] to benzaldehyde. A 1,3-Brook rearrangement then afforded **122** from **124**.[107] This ketone could be further functionalized through lithium aluminum hydride reduction followed by deprotection to afford triol **125** containing four contiguous stereocenters. Thus, the molecular complexity of silyloxyalkynes can be increased dramatically in just three operations.

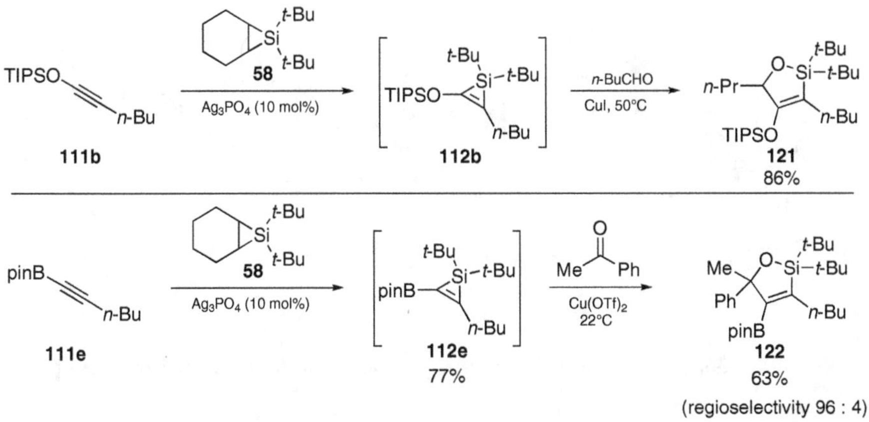

Scheme 7.34. Copper-mediated ring expansion to afford oxasilacyclopentenes.

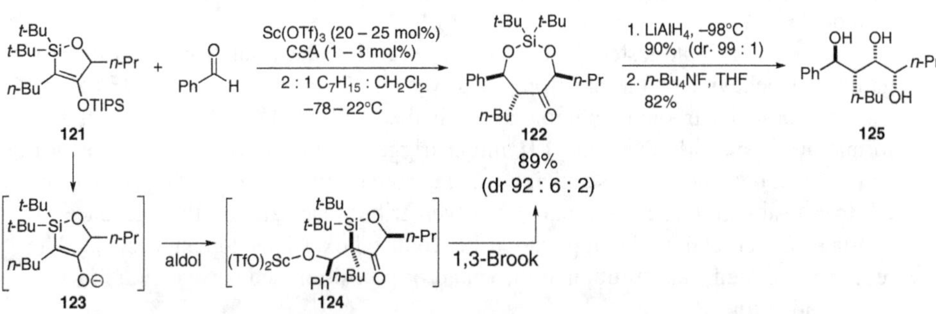

Scheme 7.35. Mukaiyama aldol functionalization of oxasilacyclopentenes.

7.6 SILVER-MEDIATED TRANSFER OF DI-*tert*-BUTYLSILYLENE TO CARBONYL COMPOUNDS

In addition to participating in [2 + 1]-cycloaddition reactions, divalent reactive intermediates can form ylides in the presence of carbonyl or other Lewis basic functionalities.[108] These ylides participate in cycloaddition or other pericyclic reactions to furnish products with dramatically increased complexity. While carbenes (or metal carbenoids) are well known to participate in these pericyclic reactions, silylenes, in contrast, were reported to react with aldehydes or ketones to form cyclic siloxanes[109,110] or enoxysilanes.[111,112] Reaction of silylene with an α,β-unsaturated ester was known to produce an oxasilacyclopentene.[109,113,114] By forming a silver silylenoid reactive intermediate, Woerpel and coworkers enabled involvement of divalent silylenes in pericyclic reactions involving silacarbonyl ylides[115] to afford synthetically useful products.[82,116,117]

While exploring the scope of silver-catalyzed silylene transfer to olefins, Woerpel and Calad observed that the electrophilic silver silylenoid species reacted preferentially with the enolizable ester group to produce vinylsilane **129** instead of a

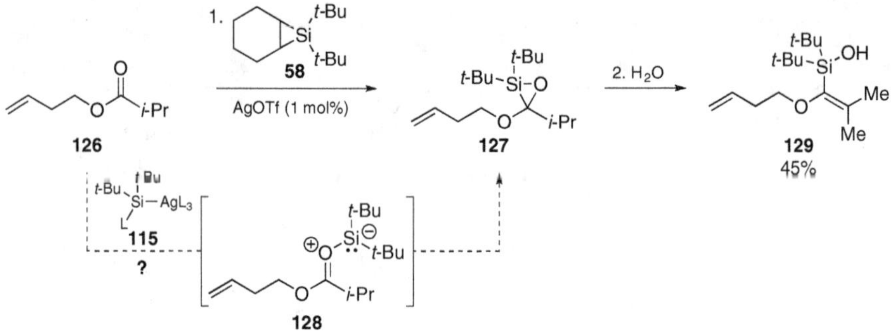

Scheme 7.36. Silver-catalyzed di-*tert*-butylsilylene transfer to an enolizable ester.

silacyclopropane (Scheme 7.36).[82] They attributed vinylsilane formation to the intermediacy of oxasilacyclopropane **127**, which might have formed in a stepwise fashion from **126** via silacarbonyl ylide **128**.[118]

Woerpel and Calad tested for the formation of the silacarbonyl ylide by interrogating the behavior of the electrophilic silver silylenoid intermediate **115** toward α,β-unsaturated carbonyl compounds (Scheme 7.37).[82] They hypothesized that formation of silacarbonyl ylide **131** might trigger a 6π-electrocyclization to form oxasilacyclopentene **132**. As anticipated, exposure of cyclohexene silacyclopropane **58** to substoichiometric amounts of silver trifluoroacetate in the presence of α,β-unsaturated carbonyl compounds **130** produced oxasilacyclopentenes **132**. The reaction tolerated a substitution at the α and/or β position and was general for both esters and ketones.

The high diastereoselectivity observed by Woerpel and Calad in the hydrolysis of oxasilacyclopentene **132** (Scheme 7.38) prompted further evaluation of these intermediates in stereoselective synthetic transformations.[82] Exposure of esters containing

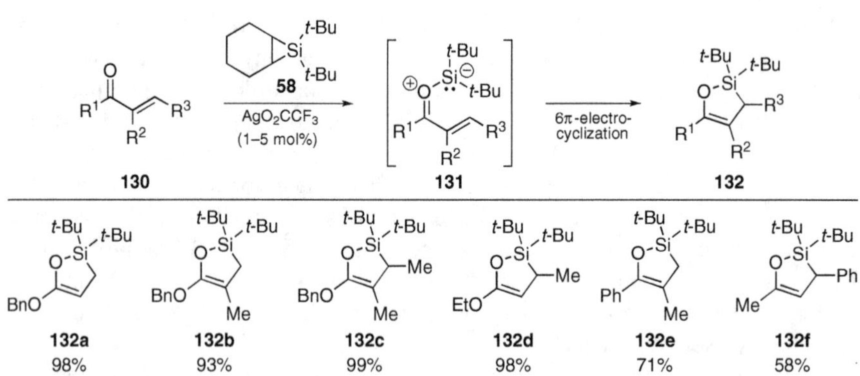

Scheme 7.37. Silver-catalyzed di-*tert*-butylsilylene transfer to α,β-unsaturated carbonyl compounds.

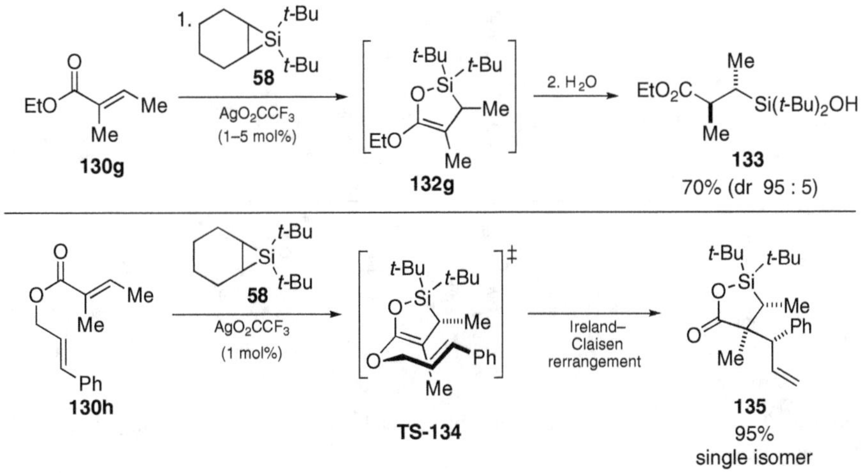

Scheme 7.38. Exploitation of oxasilacyclopentenes in diastereoselective synthetic transformations.

an allyl substituent (e.g., **130h**) to reaction conditions facilitated an Ireland–Claisen rearrangement after the initial electrocyclization. This process was highly diastereoselective; the reaction of ester **130h** produced silalactone **135** with three contiguous stereocenters as a single isomer. Woerpel and Calad rationalized the stereochemistry of the [3,3]-sigmatropic rearrangement as arising from a chair-like transition state **TS-134**, in which the allyl fragment approaches the oxasilacyclopentene from the face opposite the methyl substituent.

The synthetic utility of this cascade reaction was underscored by the facile transformation of crotyl tiglate into 1,3-diol **137** (Scheme 7.39).[82] Silver-catalyzed silylene transfer to crotyl tiglate produced silalactone **136** nearly quantitatively as a 97 : 3 mixture of diastereomers. Diol **137** was generated from silalactone after reduction of the lactone moiety with lithium aluminum hydride followed by oxidation of the C–Si bond.[65,66,81]

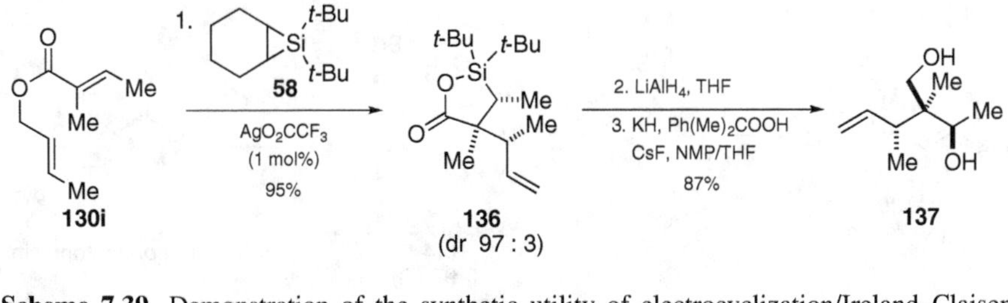

Scheme 7.39. Demonstration of the synthetic utility of electrocyclization/Ireland–Claisen rearrangement cascade reaction.

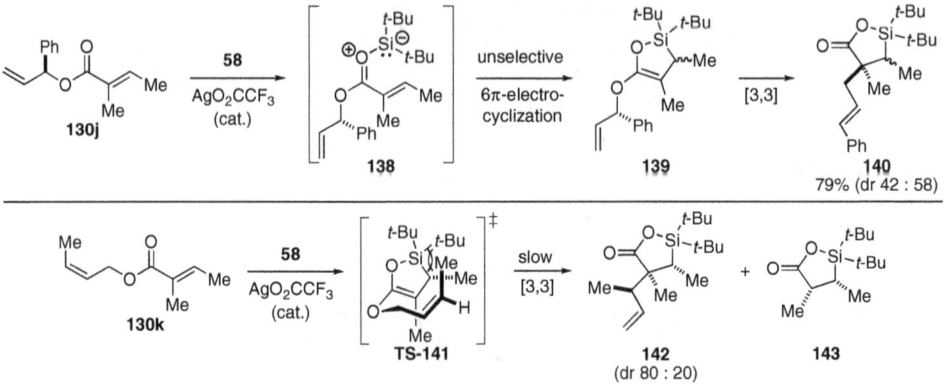

Scheme 7.40. Scope and limitations of cascade electrocyclization/[3,3]-sigmatropic rearrangement.

The limitations of each step of the cascade reaction emerged on interrogation of esters with varied substitution patterns (Scheme 7.40).[116] The reaction of ester **130j** revealed that the facial selectivity of the 6π-electrocyclization could not be controlled. Exposure of (*Z*)-allyl ester **130k** to reaction conditions produced a mixture of silalactones **142** and **143**. The formation of **143** was attributed to hydrolysis of the initially formed oxasilacyclopentene. Observation of the hydrolysis product indicates that the rate of [3,3]-sigmatropic rearrangement is much slower for (*Z*)-allyl ethers. The authors attribute this rate difference to the axial placement of the methyl substituent in transition state **TS-141**, which incurs destabilizing steric interactions with the enoxysilane moiety.

The unselective 6π-electrocyclization can be avoided with substrates that lack a β-olefin substituent (Scheme 7.41).[116] When chiral, nonracemic versions of these substrates are employed, Woerpel and Calad demonstrated that the cascade reaction efficiently transfers the chiral information. While silver salts were competent catalysts for the diastereoselective formation of silalactone **144** from **130l**, on scale-up, copper(II) triflate proved to be more efficient. The authors attribute the poorer performance by silver to product inhibition. In eight steps, silalactone **144** was transformed into the antibiotic (+)-*epi*-acetomycin, to further highlight the synthetic utility of the cascade process.

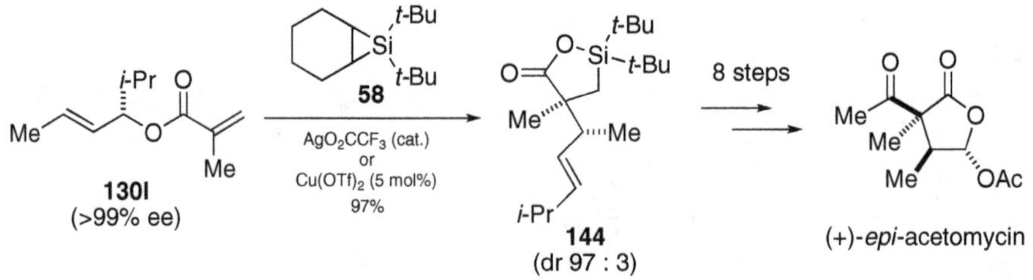

Scheme 7.41. Efficient transfer of chirality in the cascade reaction.

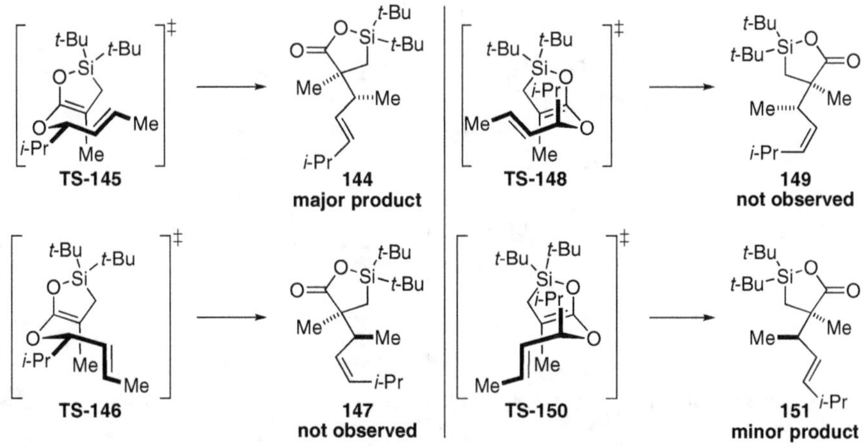

Figure 7.1. Origin of diastereoselectivity in Ireland–Claisen step of the cascade reaction.

The origin of diastereoselectivity in the cascade process was rationalized through a comparison of four different transition states (Fig. 7.1).[116] The major diastereomer was posited to originate from transition state **TS-145**, which minimizes the developing diaxial-1,3 interactions through placement of the isopropyl group in a pseudoequatorial position. Because only (*E*)-olefins were obtained, the origin of the minor diastereomer was attributed to the boat-like transition state **TS-150**.

Woerpel and Howard further extended the synthetic utility of this cascade reaction in the stereoselective synthesis of α-hydroxy acids from α-ketoesters (Scheme 7.42).[119] In every case, the α-hydroxy acid was formed with excellent diastereoselectivity (dr ≥98 : 2). Their method was tolerant of a range of substitution patterns, although diminished yields were obtained as the size of the R^1 substituent increased (compare **154a** and **154c**). In contrast, no decrease in yield was observed over a range of R^2 substituents.

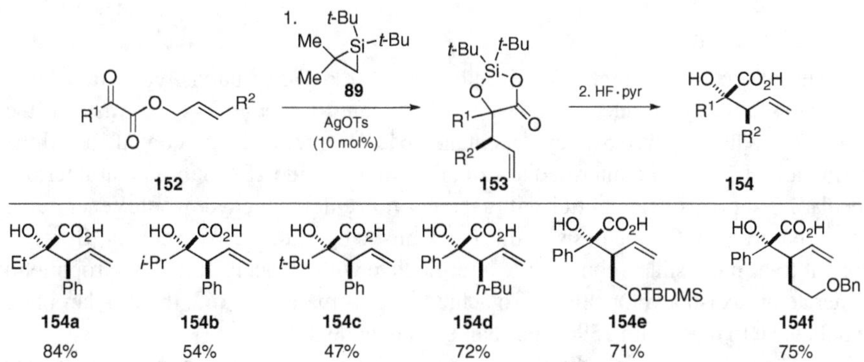

Scheme 7.42. Synthesis of α-hydroxy acids from α-ketoesters through a silver-promoted cascade reaction.

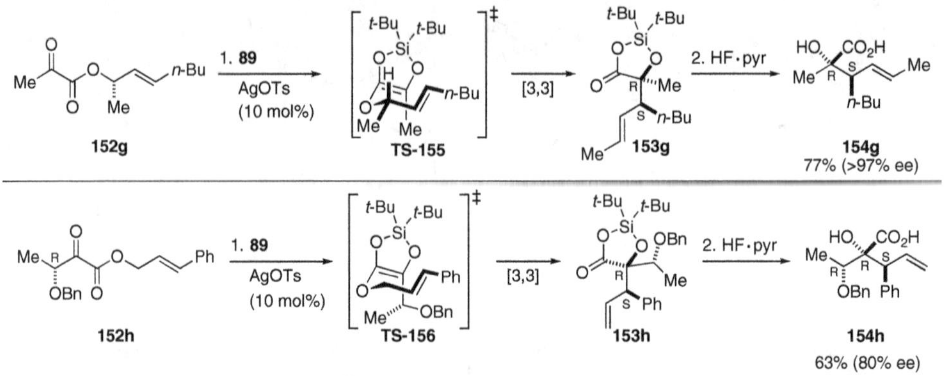

Scheme 7.43. Di-*tert*-butylsilylene transfer to chiral, nonracemic substrates.

When chiral substrates were employed in the cascade reaction, good transfer of the chiral information was observed (Scheme 7.43).[119] Excellent diastereoselectivity was obtained with **152g**, containing an allylic methyl group; and even **152h**, which contained a benzyl ether substituent that would lie outside the chairlike transition state (**TS-156**), controlled the facial selectivity of the [3,3]-sigmatropic rearrangement.

In addition to α-ketoesters, α-iminoesters were successful as substrates, allowing the synthesis of α-amino acid derivatives [equation (7.9)].[119] Exposure of **157** to *gem*-dimethylsilacyclopropane **89** and substoichiometric amounts of silver tosylate produced azasilalactone **158**. In contrast to the above–mentioned synthesis of **154**, hydrolysis of **158** was not observed on aqueous extraction. The authors attributed the enhanced robustness of **158** to the steric bulk of the anisidine group.

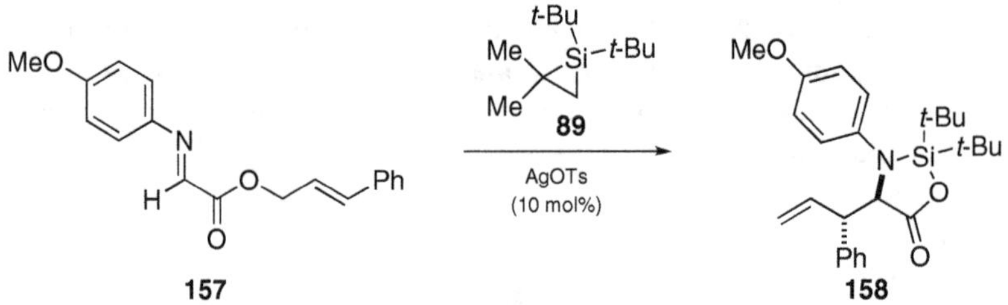

(7.9)

In the absence of an internal olefin, silver-catalyzed di-*tert*-butylsilylene transfer to aldehydes did not produce an oxasilacyclopropane (Scheme 7.44).[117] Instead, the reaction produced a dioxosilacyclopentane **161**. Woerpel and Bourque attributed the formation of **161** to the intermediacy of silacarbonyl ylide **159**. Subsequent intermolecular 1,3-dipolar cycloaddition with a second molecule of aldehyde would generate the observed product. On the basis of this mechanistic hypothesis, Woerpel and Bourque sought to trap the silacarbonyl ylide intermediate with an acetylenic dipolarophile to generate an oxasilacyclopentene. To achieve the formation of **162**, the dipolarophile would have to react with **159** faster than the aldehyde did.

To identify potential dipolarophiles, a series of competition experiments were performed (Scheme 7.45).[117] While inclusion of terminal acetylenes resulted in preferential silacyclopropene **99c** or dioxosilacyclopentane **161** formation, the desired oxasilacylopentene (**163**) was obtained when diethylacetylene dicarboxylate was added

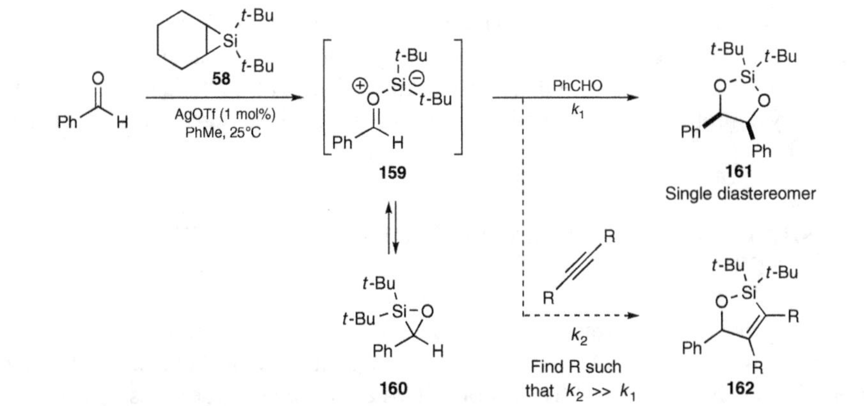

Scheme 7.44. Silver-promoted 1,3-dipolar cycloaddition reactions.

to the reaction mixture. Substitution of the carboethoxy groups with methyl and phenyl groups, unfortunately, resulted in dioxosilacyclopentane **161** formation.

The cycloaddition reaction with diethyl acetylenedicarboxylate tolerated a range of carbonyl compounds (Scheme 7.46). Facile reaction was obtained with aldehydes,

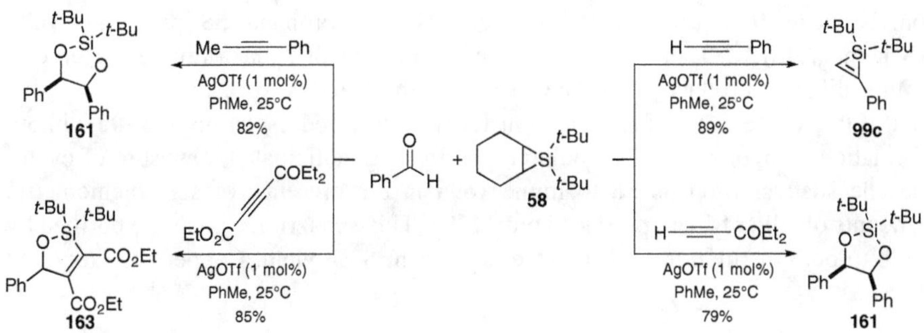

Scheme 7.45. Intermolecular competition reactions that examine the relative rates of cycloaddition and cyclization.

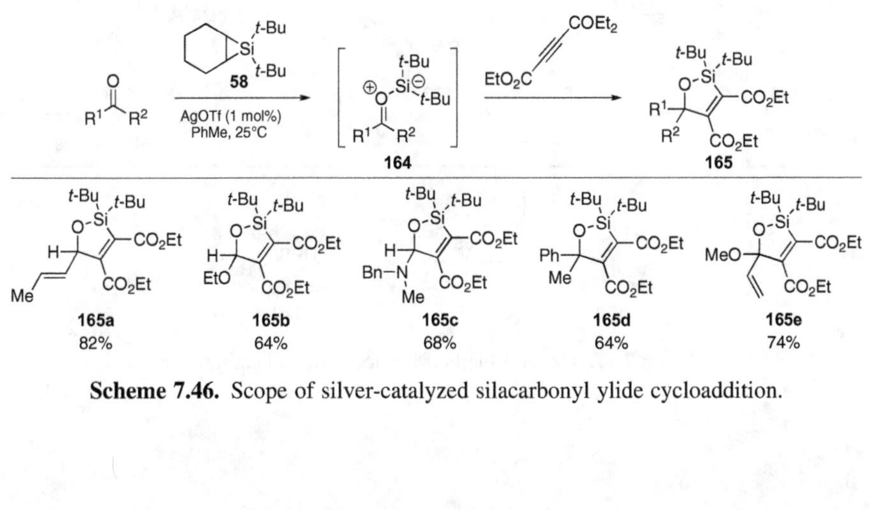

Scheme 7.46. Scope of silver-catalyzed silacarbonyl ylide cycloaddition.

formates, ketones, and formamides. The slower rate of reaction of formamide required heating and prolonged time (50°C, 12 h). 1,3-Dipolar cycloaddition was even obtained in the reaction of methyl acrylate to produce **165e**, although excess ester and silacyclopropane was required to compete with silylene transfer to the olefinic moiety.

7.7 SILVER-MEDIATED TRANSFER OF DI-*tert*-BUTYLSILYLENE TO IMINES

The nitrogen analog of oxasilacyclopropane, azasilacyclopropane (or silaziridine), has been postulated as a potential reactive intermediate in photochemical- or thermal processes involving silylenes. These strained *N*-heterocycles can be accessed from the reaction of isocyanides and photochemically generated silenes, but this method is limited to the formation of methylene silaaziridines **168** (Scheme 7.47).[120–122] Isolation of these *N*-heterocycles proved challenging, as they rapidly inserted another molecule of the isocyanide starting material on warming to room temperature.[121] Woerpel and coworkers believed that the mild nature of silver-catalyzed di-*tert*-butylsilylene transfer might be exploited to provide general access to these strained *N*-heterocycles (**170**) and enable a detailed study of their reactivity.

Toward this end, Woerpel and Nevárez examined the possibility of di-*tert*-butylsilylene transfer from cyclohexene silacyclopropane **58** to imine **169a** (Scheme 7.48).[123] Thermolysis produced a mixture of silaaziridine **170a** and an imine–dimer byproduct (**171**). The results by Brook and coworkers[120–122] suggested that if the temperature of silylene transfer were lowered, isolation of **170a** without formation of byproduct **171** would be possible. As anticipated, exposure of cyclohexene silacyclopropane **58** to imine **169a** in the presence of substoichiometric amounts of silver triflate produced only **170a**. This silaazridine could be purified by bulb-to-bulb distillation to afford the product in 80% yield. Copper salts required

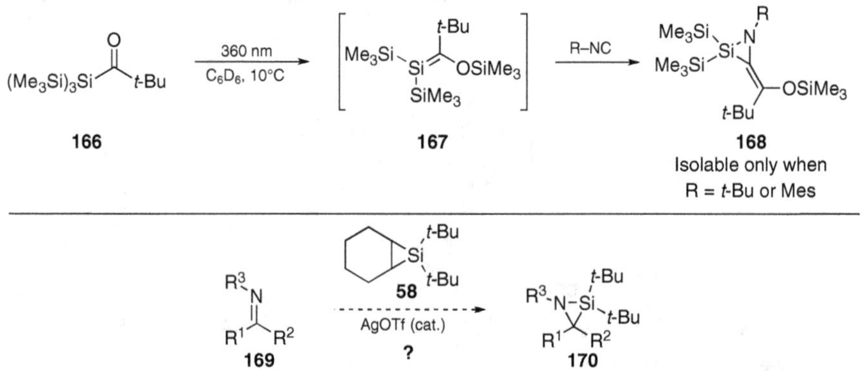

Scheme 7.47. General method for accessing silaaziridines.

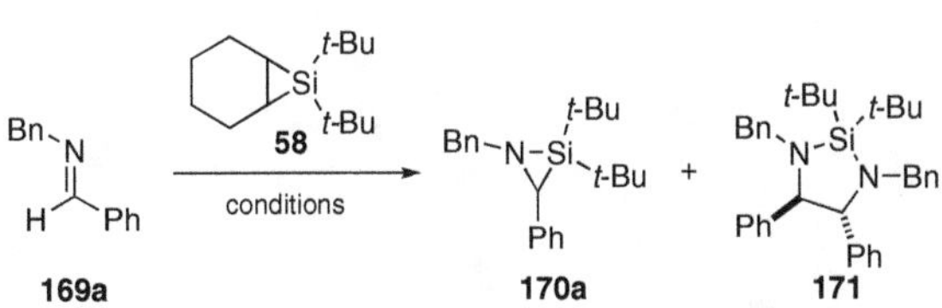

Conditions	Yield (%)	**170a** : **171**
100°C	72	63 : 37
AgOTf (cat.)	90	100 : 0

Scheme 7.48. Di-*tert*-butylsilylene transfer to imines.

higher temperatures to promote silylene transfer from **169a** to **170a**. Yields were significantly diminished using copper catalysts.

Silver-catalyzed di-*tert*-butylsilylene transfer to imines proved general (Scheme 7.49).[123] Exposure of alkyl- or arylimines with *N*-benzyl or *N*-aryl groups to cyclohexene silacyclopropane **58** in the presence of 1 mol% of silver triflate produced silaaziridines **170**. Even ketimine **169f** was tolerated as a substrate.

In contrast to the results presented above, silver-mediated silylene transfer to α,β-unsaturated imines or imines with *N*-carbamate groups did not produce silaaziridines (Scheme 7.50).[123] These substrates instead formed enamine **173** or oxazoline **175**. The authors attributed these products to a 6π-electrocyclization of ylide **172** and **174**.

While enamine **173** and oxazoline **175** were observed using ^{1}H NMR spectroscopy, they could not be isolated. Instead, dimeric **177** and silaoxazolidinone **179** were obtained (Scheme 7.51).[123] The formation of azasilacyclopentane **177** was rational-

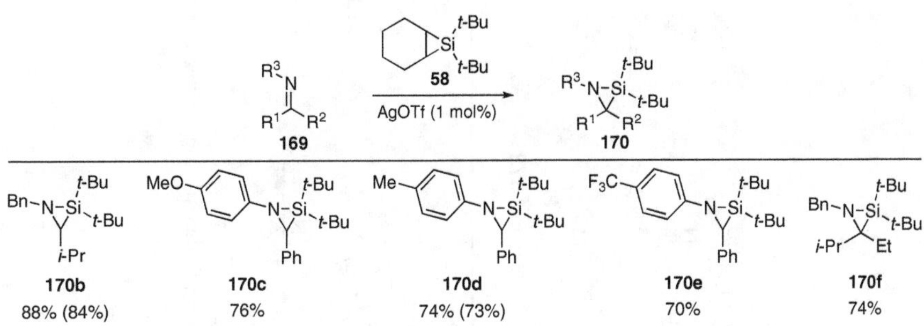

Scheme 7.49. Scope of Ag-catalyzed di-*tert*-butylsilylene transfer to aldimines and ketimines.

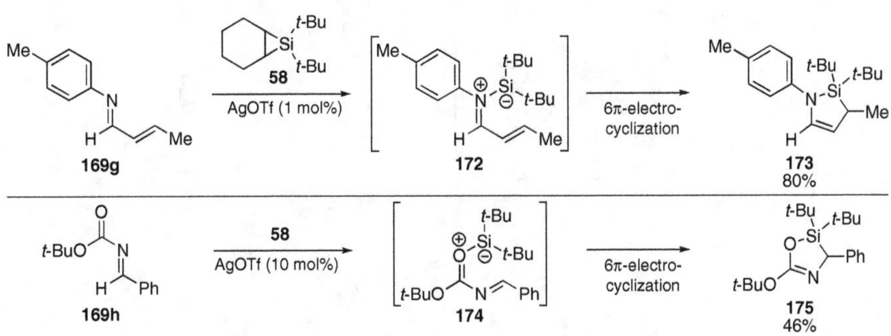

Scheme 7.50. Silver-catalyzed di-*tert*-butylsilylene transfer to imines.

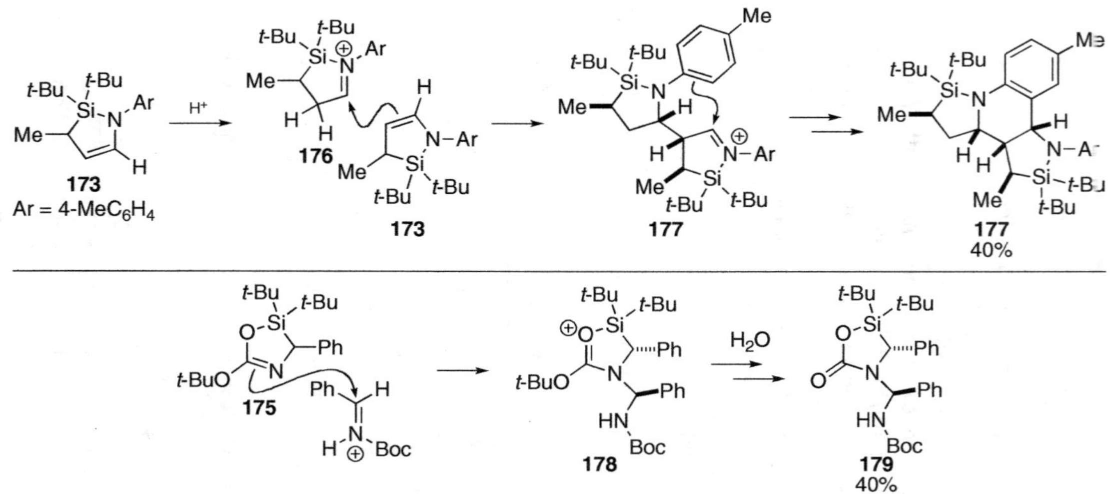

Scheme 7.51. Decomposition of silaaziridines derived from α,β-unsaturated imines or carbamate-protected imines.

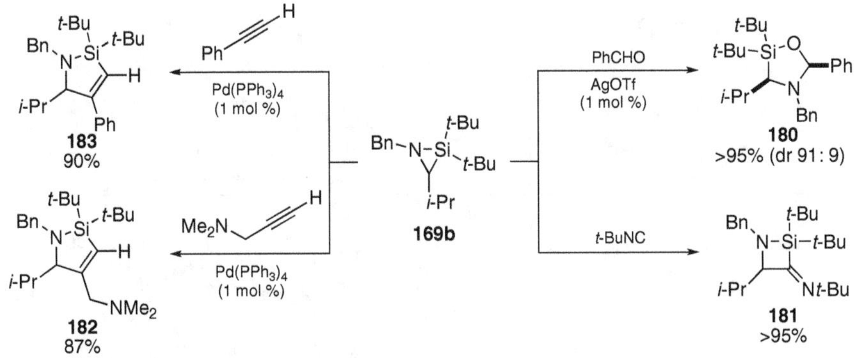

Scheme 7.52. Synthetic utility of silaziridines.

ized by the authors as occurring through a series of nucleophilic additions to iminium ions. These additions were diastereoselective as a single isomer of **177** was isolated. A similar addition mechanism was postulated to account for carbamate **179** formation; nucleophilic addition of oxazoline **175** to imine **169h** would produce **179** after hydrolysis.

Woerpel and Nevárez demonstrated the synthetic potential of silaaziridines by selective insertion reactions (Scheme 7.52).[123] Silver-catalyzed aldehyde insertion into the Si–N bond of **169b** produced the *N*,*O*-cyclic acetal **180** as the *cis* isomer. In contrast to this process, insertion of *tert*-butyl isocyanide occurred into the weaker C–Si bond to afford imine **181**. The authors rationalized the chemoselectivity for these two processes on the basis of Pearson's hard–soft acid–base theory;[124–126] the more ionic Si–N bond reacted with harder benzaldehyde electrophile, whereas the more covalent Si–C bond reacted with the softer isocyanide.

Azasilacyclopentenes could also be produced from silaaziridines through the use of a palladium(0) catalyst (Scheme 7.52).[123] This process inserted the terminal acetylene into the weaker Si–C bond to produce **182** and **183** as single regioisomers. The authors asserted that the regioselectivity of the insertion was controlled through minimization of steric interactions between the *tert*-butyl groups and the alkyne substituent. A range of functional groups were tolerated in this process, including silyl ethers, silanes, and tertiary amines. Protodesilylation reveals that azasilacyclopentenes are masked allylic amines:

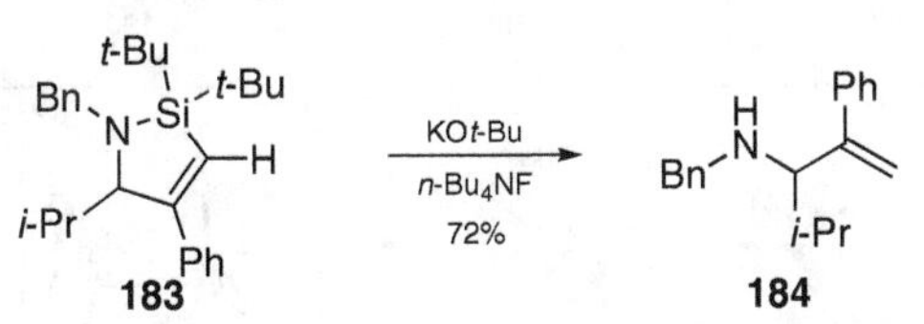

(7.10)

In contrast to the formation of an *N*,*O*-acetal in the reaction of *N*-benzylsilaaziridine **169b** with isobutyraldehyde, the reaction of *N*-arylsilaaziridines **169g**

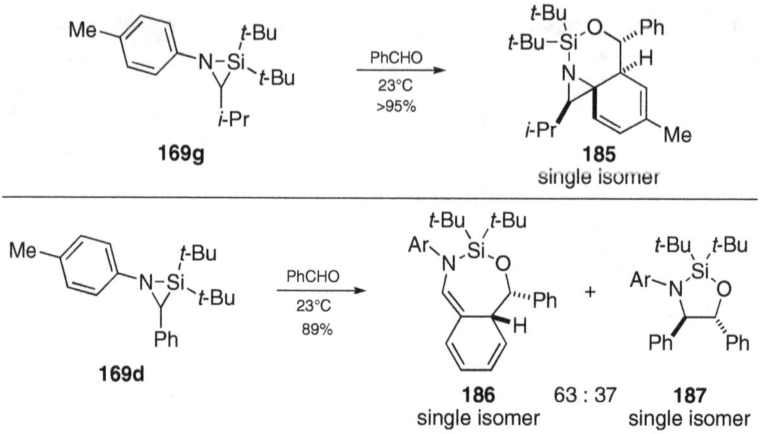

Scheme 7.53. Dearomatization reactions of *N*-arylsilaziridines and benzaldehyde.

and **169d** with benzaldehyde produced new C–C bonds through dearomatization processes (Scheme 7.53).[127] The reaction of silaziridine **169g** with benzaldehyde produced aziridine **185**, which contains four contiguous stereocenters, as a single isomer. An alternative dearomatization reaction occurred with silaaziridine **169d**; exposure to benzaldehyde resulted in the break of the aromaticity of the phenyl substituent instead of the *N*-aryl group to produce **186** as a single isomer. Insertion of benzaldehyde into **169d** to form **187** was competitive with this dearomatization mechanism.

Woerpel and Nevárez postulated a stepwise mechanism for the formation of **185** (Scheme 7.54).[127] Coordination of benzaldehyde to the Lewis acidic silicon atom of **169g**[86] triggers an attack by the pendent *N*-aryl group to form zwitterion **189**. Isomerization of **189** to azomethine ylide **190**[128] enabled a 4π-electrocyclization to produce aziridine **185**. The thermodynamic driving force for this reaction was attributed to the relief of the silaaziridine's ring strain.

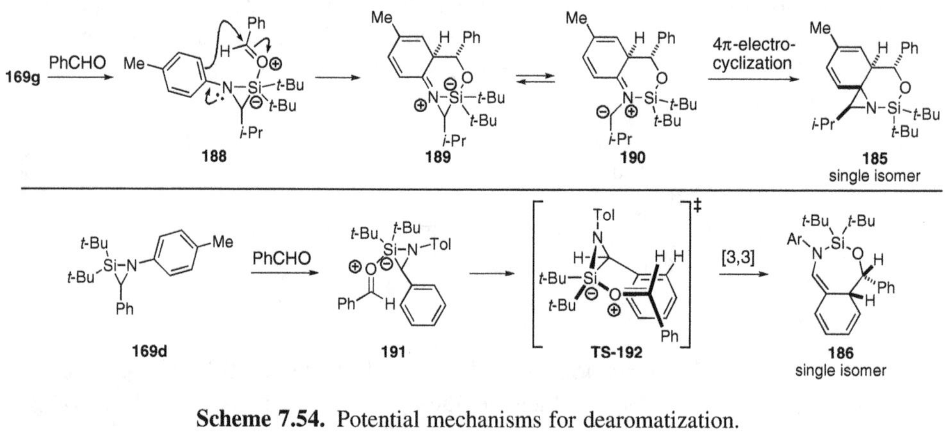

Scheme 7.54. Potential mechanisms for dearomatization.

The authors suggested a slightly different mechanism for the formation of **186**.[127] Instead of triggering an electrophilic aromatic substitution reaction, initial coordination of benzaldehyde to **169d** enabled a [3,3]-sigmatropic rearrangement through the boat-like transition state **TS-192**[129] to form the observed product with the requisite *cis*-stereochemical arrangement of hydrogens.

7.8 SILVER-MEDIATED DI-*tert*-BUTYLSILYLENE INSERTION INTO C–O BONDS

Transition-metal-catalyzed [2,3]-sigmatropic rearrangement of an oxonium ylide derived from an allylic ether is a well established method for the synthesis of new C–C bonds in a stereoselective fashion (Scheme 7.55).[130–133] Competitive [1,2]-rearrangements, however, do occur if the rate of the [2,3] rearrangement is slow (see **194–196**).[134,135] Since silver-catalyzed di-*tert*-butylsilylene transfer to Lewis basic carbonyls was established to afford ylides, Woerpel and coworkers investigated the potential of silylene transfer to allylic ethers **197**.[136] The resulting oxonium ylide (**198**) could rearrange through a [2,3]-sila-Wittig rearrangement to afford allylsilane **199** or undergo a [1,2]-Stevens rearrangement to afford allylsilane **200**. The latter rearrangement would constitute a formal insertion of silylene into a C–O bond.

While investigating trisubstituted allylic ethers, Woerpel and coworkers observed insertion of silylene into the allylic C–O bond (Scheme 7.56).[136] The product formation was dependent on the identity of the catalyst. Silver trifluoroacetate promoted the formation of allylic disilane **201**, whereas copper(I) triflate

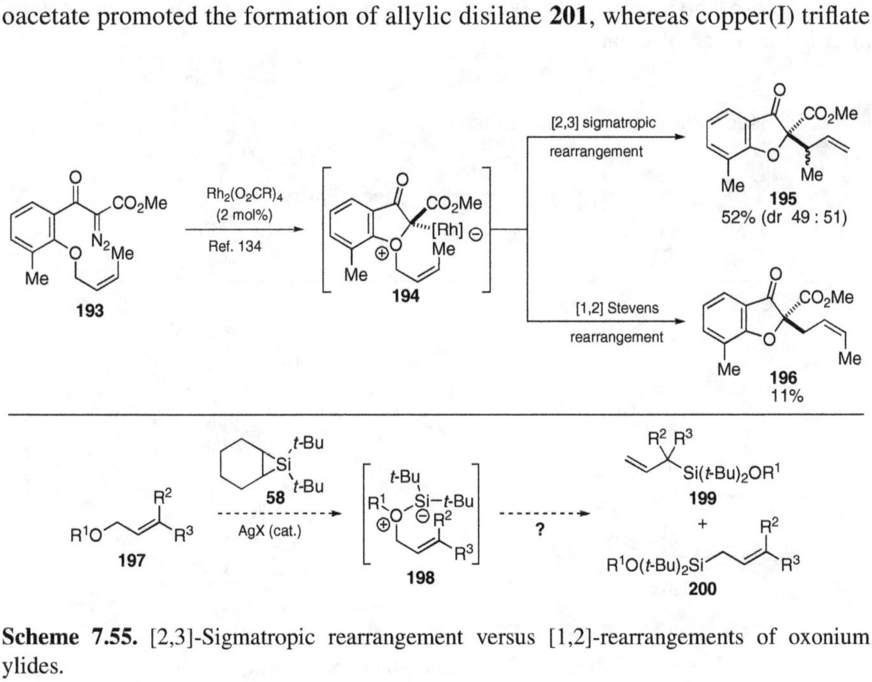

Scheme 7.55. [2,3]-Sigmatropic rearrangement versus [1,2]-rearrangements of oxonium ylides.

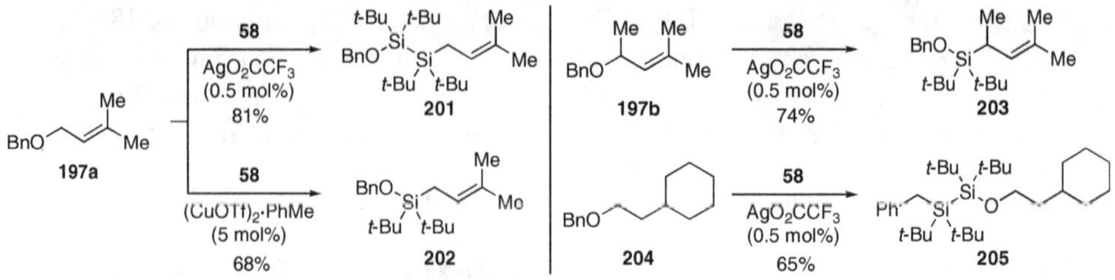

Scheme 7.56. Transition metal-catalyzed insertion of di-*tert*-butylsilylene into C–O bonds.

catalyzed the formation of allylic silane **202**. The copper-catalyzed reaction was not as efficient, requiring higher catalyst loading and increased reaction time to afford **202**. The identity of the product also depended on the steric environment of the allylic ether, silver-catalyzed silylene transfer to secondary allylic ethers afforded allylic silane **203**, the product of a single silylene insertion. The insertion of a silylene into the benzylic C–O bond of **204** was also observed to afford disilane **205**.

Insight into the mechanism for the formation of allylic disilane **201** and silane **202** was obtained from a series of control experiments (Scheme 7.57). Submission of allylic silane **202** to reaction conditions did not produce disilane **201**. Woerpel and coworkers interpreted this result to indicate that disilane formation does not occur through subsequent silylene insertions. Crossover experiments established that no dissociation of the alkoxy group occurs during the reaction to suggest that silylene insertion is intramolecular.

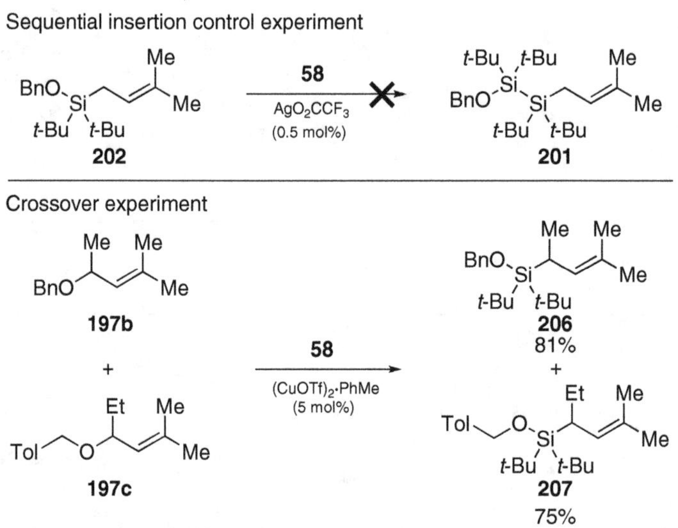

Scheme 7.57. Mechanistic experiments.

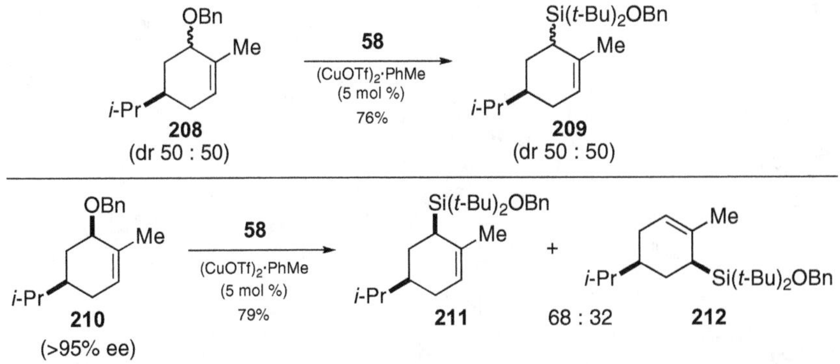

Scheme 7.58. Examination of the stereochemical course of the reaction.

Interrogation of the stereochemical course of the mechanism was obtained through submission of allylic ethers **208** and **210** (>95% ee) to reaction conditions (Scheme 7.58). The reaction of a 1 : 1 mixture of allylic ether **208** produced the allylic silane as a 1 : 1 mixture of diastereomers. Exposure of **210** to substoichiometric amounts of copper(I) triflate and cyclohexene silacyclopropane produced *cis*-substituted allyl silanes **211** and **212** to reveal that C–Si bond formation occurs on the same face as the C–O bond that is cleaved. The loss of enantiopurity, however, indicates that the rate of allylic transposition is competitive with the insertion process.

Woerpel and coworkers interpreted the results of these mechanistic experiments as evidence that the insertion of silylene into the C–O bond occurs through a [1,2]-Stevens rearrangement of oxonium ylide **198** and that a competitive [2,3]-sigmatropic rearrangement of **198** could account for allylic transposition.

7.9 CONCLUSION

Woerpel and coworkers discovered that silver salts promote di-*tert*-butylsilylene extrusion from a sacrificial silacyclopropane (**58** or **89**, Scheme 7.59). The resulting silver silylenoid reacts with a variety of different functional groups to afford three- or five-membered heterocycles. These silacycles are easily transformed into a variety of important synthetic intermediates, including polypropionates, allylic ethers, allylic amines, and α-hydroxy acids. The synthetic utility of silver-catalyzed silylene transfer was showcased in the total syntheses of 1′-*epi*-stegobinone and (+)-*epi*-acetomycin. The mild nature of silver-catalyzed silylene transfer allows the isolation of particularly reactive monosubstituted silacyclopropenes and silaaziridines to enable their use in stereoselective methods for organic synthesis. The use of silver to mediate silylene transfer reactions has resulted in the development of useful organic transformations that exploit the inherent reactivity of silanes to produce important synthetic intermediates.

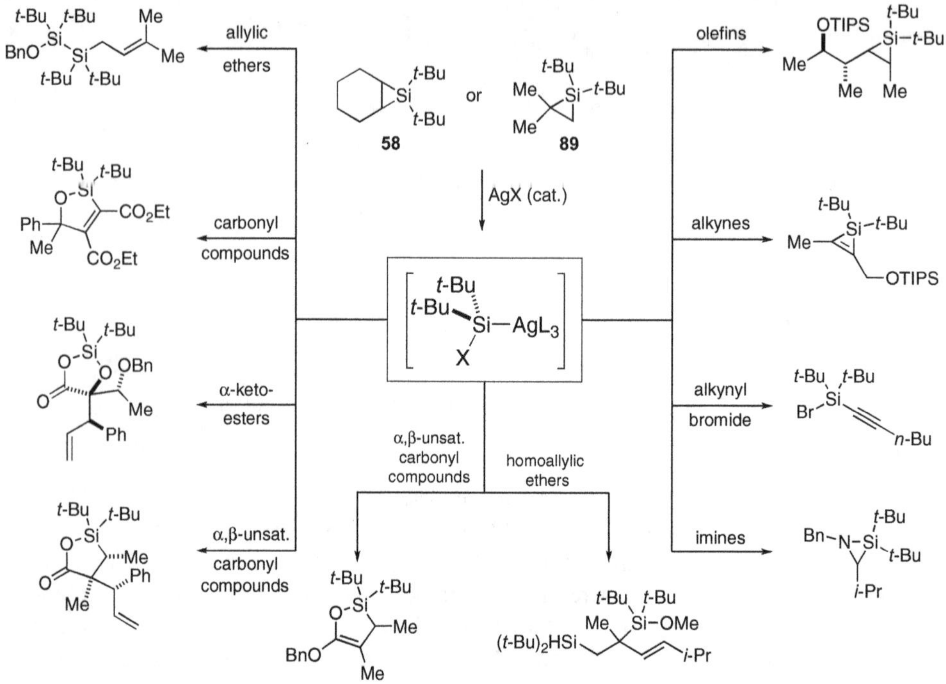

Scheme 7.59. Silver-catalyzed di-*tert*-butylsilylene transfer.

REFERENCES

1. Ye, T.; McKervey, M. A., *Chem. Rev.* **1994**, *94*, 1091–1160.
2. Doyle, M. P.; McKervey, M. A.; Ye, T., *Modern Catalytic Methods for Organic Synthesis with Diazo Compounds: From Cyclopropanes to Ylides*, Wiley, New York, **1998**.
3. Kirmse, W., *Eur. J. Org. Chem.* **2002**, 2193–2256.
4. Li, Z.; He, C., *Eur. J. Org. Chem.* **2006**, *2006*, 4313–4322.
5. For silver-promoted C–H bond functionalization or cyclopropanation reactions, see (a) Burgess, K.; Lim, H.-J.; Porte, A. M.; Sulikowski, G. A., *Angew. Chem., Int. Ed. Engl.* **1996**, *35*, 220–222; (b) Dias, H. V. R.; Browning, R. G.; Polach, S. A.; Diyabalanage, H. V. K.; Lovely, C. J., *J. Am. Chem. Soc.* **2003**, *125*, 9270–9271; (c) Dias, H. V. R.; Browning, R. G.; Richey, S. A.; Lovely, C. J., *Organometallics* **2004**, *23*, 1200–1202; (d) Lovely, C. J.; Browning, R. G.; Badarinarayana, V.; Dias, H. V. R., *Tetrahedron Lett.* **2005**, *46*, 2453–2455; (e) Urbano, J.; Belderrain, T. R.; Nicasio, M. C.; Trofimenko, S.; Díaz-Requejo, M. M.; Pérez, P. J., *Organometallics* **2005**, *24*, 1528–1532; (f) Thompson, J. L.; Davies, H. M. L., *J. Am. Chem. Soc.* **2007**, *129*, 6090–6091; (g) Dias, H. V. R.; Lovely, C. J., *Chem. Rev.* **2008**, *108*, 3223–3238.
6. Wolff, L.; *Justus Liebigs Ann. Chem.* **1902**, *325*, 129–195.
7. Bachmann, W. E.; Struve, W. S., *Org. React.* **1946**, *1*, 38.
8. Newman, M. S.; Beal, P. F., *J. Am. Chem. Soc.* **1950**, *72*, 5163–5165.
9. Lee, V.; Newman, M. S., *Org. Synth.* **1970**, *50*, 77.

10. For leading mechanistic studies on silver-mediated Wolff rearrangement, see (a) Yukawa, Y.; Tsuno, Y.; Ibata, T., *Bull. Chem. Soc. Jpn.* **1967**, *40*, 2613–2617; (b) Yukawa, Y.; Tsuno, Y.; Ibata, T., *Bull. Chem. Soc. Jpn.* **1967**, *40*, 2618–2623; (c) Takebayashi, M.; Ibata, T., *Bull. Chem. Soc. Jpn.* **1968**, *41*, 1700–1707; (d) Duggleby, P. M.; Holt, G.; Hope, M. A.; Lewis, A., *J. Chem. Soc. Perkin Trans.* **1972**, *1*, 3020–3024; (e) Agosta, W. C.; Wolff, S., *J. Org. Chem.* **1975**, *40*, 1027–1030; (f) Sudrik, S. G.; Maddanimath, T.; Chaki, N. K.; Chavan, S. P.; Chavan, S. P.; Sonawane, H. R.; Vijayamohanan, K., *Org. Lett.* **2003**, *5*, 2355–2358; (g) Julian, R. R.; May, J. A.; Stoltz, B. M.; Beauchamp, J. L., *J. Am. Chem. Soc.* **2003**, *125*, 4478–4486.
11. Ćiraković, J.; Driver, T. G.; Woerpel, K. A., *J. Am. Chem. Soc.* **2002**, *124*, 9370–9371.
12. For reviews, see (a) West, R., *Pure Appl. Chem.* **1984**, *56*, 163–173; (b) Raabe, G.; Michl, J., *Chem. Rev.* **1985**, *85*, 419–509; (c) West, R., *Angew. Chem. Int. Ed. Engl.* **1987**, *26*, 1201–1211; (d) Barrau, J.; Escudié, J.; Satgé, J., *Chem. Rev.* **1990**, *90*, 283–319; (e) Jutzi, P., *J. Organomet. Chem.* **1990**, *400*, 1–17; (f) Lappert, M. F.; Rowe, R. S., *Coord. Chem. Rev.* **1990**, *100*, 267–292; (g) Satgé, J., *J. Organomet. Chem.* **1990**, *400*, 121–147; (h) Tsumuraya, T.; Batcheller, S. A.; Masamune, S., *Angew. Chem. Int. Ed. Engl.* **1991**, *30*, 902–930; (i) Weidenbruch, M., *Coord. Chem. Rev.* **1994**, *130*, 275–300; (j) Weidenbruch, M., *Chem. Rev.* **1995**, *95*, 1479–1493; (k) Brook, A. G.; Brook, M., *Adv. Organomet. Chem.* **1996**, *39*, 71–158; (l) Hemme, I.; Klingebiel, U., *Adv. Organometal. Chem.* **1996**, *39*, 159–192; (m) Driess, M., *Adv. Organomet. Chem.* **1996**, *39*, 193–229; (n) Okazaki, R.; West, R., *Adv. Organomet. Chem.* **1996**, *39*, 231–273; (o) Baines, K. N.; Stibbs, W. G., *Adv. Organomet. Chem.* **1996**, *39*, 275–324; (p) Power, P. P., *J. Chem. Soc. Dalton Trans.* **1998**, 2939–2951; (q) Weidenbruch, M., *Eur. J. Inorg. Chem.* **1999**, 373–381; (r) Tokitoh, N.; Okazaki, R., *Coord. Chem. Rev.* **2000**, *210*, 251–277; (s) Haaf, M.; Schmedake, T. A.; West, R., *Acc. Chem. Res.* **2000**, *33*, 704–714; (t) Gaspar, P. P., *Sci. Synth.* **2002**, 135–158; (u) Ottosson, H.; Steel, P. G., *Chem. Eur. J.* **2006**, *12*, 1576–1585.
13. Waterman, R.; Hayes, P. G.; Tilley, T. D., *Acc. Chem. Res.* **2007**, *40*, 712–719.
14. Lickiss, P. D., *Chem. Soc. Rev.* **1992**, 271–279.
15. Brown-Wensley, K. A., *Organometallics* **1987**, *6*, 1590–1591.
16. Hengge, E.; Weinberger, M., *J. Organomet. Chem.* **1993**, *443*, 167–173.
17. Ojima, I.; Inaba, S.-I.; Kogure, T.; Nagai, Y., *J. Organomet. Chem.* **1973**, *55*, C7–C8.
18. Zybill, C.; Müller, G., *Angew. Chem. Int. Ed. Engl.* **1987**, *26*, 669–670.
19. Ueno, K.; Tobita, H.; Shimoi, M.; Ogino, H., *J. Am. Chem. Soc.* **1988**, *110*, 4092–4093.
20. Tobita, H.; Ueno, K.; Shimoi, M.; Ogino, H., *J. Am. Chem. Soc.* **1990**, *112*, 3415–3420.
21. Sharma, H. K.; Pannell, K. H., *Organometallics* **2001**, *20*, 7–9.
22. Pannell, K. H.; Sharma, H. K.; Kapoor, R. N.; Cervantes-Lee, F., *J. Am. Chem. Soc.* **1997**, *119*, 9315–9316.
23. Ueno, K.; Masuko, A.; Ogino, H., *Organometallics* **1999**, *18*, 2694–2699.
24. Zhang, Y.; Cervantes-Lee, F.; Pannell, K. H., *J. Organomet. Chem.* **2001**, *634*, 102–108.
25. Zhang, Y.; Cervantes-Lee, F.; Pannell, K. H., *Organometallics* **2003**, *22*, 2517–2524.
26. Zhang, Y.; Pannell, K. H., *Organometallics* **2003**, *22*, 1766–1770.
27. Zhang, Y.; Cervantes-Lee, F.; Pannell, K. H., *Organometallics* **2002**, *21*, 5859–5867.
28. Zhang, Y.; Pannell, K. H., *Organometallics* **2002**, *21*, 503–510.
29. Sato, T.; Okazaki, M.; Tobita, H.; Ogino, H., *J. Organomet. Chem.* **2003**, *669*, 189–199.
30. Tobita, H.; Sato, T.; Okazaki, M.; Ogino, H., *J. Organomet. Chem.* **2000**, *611*, 314–322.

31. Ueno, K.; Masuko, A.; Ogino, H., *Organometallics* **1997**, *16*, 5023–5026.
32. Ueno, K.; Sakai, M.; Ogino, H., *Organometallics* **1998**, *17*, 2138–2140.
33. Ueno, K.; Asami, S.; Watanabe, N.; Ogino, H., *Organometallics* **2002**, *21*, 1326–1328.
34. Feldman, J. D.; Mitchell, G. P.; Nolte, J.-O.; Tilley, T. D., *J. Am. Chem. Soc.* **1998**, *120*, 11184–11185.
35. Grumbine, S. K.; Mitchell, G. P.; Straus, D. A.; Tilley, T. D.; Rheingold, A. L., *Organometallics* **1998**, *17*, 5607–5619.
36. Peters, J. C.; Feldman, J. D.; Tilley, T. D., *J. Am. Chem. Soc.* **1999**, *121*, 9871–9872.
37. Feldman, J. D.; Peters, J. C.; Tilley, T. D., *Organometallics* **2002**, *21*, 4065–4075.
38. Straus, D. A.; Grumbine, S. D.; Tilley, T. D., *J. Am. Chem. Soc.* **1990**, *112*, 7801–7802.
39. Mitchell, G. P.; Tilley, T. D., *J. Am. Chem. Soc.* **1998**, *120*, 7635–7636.
40. West, R.; Denk, M. K., *Pure Appl. Chem.* **1996**, *68*, 785–788.
41. Schmedake, T. A.; Haaf, M.; Paradise, B. J.; Millevolte, A. J.; West, R., *J. Organomet. Chem.* **2001**, *636*, 17–25.
42. Denk, M.; Hayashi, R. K.; West, R., *J. Chem. Soc. Chem. Commun.* **1994**, 33–34.
43. Schmedake, T. A.; Haaf, M.; Paradise, B. J.; Powell, D.; West, R., *Organometallics* **2000**, *19*, 3263–3265.
44. Theil, M.; Jutzi, P.; Neumann, B.; Stammler, A.; Stammler, H.-G., *J. Organomet. Chem.* **2002**, *662*, 34–42.
45. Gehrhus, B.; Hitchcock, P. B.; Lappert, M. F.; Maciejewski, H., *Organometallics* **1998**, *17*, 5599–5601.
46. Amoroso, D.; Haaf, M.; Yap, G. P. A.; West, R.; Fogg, D. E., *Organometallics* **2002**, *21*, 534–540.
47. Woo, L. K.; Smith, D. A.; Young, V. G., *Organometallics* **1991**, *10*, 3977–3982.
48. (a) Crabtree, R. H.; in *The Organometallic Chemistry of the Transition Metals*, Wiley, New York, **1994**, pp. 270–310; (b) Bernasconi, C. F., *Chem. Soc. Rev.* **1997**, *26*, 299–307; (c) de Meijere, A.; Schirmer, H.; Duetsch, M., *Angew. Chem. Int. Ed.* **2000**, *39*, 3964–4002.
49. Probst, R.; Leis, C.; Gamper, S.; Herdtweck, E.; Zybill, C.; Auner, N., *Angew. Chem. Int. Ed. Engl.* **1991**, *30*, 1132–1135.
50. Handwerker, H.; Leis, C.; Probst, R.; Bissinger, P.; Grohmann, A.; Kiprof, P.; Herdtweck, E.; Bluemel, J.; Auner, N.; Zybill, C., *Organometallics* **1993**, *12*, 2162–2176.
51. Marquez, A.; Fernandez Sanz, J., *J. Am. Chem. Soc.* **1992**, *114*, 2903–2909.
52. Cundari, T. R.; Gordon, M. S., *Organometallics* **1992**, *11*, 3122–3129.
53. Leis, C.; Wilkinson, D. L.; Handwerker, H.; Zybill, C.; Mueller, G., *Organometallics* **1992**, *11*, 514–529.
54. Yamashita, H.; Tanaka, M.; Goto, M., *Organometallics* **1992**, *11*, 3227–3232.
55. Glaser, P. B.; Tilley, T. D., *J. Am. Chem. Soc.* **2003**, *125*, 13640–13641.
56. Calimano, E.; Tilley, T. D., *J. Am. Chem. Soc.* **2008**, *130*, 9226–9227.
57. Klei, S. R.; Tilley, T. D.; Bergman, R. G., *Organometallics* **2002**, *21*, 3376–3387.
58. Okamoto, K.; Hayashi, T., *Org. Lett.* **2007**, *9*, 5067–5069.
59. Okamoto, K.; Hayashi, T., *Chem. Lett.* **2008**, *37*, 108–109.
60. Ohmura, T.; Masuda, K.; Suginome, M., *J. Am. Chem. Soc.* **2008**, *130*, 1526–1527.

61. Palmer, W. S.; Woerpel, K. A., *Organometallics* **1997**, *16*, 4824–4827.
62. Franz, A. K.; Woerpel, K. A., *Acc. Chem. Res.* **2000**, *33*, 813–820.
63. Franz, A. K.; Woerpel, K. A., *J. Am. Chem. Soc.* **1999**, *121*, 949–957.
64. Franz, A. K.; Woerpel, K. A., *Angew. Chem. Int. Ed. Engl.* **2000**, *39*, 4295–4299.
65. For reviews, see (a) Tamao, K., in *Advances in Silicon Chemistry*, JAI Press, Greenwich, CT, **1996**, Vol. 3, pp. 1–62; (b) Fleming, I., *Chemtracts Org. Chem.* **1996**, *9*, 1–64; (c) Jones, G. R.; Landais, Y., *Tetrahedron* **1996**, *52*, 7599–7662.
66. Smitrovich, J. H.; Woerpel, K. A., *J. Org. Chem.* **1996**, *61*, 6044–6046.
67. Boudjouk, P.; Samaraweera, U.; Sooriyakumaran, R.; Chrusciel, J.; Anderson, K. R., *Angew. Chem. Int. Ed. Engl.* **1988**, *27*, 1355–1356.
68. Seyferth, D.; Annarelli, D. C., *J. Am. Chem. Soc.* **1975**, *97*, 7162–7164.
69. Schäfer, A.; Weidenbruch, M.; Peters, K.; von Schnering, H.-G., *Angew. Chem. Int. Ed. Engl.* **1984**, *23*, 302–303.
70. Boudjouk, P.; Black, E.; Kumarathasan, R., *Organometallics* **1991**, *10*, 2095–2096.
71. Belzner, J.; Ihmels, H.; Kneisel, B. O.; Gould, R. O.; Herbst-Irmer, R., *Organometallics* **1995**, *14*, 305–311.
72. Belzner, J.; Dehnert, U.; Ihmels, H., *Tetrahedron* **2001**, *57*, 511–517.
73. Driver, T. G.; Franz, A. K.; Woerpel, K. A., *J. Am. Chem. Soc.* **2002**, *124*, 6524–6525.
74. Ćiraković, J.; Driver, T. G.; Woerpel, K. A., *J. Org. Chem.* **2004**, *69*, 4007–4012.
75. Franz, A. K.; Woerpel, K. A., *Angew. Chem.* **2000**, *112*, 4465–4469.
76. Calad, S. A.; Ćiraković, J.; Woerpel, K. A., *J. Org. Chem.* **2007**, *72*, 1027–1030.
77. Kuwahara, Y.; Fukami, H.; Howard, R.; Ishii, S.; Matsumura, F.; Burkholder, W. E., *Tetrahedron* **1978**, *34*, 1769–1774.
78. Shaw, J. T.; Woerpel, K. A., *J. Org. Chem.* **1997**, *62*, 442–443.
79. Shaw, J. T.; Woerpel, K. A., *Tetrahedron* **1997**, *53*, 16597–16606.
80. Shaw, J. T.; Woerpel, K. A., *Tetrahedron* **1999**, *55*, 8747–8756.
81. Peng, Z.-H.; Woerpel, K. A., *Org. Lett.* **2002**, *4*, 2945–2948.
82. Calad, S. A.; Woerpel, K. A., *J. Am. Chem. Soc.* **2005**, *127*, 2046–2047.
83. Driver, T. G.; Woerpel, K. A., *J. Am. Chem. Soc.* **2004**, *126*, 9993–10002.
84. Lawrance, G. A., *Chem. Rev.* **1986**, *86*, 17–33.
85. Cleary, P. A.; Woerpel, K. A., *Org. Lett.* **2005**, *7*, 5531–5533.
86. Damrauer, R.; Crowell, A. J.; Craig, C. F., *J. Am. Chem. Soc.* **2003**, *125*, 10759–10766.
87. Belzner, J.; Ihmels, H., *Adv., Organomet. Chem.* **1999**, *43*, 1–42.
88. Belzner, J.; Dehnert, U.; Ihmels, H.; Hübner, M.; Müller, P.; Usón, I., *Chem. Eur. J.* **1998**, *4*, 852–863.
89. DePuy, C. H.; Damrauer, R.; Bowie, J. H.; Sheldon, J. C., *Acc. Chem. Res.* **1987**, *20*, 127–133.
90. Palmer, W. S.; Woerpel, K. A., *Organometallics* **1997**, *16*, 1097–1099.
91. Palmer, W. S.; Woerpel, K. A., *Organometallics* **2001**, *20*, 3691–3697.
92. Ishikawa, M.; Sugisawa, H.; Harata, O.; Kumada, M., *J. Organomet. Chem.* **1981**, *217*, 43–50.
93. Seyferth, D.; Shannon, M. L.; Vick, S. C.; Lim, T. F. O., *Organometallics* **1985**, *4*, 57–62.

94. Seyferth, D.; Vick, S. C.; Shannon, M. L., *Organometallics* **1984**, *3*, 1897–1905.

95. Clark, T. B.; Woerpel, K. A., *J. Am. Chem. Soc.* **2004**, *126*, 9522–9523.

96. Goller, A.; Heydt, H.; Clark, T., *J. Org. Chem.* **1996**, *61*, 5840–5846.

97. Gordon, M. S., *J. Am. Chem. Soc.* **1980**, *102*, 7419–7422.

98. Shaw, J. T.; Woerpel, K. A., *J. Org. Chem.* **1997**, *62*, 6706–6707.

99. Anderson, L. L.; Woerpel, K. A., *Org. Lett.* **2009**, *11*, 425–428.

100. For the analogous photochemical reaction of silacyclopropenes with nitriles, see Sakurai, H.; Kamiyama, Y.; Nakadaira, Y., *J. Chem. Soc. Chem. Commun.* **1978**, 80–81.

101. Clark, T. B.; Woerpel, K. A., *Organometallics* **2005**, *24*, 6212–6219.

102. The sulfonyl oxygen has been demonstrated to be nucleophilic. For leading reports, see (a) Braverman, S.; Duar, Y., *J. Am. Chem. Soc.* **1983**, *105*, 1061–1063; (b) Carretero, J. C.; Garcia Ruano, J. L.; Martinez, M. C.; Rodriguez, J. H., *Tetrahedron* **1987**, *43*, 4417–4423; (c) Lucchini, V.; Modena, G.; Pasquato, L., *J. Chem. Soc. Chem. Commun.* **1992**, 293–294; (d) Saginova, L. G.; Polyakova, O. V.; Shabarov, Y. S., *J. Org. Chem. USSR* **1992**, *28*, 938–942; (e) Leca, D.; Song, K.; Amatore, M.; Fensterbank, L.; Lacote, E.; Malacria, M., *Chem. Eur. J.* **2004**, *10*, 906–916.

103. For the bond dissociation enthalpies of S=N, S=O, Si–N, and Si–O, see (a) Parsons, S.; Passmore, J., *Inorg. Chem.* **1992**, *31*, 526–528; (b) Roux, M. V.; Temprado, M.; Jimenez, P.; Davalos, J. Z.; Notario, R.; Guzman-Mejia, R.; Juaristi, E., *J. Org. Chem.* **2003**, *68*, 1762–1770; (c) Johnson, E. R.; Clarkin, O. J.; DiLabio, G. A., *J. Phys. Chem. A* **2003**, *107*, 9953–9963.

104. Clark, T. B.; Woerpel, K. A., *Org. Lett.* **2006**, *8*, 4109–4112.

105. For leading reviews on the Mukaiyama aldol reaction, see (a) Carreira, E. M., in *Comprehensive Asymmetric Catalysis I–III*; Jacobsen, E. N.; Pfaltz, A.; Yamamoto, H., eds., Springer-Verlag, Berlin, **1999**, pp. 997–1065; (b) Mahrwald, R., *Chem. Rev.* **1999**, *99*, 1095–1120.

106. (a) Lin, S.; Bondar, G. V.; Levy, C. J.; Collins, S., *J. Org. Chem.* **1998**, *63*, 1885–1892; (b) Hollis, T. K.; Bosnich, B., *J. Am. Chem. Soc.* **1995**, *117*, 4570–4581; (c) Carreira, E. M.; Singer, R. A., *Tetrahedron Lett.* **1994**, *35*, 4323–4326.

107. For examples of 1,3-Brook rearrangements, see (a) Chuprakov, S.; Malyshev, D. A.; Trofimov, A.; Gevorgyan, V., *J. Am. Chem. Soc.* **2007**, *129*, 14868–14869; (b) Smith, A. B.; Pitram, S. M.; Boldi, A. M.; Gaunt, M. J.; Sfouggatakis, C.; Moser, W. H., *J. Am. Chem. Soc.* **2003**, *125*, 14435–14445; (c) Naganuma, K.; Kawashima, T.; Okazaki, R., *Chem. Lett.* **1999**, *28*, 1139–1140; (b) Shinokubo, H.; Miura, K.; Oshima, K.; Utimoto, K., *Tetrahedron* **1996**, *52*, 503–514; (c) Yamamoto, K.; Kimura, T.; Tomo, Y., *Tetrahedron Lett.* **1985**, *26*, 4505–4508; (d) Wilson, S. R.; Georgiadis, G. M., *J. Org. Chem.* **1983**, *48*, 4143–4144; (e) Hudrlik, P. F.; Hudrlik, A. M.; Kulkarni, A. K., *J. Am. Chem. Soc.* **1982**, *104*, 6809–6811.

108. For reviews involving carbenes and metallo carbenoids, see (a) Padwa, A.; Hornbuckle, S. F., *Chem. Rev.* **1991**, *91*, 263–309; (b) Padwa, A.; Weingarten, M. D., *Chem. Rev.* **1996**, *96*, 223–270; (c) Padwa, A.; in *Topics in Current Chemistry*, Springer, Berlin/Heidelberg, **1997**, Vol. 189, pp. 121–158; (d) Padwa, A., *Pure Appl. Chem.* **2003**, *75*, 47–62.

109. Gehrhus, B.; Hitchcock, P. B.; Lappert, M. F., *Organometallics* **1997**, *16*, 4861–4864.

110. (a) Ando, W.; Ikeno, M.; Sekiguchi, A., *J. Am. Chem. Soc.* **1978**, *100*, 3613–3615; (b) Seyferth, D.; Lim, T. F. O., *J. Am. Chem. Soc.* **1978**, *100*, 7074–7075; (c) Belzner, J.; Ihmels, H.; Pauletto, L.; Noltemeyer, M., *J. Org. Chem.* **1996**, *61*, 3315–3319; (d) Jutzi, P.;

Eikenberg, D.; Bunte, E.-A.; Möhrke, A.; Neumann, B.; Stammler, H. -G., *Organometallics* **1996**, *15*, 1930–1934; (e) Sakai, N.; Fukushima, T.; Minakata, S.; Ryu, I.; Komatsu, M., *Chem. Commun.* **1999**, 1857–1858.

111. Schäfer, A.; Weidenbruch, M.; Pohl, S., *J. Organomet. Chem.* **1985**, *282*, 305–313.
112. (a) Ando, W.; Ikeno, M.; Sekiguchi, A., *J. Am. Chem. Soc.* **1977**, *99*, 6447–6449; (b) Ishikawa, M.; Nakagawa, K. -I.; Kumada, M., *J. Organomet. Chem.* **1977**, *135*, C45–C49.
113. Heinicke, J.; Gehrhus, B., *J. Organomet. Chem.* **1992**, *423*, 13–21.
114. Gehrhus, B.; Lappert, M. F., *J. Organomet. Chem.* **2001**, *617–618*, 209–223.
115. Ando, W.; Hagiwara, K.; Sekiguchi, A., *Organometallics* **1987**, *6*, 2270–2271.
116. Calad, S. A.; Woerpel, K. A., *Org. Lett.* **2007**, *9*, 1037–1040.
117. Bourque, L. E.; Woerpel, K. A., *Org. Lett.* **2008**, *10*, 5257–5260.
118. For the isolation of oxasilacyclopropane-derived ketones, see Ando, W.; Hamada, Y.; Sekiguchi, A.; Ueno, K., *Tetrahedron Lett.* **1982**, *23*, 5323–5326.
119. Howard, B. E.; Woerpel, K. A., *Org. Lett.* **2007**, *9*, 4651–4653.
120. Brook, A. G.; Kong, Y. K.; Saxena, A. K.; Sawyer, J. F., *Organometallics* **1988**, *7*, 2245–2247.
121. Brook, A. G.; Saxena, A. K.; Sawyer, J. F., *Organometallics* **1989**, *8*, 850–852.
122. Brook, A. G.; Azarian, D.; Baumegger, A.; Hu, S. S.; Lough, A. J., *Organometallics* **1993**, *12*, 529–534.
123. Nevárez, Z.; Woerpel, K. A., *Org. Lett.* **2007**, *9*, 3773–3776.
124. Pearson, R. G.; Songstad, J., *J. Am. Chem. Soc.* **1967**, *89*, 1827–1836.
125. Pearson, R. G., *J. Org. Chem.* **1989**, *54*, 1423–1430.
126. Pearson, R. G., *Acc. Chem. Res.* **1993**, *26*, 250–255.
127. Nevárez, Z.; Woerpel, K. A., *J. Org. Chem.* **2008**, *73*, 8113–8115.
128. Nielsen, I. M. B., *J. Phys. Chem. A* **1998**, *102*, 3193–3201.
129. For leading studies on the boat-like nature of [3,3]-sigmatropic rearrangements of divinyl-substituted three- and four-membered rings, see (a) Hammond, G. S.; DeBoer, C. D., *J. Am. Chem. Soc.* **1964**, *86*, 899–902; (b) Trecker, D. J.; Henry, J. P., *J. Am. Chem. Soc.* **1964**, *86*, 902–905;(c) Piers, E., in *Comprehensive Organic Synthesis*, Trost, B. M.; Fleming, I.; Paquette, L. A., eds.; Pergamon Press, Oxford, **1991**, Vol. 5, pp. 971–998; (d) Bronson, J. J.; Danheiser, R. L., in *Comprehensive Organic Synthesis*, Trost, B. M.; Fleming, I.; Paquette, L. A., eds.; Pergamon Press, Oxford, **1991**, Vol. 5, pp. 999–1035; (e) Özkan, İ.; Zora, M., *J. Org. Chem.* **2003**, *68*, 9635–9642; (f) Zora, M., *J. Org. Chem.* **2005**, *70*, 6018–6026.
130. Roskamp, E. J.; Johnson, C. R., *J. Am. Chem. Soc.* **1986**, *108*, 6062–6063.
131. Doyle, M. P.; Bagheri, V.; Harn, N. K., *Tetrahedron Lett.* **1988**, *29*, 5119–5122.
132. Doyle, M. P.; Forbes, D. C.; Vasbinder, M. M.; Peterson, C. S., *J. Am. Chem. Soc.* **1998**, *120*, 7653–7654.
133. Clark, J. S.; Fretwell, M.; Whitlock, G. A.; Burns, C. J.; Fox, D. N. A., *Tetrahedron Lett.* **1998**, *39*, 97–100.
134. Kitagaki, S.; Yanamoto, Y.; Tsutsui, H.; Anada, M.; Nakajima, M.; Hashimoto, S., *Tetrahedron Lett.* **2001**, *42*, 6361–6364.
135. Kawachi, A.; Doi, N.; Tamao, K., *J. Am. Chem. Soc.* **1997**, *119*, 233–234.
136. Bourque, L. E.; Cleary, P. A.; Woerpel, K. A., *J. Am. Chem. Soc.* **2007**, *129*, 12602–12603.

8

SILVER CARBENOIDS

CARL J. LOVELY

Department of Chemistry and Biochemistry, The University of Texas at Arlington, Arlington, Texas

8.1 INTRODUCTION

Although the other coinage metals, silver and gold, are in the same group as copper, which has a rich history in organic synthesis, the utility of Ag and Au as homogeneous catalysts in organic synthesis has only just begun to be realized since the early 2000s.[1] Silver salts and complexes have seen extensive use in organic synthesis as oxidants, as halide sinks, and as heterogeneous catalysts in the Wolff rearrangement.[2] Since the late 1990s, silver salts have begun to emerge as viable homogeneous catalysts for a variety of transformations, few of which are feasible

Silver in Organic Chemistry Edited by Michael Harmata

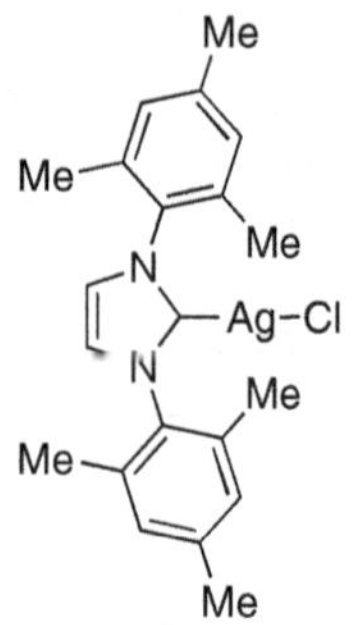

1

Figure 8.1. NHC silver complex.

with other transition metal complexes, and as a result provide new avenues for investigation, in particular mechanistic issues and derived applications. One might anticipate, on the basis of periodic trends, that copper and silver would exhibit some similarities in their chemistries, and therefore it is not surprising that the potential of silver salts to function as group transfer catalysts has been explored.[3] This chapter summarizes the current status of net silver-catalyzed carbene transfer processes primarily with diazo precursors.[4,5] The term *net* here refers to the fact there is evidence to suggest that in some cases the reaction does not proceed via a silver carbene, but rather that the silver complex functions in a different capacity, usually as a Lewis acid. In other situations, the reactivity profile suggests that silver carbenes are the reactive intermediates, but this has not been established unequivocally. It should be noted that this discussion does not include stable silver carbenes (NHC = *N*-heterocylic carbene) of the type exemplified in Figure 8.1,[6,7] which have been used to transfer the imidazolylidene fragment to other metal complexes.[8]

8.2 WOLFF REARRANGEMENT

The Wolff rearrangement constitutes the first step of the well-known Arndt–Eistert homologation of carboxylic acids, which can be performed under a number of conditions, many of which include the presence of silver salts.[2] While the mechanistic details of these catalyzed reactions remain to be fully elucidated, there is some evidence to suggest that silver carbenes may serve as intermediates in certain circumstances. In a combined mass spectrometry and computational study, Beauchamp and Stoltz have demonstrated the formation and rearrangement of a silver carbene species derived from dimethyl diazomalonate.[9] In this study, the first stable species observed was the β-diketo complex **2** (for comparison see complex **21** in Fig. 8.2), which, on activation using low-energy collision-activated dissociation (CAD), loses the acetonitrile ligand, producing **3** (Scheme 8.1). Further activation leads to loss of nitrogen and a mole of carbon monoxide, leading to the formation of a species with a molecular mass consistent with the silver carbene **4**. An additional round of CAD leads to the expulsion of an additional mole of carbon monoxide and the formation of a second silver carbene species **5**. Additional support for this

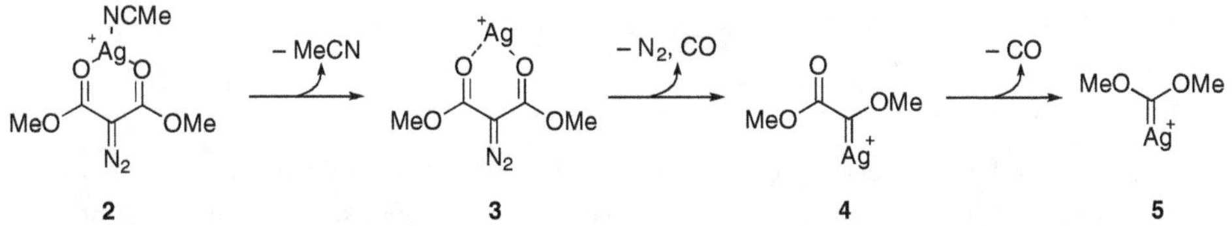

Scheme 8.1

decomposition pathway was provided through the use of labeled substrates (deuterium or ^{13}C), although these experiments were performed with the corresponding copper system, which underwent similar chemistry.[9]

Chavan and coworkers provide evidence that the Wolff rearrangement is facilitated by the formation of silver nanoclusters, which initiate electron transfer to the diazo compound providing **8**. While the precise fate of this species remains to be firmly established, they suggest a multicycle process involving the intermediacy of a silver carbene **10** (Scheme 8.2).[10–12] Decomposition of the silver carbene to the free carbene **14** precedes rearrangement to ketene **13**, which is then trapped with water to provide the carboxylic acid **15** (Scheme 8.2).

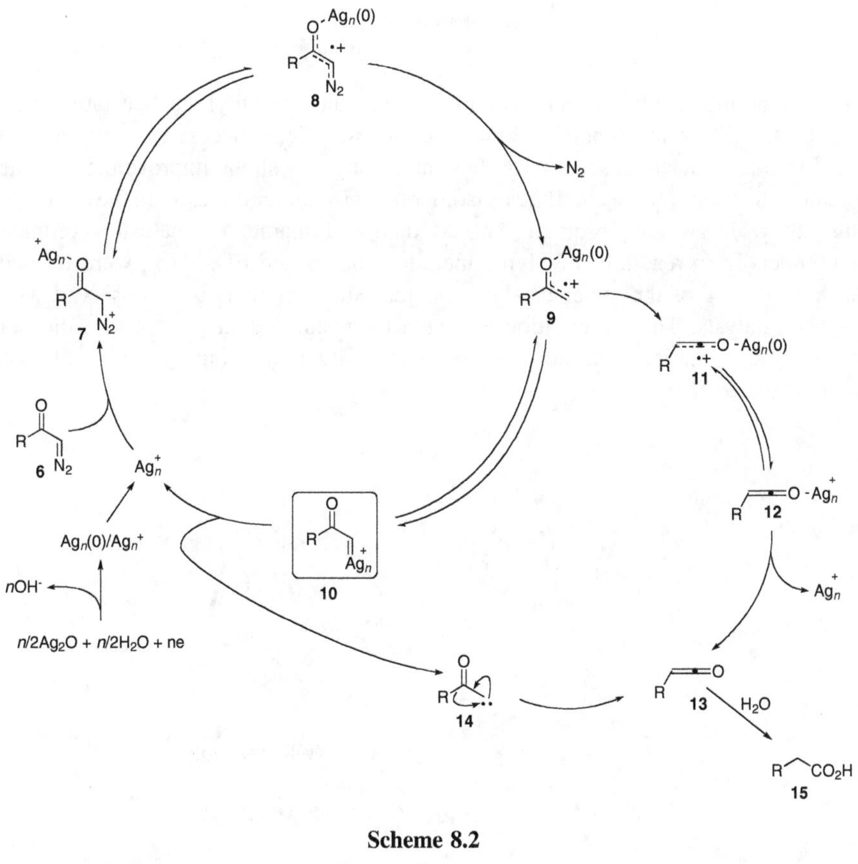

Scheme 8.2

8.3 CARBENE TRANSFER REACTIONS TO π BONDS

8.3.1 Aziridination

Jørgensen and coworkers reported the preparation of *N*-tosyl aziridines **19–20** by the net carbene addition (via a diazo compound) to *N*-tosyl iminoesters with either copper or silver catalysts.[13,14] It was noted that the copper catalysts were generally superior, although a catalyst derived from $AgSbF_6$ and (*R*)-Tol-BINAP provided the corresponding aziridine **19** from **16** and trimethylsilyl diazomethane **17** (R = TMS) in excellent chemical yield with high levels of diastereoselectivity, but unfortunately the enantioselectivity was poor (Scheme 8.3). This success with trimethylsilyldiazo-

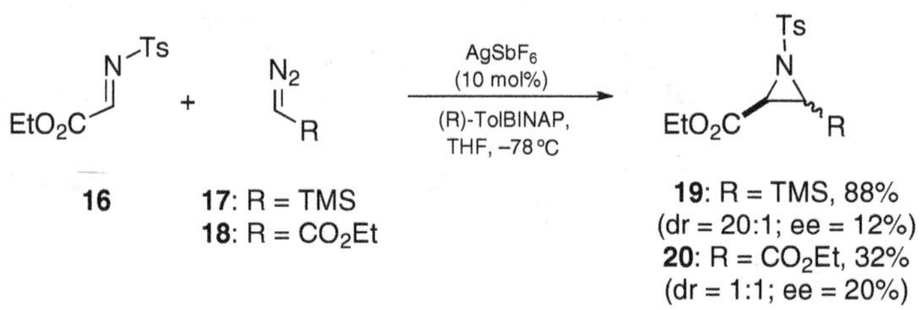

Scheme 8.3

methane prompted Jørgensen to extend this reaction with ethyl diazoacetate (EDA, **18**, R = CO_2Et). Unfortunately, these reactions were less successful, with reduced yield, reduced diastereoselectivity, but interestingly a slight improvement in the enantioselectivity for the chiral *trans* isomer (the *cis*-isomer is *meso*). In the course of this study, Jørgensen observed that no diethyl fumarate or maleate, common byproducts from reactions involving metallocarbenes and EDA (**18**), were formed in the presence of the silver catalyst, whereas such products were observed with copper catalysts. This observation led them to speculate that the reaction did not involve the formation of a metallocarbene with silver salts, but was in fact Lewis acid-catalyzed.[14]

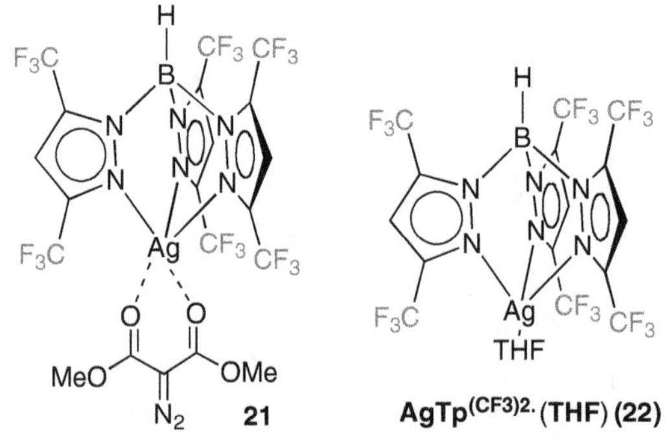

Figure 8.2. Silver tris(pyrazolyl)borato complexes.

8.3.2 Cyclopropanation

The Dias group has developed an interest in the synthesis and characterization of putative reactive intermediates in a variety of transition-metal-mediated processes, particularly those involving the coinage metals.[4,15–17] As a part of this study, the stable AgTp–diazomalonate [Tp = tris(pyrazolyl)borato] complex **21** was prepared in an attempt to obtain a discrete, and characterizable, silver carbene complex.[18] However, rather than the desired carbene complex, the β-diester complex **21** (Fig. 8.2) was isolated and characterized by X-ray crystallography. It is of note that coordination occurs via the ester carbonyl oxygens, and thus this may be considered akin to nonproductive resting state in carbene transfer reactions in which diazomalonate participates. A related species was identified in the Stoltz–Beauchamp mass spectrometry investigation discussed above (see compound **3**, Scheme 8.1) as formed in high abundance, which then formed the silver carbene on collisional activation. To further probe this issue, an attempt was made to prepare and characterize the corresponding complex with the more reactive EDA (**18**) and **22**; however, no characterizable silver complex was obtained. Interestingly, the evolution of a gas was observed, indicating that a reaction occurred, presumably the formation of a reactive silver–carbene with the release of nitrogen gas. On the basis of these observations, the experiment was repeated in the presence of a carbene trap to intercept the putative silver carbene. An initial attempt at cyclopropanation with styrene **23** was not successful in the clean formation of the corresponding cyclopropàne **24**. Instead, a rather complex mixture of products was obtained containing the expected cyclopropane **24** (Scheme 8.4), the ring expanded product **25**, the solvent insertion product **26** and other products (possibly **27**).[17] This is in contrast to the corresponding copper complex, which smoothly cyclopropanoates a variety of olefins, including styrene.[19]

Ultimately it was found that by conducting the reaction with benzene in CH_2Cl_2 or in neat benzene, the ring expansion product **30a**, via the norcaradiene **29a** (Büchner reaction), was obtained in good yield (Scheme 8.5).[17] Evaluation of the substrate

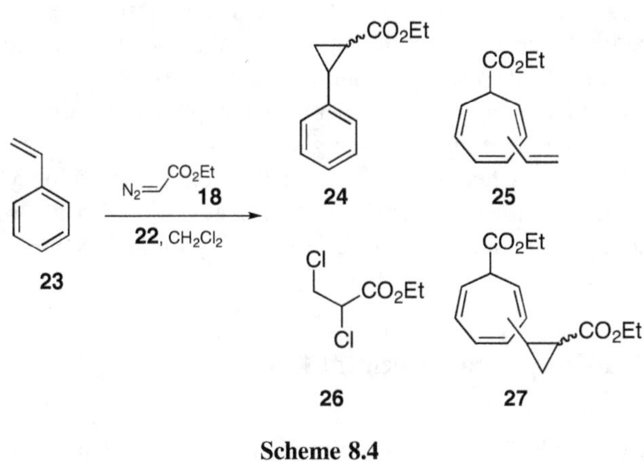

Scheme 8.4

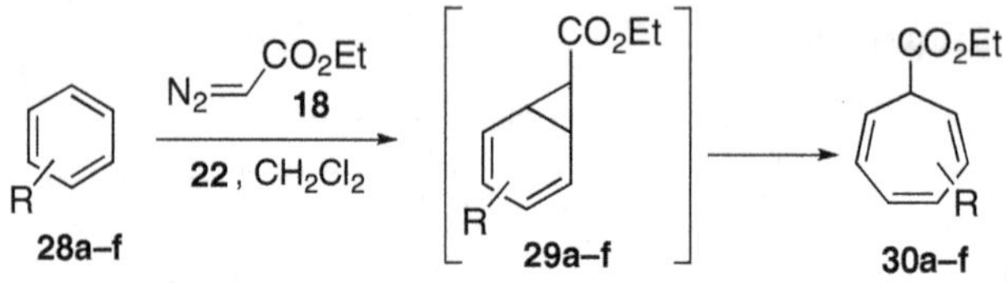

Scheme 8.5

scope revealed that other aromatic derivatives, **28b–f**, participated in the rearrangement, but generally the yields of the ring-expanded product **30b–f** were lower than with benzene (Table 8.1). In some cases (**30b** and **30f**), competitive benzylic C–H insertion was observed, providing **31** and **32,** respectively (Fig. 8.3). Perez and coworkers have observed similar Büchner chemistry, as well as cyclopropanation of styrene with the polybrominated Tp–silver complex **33** (Fig. 8.3), although no chemical yields were reported in either case.[20] The corresponding Cu–scorpionate complexes function well as cyclopropanation catalysts.[19,21–23]

Davies and Thompson have demonstrated that several donor-acceptor diazo compounds (**35** and **37**) participate in selective silver-catalyzed cyclopropanation of a variety of olefins with $AgSbF_6$ as catalyst at reflux in CH_2Cl_2 (Schemes 8.6 and 8.7).[24] Interestingly, it had been reported previously that this same group of olefins undergoes predominantly C–H insertion if $Rh_2(OAc)_4$ is employed as the catalyst. Very high levels of chemoselectivity were obtained, with typically $\geq 15:1$ selectivity for cyclopropanation over C–H insertion (Tables 8.2 and 8.3). Lower selectivity was observed in only one case, the reaction of styryl diazoacetate (**37**) and cyclohexadiene (Table 8.3, entry 2). In addition to high levels of chemoselectivity, excellent diastereoselectivities were observed, with additional isomers undetected by ^{1}H NMR spectroscopy.[24]

8.4 FORMATION AND REACTIONS OF YLIDES

Metallocarbenes derived from diazoacetates and the appropriate transition metal are generally thought to be electrophilic in nature; that is, they are susceptible to nucleophilic addition. This reactivity manifold has been exploited with the formation of ylide species on reaction with a number of nucleophiles providing intermediates that can undergo either rearrangements (e.g., **41** $\rightarrow$ **42**, Scheme 8.8)[25] or cycloadditions (e.g., **43** $\rightarrow$ **44**, Scheme 8.8)[26] depending on the type of ylide formed (Scheme 8.8). In the case of silver-derived carbenes, this area has only recently begun to attract attention, and therefore there are only limited examples reported in the literature.

8.4.1 C–Hal Addition–Rearrangement Reactions

In the course of our investigation of the cyclopropanation of olefins, we observed the formation of a by-product, ethyl-2,3-dichloropropanoate (**26**, Scheme 8.4), when the reaction was conducted in CH_2Cl_2, resulting from the net insertion of the carbene

TABLE 8.1. Yields and Isomer Ratios of the Silver-Catalyzed Büchner Reaction of Benzene Derivatives with EDA

Substrate	Product	Yield (%)[a,b]	Isomer Ratio (2 : 3 : 4)
 **28a**	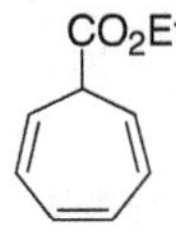 **30a**	74 75[c]	NA NA
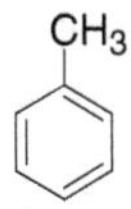 **28b**	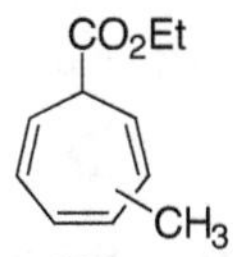 **30b**	64[d] 62[c,e]	1 : 1.8 : 2.4
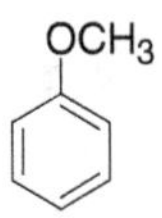 **28c**	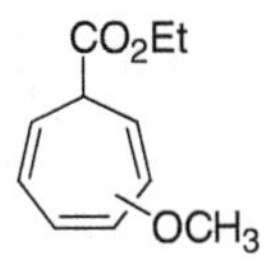 **30c**	40	1 : 0 : 19
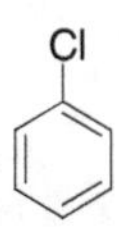 **28d**	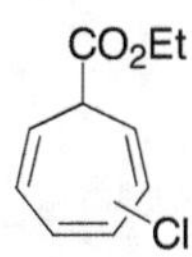 **30d**	49[f]	1 : 8.5 : 18
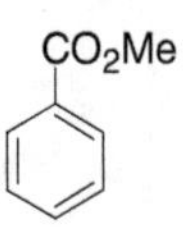 **28e**	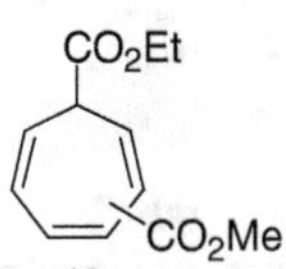 **30e**	14[f]	0 : 1 : 1
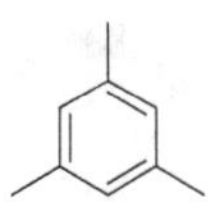 **28f**	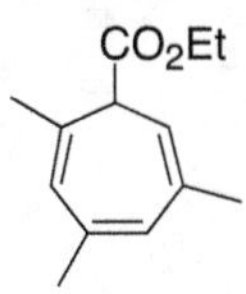 **30f**	35[g]	NA

[a] The yields are the average of at least two runs and refer to isolated material and, unless noted, were conducted in CH_2Cl_2.
[b] In the reactions conducted in CH_2Cl_2, there 15–25% of the C–Cl insertion product **26** was observed in the ^1H NMR spectrum of the crude reaction mixture.
[c] These reactions were conducted in the neat arene.
[d] The C–H insertion product **31** (~ 4%) was observed in ^1H NMR spectrum of crude reaction mixture.
[e] The C–H insertion product **31** (~ 5%) was observed in ^1H NMR spectrum of crude reaction mixture.
[f] The purified material was contaminated with ~ 5% of the C–Cl insertion product.
[g] The C–H insertion product **32** (~ 28%) was observed in ^1H NMR spectrum of crude reaction mixture.

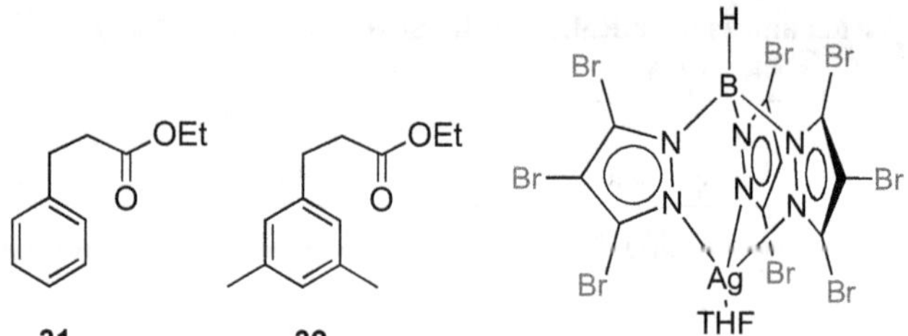

31 32 $Tp^{Br3}Ag(THF)$ (33)

Figure 8.3. C–H insertion products and the Perez catalyst.

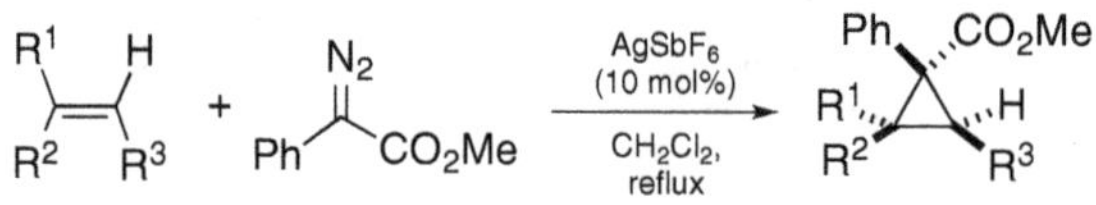

34 35 36

Scheme 8.6

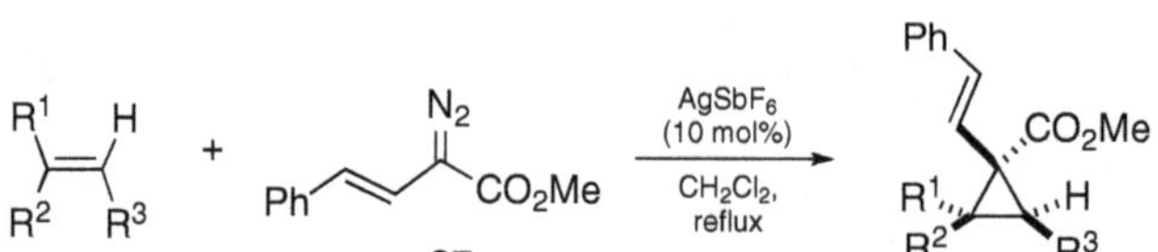

34 37 38

Scheme 8.7

TABLE 8.2. Silver-Catalyzed Cyclopropanation with 35[a]

Entry	R^1	R^2	R^3	Yield (%)
1	H	-$(CH_2)_4$-		88
2	H	-($CH{=}CHCH_2CH_2$)-		79
3	H	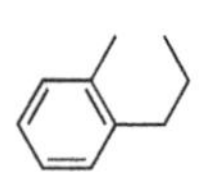		80
4	H	H	Ph	96
5	Ph	Ph	H	82
6	Me	H	Ph	80
7	H	Me	Ph	86
8	Ph	H	Ph	84
9	H	Ph	Ph	54

[a] Approximately 5–10 equiv of the olefin was employed.

TABLE 8.3. Silver-Catalyzed Cyclopropanation with 37[a]

Entry	R^1	R^2	R^3	Yield (%)
1	H	$-(CH_2)_4-$		43
2	H	$-(CH{=}CHCH_2CH_2)-$		67[b]
3	H	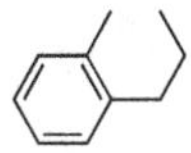		57
4	H	H	Ph	82
5	Ph	Ph	H	56[c]
6	Me	H	Ph	65
7	Ph	H	Ph	34

[a] Approximately 5–10 equiv of the olefin was employed.
[b] Includes 15% of insertion product.
[c] Reaction conducted at room temperature.

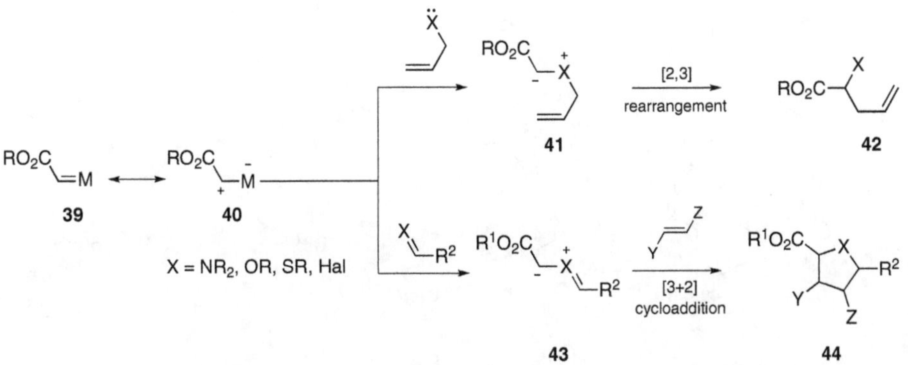

Scheme 8.8

fragment into the C–Cl bond. The participation of halogenated solvents with metallocarbenes is rare, but has been observed previously.[27,28] Similarly, the reaction of free carbenes with haloalkanes and allyl halides has also been described previously in the literature, but these processes are quite unselective.[29–34]

To further explore the formation of **26**, the reaction was repeated in CH_2Cl_2 only, and in this case the insertion product was obtained in 26% yield (Scheme 8.9; Table 8.4, entry 1).[35] On the basis of this result, an examination of the scope and

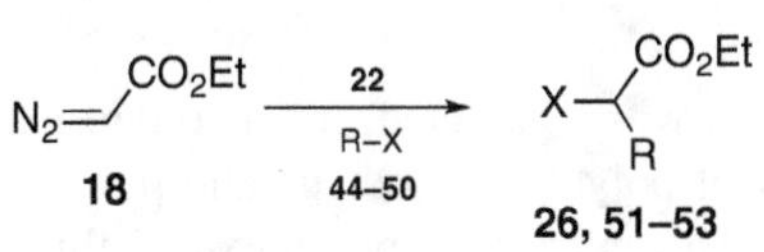

Scheme 8.9

TABLE 8.4. Products and Yields of the Carbene Insertion into Haloalkanes

Entry	Substrate (a = Cl, b = Br)	Product (a = Cl, b = Br)	Yield (X = Cl) (%)[a,b]	Yield (X = Br) (%)[a,b]
1	CH_2X_2 **44a-b**	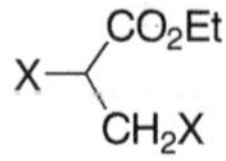 **26a–b**	26	65
2	CHX_3 **45a-b**	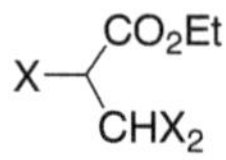 **51a–b**	60	29[c,d]
3	CX_4 **46a-b**	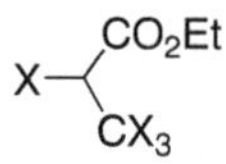 **52a–b**	62	48[c,e]
4	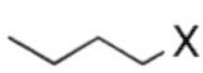 **47a–b**	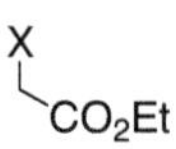 **53a–b**	39	78
5	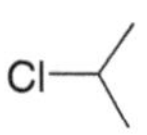 **48a**	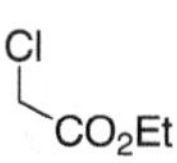 **53a**	70	—
6	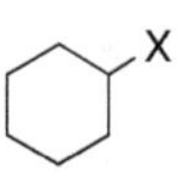 **49a–b**	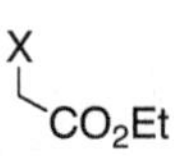 **53a–b**	63[f,g]	60[f,h]
7	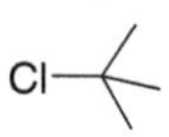 **50a**	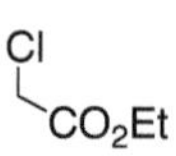 **53a**	73	—

[a] These isolated yields are based on the average of at least two experiments and on the amount of EDA used.
[b] The material balance is accounted for by dimerization of EDA (fumarate and maleate).
[c] Reaction performed in CH_2Cl_2.
[d] In addition, **26a** was formed (~ 9%).
[e] In addition, **26a** was formed (~ 3%).
[f] Yield determined by GC.
[g] In addition, cyclohexene was formed (~ 47%, by GC).
[h] In addition, cyclohexene was formed (~ 57%, by GC).

limitations of this reaction was conducted. It was found that similar products were obtained with a number of polychloromethane and polybromomethane derivatives (Table 8.4, entries 1–3). However, attempts to extend this addition–rearrangement sequence to other haloalkanes was not successful in the expected sense, providing instead the α-haloacetate derivative **53a-b** via a net 1,1-hydrohalogenation/elimination process and the olefin (Table 8.4, entries 4–7).[35] In the case of the reaction with halocyclohexanes, the formation of cyclohexene was established and quantified using GC analysis, which revealed that an equivalent amount (within experimental error) of olefin was formed.

On the basis of these results, a mechanism (Scheme 8.10) involving the intermediacy of a silver–carbene **54** was proposed in which the insertion product arises from the formation of the halonium ylide **55**, followed by a 1,2 shift (**55** → **26**, or **51** or **52**). Alternatively, if the substrate and thus the halonium ylide **56** contain a β-hydrogen, this could be removed by an intramolecular deprotonation with concomitant loss of halide resulting in formation of the olefin **57** and the α-haloacetate **53**. At this stage, no independent evidence has been obtained to support this pathway; thus this mechanism is purely speculative (see text below). Indeed, although the pathway has been depicted as involving metal-free intermediates, it is quite likely that this is not the case, but this awaits independent experimental verification.

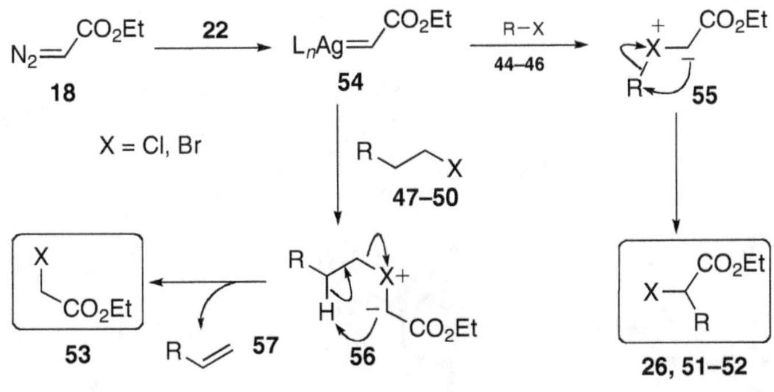

Scheme 8.10

The interception of metallocarbenes (Cu or Rh) with allylic substituted amines, ethers, thioethers, and selenides and their subsequent [2,3] rearrangement is well documented.[36] On the other hand, fewer reports exist of the corresponding process with allylic halides.[37–39] Very recently it has been reported by our lab that the AgTp complex **22** does, in fact, catalyze this reaction in excellent efficiencies, although the diastereoselectivities are very low at this point (Table 8.5, Schemes 8.11 and 8.12).[15] In this case, allylic halides **58–62** were reacted with ethyl- or *tert*-butyl diazoacetate (BDA) in the presence of **22** (Scheme 8.11), leading to the formation of homoallylic haloacetates **64a,b–68a,b** (Table 8.5, entries 1–5). Propargylic substrates **69–70** provided similar rearrangement products, although in this case allenes **72b–73a,b** were obtained (Scheme 8.12). A mechanism involving the formation and rearrangement of the halonium ylide (**63** or **71**) was proposed, again via the intermediacy of a silver carbene.[15] Similarly to the 1,2 shift described above, no detailed mechanistic studies have been performed, and so this pathway should be considered a working hypothesis, pending additional experimental investigation.

TABLE 8.5. Products and Yields from the Rearrangement[a]

			R/Yield (%)	
Entry	Substrate	Product	Et	*t*-Bu
1	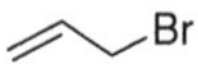 **58**	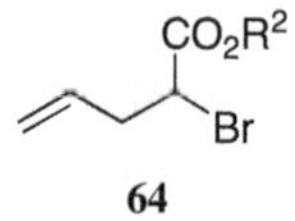 **64**	75	65
2	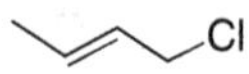 **59**	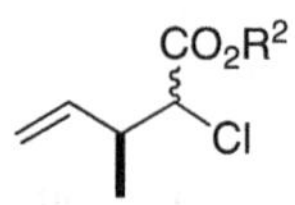 **65**	86 (1 : 1)[b]	96(1 : 1)[b]
3	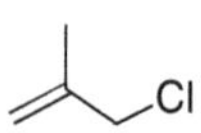 **60**	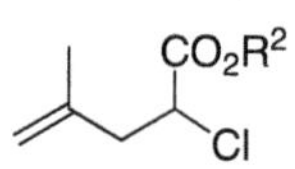 **66**	59	84
4	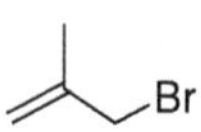 **61**	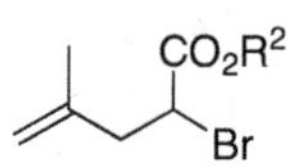 **67**	57	89
5	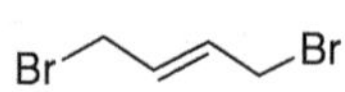 **62**	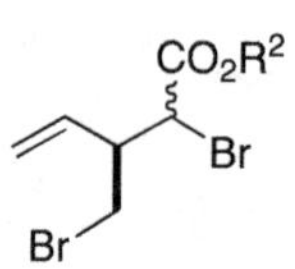 **68**	70 (1 : 1)[b]	80 (1 : 1)[b]
6	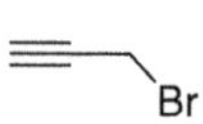 **69**	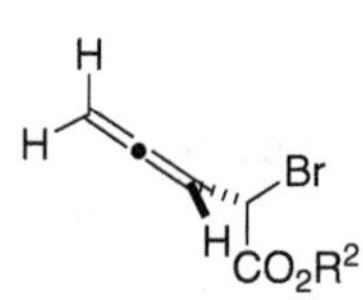 **72**	—	66
7	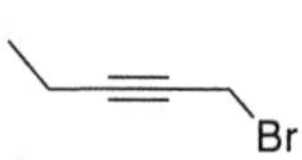 **70**	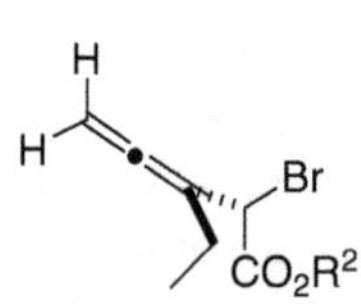 **73**	71	74

[a] The reactions are conducted in neat halide with the exception of entry 5, in which the substrate is dissolved in CH_2Cl_2.

[b] *Syn*/*anti* ratio, determined by ^{1}H NMR spectroscopy.

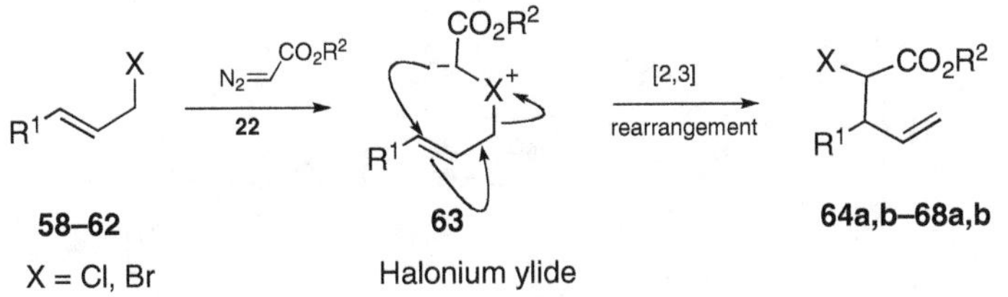

Scheme 8.11

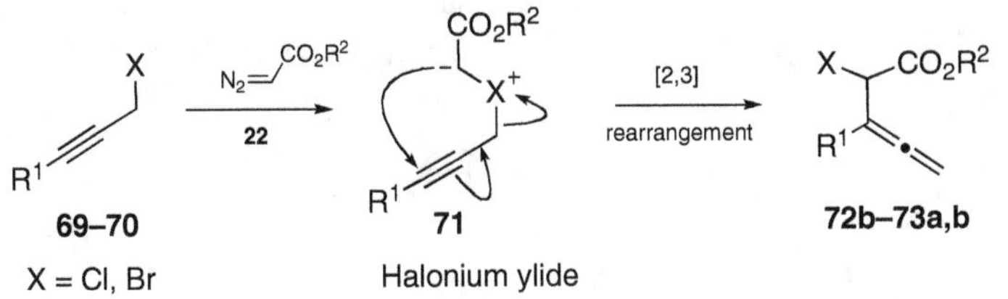

Scheme 8.12

Most of the substrates evaluated in this reaction were primary halides; however, one example of a secondary substrate **74** was reported to successfully engage in this rearrangement (Scheme 8.13, **74** → **75**).[15]

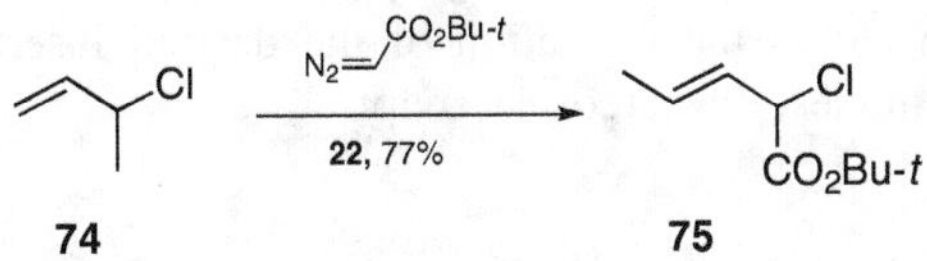

Scheme 8.13

Of note in this chemistry is that in both the C-Hal insertion and C-Hal rearrangement reactions, new carbon–carbon bonds are formed as a result of the *participation* of the halogen, but the halogen is *retained* in the product. This is quite unusual as in most cases, metal-catalyzed reactions involving halogens result in the consumption of the halogen for the ensuing bond formation. This process provides substrates with multiple functional groups that can then be used in future manipulations.

8.4.2 C-S Addition–Rearrangement Reactions

Very recently, Davies and coworkers have reported a silver-catalyzed Doyle–Kirmse reaction of allylic and propargylic sulfides with ethyl diazophenylacetate (**35**). A selection of examples was investigated providing the rearranged products, either homoallylic thioethers (**76** → **77**, Scheme 8.14a) or allenyl thioethers (**78** → **79**, Scheme 8.14b) in good yields, and with good overall scope.[40]

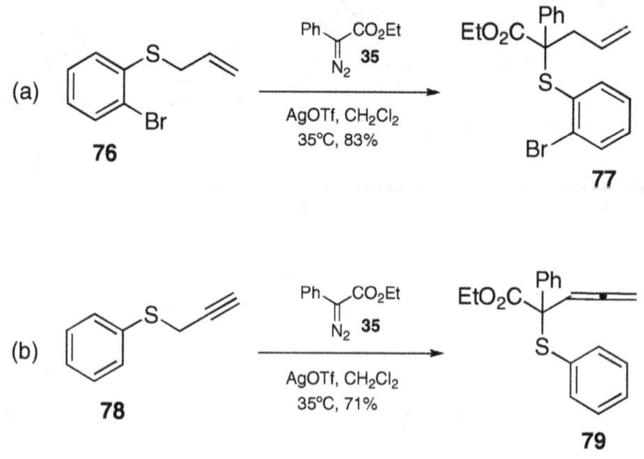

Scheme 8.14

8.5 C–H INSERTION

Several silver complexes have been investigated in C–H insertion reactions of carbenes. One of the earliest reports by Sulikowski and Burgess describes an intramolecular C–H insertion of a chiral aryl diazoacetate **80** as part of a study on the total synthesis of mitomycin family of natural products (Scheme 8.15).[41] Among several catalysts that were screened was $AgSbF_6$ in the presence of bis(isoxazoline)-type ligands, which led to the formation of the desired insertion product **81** in moderate yield and modest diastereoselectivity.

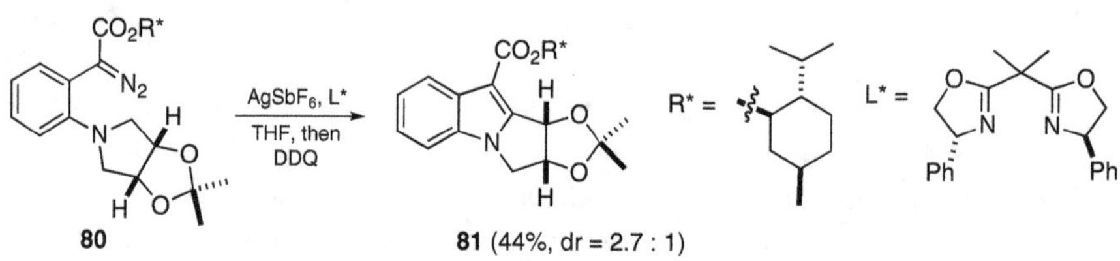

Scheme 8.15

More recently, both our group and Perez's group have demonstrated that Ag complexes **22** and **33** containing scorpionate ligands are effective for C–H insertion with diazoacetates (Scheme 8.16).[17,20,42] Complex **22** efficiently effects insertion of

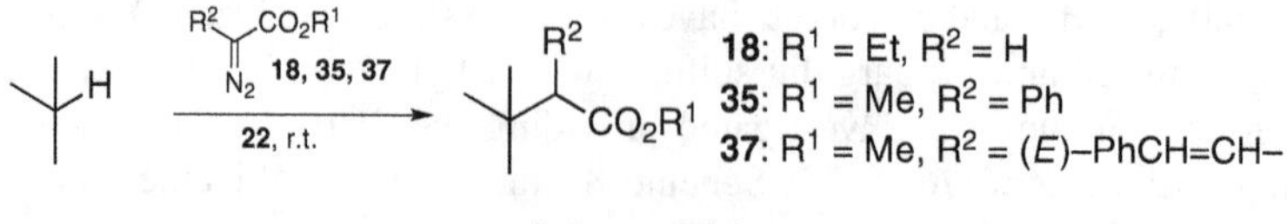

Scheme 8.16

EDA into cycloalkanes **82** and **83** (Table 8.6, entries 1 and 2), but in the case of oxygen-containing substrates **84–88**, the C–H insertion yields were rather modest (Table 8.6, entries 3–5), or insertion did not occur at all (Table 8.6, entries 6 and 7).[42] Interestingly, some of the same substrates were evaluated with the corresponding $CuTp^{(CF_3)_2}$ complex, and in this case the oxygen-containing substrates appear to react more efficiently.[19] This reaction has been extended to include donor–acceptor carbenes, and it was found to provide the expected products in most cases.[43] The $Tp^{(3,5\text{-}CF_3)_2}Ag$ complex **22** and the related silver complex **33** were examined in C–H insertion reactions with several acyclic alkanes.[17,20,42] The chemical yields were excellent, and interestingly, both complexes exhibited relatively high proportions of C–H insertion at the primary carbon (Table 8.7). Perez and coworkers speculated that this increase selectivity for insertion at primary sites was due to increased electrophilic character at the metal center rather than steric factors.[20] The Perez lab has reported a new set of silver complexes containing fluorinated indazolylborates (e.g., **108**; Fig. 8.4).[44] These complexes perform similar C–H insertion reactions to the pyrazolylborates, but at very low catalyst loadings. Perhaps not unexpectedly, the selectivities closely mirrored those obtained for the corresponding fluorinated pyrazolyl complexes.

8.6 N-H INSERTION

Both copper and rhodium complexes have been shown to catalyze carbene insertion into N–H bonds, providing a convenient method for amino acid synthesis.[45] Accordingly, the corresponding silver-catalyzed process has attracted attention. Jørgensen and coworkers reported the reaction of aniline with substituted diazoacetate derivatives and several simple silver salts admixed with bis(isoxazoline) ligands (Scheme 8.17).[46] Generally the yields were quite moderate, as were the enantioselectivities (Table 8.8). Parallel investigations with copper-based catalysts showed improved chemical yields, but the enantioselectivities were still poor. The authors suggested that the silver-mediated reactions proceeded via a Lewis acid–catalyzed pathway, rather than the expected carbene transfer mechanism, which was thought to be operating in the case of the copper catalysts.[46]

Hu and coworkers have examined N–H (and O–H and S–H) insertions in the presence of silver salts as well as with copper or rhodium catalysts with styryl diazoacetates (Schemes 8.18 and 8.19, Tables 8.9 and 8.10).[47] Two possible products (**114/117** and **115/118**) were obtained that are derived from either direct insertion or insertion with net transposition (Schemes 8.18 and 8.19). Silver and copper salts tended to favor transposition (Table 8.9, entries 2–5; Table 8.10, entries 2–9), whereas rhodium favored direct insertion (Table 8.9, entry 1; Table 8.10, entry 1). The selectivity differences between the two products were again rationalized in terms of two mechanistic pathways. In the case of rhodium-based catalysts, it was proposed that the reaction occurs via a metallocarbene, whereas with copper and silver catalysts the reaction was interpreted as proceeding by Lewis acid activation.

TABLE 8.6. C–H Insertion into Cyclic Substrates of Diazoacetates Catalyzed by AgTp Complex 22

Entry	Substrate	Products	Yield (%)	Products	Yield (%)	Products	Yield (%)
1	 **82**	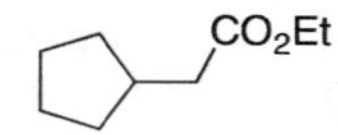 **89**	88	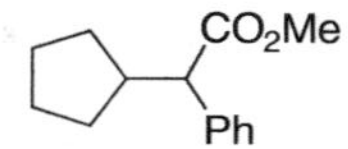 **96**	87	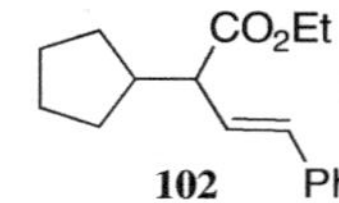 **102**	84
2	 **83**	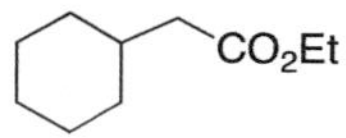 **90**	88	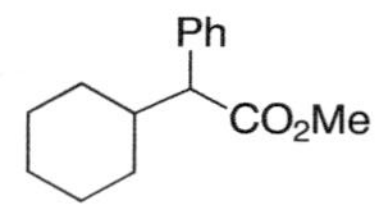 **97**	76	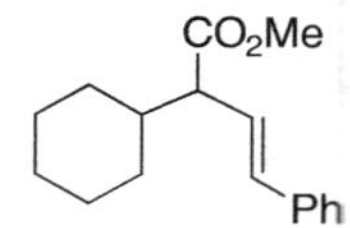 **103**	77
3	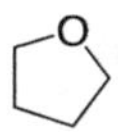 **84**	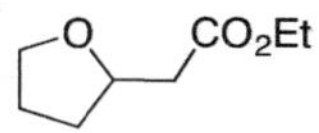 **91**	24	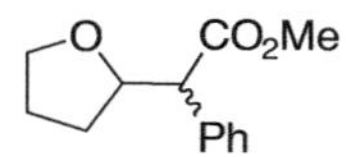 **98**	91 (3 : 2)	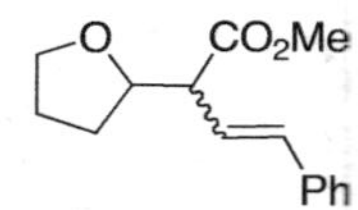 **104**	0
4	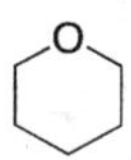 **85**	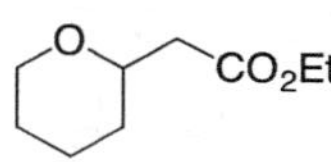 **92**	41	—	—	—	—

5	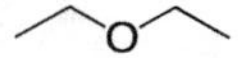 **86**	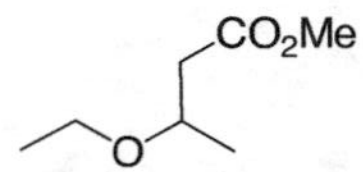 **93**	5	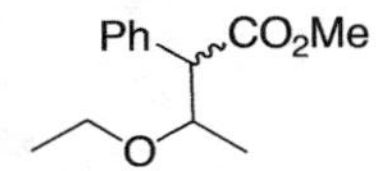 **99**	71 (3 : 2)	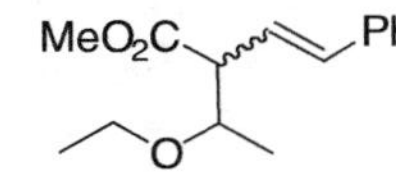 **105**	88 (3 : 2)
6	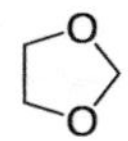 **87**	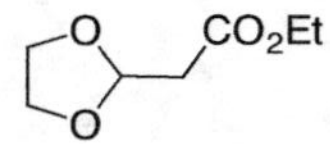 **94**	0	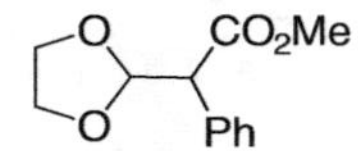 **100**	0	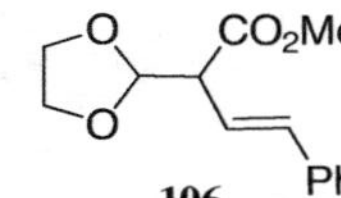 **106**	0
7	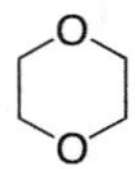 **88**	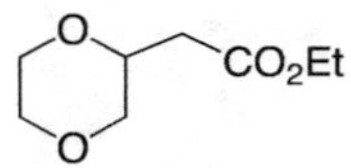 **95**	0	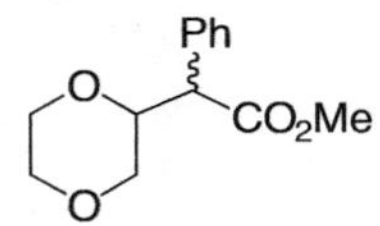 **101**	0	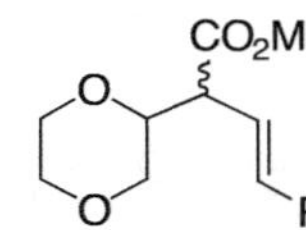 **107**	0

TABLE 8.7. C–H Insertions into Acyclic Alkanes Catalyzed by AgTp Complexes 22, 33, and 108

Catalyst	Substrate	Primary Site (%)		Secondary Site (%)		Tertiary Site (%)
	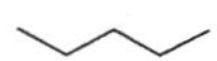	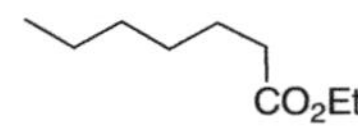		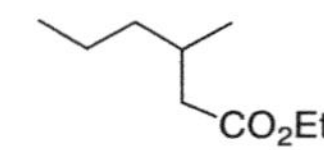	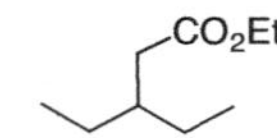	
22		41		47	12	
33[a]		29		57	14	
108		36		48	5	
	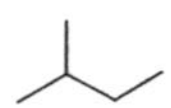	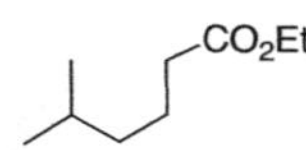	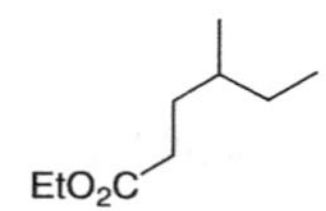	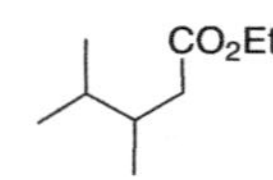		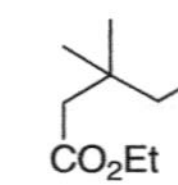
22		59 (both)		27		14
33[a]		42 (both)		37		21
108		56 (both)		26		18
	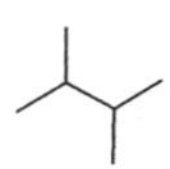	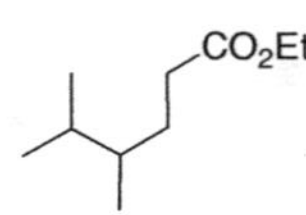				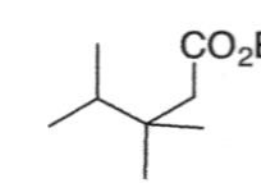
22		80				20
33[a]		60				40
108		75				25

	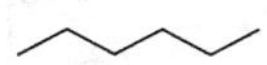	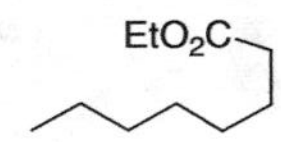		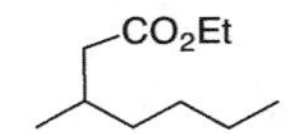	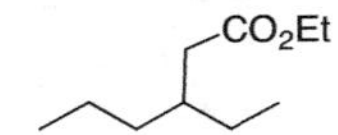	
33[a]		22		56	21	
108		20		45	35	
	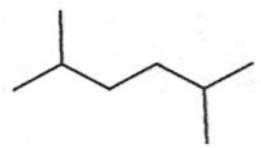	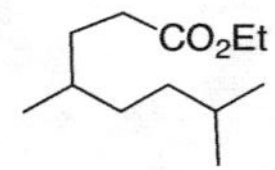		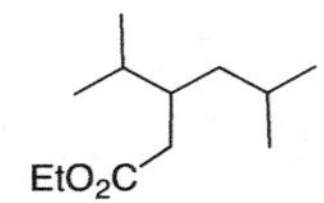		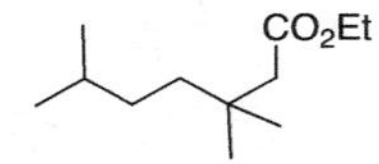
33[a]		46		17		37
	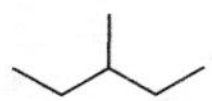	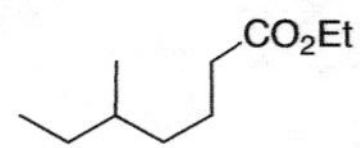	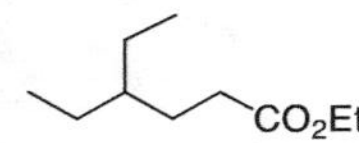	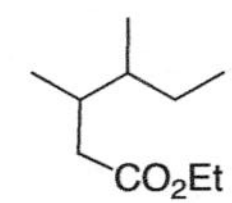		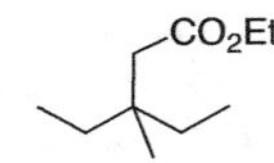
33[a]		18	48	33		11
	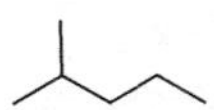	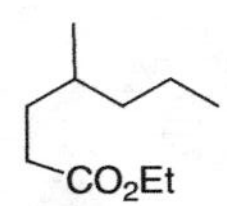	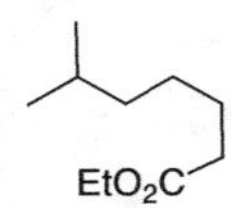	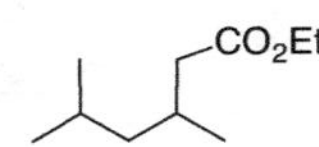		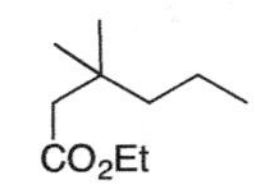
33[a]		50 (both)		39		21
108		66 (both)		24		10
	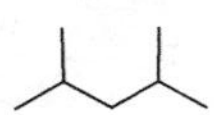	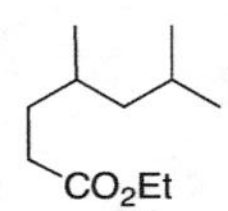		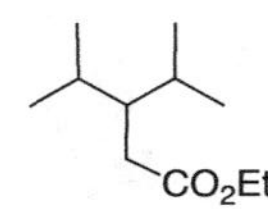		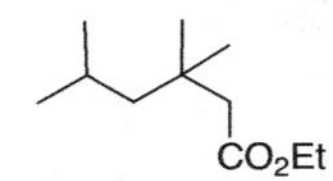
55[a]		81		6		13

[a]These values have been normalized to account for C–H insertion products only.

TABLE 8.8. Silver(I)-Catalyzed N–H Insertion

Entry	Ag Salt	Ligand	R^1	**112** Yield (%)	ee (%)
1	$AgSbF_5$	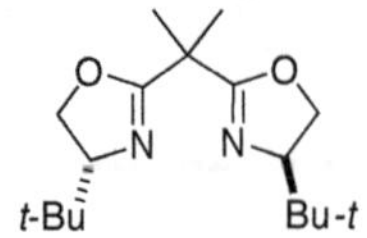	Me	8	25
2	AgOTf	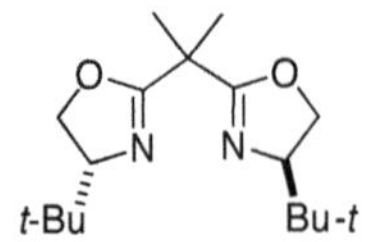	Me	5	48
3	AgOTf	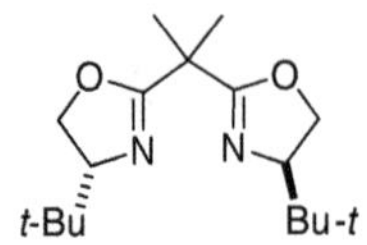	Me	4	20
4	$AgClO_4$	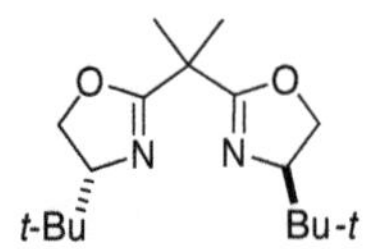	Me	5	48
5	$AgSbF_5$	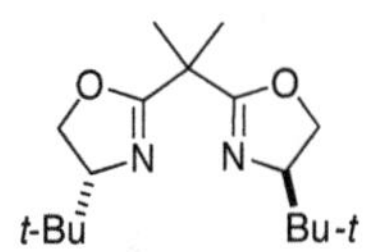	Ph	33	13
6	AgOTf	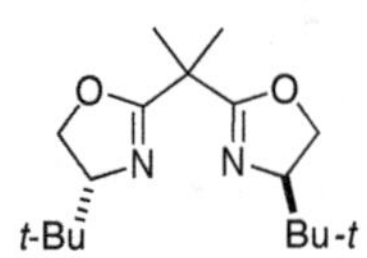	Ph	49	9
7	AgOTf	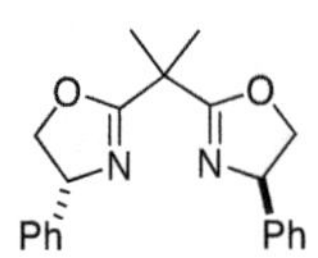	Ph	58	9
8	AgOTf	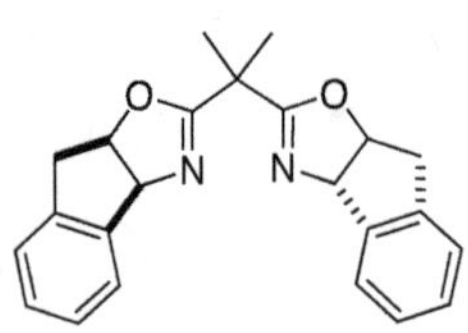	Ph	58	(±)

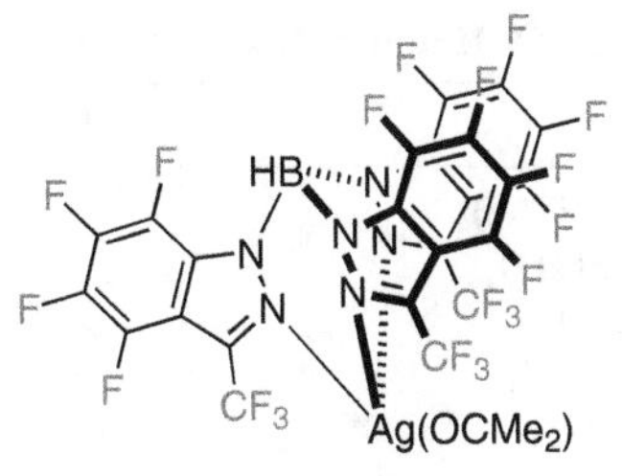

108

Figure 8.4. Polyfluorinated indazolyl silver complex (**108**), reported in 2008.

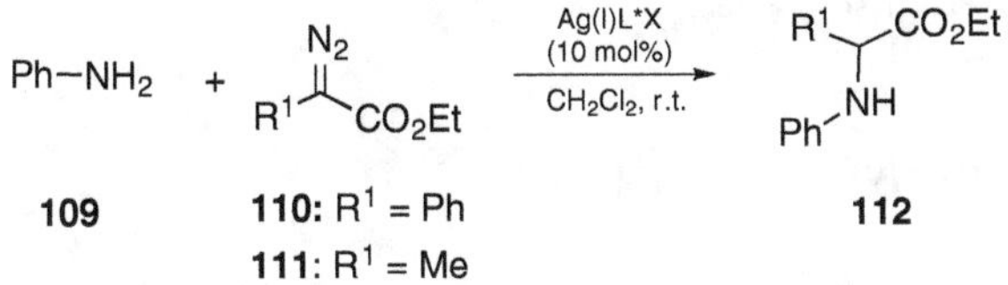

Scheme 8.17

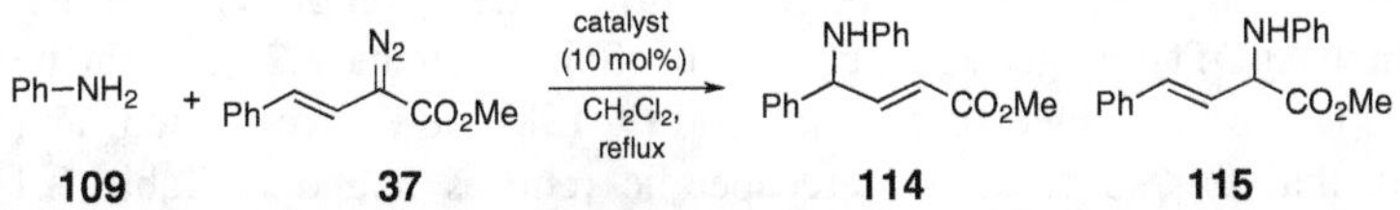

Scheme 8.18

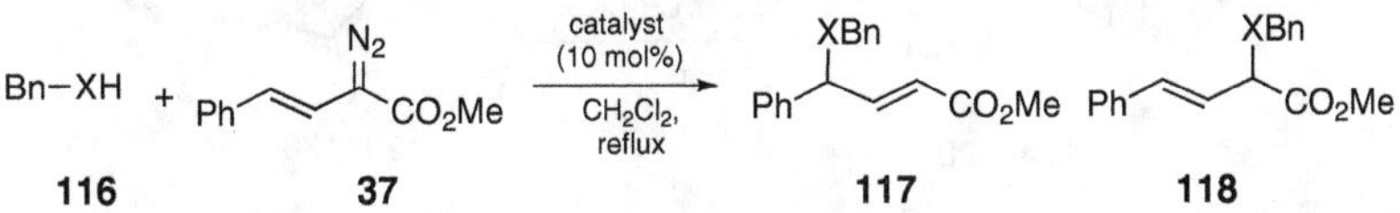

Scheme 8.19

TABLE 8.9. Metal-Catalyzed N-H Insertion

Entry	Catalyst	Yield [**114** + **115**(%)]	**114** : **115**
1	$Rh_2(OAc)_4$[a]	79	<2 : 98
2	$Cu(OTf)_2$	72	85 : 15
3	$AgBF_4$	67	73 : 27
4	$AgClO_4$	64	67 : 33
5	AgOTf	66	65 : 35

[a] In this case 1 mol% catalyst was used.

TABLE 8.10. Metal-Catalyzed O–H and S–H Insertions

Entry	X	Catalyst	Yield [**117** + **118** (%)]	**117**: **118**
1	O	$Rh_2(OAc)_4$[a]	78	<2 : 98
2	O	$Cu(OTf)_2$	53	72 : 28
3	O	$AgBF_4$	67	>98 : 2
4	O	$AgClO_4$	71	>98 : 2
5	O	AgOTf	63	>98 : 2
6	S	$Cu(OTf)_2$	68	78 : 22
7	S	$AgBF_4$	69	69 : 31
8	S	$AgClO_4$	71	68 : 32
9	S	AgOTf	59	72 : 28

[a] In this case 1 mol% catalyst was used.

8.7 RING EXPANSION REACTIONS

It was demonstrated previously that cyclopropyl carbenes, generated by the thermal decomposition of the corresponding diazo compound, provide cyclobutenes, but this process can be nonselective and inefficient.[48,49] Tang and coworkers have reported a transition-metal-mediated variant in which significant improvements in the yields and selectivities were observed.[50] Several transition metal complexes were evaluated, from which AgOTf emerged as the catalyst of choice (Scheme 8.20). As can be seen in Table 8.11, a broad range of substrates (**119–128**) were investigated, and it was observed that this reaction is stereospecific (entries 5 and 6, Table 8.11) and regioselective (entries 7–10, Table 8.11).

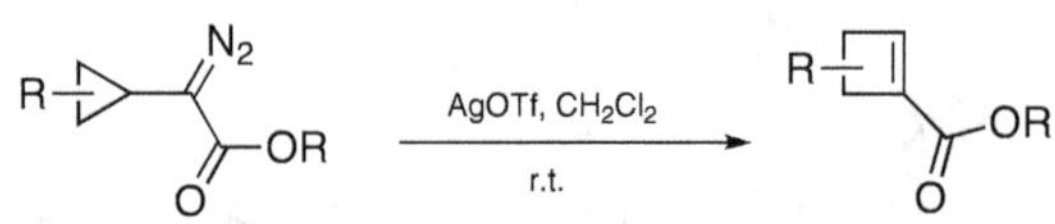

124–128 **129–138**

Scheme 8.20

In addition to examining the scope of the reaction, the authors observed some interesting selectivity issues with respect to the migrating bond, noting both a catalyst dependence (Scheme 8.21a) and a substrate dependence (Scheme 8.21b), which, in principle, provides substantial flexibility in the utility of the reaction.

8.8 INTERMEDIACY OF SILVER CARBENES

At this point, although there is no independent experimental evidence for the formation of silver carbenes in the reactions described above, and in some cases

TABLE 8.11. Scope of Silver-Catalyzed Cyclopropyldiazoacetate Ring Expansion

Entry	Cyclopropane	Cyclobutane	Yield (%)	Ratio
1	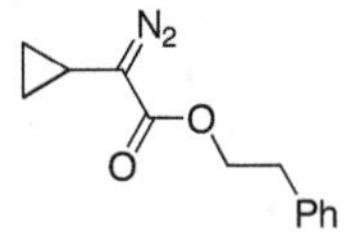 **119**	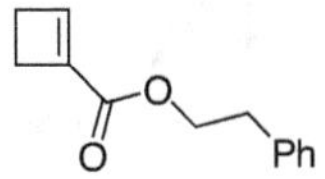 **129**	91	—
2	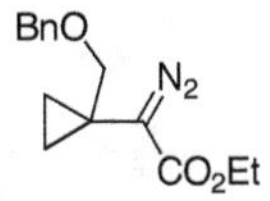 **120**	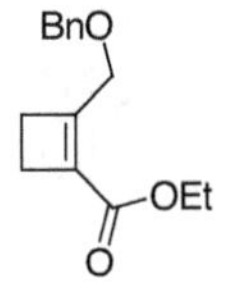 **130**	77	—
3	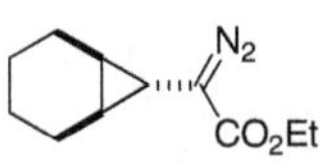 **121**	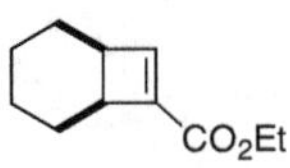 **131**	72	—
4	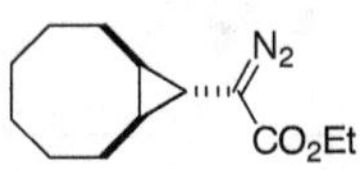 **122**	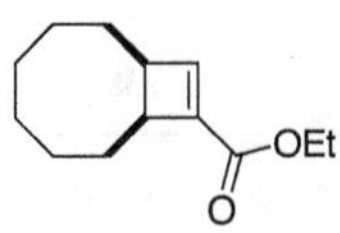 **132**	90	—
5	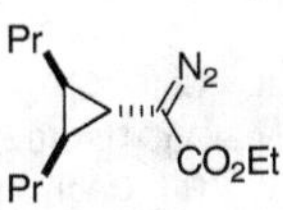 **123**	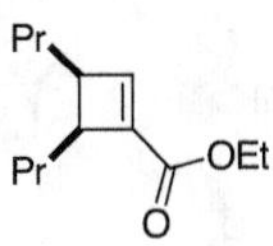 **133**	71	Single isomer
6	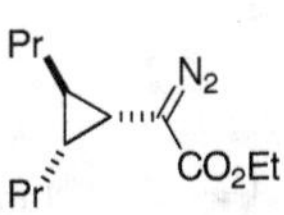 **124**	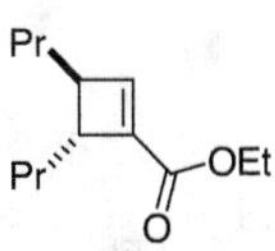 **134**	92	Single isomer
7	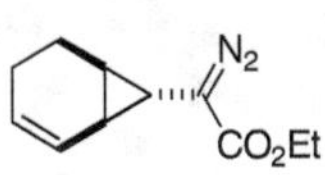 **125**	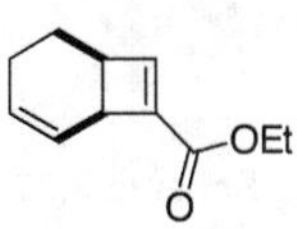 **135**	73	10 : 1

(Continued)

TABLE 8.11 *(Continued)*

Entry	Cyclopropane	Cyclobutane	Yield (%)	Ratio
8	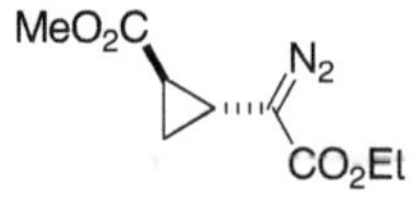 **126**	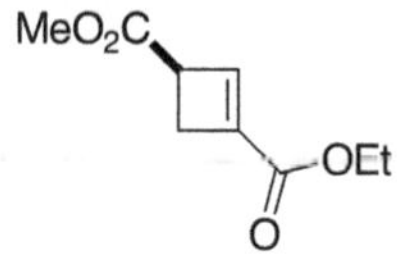 **136**	70	10 : 1
9	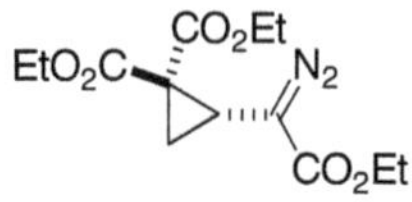 **127**	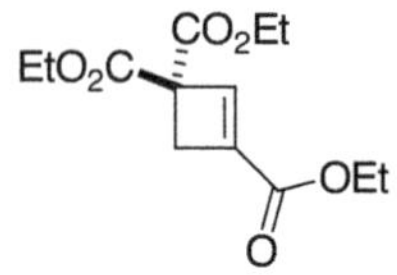 **137**	87	Single isomer
10	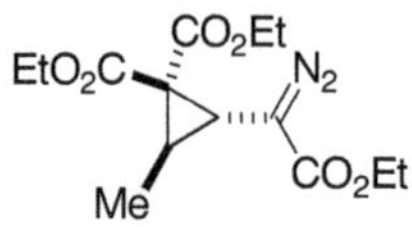 **128**	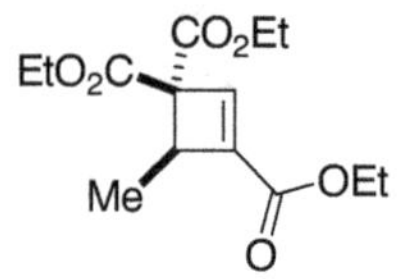 **138**	77	Single isomer

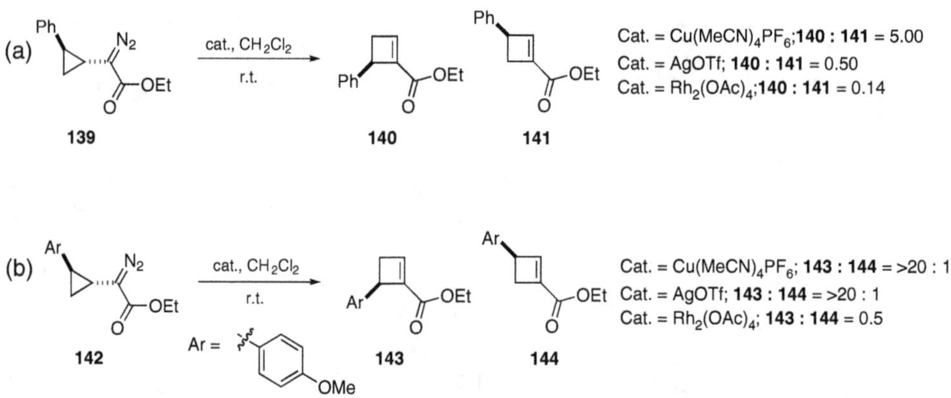

Scheme 8.21

alternative mechanistic pathways have been posited in which the silver complex behaves as a Lewis acid, the overall behavior of the intermediate in many of the other transformations discussed is *consistent* with the formation of a silver carbene. A DFT study by Perez and coworkers on the mechanism of Ag-catalyzed C–H insertion of diazoesters has appeared detailing and supporting the role of silver carbenes as intermediates in this chemistry.[51]

8.9 MISCELLANEOUS REACTIONS INVOLVING SILVER CARBENOIDS

All the examples described above involved the reaction of diazoacetate derivatives with silver salts to initiate the formation of a putative silver carbene; however, other pathways exist. For example, Porcel and Echavarren have reported an intramolecular cyclization of an allylstannane to a pendent alkyne (Scheme 8.22) that involves the intermediacy of a silver carbene.[52] As can be seen in Table 8.12, the reactions proceeded in moderate to excellent yield, providing the dienylstannane, while in some cases, reductive destannylation occurred. Several asymmetric reactions were reported with substrate (*E*)-**145d**, leading to the formation of the expected adduct in reasonable enantioselectivities (ee = 73–78%) in a preliminary screen with a number of different ligands.

Mechanistically, most alkyne cyclizations of this type have been interpreted in terms of the catalyst serving as a Lewis acid, activating the π bond to nucleophilic attack,[53–57] and while this pathway is feasible, the authors preferred one involving the intermediacy of a silver carbene.[52] Control studies ruled out a pathway involving

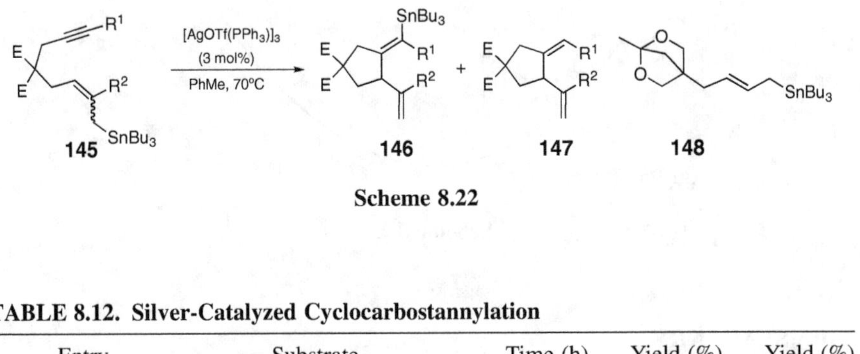

Scheme 8.22

TABLE 8.12. Silver-Catalyzed Cyclocarbostannylation

	Entry	Substrate	Time (h)	Yield (%)	Yield (%)
1	(*E*)-**145a**	E = CO_2Me, R^1 = H, R^2 = Me	0.5	**146a** (87)	**147a** (3)
2	(*E*)-**145b**	E = CO_2Me, R^1 = R^2 = H	5.0	**146b** (71)	**147b** (6)
3	(*Z*)-**145c**	E = SO_2Ph, R^1 = R^2 = H	3.5	**146c** (72)	**147b** (9)
4	(*E*)-**145d**	E = SO_2Ph, R^1 = H, R^2 = Me	0.5	**146d** (90)	
5	(*Z*)-**145d**	E = SO_2Ph, R^1 = H, R^2 = Me	0.5	**146d** (91)	
6	(*E*)-**145e**	E = CH_2OAc, R^1 = R^2 = H	2.5	**146e** (69)	**147e** (12)
7	(*E*)-**145f**	E = $CH_2OTBDPS$, R^1 = R^2 = H	2.0	**146f** (93)	
8	(*E*)-**145g**	E = CH_2OH, R^1 = R^2 = H	0.2[a]	**146g** (31)	**147g** (11)

[a] Compound **148** was obtained in 30% yield in addition to the expected products **146g** and **147g**.

transmetallation, along with the participation of silver acetylides; therefore, the pathway depicted in Scheme 8.23 was offered as a possibility. Porcel and Echavarren proposed that the silver complex selectively activates the alkyne to form the cyclopropyl carbene–silver(I) complex **149**. Fragmentation leads to the formation of alkenylsilver complex **150**, and reaction with Bu_3SnOTf then leads to the formation of the final cyclization product and regeneration of the silver catalyst.

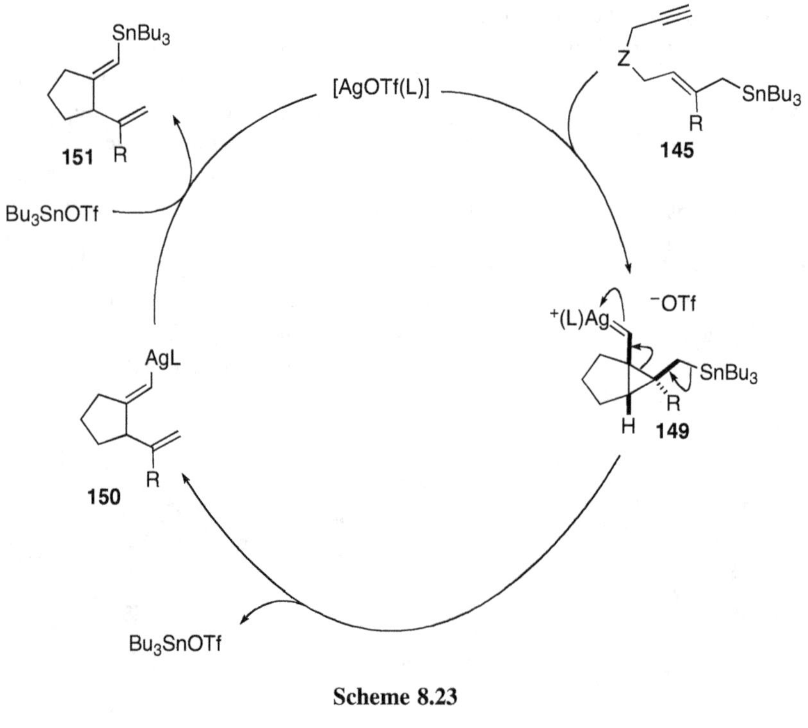

Scheme 8.23

8.10 SUMMARY

It is clear from the discussion above that the utility of silver complexes as carbene transfer catalysts has come a long way, but much remains to be done. There is a far from complete picture with respect to the precise mechanisms operating in many of these reactions, which hinders the design of more efficient second-generation catalysts. This is particularly relevant in cases where little or no precedent exists for a reaction, for example, the C-Hal insertion reactions reported by us.[15,35] A second issue that demands attention concerns the development of asymmetric variants of many of these processes, which will also rely on a more complete mechanistic understanding of the design of appropriate supporting ligands.

ACKNOWLEDGMENTS

Our own endeavors in this area have been possible because of the hard work and valuable intellectual contributions of a number of coworkers, whose names appear in the references. Financial support has generously been provided by The Robert A. Welch Foundation (Y-1362).

REFERENCES

1. Issue number 8 of *Chemical Reviews* (*Chem. Rev.* **2008**, *108*) was devoted to the utility of coinage metals in synthesis.
2. Kirmse, W., *Eur. J. Org. Chem.* **2002**, 2193–2256.
3. Li, Z.; He, C., *Eur. J. Org. Chem.* **2006**, 4313–4322.
4. Dias, H. V. R.; Lovely, C. J., *Chem. Rev.* **2008**, *108*, 3223.
5. Zhang, Z.; Wang, J., *Tetrahedron* **2008**, *64*, 6577–6605.
6. Arduengo, A. J.; Dias, H. V. R.; Calabrese, J. C.; Davidson, F., *Organometallics* **1993**, *12*, 3405–3409.
7. Ramnial, T.; Abernethy, C. D.; Spicer, M. D.; McKenzie, I. D.; Gay, I. D.; Clyburne, J. A. C., *Inorg. Chem.* **2003**, *42*, 1391–1393.
8. Ahrens, S.; Zeller, A.; Taige, M.; Strassner, T., *Organometallics* **2006**, *25*, 5409–5415.
9. Julian, R. R.; May, J. A.; Stoltz, B. M.; Beauchamp, J. L., *J. Am. Chem. Soc.* **2003**, *125*, 4478–4486.
10. Sudrik, S. G.; Maddanimath, T.; Chaki, N. K.; Chavan, S. P.; Chavan, S. P.; Sonawane, H. R.; Vijayamohanan, K., *Org. Lett.* **2003**, *5*, 2355–2358.
11. Sudrik, S. G.; Sharma, J.; Chavan, V. B.; Chaki, N. K.; Sonawane, H. R.; Vijayamohanan, K. P., *Org. Lett.* **2006**, *8*, 1089–1092.
12. Sudrik, S. G.; Chaki, N. K.; Chavan, V. P.; Chavan, S. P.; Chavan, S. P.; Sonawane, H. R.; Vijayamohanan, K., *Chem. Eur. J.* **2006**, *12*, 859–864.
13. We use the term *carbene transfer* to represent the net chemical transformation and not necessarily to imply that this is a reactive intermediate (either a free carbene or a metallocarbene).
14. Juhl, K.; Hazell, K. A.; Jorgensen, K. A., *J. Chem. Soc. Perkin Trans.* **1999**, *1*, 2293–2297.
15. Krishnamoorthy, P.; Browning, R. G.; Singh, S.; Sivappa, R.; Lovely, C. J.; Dias, H. V. R., *Chem. Commun.* **2007**, 731–732.
16. Dias, H. V. R.; Lu, H. -L.; Kim, H. -J.; Polach, S. A.; Goh, T. K. H. H.; Browning, R. G.; Lovely, C. J., *Organometallics* **2002**, *21*, 1466–1473 and the references cited therein.
17. Lovely, C. J.; Browning, R. G.; Badarinarayana, V.; Dias, H. V. R., *Tetrahedron Lett.* **2005**, *46*, 2453–2455.
18. Dias, H. V. R.; Polach, S. A., *Inorg. Chem.* **2000**, *39*, 4676–4677.
19. Browning, R. G.,Ph.D. dissertation, Univ. Texas, Arlington, **2003**.
20. Urbano, J.; Belderrain, T. R.; Nicasio, M. C.; Trofimenko, S.; Diaz-Requejo, M. M.; Perez, P. J., *Organometallics* **2005**, *24*, 1528–1532.
21. Morilla, M. E.; Díaz-Requejo, M. M.; Belderrain, T. R.; Nicasio, M. C.; Trofimenko, S.; Pérez, P. J., *Organometallics* **2004**, *23*, 293–295.

22. Díaz-Requejo, M. M.; Caballero, A.; Belderraín, T. R.; Nicasio, M. C.; Trofimenko, S.; Pérez, P. J., *J. Am. Chem. Soc.* **2002**, *124*, 978–983.
23. Díaz-Requejo, M. M.; Belderraín, T. R.; Trofimenko, S.; Pérez, P. J., *J. Am. Chem. Soc.* **2001**, *123*, 3167–3168.
24. Thompson, J. L.; Davies, H. M. L., *J. Am. Chem. Soc.* **2007**, *129*, 6090–6091.
25. Sweeney, J. B., *Chem. Soc. Rev.* **2009**, *38*, 1027–1038.
26. Padwa, A., *Chem. Soc. Rev.*, **2009**, 3072–3081, DOI: 10.1039/b816701j.
27. Pirrung, M. C.; Zhang, J.; McPhail, A. T., *J. Org. Chem.* **1991**, *56*, 6269–6271.
28. Pirrung, M. C.; Zhang, J.; Lackey, J.; Sternbach, D. D.; Brown, F., *J. Org. Chem.* **1995**, *60*, 2112–2125.
29. Migita, T.; Ando, W.; Kondo, S.; Matsuyama, H.; Kosugi, M., *Nippon Kagaku Zasshi* **1970**, *91*, 374–377.
30. Urry, W. H.; Bilow, N., *J. Am. Chem. Soc.* **1964**, *86*, 1815–1819.
31. Urry, W. H.; Eiszner, J. R.; Wilt, J. W., *J. Am. Chem. Soc.* **1957**, *79*, 918–922.
32. Urry, W. H.; Wilt, J. H., *J. Am. Chem. Soc.* **1954**, *76*, 2594–2595.
33. Urry, W. H.; Eiszner, J. R., *J. Am. Chem. Soc.* **1952**, *74*, 5822–5826.
34. Urry, W. H.; Eiszner, J. R., *J. Am. Chem. Soc.* **1951**, *73*, 2977.
35. Dias, H. V. R.; Browning, R. G.; Polach, S. A.; Diyabalanage, H. V. K.; Lovely, C. J., *J. Am. Chem. Soc.* **2003**, *125*, 9270–9271.
36. Hodgson, D. M.; Pierard, F. Y. T. M.; Stupple, P. A., *Chem. Soc. Rev.* **2003**, *30*, 50–61.
37. Simonneaux, G.; Galardon, E.; Paul-Roth, C.; Gulea, M. M. S., *J. Organomet. Chem.* **2001**, *617–618*, 360–363.
38. Doyle, M. P.; Forbes, D. C.; Vasbinder, M. M.; Peterson, C. S., *J. Am. Chem. Soc.* **1998**, *120*, 7653–7654.
39. Doyle, M. P.; Tamblyn, W. H.; Bagheri, V., *J. Org. Chem.* **1981**, *46*, 5094–5102.
40. Davies, P. W.; Albrecht, S. J. -C.; Assanelli, G., *Org. Biomol. Chem.* **2009**, *7*, 1276–1279.
41. Burgess, K.; Lim, H. -J.; Porte, A. M.; Sulikowski, G. A., *Angew. Chem. Int. Ed. Engl.* **1996**, *35*, 220–222.
42. Dias, H. V. R.; Browning, R. G.; Richey, S. A.; Lovely, C. J., *Organometallics* **2004**, *23*, 1200–1204.
43. Lovely, C. J.; Flores, J. A.; Meng, X.; Dias, H. V. R., *Synlett* **2009**, 129.
44. Despagnet-Ayoub, E.; Jacob, K.; Vendier, L.; Etienne, M.; Alvarez, C.; Diaz-Requejo, M. M.; Perez, P. J., *Organometallics* **2008**, *27*, 4779–4787.
45. Moody, C. J., *Angew. Chem. Int. Ed.* **2007**, *46*, 9148–9150.
46. Bachmann, S.; Fielenbach, D.; Jorgensen, K. A., *Org. Biomol. Chem.* **2004**, *2*, 3044–3049.
47. Yue, Y.; Wang, Y.; Hu, W., *Tetrahedron Lett.* **2007**, *48*, 3975–3977.
48. Gallucci, R. R.; Jones, M., *J. Am. Chem. Soc.* **1976**, *98*, 7704–7711.
49. Friedman, L.; Shechter, H., *J. Am. Chem. Soc.* **1960**, *82*, 1002–1003.
50. Xu, H.; Zhang, W.; Shu, D.; Werness, J. B.; Tang, W., *Angew. Chem. Int. Ed.* **2008**, *47*, 8933–8936.
51. Braga, A. A. C.; Maseras, F.; Urbano, J.; Caballero, A.; Diaz-Requejo, M. M.; Perez, P. J., *Organometallics* **2006**, *25*, 5292–5300.
52. Porcel, S.; Echavarren, A. M., *Angew. Chem. Int. Ed.* **2007**, *46*, 2672–2676.

53. Zhao, J.; Hughes, C. O.; Toste, F. D., *J. Am. Chem. Soc.* **2006**, *128*, 7436–7437.

54. Sun, J.; Kozmin, S. A., *Angew. Chem. Int. Ed.* **2006**, *45*, 4991–4993.

55. Harrison, T. J.; Kozak, J. A.; Corbella-Pané, M.; Dake, G. R., *J. Org. Chem.* **2006**, *71*, 4525–4529.

56. Sweis, R. F.; Schramm, M. P.; Kozmin, S. A., *J. Am. Chem. Soc.* **2004**, *126*, 7442–7443.

57. Harrison, T. J.; Dake, G. R., *Org. Lett.* **2004**, *6*, 5023–5026.

9

ALDOL AND RELATED PROCESSES

MASANORI KAWASAKI
Medicinal Chemistry Research Laboratories, Pharmaceutical Research Division, Takeda Pharmaceutical Company Limited, Osaka, Japan

HISASHI YAMAMOTO
Department of Chemistry, University of Chicago, Chicago, Illinois

9.1 INTRODUCTION

Carbon–carbon bond-forming reactions are essential reactions in organic synthesis. Among them, the allylation reaction of the carbonyl group, which leads to homoallylic alcohols, is one of the most important transformations, since the resulting homoallylic alcohols are useful building blocks in synthetic organic chemistry. For this transformation, allylic organometallic reagents can be used with or without catalysts. Since the 1990s, the enantioselective version of this reaction was developed using chiral allylic organometallic reagents or chiral catalysts. The reactions of chiral allylic borane or allylic titanium as allylic reagents with carbonyls afford homoallylic alcohols with high enantioselectivity and diastereoselectivity.[1] Because of its high

Silver in Organic Chemistry Edited by Michael Harmata

selectivity, this reaction has been applied to the synthesis of various natural and unnatural products. Although excellent results have been obtained through this approach, a stoichiometric amount of the chiral source is necessary. In the past few decades, the catalytic enantioselective version of this reaction has been developed to resolve that issue. In this chapter, we would like to focus on allylation reactions of carbonyls using silver catalysts.

9.2 ALLYLATION REACTION USING ALLYLTRIBUTYLTIN

In 1996, Yamamoto and Yanagisawa reported the allylation reaction of aldehydes with allytributyltin in the presence of a chiral silver catalyst.[2] They found that the combination of silver and a phosphine ligand accelerates the allylation reaction between aldehydes and allyltributyltin. After this discovery, they screened several chiral phosphine ligands and found that chiral silver–diphosphine catalysts can effect the reaction in an enantioselective fashion (Table 9.1).[2] For example, when benzaldehyde and allyltributyltin were mixed in the presence of 5 mol% of AgOTf and (*S*)-2,2′-bis(diphenylphosphino)-1,1′-binaphthyl (BINAP), the corresponding homoallyl alcohol was obtained with 96% ee and 88% yield (Table 9.1). Generally, the reaction with aromatic aldehydes afforded the corresponding homoallyl alcohols in excellent

TABLE 9.1. Catalyst Screening for the Allylation Reaction with Aldehydes and Allyltributyltin

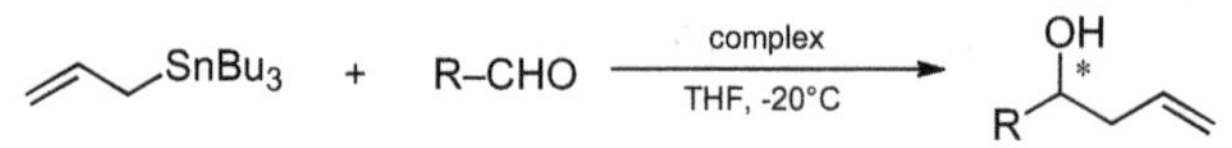

Entry	Complex[a]	Yield (%)	ee (%)
1	(*S*)-BINAP–$AgOCOCF_3$	47	40
2	(*S*)-BINAP–$AgClO_4$	1	26
3	(*S*)-BINAP–$AgNO_3$	26	53
4	(*S*)-BINAP–AgOTf	88	96
5	(*R*,*R*)-Chiraphos–AgOTf	97	2
6	(*S*,*S*)-Me-Duphos–AgOTf	4	48
7	(*S*,*S*)-Et-Duphos–AgOTf	13	3

[a] Structures ligands:

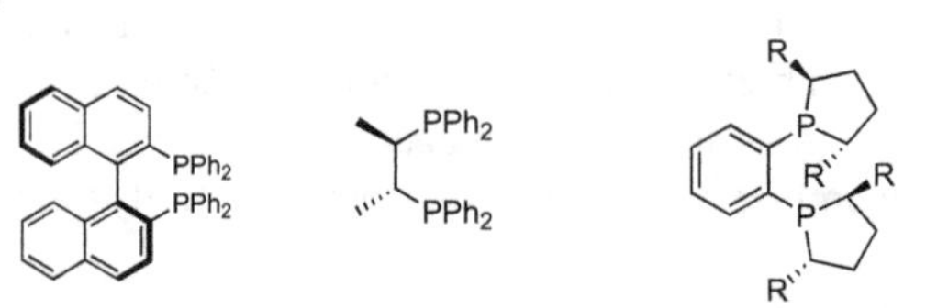

(S)-BINAP (R,R)-Chiraphos R = Me: (S,S)-Me-Duphos
R = Et : (S,S)-Et-Duphos

TABLE 9.2. Allylation Reaction with Various Aldehydes

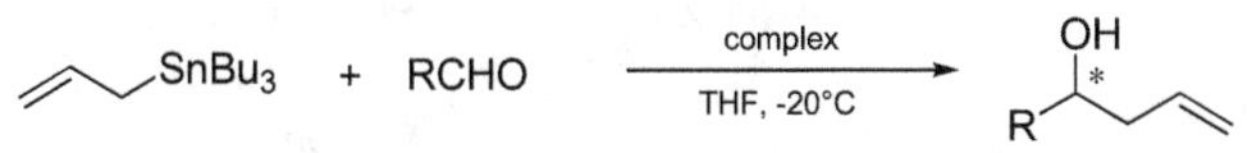

Entry	RCHO	Yield (%)	ee (%)
1	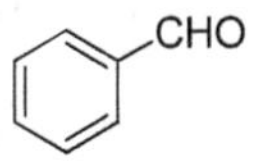	88	96 (*S*)
2	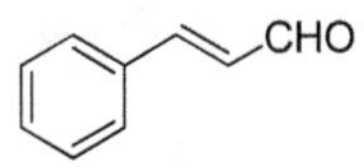	83	88 (*S*)
3	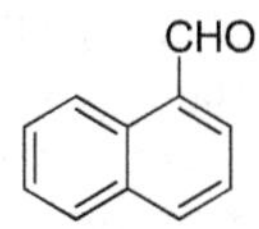	89	97
4	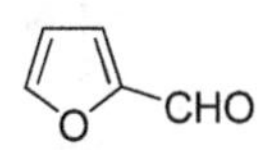	94	93
5	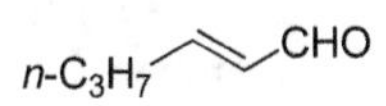	72	93
6	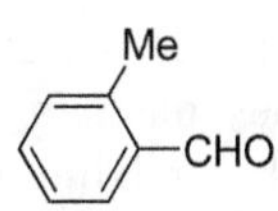	85	97
7	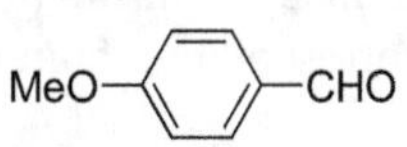	59	97
8	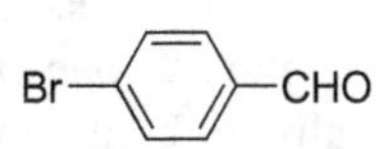	95	96
9	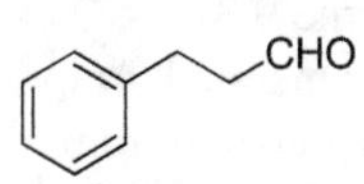	47	88

enantioselectivity. On the other hand, when aliphatic aldehydes were used as substrates, the enantioselectivity was slightly decreased (Table 9.2).

In addition, this silver–BINAP catalyst has been applied to the crotylation reaction between aldehydes and 2-butenylstannanes (Scheme 9.1).[3] The addition of (*E*)-2-butenylstannane to benzaldehyde in the presence of 20 mol% AgOTf/BINAP gave the corresponding adduct with 94% ee with a high *anti* : *syn* ratio. In contrast to other

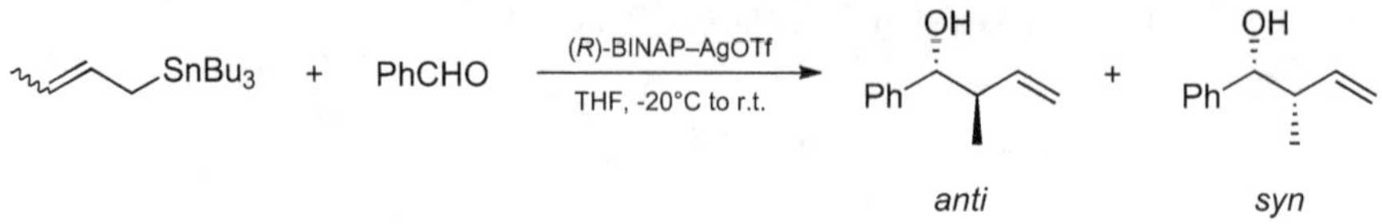

E : *Z* ratio of 2-butenylstannane	Yield (%)	*anti* (% ee) : *syn* (% ee)
95 : 5	56	85 (94) : 15 (64)
2 : 98	72	85 (91) : 15 (50)
53 : 47	45	85 (94) : 15 (57)

Scheme 9.1. Crotylation with benzaldehyde.

Lewis acid-catalyzed allylation reactions, the *syn* : *anti* ratio observed here did not depend on the *E* : *Z* ratio of the starting stannane.[4] Although the reaction mechanism was not clear, this result would suggest that the reaction took place via transmetallation from stannane to silver to generate an allylsilver species.

This silver–BINAP catalyst can be applied for the reaction of 2,4-pentadienylstannane and aldehydes.[5] Generally, the reaction of 2,4-pentadienylstannane and aldehydes in the presence of a Lewis acid catalyst afforded the corresponding ε adduct. On the other hand, the reaction using the silver catalyst gave the corresponding γ adduct predominantly (Table 9.3). These results suggested that the reaction catalyzed by silver could proceed through a six-membered cyclic transition state (Fig. 9.1).

In 2000, Loh and coworkers reported the enantioselective allylation reaction in the presence of $AgNO_3$–Tol–BINAP in an aqueous medium (Scheme 9.2).[6] They surveyed the reaction media and found that a 1 : 9 mixture of water : ethanol was optimal. The reaction of allyltributyltin and aldehyde in the presence of $AgNO_3$–Tol–BINAP in water and ethanol gave the corresponding homoallyl alcohols with moderate to good enantioselectivities.

Cesarotti and coworkers have developed an allylation reaction using allyltributyltin in the presence of an AgOTf–BITIANP complex.[7] In one example, the reaction was conducted with 4-bromobenzaldehyde and allyltributyltin in the presence of the catalyst to give the adduct with 78% ee (Scheme 9.3).

Shi and Wang reported that the catalyst generated from AgOTf and a chiral diphenylthiophosphoramide ligand, which was prepared from chiral 1,1′-binaphthyl-2,2′-diamine, could promote the allylation reaction of allyltributyltin and aromatic aldehydes (Table 9.4).[8] Thus, the reaction of allyltributyltin and aldehyde in the presence of AgOTf and chiral diphenylthiophosphoramide gave the corresponding adducts with up to 98% ee.

More recently, Benaglia and coworkers reported that the allylation reaction of α-iminoesters proceeded to give the homoallyl alcohols with moderate enantioselectivities (Scheme 9.4).[9] The chiral diimine that has a chiral 1,2-cyclohexyldiamine framework was used as the ligand. The reaction of α-iminoesters and allyltributyltin in the presence of AgOTf and diimine ligand proceeded to afford chiral amino acid derivatives with moderate enantioselectivities.

TABLE 9.3. Dienylation of Aldehydes with Organostannanes

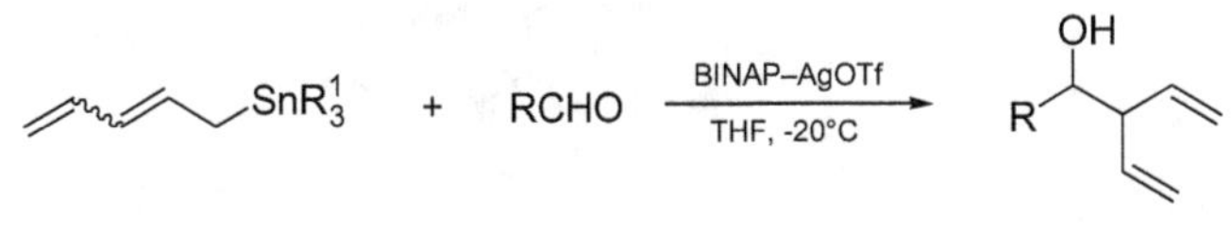

Entry	SnR^1_3	RCHO	Yield (%)	ee (%)
1	$SnBu_3$	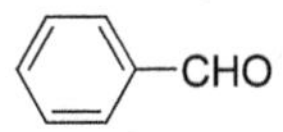	61	90
2	$SnMe_3$		68	89
3	$SnMe_3$	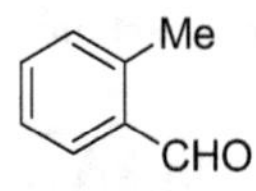	57	90
4	$SnMe_3$	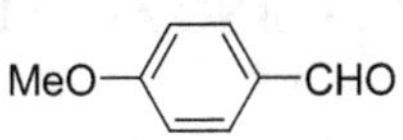	41	87
5	$SnMe_3$	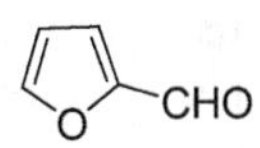	62	89
6	$SnBu_3$	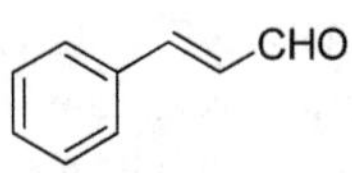	73	58
7	$SnMe_3$		68	58
8	$SnBu_3$	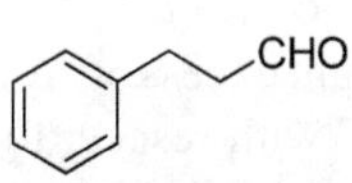	52	71
9	$SnBu_3$	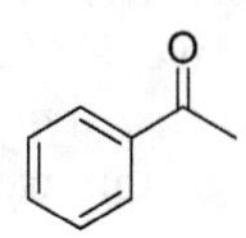	<1	—

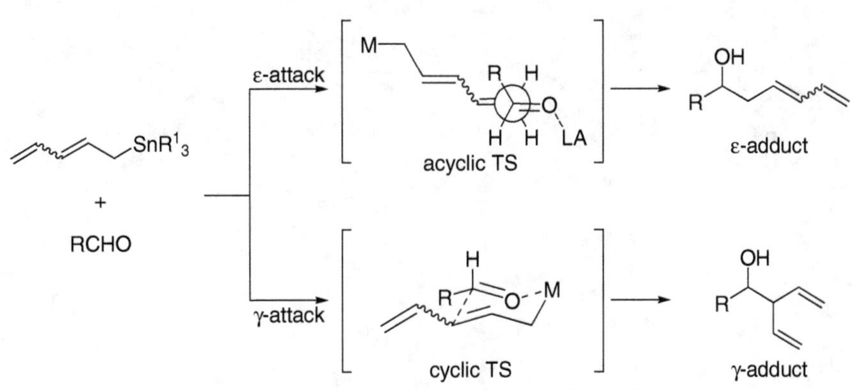

Figure 9.1. Proposed transition states.

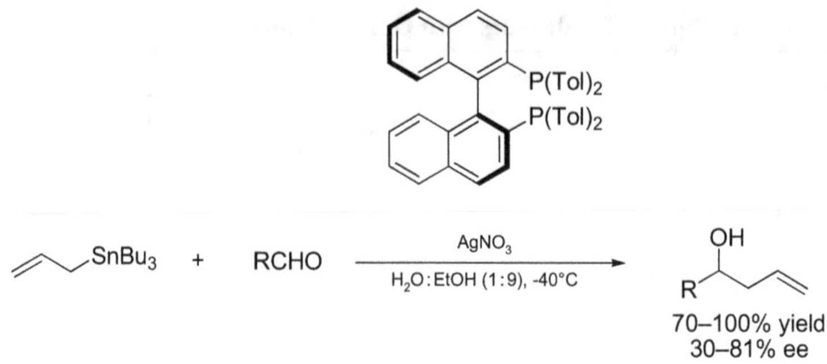

Scheme 9.2. Allylation reaction in aqueous media.

9.3 ALLYLATION REACTION USING ALLYLSILANES

Allylsilanes are more attractive allylation reagents than allylstannanes from an environmental perspective. However, the use of allylsilanes for organic synthesis is limited since the reactivity of allylsilanes is typically low compared to allyltributyltin. In 1999, Yamamoto and coworkers investigated the allylation reaction using allylsilanes and found that BINAP–AgF catalyzes enantioselective Sakurai–Hosomi-type reaction using allyltrimethoxysilanes (Table 9.5).[10] The reaction was conducted with aromatic aldehydes and allyltrimethoxysilanes in the presence of 6 mol% BINAP and 10 mol% AgF to afford the corresponding adducts with high enantioselectivities. The use of a slight excess of AgF was essential for this reaction, because when slight excess of AgF with respect to BINAP was used, the catalytically active 1 : 1 complex of AgF and BINAP was formed predominantly. It was also established that this catalytic system could be used for asymmetric crotylation reactions (Scheme 9.5). The high *anti* selectivity was observed regardless of the geometry of the crotylsilane. Although this catalytic system was effective for

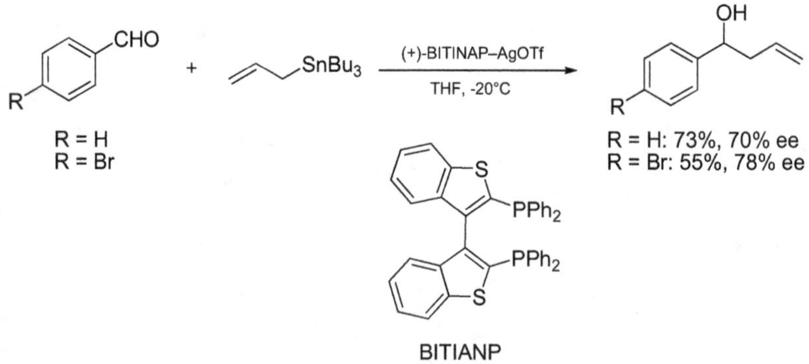

Scheme 9.3. Allylation using BITIANP ligand.

TABLE 9.4. Allylation Using a Thiophosphoramide Ligand

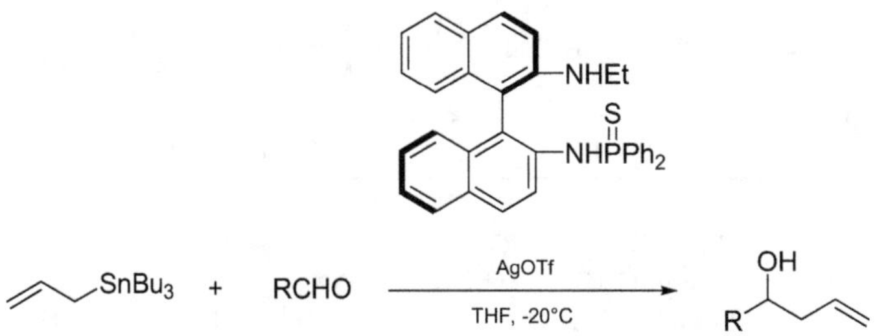

Entry	RCHO	Yield (%)	ee (%)
1	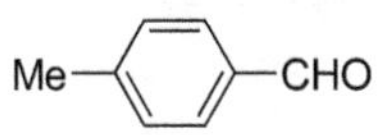	75	98
2	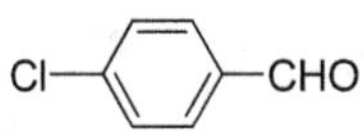	74	96
3		56	80
4	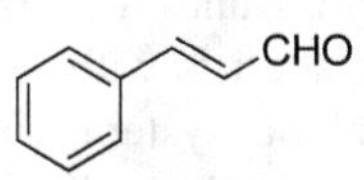	80	68

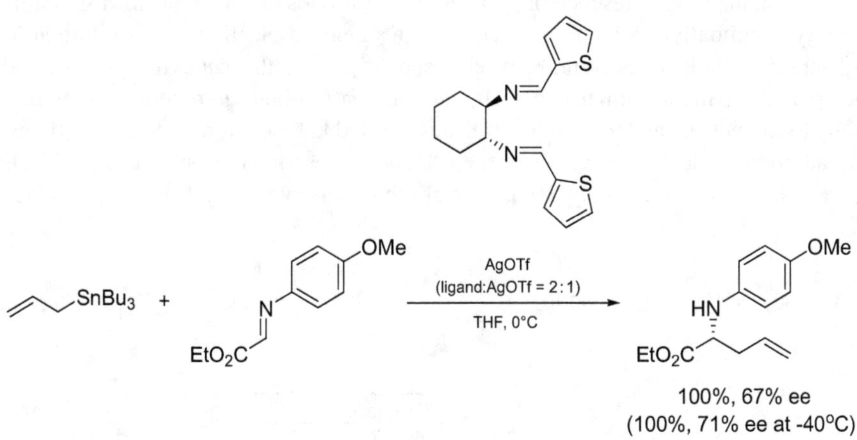

Scheme 9.4. Allylation using a diimine ligand.

TABLE 9.5. Allylation of Aldehydes with Allyltrimethoxysilane

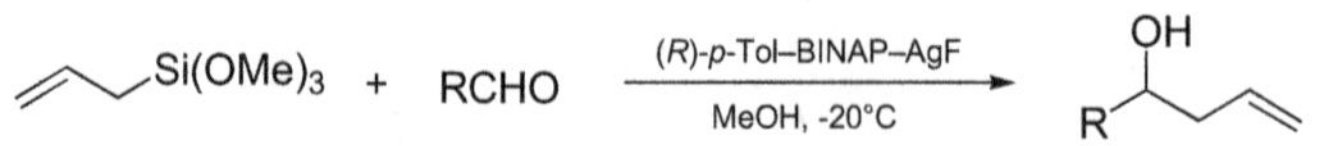

Entry	RCHO	Yield (%)	ee (%)[a]
1	PhCHO	80	94 (*R*)
2	(*E*)-PhCH=CHCHO	93	78 (*R*)
3	2-Furyl-CHO	70	83 (*R*)
4	1-Naphthyl-CHO	81	92 (*R*)
5	4-Me-C_6H_4CHO	67	93 (*R*)
6	4-Br-C_6H_4CHO	90	93

[a] Configuration.

various aromatic aldehydes, aliphatic aldehydes did not react in the presence of this catalyst.

To overcome this limitation, Yamamoto and coworkers investigated the reaction conditions and found that the use of KF and 18-*crown*-6 in the presence of AgOTf and BINAP was effective in promoting the allylation reaction (Table 9.6).[11] Thus, the reaction was conducted with aliphatic aldehydes and allytrimethoxysilane in the presence of catalyst to give the corresponding adducts with satisfactory yields and high enantioselectivities. Moreover, this catalytic system can be applied to the reaction between aromatic aldehydes and allyltrimethoxysilane.

The allylation reaction between ketones and allylsilanes was achieved in 2005. Yamamoto and Wadamoto developed the asymmetric allylation reaction in the presence of AgF-Difluorphos (Scheme 9.6).[12] The reaction of ketones and allyltrimethoxysilane in the presence of AgF and Difluorophos afforded the corresponding tertiary homoallylic alcohols with high enantioselectivities. Additionally, α,β-unsaturated ketones could be used as substrates, and this catalytic system could be applied for the asymmetric crotylation reaction to obtain *anti* adducts preferentially (Schemes 9.7 and 9.8). When α,β-unsaturated ketones were used as substrates, 1,2-addition products were obtained exclusively. As described before, the *anti* adducts were obtained predominately, regardless of the geometry of crotyltrimethoxysilane.

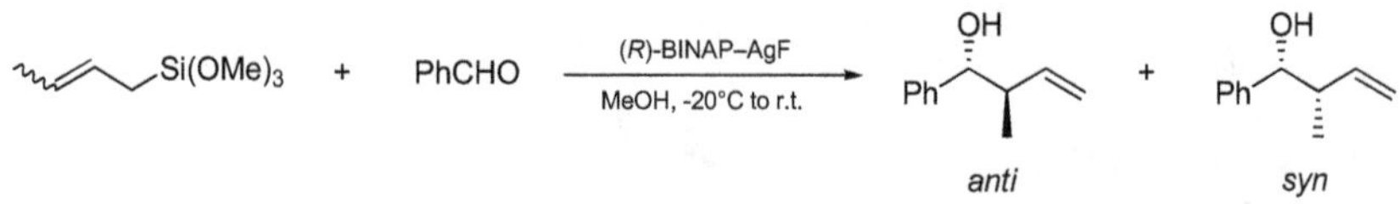

E : *Z* ratio	Yield (%)	*anti* (% ee) : *syn* (% ee)
83 : 17	77	92 (96) : 8 (62)
1< : 99	82	94 (94) : 6 (60)
45 : 55	99	93 (94) : 7 (60)

Scheme 9.5. Crotylation of benzaldehyde.

TABLE 9.6. Allylation of Various Aldehydes

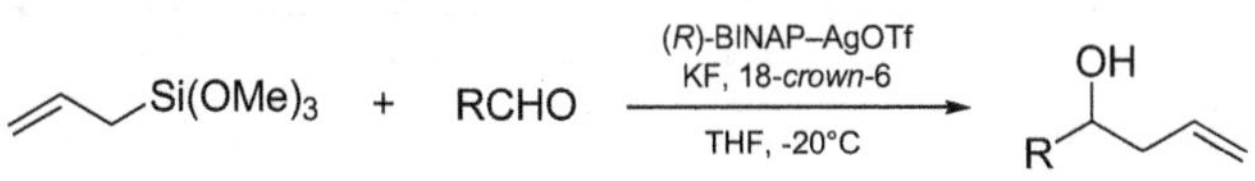

Entry	RCHO	Yield (%)	ee (%)[a]
1	PhCHO	91	95 (*R*)
2	(*E*)-PhCH=CHCHO	81	87 (*R*)
3	2-Furyl-CHO	57	95 (*R*)
4	1-Naphthyl-CHO	95	92 (*R*)
5	4-Me-C_6H_4CHO	61	95 (*R*)
6	4-Br-C_6H_4CHO	95	96 (*R*)
7	2-Me-C_6H_4CHO	82	97 (*R*)
8	*c*-C_6H_{11}CHO	62	93 (*R*)
9	$PhCH_2CH_2CHO$	76	86 (*R*)

[a] Configuration.

More recently, the catalyst that was prepared from AgOTf and Tol–BINAP was applied to the synthesis of 2,3-dihydrobenzofurans. The reaction of 2,3-dihydrobenzoxasilepine and an aromatic aldehyde was carried out in the presence of the silver catalyst, KF, and 18-*crown*-6 to give *trans*-2,3-disubstituted 2,3-dihydrobenzofuran

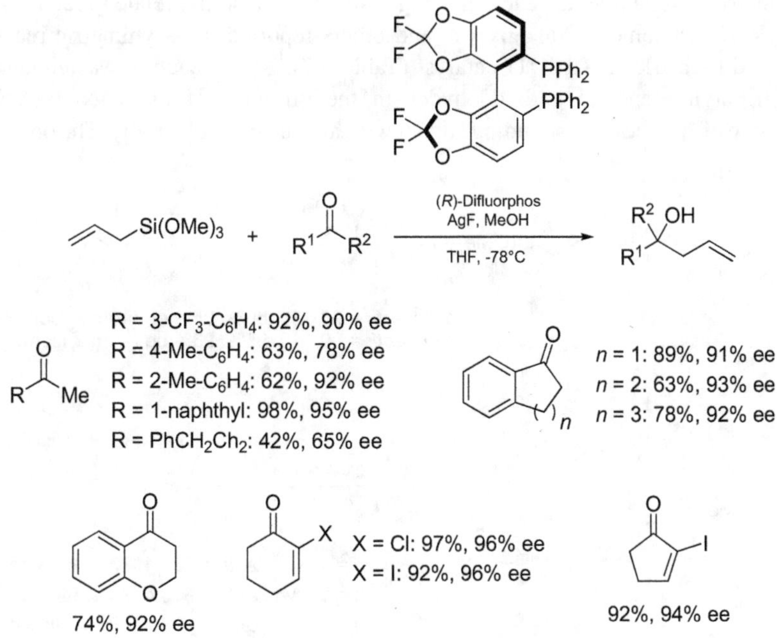

Scheme 9.6. Allylation of ketones.

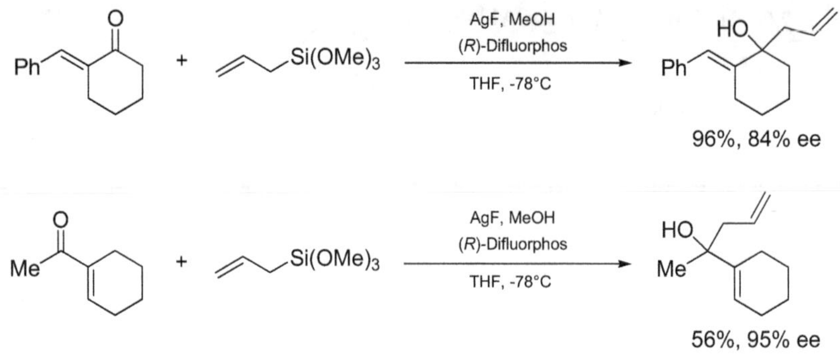

Scheme 9.7. Allylation of α,β-unsaturated ketones.

in a diastereoselective manner (Scheme 9.9).[13] The acyl group at the *ortho* position of aromatic ring was essential for a successful reaction.

9.4 ALDOL REACTION USING TIN ENOLATES

The aldol reaction is one of the most useful and classical reactions in organic synthesis, since the resulting β-hydroxycarbonyl compounds are highly useful building blocks for pharmaceuticals and natural products.[14] The asymmetric version of this reaction has been developed to a significant extent. We focus here on the asymmetric classical aldol reaction using a silver catalyst and related reactions.[15]

In 1997, Yamamoto, Yanagisawa, and others reported the asymmetric reaction catalyzed by a BINAP–AgOTf catalyst (Table 9.7).[16] The reaction was conducted with tributyltin enolate and aldehydes in the presence of the BINAP–AgOTf catalyst to afford the corresponding adduct with high enantioselectivity. The observed

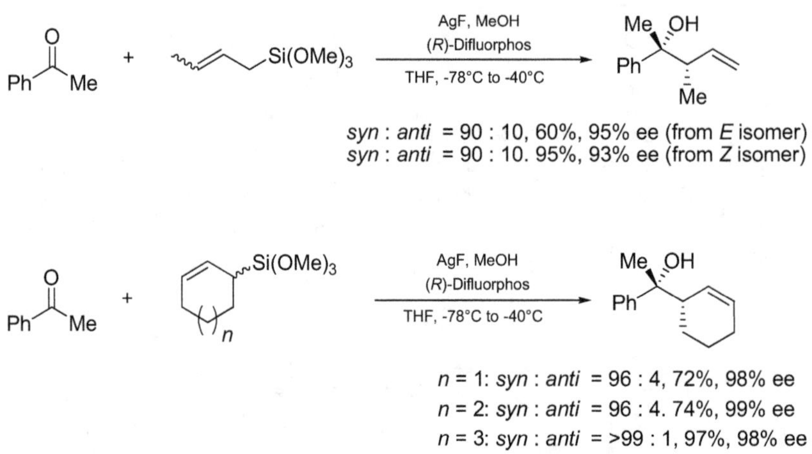

Scheme 9.8. Crotylation of ketones.

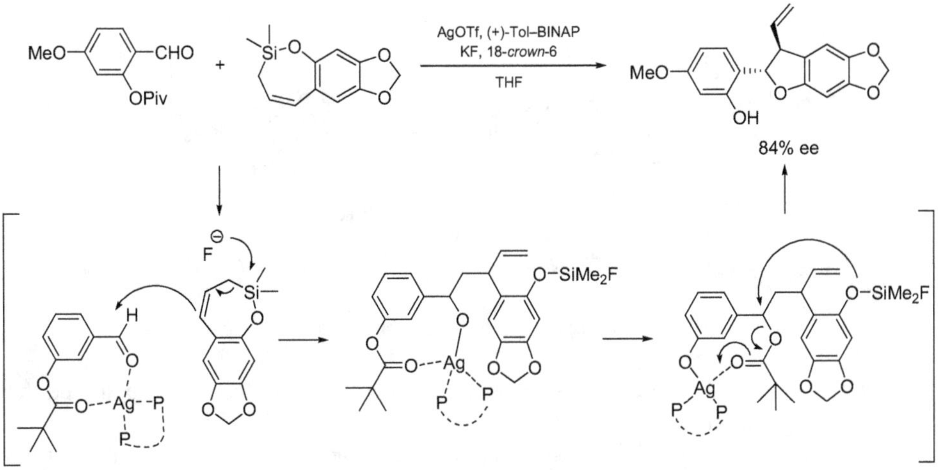

Scheme 9.9. Synthesis of a dihydrobenzofuran.

TABLE 9.7. Aldol Reaction with Tin Enolates

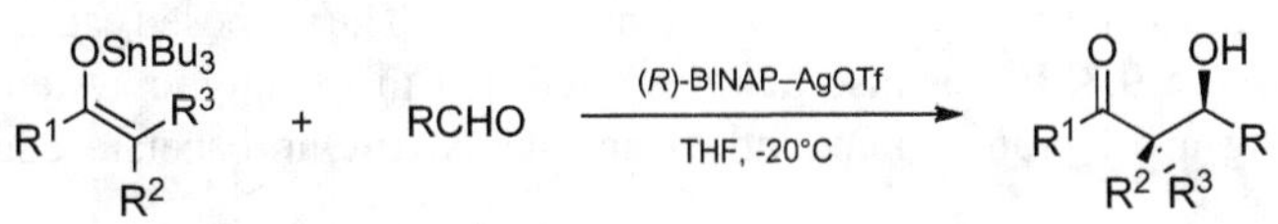

Entry	Tin Enolate	RCHO	Yield (%)	*anti* : *syn*	ee (%)
1	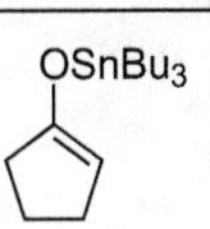	PhCHO	92	89 : 11	92
2	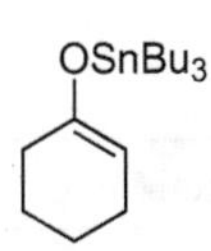	PhCHO	94	92 : 8	93
3		PhCHO	95	93 : 7	94
4	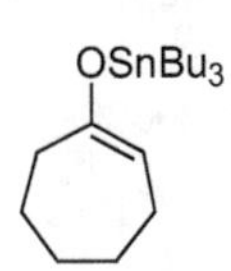	PhCHO	90	85 : 15	96
5	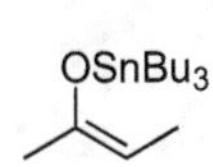	PhCHO	81	<1 : 99	95
6		$PhCH_2CH_2CHO$	77	<1 : 99	95
7	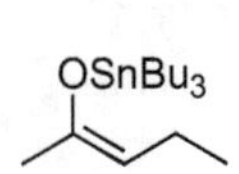	PhCHO	98	<1 : 99	91

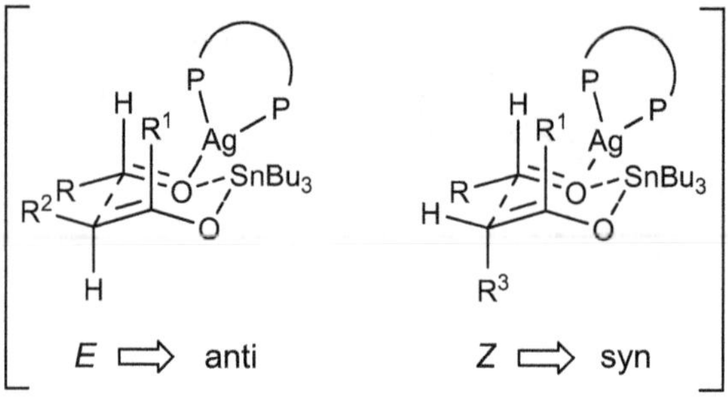

Figure 9.2. Proposed transition state.

anti : *syn* ratio depended on the $E:Z$ ratio of enolate, and this result suggested the cyclic transition state (Fig. 9.2).

Since the use of stoichiometric amounts of organostannane is unfavorable for the environment, they have established reaction conditions for a process that is catalytic in organostannane. Thus, the reaction with an alkenyl trichloroacetate as the precursor of the tin enolate in the presence of the catalytic amount of Me_3SnOMe and the appropriate aldehydes afforded the corresponding aldol adducts with high enantio- and diastereoselectivities (Table 9.8).[17] However, when the aliphatic aldehydes were used as substrates, the reaction did not take place at all. The proposed catalytic cycle is shown in Figure 9.3. It was also demonstrated that this catalytic system could be applied using a β-lactone compound as an alkenyl trichloroacetate equivalent to

TABLE 9.8. Aldol Reaction Using Catalytic Amounts of Me_3SnOMe

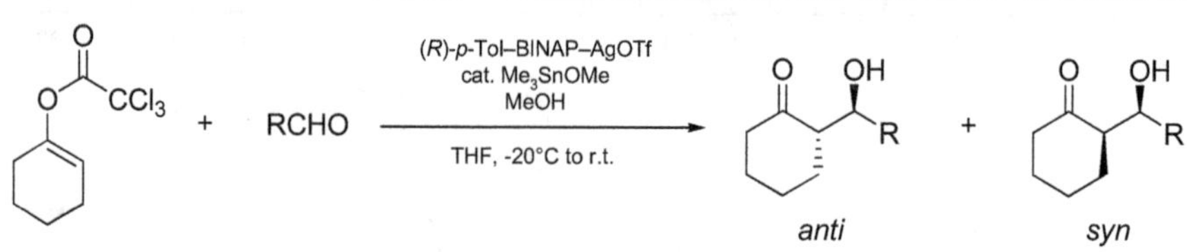

Entry	RCHO	Yield (%)	anti : syn	ee (%)
1	PhCHO	86	94 : 6	96
2	4-Me-C_6H_4CHO	66	94 : 6	95
3	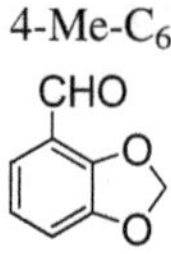	49	80 : 20	77
4	1-Naphthyl	74	96 : 4	92
5	(*E*)-PhCH=CHCHO	76	78 : 22	90
6	(*E*)-$CH_3(CH_2)_2$CH=CHCHO	61	76 : 24	83
7	$PhCH_2CH_2CHO$	29	84 : 16	79
8	*n*-C_5H_{11}CHO	<1	—	—
9	*i*-PrCHO	<1	—	—

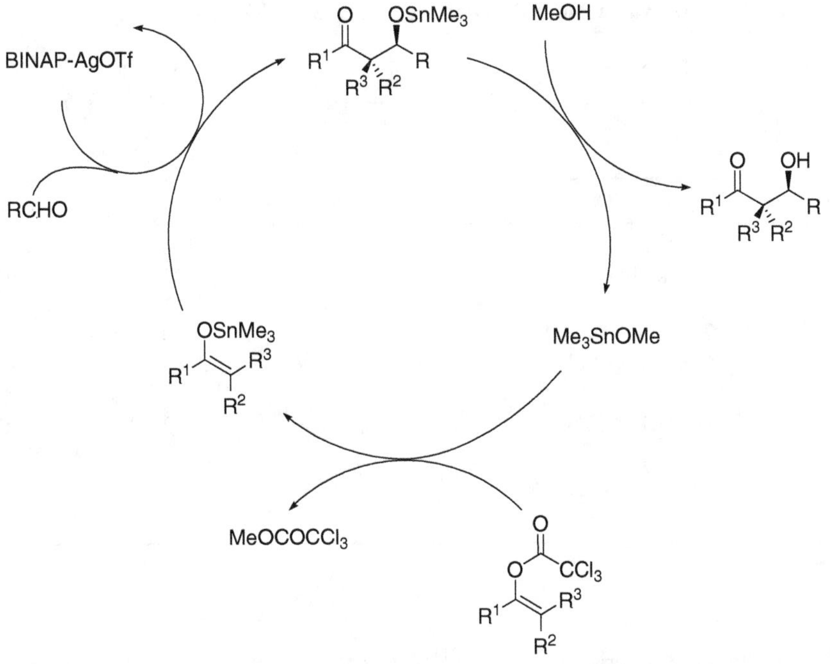

Figure 9.3. Proposed catalytic cycle.

generate a β-ketoester with high enantioselectivity (Scheme 9.10). The alkenyl trichloroacetate reacted with Me_3SnOMe to form a tin enolate, which reacted with the aldehyde in the presence of the chiral silver catalyst in a highly enantioselective manner. The resulting tin oxide was protonated by MeOH to generate the product and regenerate Me_3SnOMe.

9.5 ALDOL REACTION USING SILYL ENOL ETHERS

Because these asymmetric aldol reactions are ideal methods for constructing β-hydroxy carbonyl compounds in optically active form, the development of an asymmetric aldol reaction without the use of an organostannane would be advantageous. Yamagishi and coworkers have reported the Mukaiyama aldol reaction using trimethylsilyl enol ethers in the presence of the BINAP–$AgPF_6$ complex to afford the adducts with moderate enantioselectivities (Table 9.9).[18] They have also assigned

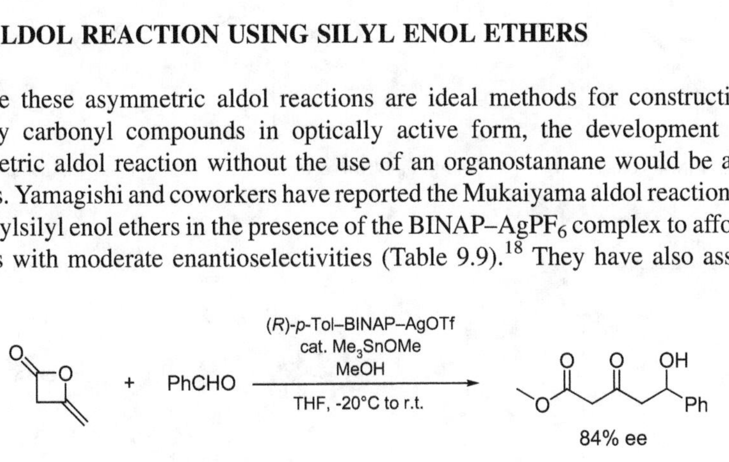

Scheme 9.10. Aldol reaction of diketene.

TABLE 9.9. Mukaiyama Aldol Reaction with Various Aldehydes

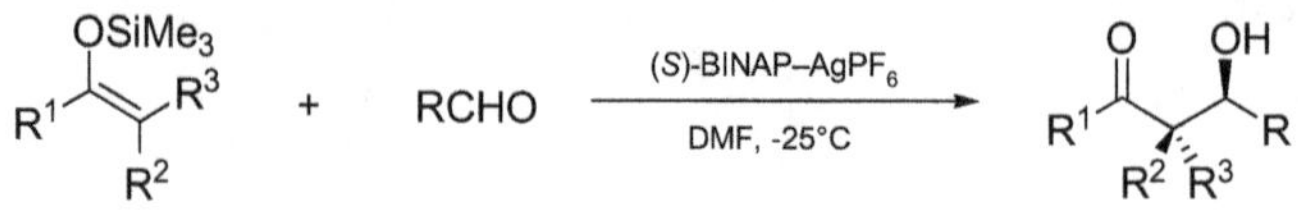

Entry	RCHO	Silyl enol ether	Yield (%)	ee (%)
1	PhCHO	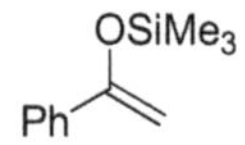	100	69
2	1-Naphthyl-CHO		36	80
3			82	54
4	2-Napththyl-CHO		100	58
5	Cyclohexyl-CHO		83	47
6	PhCHO	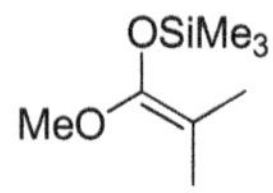	0	0

structures to three silver species using ^{31}P NMR studies and X-ray analysis. The 2 : 1 complex of BINAP and $AgPF_6$ was found to be a major complex in DMF and could not catalyze Mukaiyama aldol reactions (Fig. 9.4). The 1 : 1 complex of BINAP and $AgPF_6$ and the 1 : 2 complex of BINAP and $AgPF_6$ were found to be minor complexes. The ^{31}P NMR spectra suggested that an equilibrium existed between these two complexes and $AgPF_6$. In addition, they have demonstrated that the 1 : 1

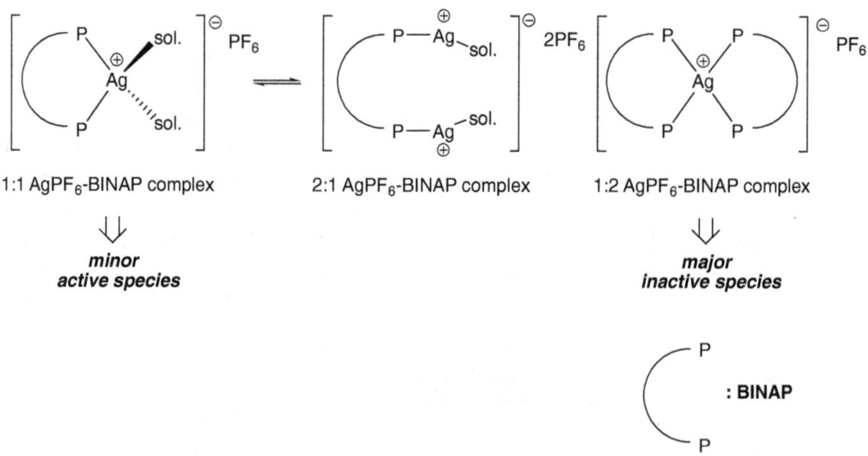

Figure 9.4. Structures of silver–BINAP complexes.

complex coordinates an aldehyde and catalyzes the reaction via a cyclic transition state.

Independently, Yamamoto, Yanagisawa, and others reported the asymmetric aldol reaction using trimethoxysilyl enol ethers.[19] The reaction was conducted with aldehydes and trimethoxysilyl enol ethers in the presence of Tol–BINAP–AgF to give the corresponding adducts with high enantioselectivities and diastereoselectivities. They obtained *syn*-aldol adducts as major products even when silyl enol ethers derived from cyclic ketones were used. Moreover, when α,β-unsaturated aldehydes were employed as substrates, 1,2 adducts were obtained exclusively (Table 9.10). From an NMR study and correlation between the $E:Z$ ratio of the enol ethers and diastereoselectiviy, they proposed a cyclic transition state (Fig. 9.5). Thus, the reaction of E enol ethers proceeded via a boat form, whereas the reaction of Z enol ethers took place via a chair form.

In addition, the combination of KF and 18-*crown*-6 in the presence of BINAP–AgOTf, which was effective in accelerating the allylation reaction using allyltrimethoxysilane, could be used for the aldol reaction (Scheme 9.11).[20] Thus, the reaction with the silyl enol ether derived from cyclohexanone and benzaldehyde in the

TABLE 9.10. Aldol Reaction Using Trimethoxysilyl Enol Ethers

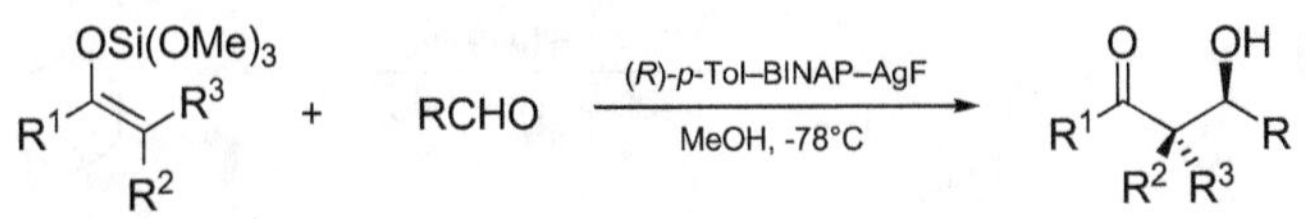

Entry	Silyl enol ether	RCHO	Yield (%)	*syn*:*anti*	ee (%)
1	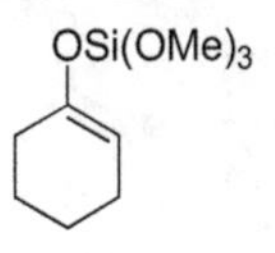	PhCHO	78	84:16	87
2		4-MeO–C_6H_4CHO	86	75:25	92
3		4-Br-C_6H_4CHO	87	76:24	90
4		1-naphthyl-CHO	68	27:73	76
5		(*E*)-PhCH=CHCHO	81	81:19	68
6		$PhCH_2CH_2CHO$	<1	—	—
7		PhCOMe	<1	—	—
8	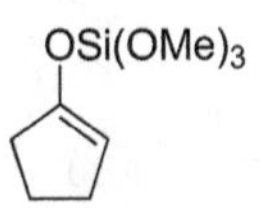	PhCHO	18	75:25	52
9	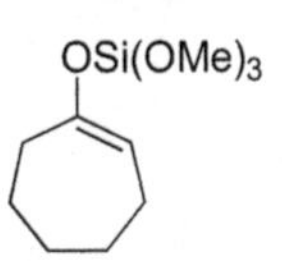	PhCHO	67	81:19	78
10	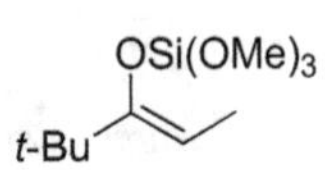	PhCHO	84	>99:1	97
11		4-MeO–C_6H_4CHO	76	>99:1	96
12		1-naphthyl-CHO	63	94:6	95
13		$PhCH_2CH_2CHO$	<1	—	—

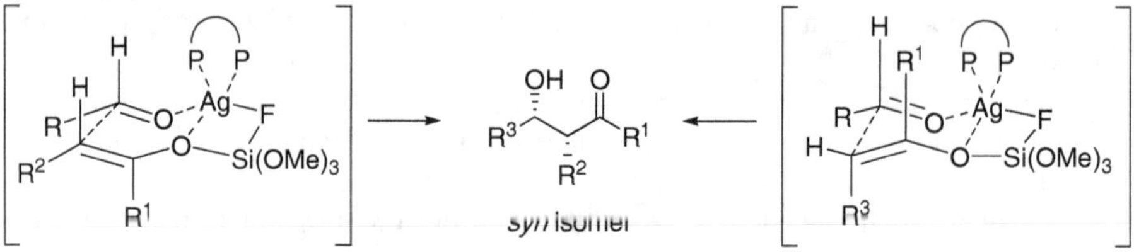

Figure 9.5. Proposed transition state.

presence of catalyst proceeded to afford an adduct with high enantioselectivity and diastereoselectivity. Remarkably, the *anti* adduct was obtained as the major product, in contrast to the BINAP–AgF catalyst system.

This catalytic system could be applied to the enantioselective hydroxymethylation of silyl enol ethers with aqueous formalin (Scheme 9.12).[21] Doyle and coworkers have successfully applied the catalyst system of BINAP–AgOTf, KF, and 18-*crown*-6 for Mukaiyama aldol reaction of α-diazo silyl enol ether (Scheme 9.13).[22]

Hoveyda and coworkers reported that the silver complex generated from AgF_2 and an amino acid–based ligand could be used for the asymmetric aldol reaction with silyl enol ethers and α-ketoesters (Table 9.11).[23]

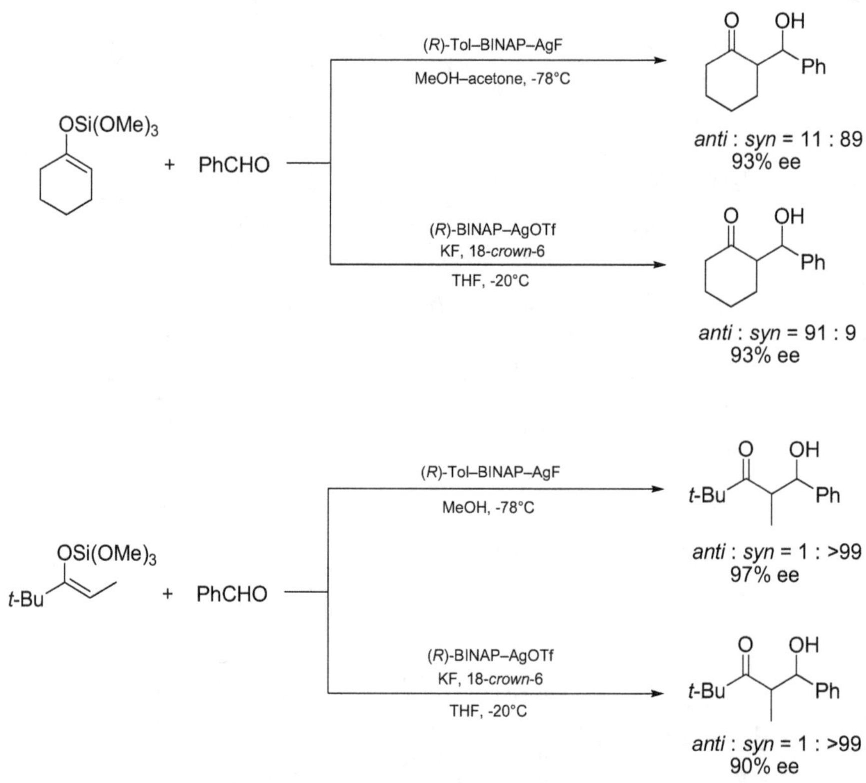

Scheme 9.11. Diastereoselectivity of aldol reaction with a different catalyst system.

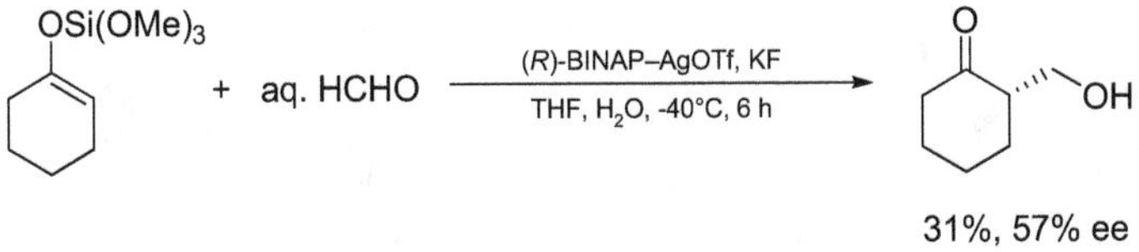

Scheme 9.12. Hydroxymethylation of a trimethoxysilyl enol ether.

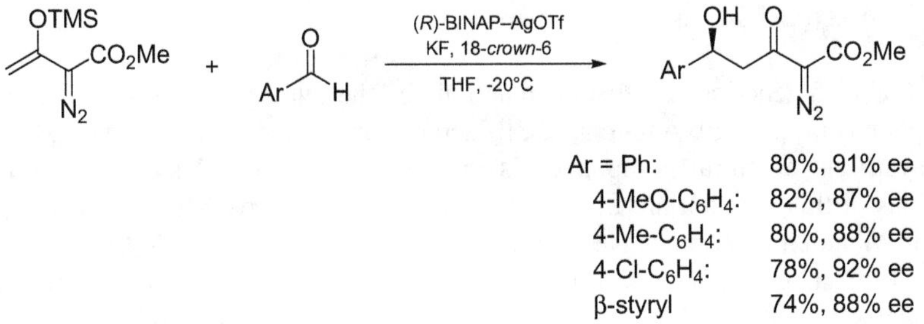

Scheme 9.13. Aldol reaction of an α-diazo trimethylsilyl enol ether.

TABLE 9.11. Aldol Reaction of α-Ketoesters

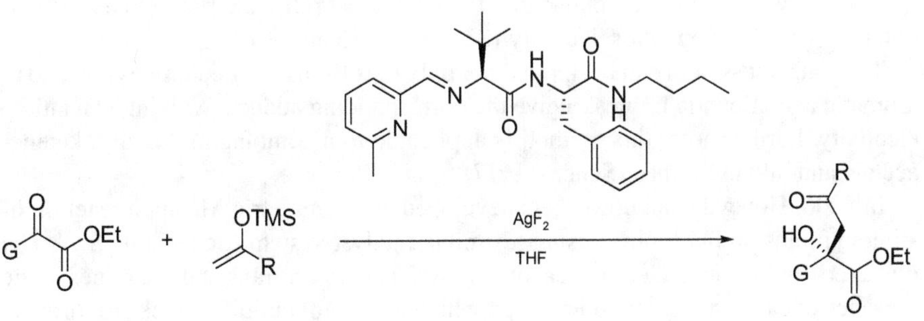

Entry	G	R	Yield (%)	ee (%)
1	CH_2CH_2Ph	Ph	92	86
2	$CH_2CH_2CO_2Me$	Ph	95	92
3	CH_2*i*-Pr	Ph	95	87
4	*i*-Pr	Ph	93	95
5	*i*-Pr	*t*-Bu	61	92
6	*i*-Pr	Me	>98	88
7	Cy	Ph	98	95
8	Cy	Me	97	90
9	Cyclopropyl	Ph	90	96
10	$CH_2{=}CH(Me)$	Ph	98	90
11	Ph	Ph	93	60
12	2-Thienyl	Ph	95	72

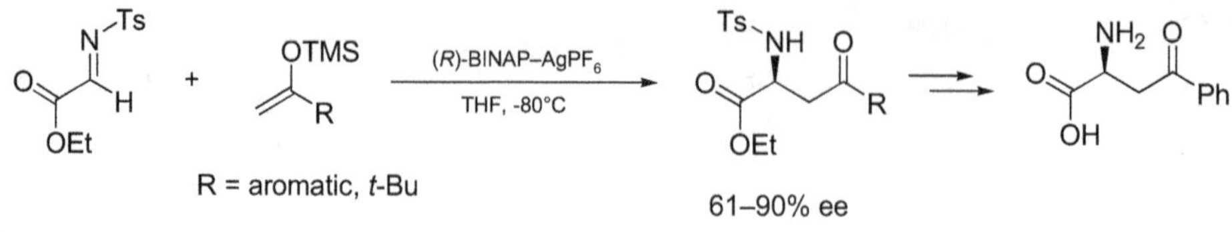

Scheme 9.14. Mannich reaction of α-iminoester with various enol ethers.

9.6 MANNICH REACTION

The Mannich reaction is a useful reaction for the synthesis of β-aminocarbonyl compounds including β-amino acids, β-lactams, and related compounds.[24] Because the synthesis of chiral compounds is required for biological study, asymmetric versions of this reaction are in high demand. Several approaches to achieve asymmetric Mannich reactions have been reported.

In 1998, Lectka and coworkers reported the asymmetric Mannich reaction of α-iminoesters catalyzed by a BINAP–Ag(I) complex (Scheme 9.14).[25]

Moreover, Hoveyda and coworkers have demonstrated that Mannich reactions between silyl enol ethers and aldimines are promoted by the chiral complex that was generated from AgOAc and an *iso*-Leu-derived phosphine ligand.[26] When the reaction was conducted with trimethylsilyl enol ether and aldimine in the presence of AgOAc, *iso*-Leu-derived phosphine ligand and 2-PrOH, the β-aminoketone was obtained with high enantioselectivity (Schemes 9.15 and 9.16).

This catalyst system can be applied not only to α,β-unsaturated imines but also to aromatic and aliphatic imines to give the corresponding adducts with high enantioselectivity. Furthermore, this method is applicable to a combination of silyl ketene-acetals and alkynylimines (Scheme 9.17).[27]

In 2006, Hoveyda and coworkers developed an asymmetric Mannich reaction of silyloxyfurans and aldimines using a similar catalyst system (Scheme 9.18).[28] The diastereo- and enantioselective reaction between silyloxyfurans and aldimines in the presence of catalyst gave γ-butenolides, which are useful building blocks for organic synthesis. They also reported a mechanistic study of this reaction.[28b]

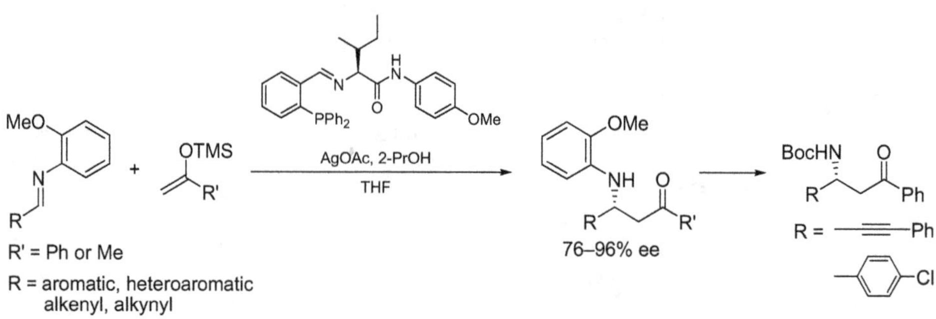

Scheme 9.15. Mannich reaction of various aldimines.

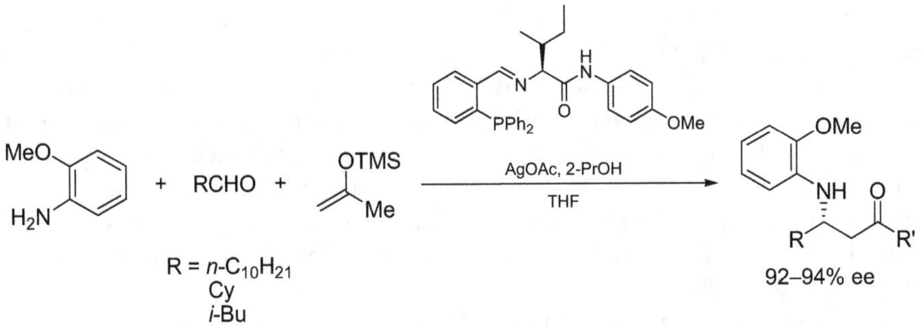

Scheme 9.16. A one-pot, three-component reaction.

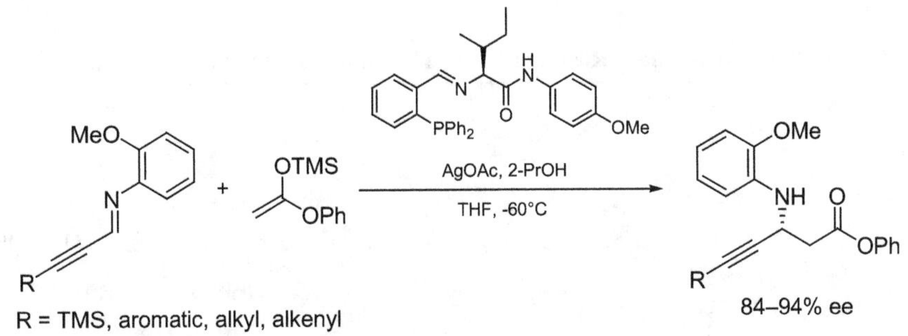

Scheme 9.17. Mannich reaction of alkynyl aldimines with a silylketeneacetal.

9.7 NITROSOALDOL REACTION

In 2002, Yamamoto and Momiyama reported an unusual aldol-like reaction with silyl enol ethers and nitrosobenzene in the presence of a catalytic amount of TESOTf (nitrosoaldol reaction).[29] Usually, nucleophiles react with nitrosobenzene without Lewis acid to give the *N* adduct predominantly. In contrast, they reported that the reaction of silyl enol ethers and nitrosobenzene catalyzed by TESOTf afforded the

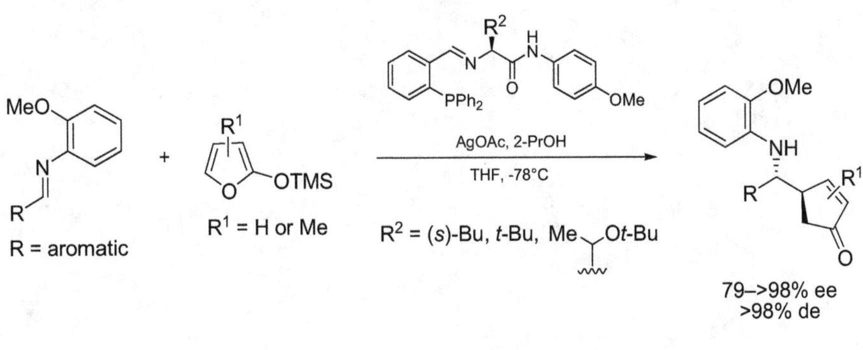

Scheme 9.18. Vinylogous Mannich reaction with silyloxyfurans.

corresponding *O* adducts exclusively. They have extended this reaction to a regio- and enantioselective version.[30] The reaction was conducted with nitrosobenzene and tin enolates in the presence of the BINAP–AgOTf catalyst to give corresponding *N* adducts or *O* adducts selectively with high enantioselectivity (Tables 9.12 and 9.13). It is noteworthy that the regioselectivity (*N* vs. *O*) can be controlled by changing the ratio of BINAP to silver salt. For example, when the reaction was conducted with trimethyltin enol ether and nitrosobenzene in the presence of a 1 : 1 mixture of a Tol–BINAP and AgOTf in THF, the *O* adduct was obtained as major product with high enantioselectivity (Table 9.12).

The *O* adducts were converted into chiral alcohols with a catalytic amount of $CuSO_4$ without the loss of enantioselectivity (Scheme 9.19).

TABLE 9.12. *O*-Nitrosoaldol Reaction Using Tin Enolates and Nitrosobenzene

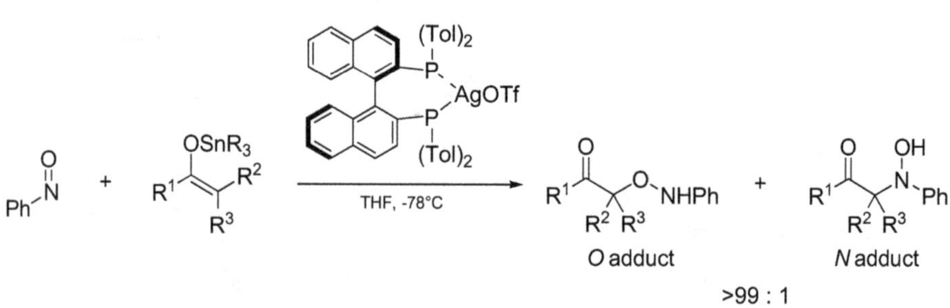

Entry	Enolate	Yield/ee (%)	Entry	Enolate	Yield/ee (%)
1	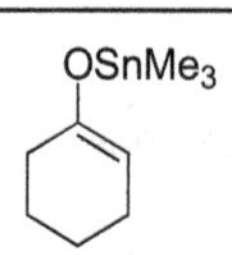	88/99	5	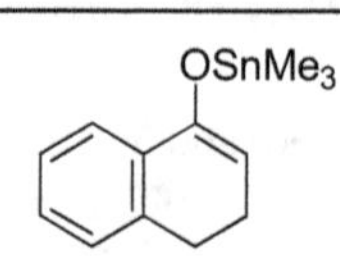	96/85
2	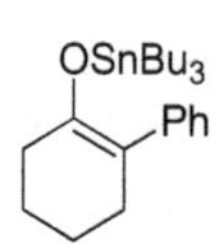	97/88	6	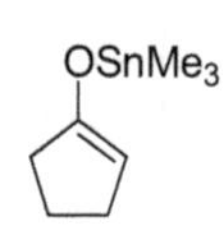	93/92
3	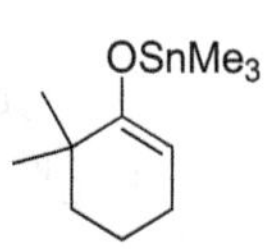	94/87	7	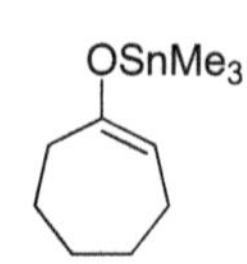	95/82
4	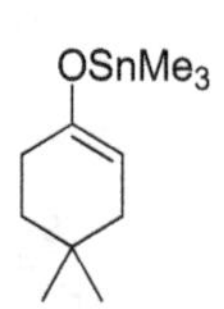	92/90	8	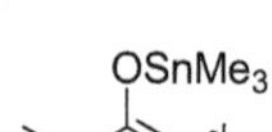	92/94

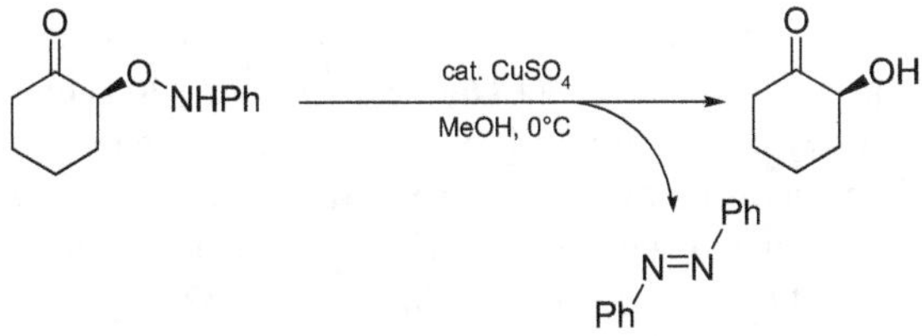

Scheme 9.19. Cleavage of the N–O bond.

In contrast, the reaction of tributyltin enol ethers and nitrosobenzene in the presence of a 1 : 2 mixture of BINAP and AgOTf in ethylene glycol diethyl ether afforded the *N* adduct predominantly with high enantioselectivity (Table 9.13). Momiyama and Yamamoto have determined the structures of silver–BINAP complex by an X-ray analysis and a ^{31}P NMR study[30b].

After their discovery, several other groups reported nitrosoaldol reactions using different catalysts including acid catalysts and organocatalysts.[31]

Although the asymmetric nitrosoaldol reaction was achieved with high enantioselectivity, the use of a stoichiometric amount of organostannane was unacceptable. In

TABLE 9.13. *N*-Nitrosoaldol Reaction Using Tin Enolates and Nitrosobenzene

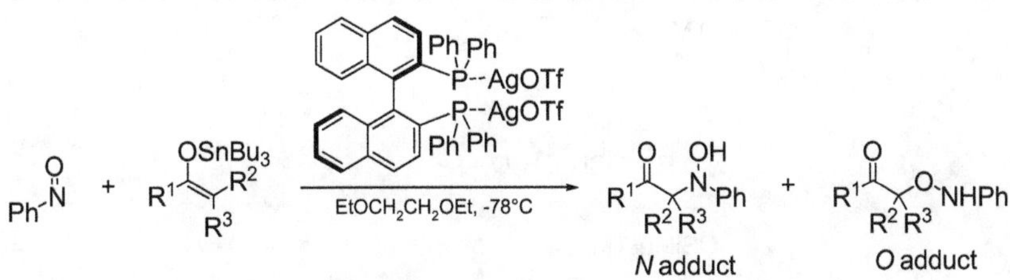

Entry	Enolate	Yield (%)	*N* : *O*	ee (%)
1	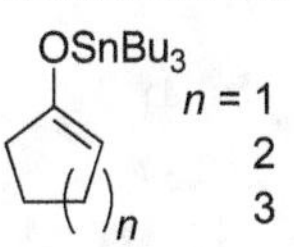	90	97 : 3	86
2		95	96 : 1	>99
3		96	>99 : 1	97
4	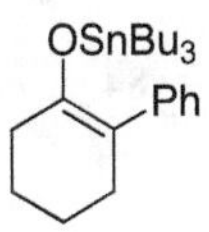	94	>99 : 1	77
5	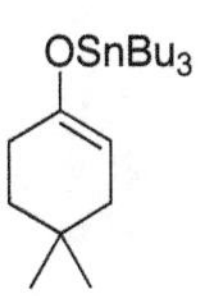	97	>99 : 1	98

2008, Yamamoto et al. reported the asymmetric *O*-nitrosoaldol reaction using silyl enol ethers in the presence of the silver catalyst.[32] In order to achieve this reaction, they developed a novel combination of silver and a chiral phosphite derived from BINOL. The disilanyl enol ether was used to ensure high yield and enantioselectivity. The reaction was conducted with disilanyl enol ether and nitrosobenzene in the presence of $AgBF_4$ and the chiral phosphite ligand in THF to produce the *O* adduct with high regio- and enantioselectivity (Table 9.14). In addition, a chiral silyl enol ether could be used as a substrate. The reaction was conducted with chiral silyl enol

TABLE 9.14. *O*-Nitrosoaldol Reaction Using Silyl Enol Ethers and Nitrosobenzene

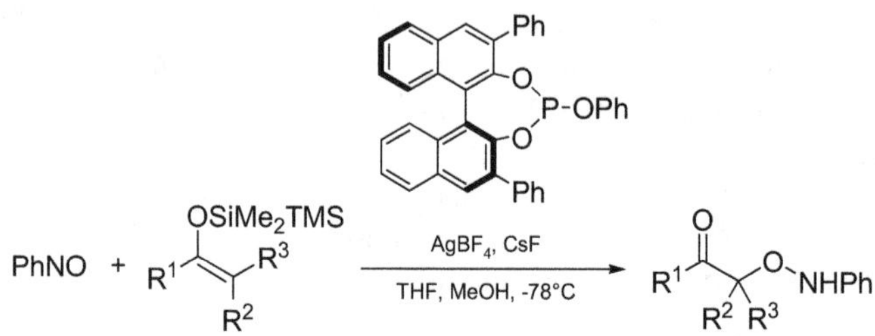

Entry	Silyl enol ether	Yield (%)	ee (%)
1	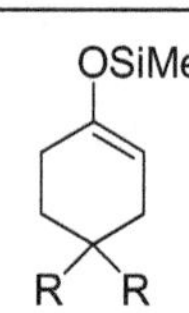 R = H	85	95
2	R = Me	72	98
3	R = OCH_2CH_2O	85	96
4	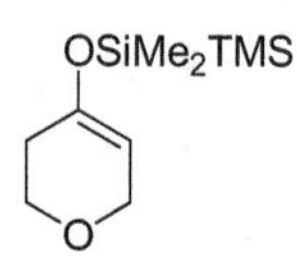	66	97
5	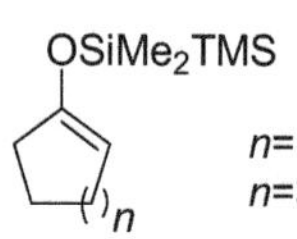 *n*=1	41	90
6	*n*=3	37	64
7	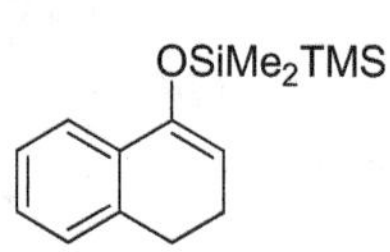	84	92
8	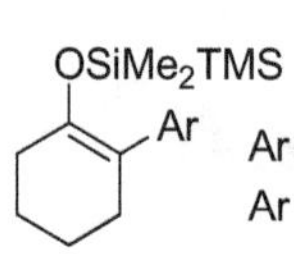 Ar = Ph	99	79
9	Ar = 2-naphthyl	80	76

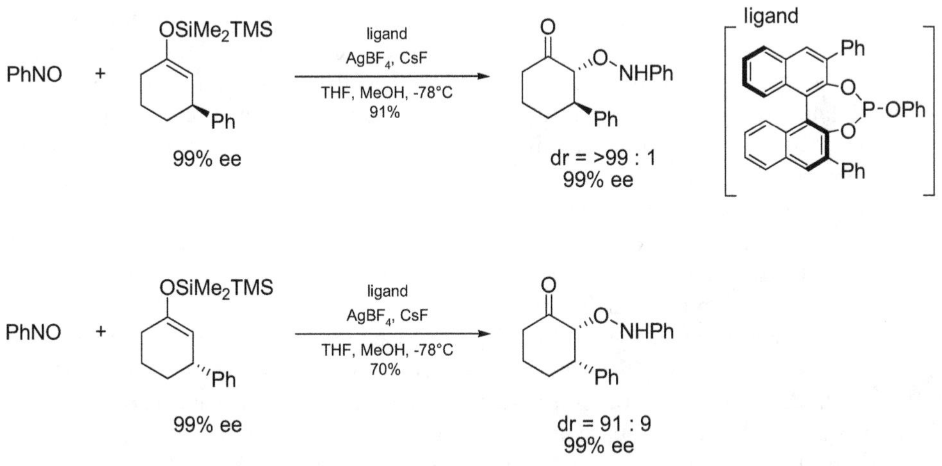

Scheme 9.20. Diastereoselective *O*-nitrosoaldol reaction.

ether in the presence of the catalyst to give *O* adduct with high diastereoselectivity. Remarkably, the stereochemical outcome of nitrosoaldol reaction could be controlled by the catalyst regardless of the configuration of the stereocenter in silyl enol ether (Scheme 9.20).

9.8 ALDOL REACTION WITH AZODICARBOXYLATE

Azodicarboxylates are attractive aminating reagents, since α-aminocarbonyl compounds are not only useful building blocks in organic synthesis, but also essential tools for biological studies.[33] Hence, the development of general and efficient methods for the preparation of these compounds is an important issue in organic synthesis.

Kobayashi and coworkers have developed an amination reaction using a silyl enol ether and an azodicarboxylate (Scheme 9.21).[34] The reaction was conducted with trimethylsilyl enol ether and azodicarboxylate in the presence of the $AgClO_4$–BINAP complex to afford the corresponding adducts with high enantioselectivities.

9.9 CONCLUSION

The allylation reaction of carbonyl compounds is a classical but excellent method for constructing highly functionalized organic molecules. More recently, allylation

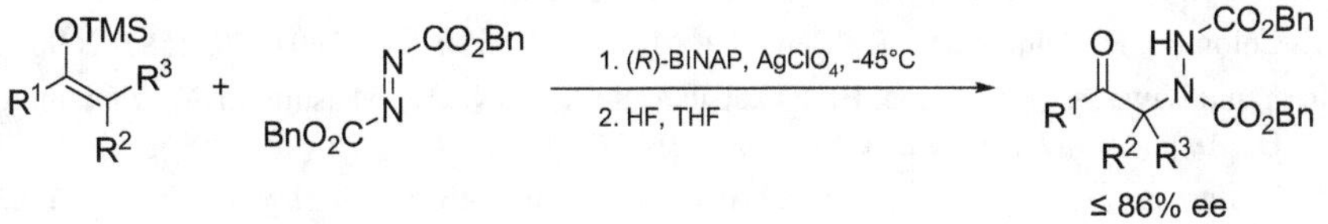

Scheme 9.21. Amination reaction using azodicarboxylate.

reactions have been rapidly extended to asymmetric versions by a combination of various catalysts and reagents. Those catalysts can also be applied for the crotylation reaction in a highly diastereo- and enantioselective manner. In addition to the allylation reaction, we have also described the aldol reaction. The aldol reaction is one of the most powerful methods for preparing 1,3-oxygenated organic compounds from carbonyl compounds, and the aldol reaction has been applied to the synthesis of numerous natural products and drugs. This reaction has also been broadened to useful variants that use nitroso compounds or azodicarboxylates instead of carbonyl compounds. The study of these reactions is just beginning, and improvements are to be expected in the near future.

Although many catalysts have been developed for these reactions, most of them are Lewis acid catalysts that are typically sensitive to moisture. The silver catalysts we mentioned above might become a preferable choice since silver catalysts can be used in water and many silver salts are commercially available and inexpensive. The chemistry of silver catalysts in this context can be expected to expand significantly in the future.

REFERENCES

1. (a) Yamamoto, Y.; Asao, N., *Chem. Rev.* **1993**, *93*, 2207–2293; (b) Roush, W. R., in *Methoden der Organischen Chemie (Houben-Weyl)*, Helmchen, G.; Hoffmann, R. W.; Mulzer, J.; Shaumann, E., eds., Thieme, Stuttgart, **1996**, Vol. 3 E21, pp. 1410–1486; (c) Bubnov, Y., in Science of Synthesis, Kaufmann, D. E.; Matteson, D. S., eds., Thieme, Stuttgart, **2004**, Vol. 6, pp. 945–1072; (d) Ramachandran, P. V., *Aldrichim. Acta* **2002**, *35*, 23–35; (e) Brown, H. C.; Ramachandran, P. V., *J. Organomet. Chem.* **1995**, *500*, 1–19; (f) Ishihara, K., in *Lewis Acids in Organic Synthesis*, Yamamoto, H., ed., Wiley-VCH, Weinheim, **2000**, Vol. 1, pp. 176–179; (g) Mikami, K.; Terada, M., in *Lewis Acids in Organic Synthesis*, Wiley-VCH, Weinheim, **2000**, Vol. 2, pp. 800–805.
2. (a) Yanagisawa, A.; Nakashima, H.; Ishiba, A.; Yamamoto, H., *J. Am. Chem. Soc.* **1996**, *118*, 4723–4724; (b) Yanagisawa, A.; Nakashima, H.; Nakatsuka, Y.; Ishiba, A.; Yamamoto, H., *Bull. Chem. Soc. Jpn.* **2001**, *74*, 1129–1137.
3. Yanagisawa, A.; Ishiba, A.; Nakashima, H.; Yamamoto, H., *Synlett* **1997**, 88–90.
4. (a) Yamamoto, Y.; Yatagai, H.; Naruta, Y.; Maruyama, K., *J. Am. Chem. Soc.* **1980**, *102*, 7107–7109; (b) Yamamoto, Y.; Yatagai, H.; Ishihara, Y.; Maeda, N.; Maruyama, K., *Tetrahedron* **1984**, *40*, 2239–2246.
5. Yanagisawa, A.; Nakatsuka, H.; Nakashima, H.; Yamamoto, H., *Synlett* **1997**, 933–934.
6. Loh, T.-P.; Zhou, J.-R., *Tetrahedron Lett.* **2000**, *41*, 5261–5264.
7. Cesarotti, E.; Araneo, S.; Rimoldi, I.; Tassi, S., *J. Mol. Catal. A.* **2003**, *204–205*, 221–226.
8. Wang, C.-J.; Shi, M., *Eur. J. Org. Chem.* **2003**, 2823–2828.
9. Colombo, F.; Annunziata, R.; Benaglia, M., *Tetrahedron Lett.* **2007**, *40*, 2687–2690.
10. Yanagisawa, A.; Kageyama, H.; Nakatsuka, Y.; Asakawa, K.; Matsumoto, Y.; Yamamoto, H., *Angew. Chem. Int. Ed.* **1999**, *38*, 3701–3703.
11. Wadamoto, M.; Ozasa, N.; Yanagisawa, A.; Yamamoto, H., *J. Org. Chem.* *68*, **2003**, 5593–5601.

12. Wadamoto, M.; Yamamoto, H., *J. Am. Chem. Soc.* **2005**, *127*, 14556–14557.
13. Gonzalez, L. J-.; Munoz, S. G-.; Corral, M. A-.; Dorado, M. M-.; Garcia, I. R., *Chem. Eur. J.* **2006**, *12*, 8762–8769.
14. Mahrwald, R., ed., *Modern Aldol Reactions*; Wiley-VCH, Weinheim, **2004**, Vols.1, 2.
15. (a) Gennari, C., in *Comprehensive Organic Synthesis*, Trost, B. M.; Fleming, I.; Heathcock, C. H., eds., Pergamon Press, Oxford, **1991**, Vol. 2, p. 629; (b) Hollis, T. K.; Bosnich, B., *J. Am. Chem. Soc.* **1995**, *117*, 4570–4581; (c) Braun, M., in *Methoden der Organischen Chemie (Houben-Weyl)*, Helmchen, G.; Hoffmann, R. W.; Mulzer, J.; Shaumann, E., eds., Thieme, Stuttgart, **1995**, Vol. E21, pp. 1410–1486; (d) Nelson, S. G., *Tetrahedron: Asymmetry* **1998**, *9*, 357–389; (e) Gröger, H.; Vogl, E. M.; Shibasaki, M., *Chem. Eur. J.* **1998**, *4*, 1137–1141; (f) Mahwald, R., *Chem. Rev.* **1999**, *99*, 1095–1120; (g) Carreira, E. M., in *Comprehensive Asymmetric Catalysis*, Jacobsen, E. N.; Pfaltz, A.; Yamamoto, H.;eds., Springer, Heidelberg, **1999**, Vol. 3, p. 997; (h) Arya, P.; Qin, H., *Tetrahedron* **2000**, *56*, 917–947; (i) Machajewski, T. D.; Wong, C. -H., *Angew. Chem. Int. Ed.,* **2000**, *39*, 1353–1357; (j) Carreira, E. M., in *Modern Carbonyl Chemistry*, Otera, J., ed., Wiley-VCH, Weinheim, **2000**, Chapter 8, p. 227.
16. Yanagisawa, A.; Matsumoto, Y.; Nakashima, H.; Asakawa, K.; Yamamoto, H., *J. Am. Chem. Soc.* **1997**, *119*, 9319–9320.
17. (a) Yanagisawa, A.; Matsumoto, Y.; Asakawa, K.; Yamamoto, H., *J. Am. Chem. Soc.* **1999**, *121*, 892–893; (b) Yanagisawa, A.; Matsumoto, Y.; Asakawa, K.; Yamamoto, H., *Tetrahedron* **2002**, *58*, 8331–8339.
18. Ohkouchi, M.; Masui, D.; Yamaguchi, M.; Yamagishi, T., *J. Mol. Catal. A Chem.* **2001**, *170*, 1–15.
19. (a) A. Yanagisawa, A.; Y. Nakatsuka, Y.; A. Asakawa, A.; H. Kageyama, H.; H. Yamamoto, H., *Synlett* **2001**, 69–72; (b) Yanagisawa, A.; Nakatsuka, Y.; Asakawa, K.; Wadamoto, M.; Kageyama, H.; Yamamoto, H., *Bull. Chem. Soc. Jpn.* **2001**, *74*, 1477–1484.
20. Wadamoto, M.; Ozasa, N.; Yanagisawa, A.; Yamamoto, H., *J. Org. Chem.* **2003**, *68*, 5593–5601.
21. Ozasa, N.; Wadamoto, M.; Ishihara, K.; Yamamoto, H., *Synlett* **2003**, 2219–2221.
22. Kundu, K.; Doyle, M. P., *Tetrahedron: Asymmetry* **2006**, *17*, 574–577.
23. Akullian, L. C.; Snapper, M. L.; Hoveyda, A. H., *J. Am. Chem. Soc.* **2006**, *128*, 6532–6533.
24. (a) Kobayashi, S.; Ishitani, H., *Chem. Rev.* **1999**, *99*, 1069–1094; (b) Córdova, A., *Acc. Chem. Res.* **2004**, *37*, 102–112; (c) Friestad, G. K.; Mathies, A. K., *Tetrahedron 63*, **2007**, 2541–2569.
25. (a) Ferraris, D.; Young, B.; Cox, C.; Dudding, T.; Drury, W. J. III; Ryzhkov, L.; Taggi, A. E.; Lectka, T., *J. Am. Chem. Soc.* **2002**, *124*, 67–77; (b) Ferraris, D.; Young, B.; Dudding, T.; Lectka, T., *J. Am. Chem. Soc.* **1998**, *120*, 4548–4549.
26. Josephsohn, N. S.; Snapper, M. L.; Hoveyda, A. H., *J. Am. Chem. Soc.* **2004**, *126*, 3734–3735.
27. Josephsohn, N. S.; Carswell, E. L.; Snapper, M. L.; Hoveyda, A. H., *Org. Lett.* **2005**, *7*, 2711–2713.
28. (a) Carswell, E. L.; Snapper, M. L.; Hoveyda, A. H., *Angew. Chem. Int. Ed.* **2006**, *45*, 7230–7233; (b) Wieland, L. C.; Vieira, E. M.; Snapper, M. L.; Hoveyda, A. H., *J. Am. Chem. Soc.* **2009**, *131*, 570–576.
29. Momiyama, N.; Yamamoto, H., *Angew. Chem. Int. Ed.* **2002**, *41*, 2986–2988.

30. (a) Momiyama, N.; Yamamoto, H., *J. Am. Chem. Soc.* **2003**, *125*, 6038–6039; (b) Momiyama, N.; Yamamoto, H., *J. Am. Chem. Soc.* **2004**, *126*, 5360–5361.

31. (a) Yamamoto, H.; Momiyama, N., *Chem. Commun.* **2005**, 3514–3525; (b) Janey, J. M., *Angew. Chem. Int. Ed.* **2005**, *44*, 4292–4300; (c) Yamamoto, H.; Kawasaki, M., *Bull. Chem. Soc. Jpn.* **2007**, *80*, 595–607.

32. Kawasaki, M.; Li, P.; Yamamoto, H., *Angew. Chem. Int. Ed.* **2008**, *47*, 3795–3797.

33. (a) Greck, C.; Drouillat, B.; Thomassingny, C., *Eur. J. Org. Chem.* **2004**, 1377–1385; (b) Erdik, E., *Tetrahedron* **2004**, *60*, 8747–8782.

34. Yamashita, Y.; Ishitani, H.; Kobayashi, S., *Can. J. Chem.* **2000**, *78*, 666–672.

10

COUPLING REACTIONS PROMOTED BY SILVER

JEAN-MARC WEIBEL, AURÉLIEN BLANC, AND PATRICK PALE

Laboratory for Organic Synthesis and Reactivity Institute of Chemistry, University of Strasbourg, France

Silver in Organic Chemistry Edited by Michael Harmata

10.1 INTRODUCTION

Carbon–carbon bond formation is clearly a key transformation in organic synthesis. Since the early times of organic chemistry, numerous methods have been developed to establish C–C bonds. Among them, those using organometallic compounds constitute the vast majority.

Originating near the end of the nineteenth century with Frankland and organozincs,[1] and continuing into the twenthieth century with Grignard and his reagents, methods based on stoichiometric amounts of organometallics established themselves as key tools in organic synthesis.[2] Looking for milder conditions and for modulation in reactivity, chemists explored other metals, especially transition metals, usually through transmetalation. During this quest, the peculiar behavior of group VIII metals, especially palladium, was noticed; and catalytic reactions were developed, leading to a burst in this chemistry in the early 1980s.[3] More recently, other metals of this group, especially coinage metals,[4] also proved useful in carbon–carbon bond formation, either as such or in combination with other metals, mainly palladium.

Although among the first organometallics ever produced,[5] organosilver compounds were applied to organic chemistry only sparingly.[6] However, these reagents exhibit an extreme mildness and a very low basicity, rendering them very useful in various transformations.[6–8] Indeed, since the end of the twentieth century, silver salts have increasingly garnered interest as catalysts or cocatalysts in various reactions in which C–C bonds are created.[7a,8] In these reactions, preformed organosilver species can be directly used as reagents, and organosilvers are often produced as transient intermediates in catalyzed reactions, but silver could also play different roles, such as a Lewis acid or halide scavenger.

In this chapter, the Ag-promoted formation of C–C bonds is described, with the presentation organized according to the hybridization of the carbons involved in the bond formation.

10.2 sp^3–sp^3 COUPLING REACTIONS PROMOTED BY SILVER SALTS

One of the earliest, if not the first, uses of silver in C–C bond formation was the homocoupling reaction originally described by Gardner et al. and further investigated by Kharash and Reinmuth and by Kochi et al.[9–11] In this reaction, Grignard reagents in THF solution reacted with stoichiometric amounts of various silver salts leading to homocoupling of alkyl groups (Scheme 10.1).

Interestingly, this reaction could be performed with catalytic amounts of silver provided that the nitrate counterion was present. The latter could be obtained from silver nitrate or by addition of lithium nitrate to silver bromide. Mixtures of alkanes were obtained starting from two different organomagnesiums, suggesting radical formation.

$$\text{R–MgX} + \text{AgY} \longrightarrow \text{R–Ag} + \text{MgXY}$$

$$2\ \text{R–Ag} \longrightarrow \text{R–R} + 2\ \text{Ag}$$

Scheme 10.1

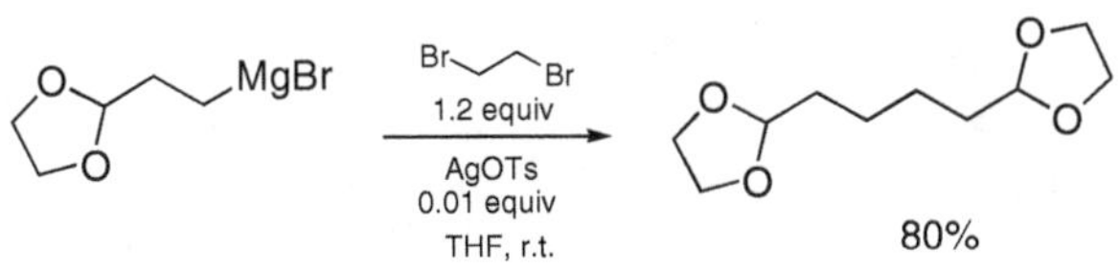

Scheme 10.2

However, their nonstatistical distribution as well as retention of stereochemistry if alkenyl organometallics were used (see Scheme 10.14),[12] suggested that free radicals were not intermediates in such reactions. Such Wurtz-type couplings therefore probably proceeded by transmetallation and then recombination (Scheme 10.1).

More recently, this reaction was slightly modified using the more soluble silver tosylate as catalyst (1 mol%) and 1,2-dibromoethane as reoxidant (Scheme 10.2).[13]

In a more interesting way, nonsymmetric couplings have been achieved much more recently.[14] Tertiary and secondary alkyl halides readily reacted with benzyl or allyl Grignard reagents in the presence of silver nitrate, giving the corresponding benzyl or allyl derivatives in high yields (Scheme 10.3). In competitive experiments, tertiary alkyl halides proved more reactive than the secondary ones. Electronic effects on benzyl Grignard reagents did not modify reactivity. Experimental evidence suggested radicals as intermediates in this reaction.

Instead of starting from Grignard reagents, Johnson[15] and then Brown et al.[16–18] showed that alkylboranes, readily accessible through hydroboration of the corresponding alkenes, also reacted in the presence of silver nitrate or oxide and produced alkanes resulting from homocoupling. The presence of hydroxide brought a very rapid reaction at room temperature or below. In such reactions, alkylboranes are probably transmetallated to silver and the thus-formed alkylsilver reacts to give the corresponding homocoupling product (Scheme 10.4).

More functionalized boranes have also been coupled in the presence of silver oxide, but the reaction proved more efficient when chromous chloride was added as catalyst (Scheme 10.5).[19]

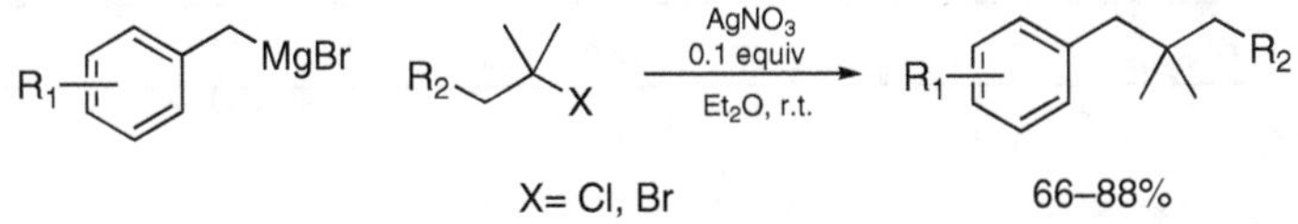

Scheme 10.3

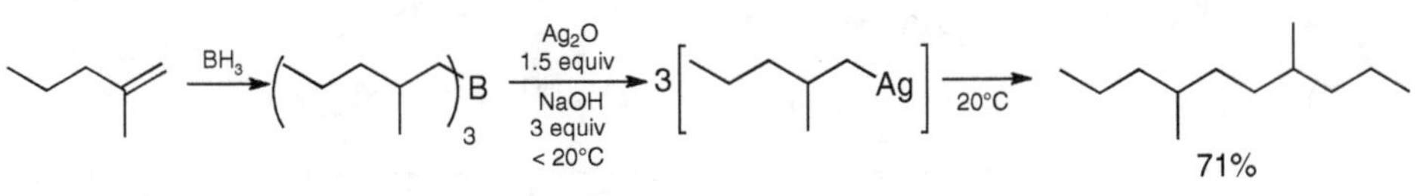

Scheme 10.4

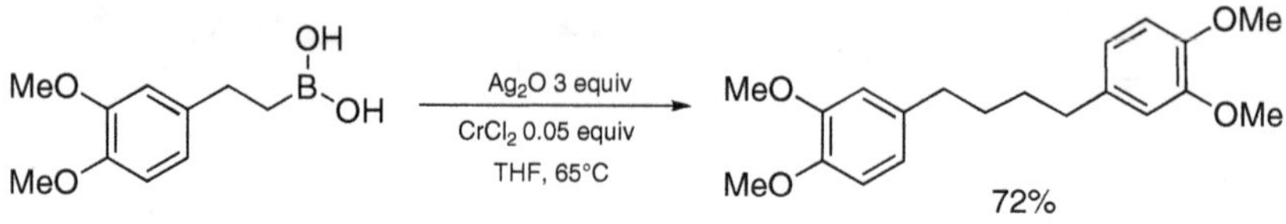

Scheme 10.5

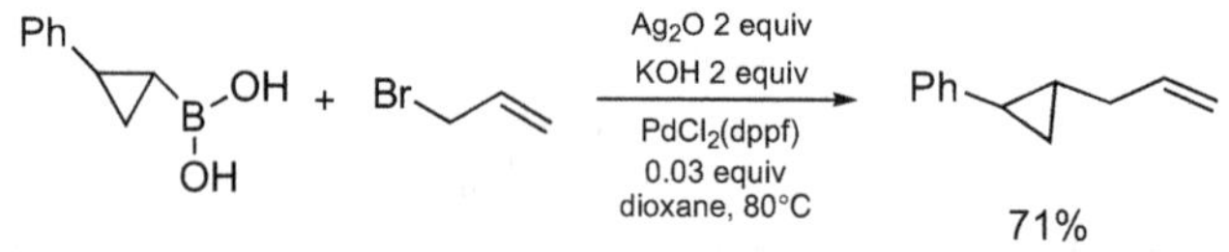

Scheme 10.6

As before, nonsymmetric couplings could also be realized from boronic acids, but using a palladium catalyst instead of chromium(II) chloride. Silver oxide was used as an activator for the Pd-catalyzed coupling of cyclopropylboronic acids to allyl bromides (Scheme 10.6).[20]

In a related reaction, Caddick et al. expanded this Suzuki-type coupling to other alkyl–alkyl couplings.[21] Although the yields were still modest, the reaction seemed compatible with various functional groups and required only mild conditions (Table 10.1).

TABLE 10.1. Suzuki–Miyaura Cross-Coupling of Alkyl Bromides

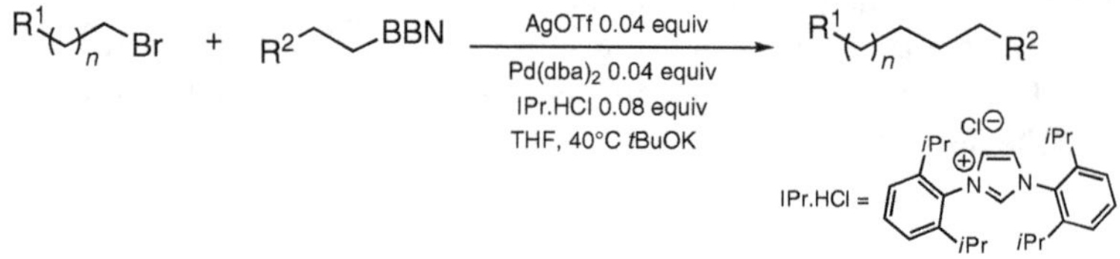

n	R^1	R^2	Product	Yield (%)
10	CH_3	*n*-Hex	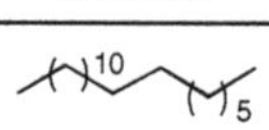	56
10	CH_3	CH_2Ph-*p*OMe	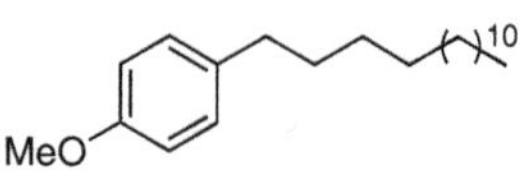	46
4	COOEt	*n*-Bu	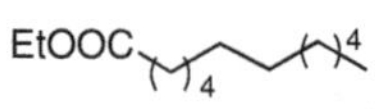	53
4	CN	*n*-Bu	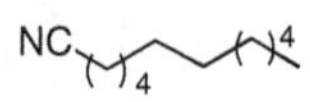	52

10.3 sp^3–sp^2 COUPLING REACTIONS PROMOTED BY SILVER SALTS

No coupling reactions connecting sp^2 and sp^3 carbons directly promoted by silver salts seem to have been reported. However, silver salts have been used as activators or cocatalysts in a few palladium-catalyzed couplings.

The Caddick protocol set up for alkyl–alkyl coupling (see Section 10.2) could also be used for the coupling of a vinylborane with an alkyl bromide (Scheme 10.7). The yield was, however, modest.[17]

The synthesis of medicinally relevant 4-substituted 2(5*H*)-furanones has been achieved by a coupling reaction of the triflate derived from β-tetronic acid with alkyl- and cyclopropylboronic acids (Scheme 10.8).[22] This reaction was mediated by bis(acetonitrile)dichloropalladium and silver oxide. Interestingly, optically active cyclopropyl boronic acids were not racemized during this coupling reaction.

In 2006, Yu et al. combined pyridinyl-directed C–H activation and C–C bond formation with alkylboronic acids (see Section 10.5.4.2).[23] The success of this transformation relied on the combination of palladium acetate (10 mol%), benzoquinone (1 equiv), and silver oxide or carbonate (0.5 equiv) in a protic solvent, but an excess of boronic acid (3 equiv) was required (Scheme 10.9). Interestingly, in this reaction silver oxide played a dual role as promoter for the transmetallation step and as cooxidant with benzoquinone.

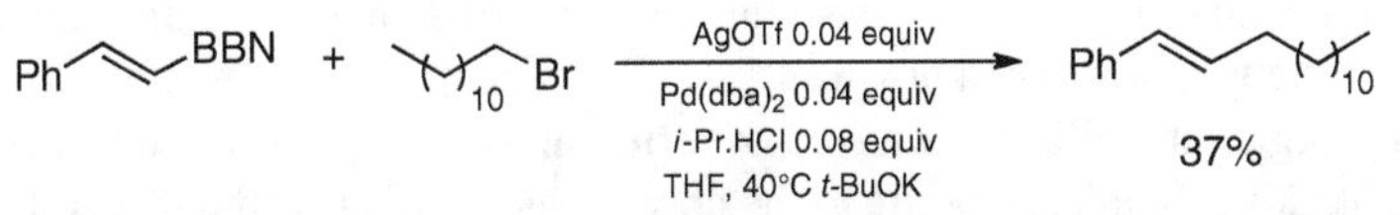

Scheme 10.7

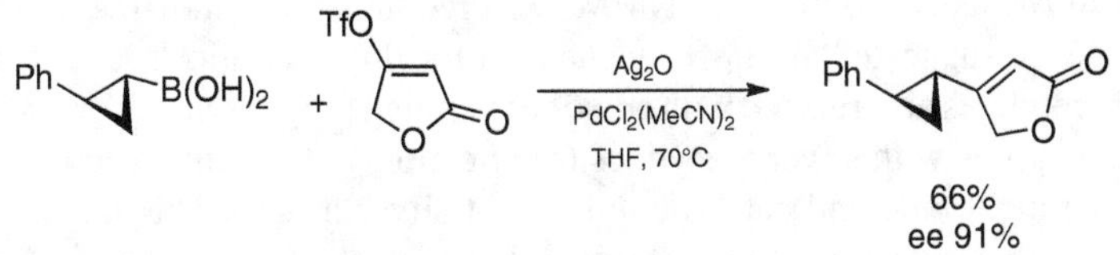

Scheme 10.8

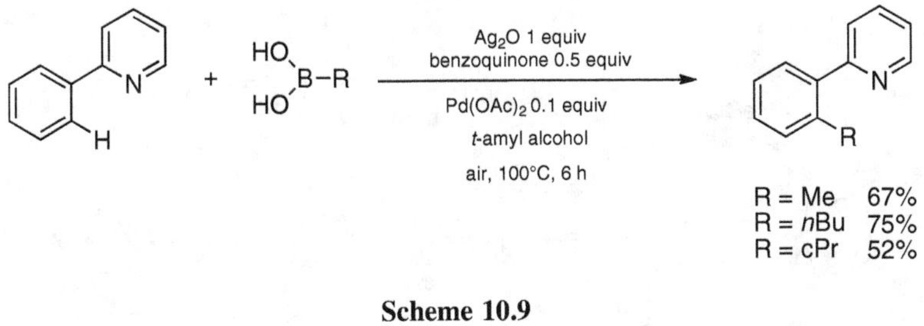

Scheme 10.9

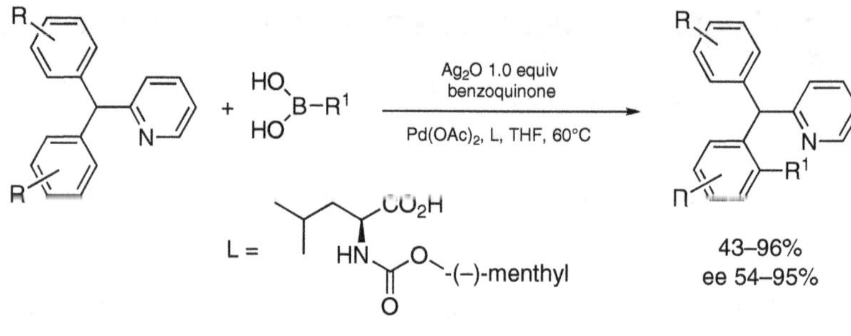

Scheme 10.10

In 2008, the same group developed an asymmetric version of this reaction (Scheme 10.10).[24] Run under similar conditions, but with more silver oxide (1 equiv) and thus less reoxidant (benzoquinone 0.5 equiv) and in the presence of catalytic amounts of chiral ligand (20 mol%), the best enantiomeric excesses and yields were obtained with menthyl-L-leucine ester as the chiral ligand.

10.4 *sp*^{3}–*sp* COUPLING REACTIONS PROMOTED BY SILVER SALTS

The title reactions refer to alkylation reactions using alkynyl silvers as reagents. It seems that no catalytic version has so far been reported, and only a few examples of stoichiometric reactions are known.[8]

As any organometallic derivatives of terminal alkynes, silver acetylides readily reacted with alkyl halides, leading to the corresponding disubstituted acetylenes. For example, methyl iodide gave the methylacetylene derivative (Scheme 10.11, top).[25] Surprisingly, silver arylacetylides react with adamantyl bromide or iodide, although refluxing in *N*-methylmorpholine (NMM) was required (Scheme 10.11, bottom).[26] Some evidence suggested an $S_{RN}1$ mechanism for this alkylation.

Silver acetylides also reacted with epoxides yielding propargylic alcohols.[27] As for related alkylations with silver acetylides (see Section 10.6.1), this reaction required zirconocene dichloride and catalytic amount of silver triflate. This method proved useful for both electron-rich and electron-deficient alkynes and compatible with various acid- and base-sensitive functional groups (Table 10.2).

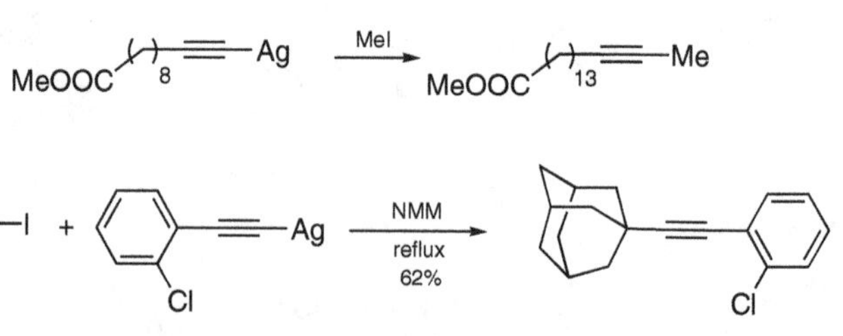

Scheme 10.11

TABLE 10.2. Epoxide Opening with Silver Acetylides

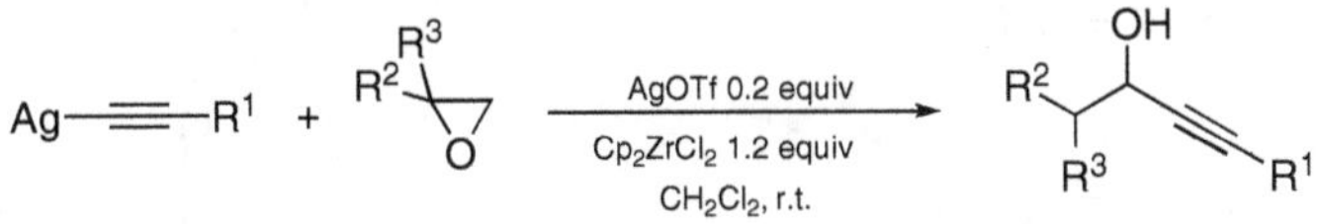

R^1	Epoxide	Product	Yield (%)
CH_2OTHP	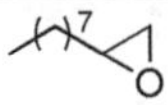	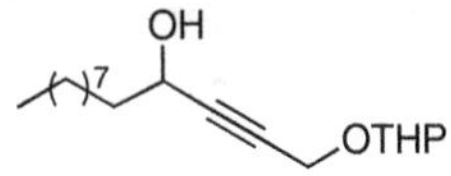	82
n-Bu	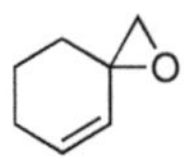	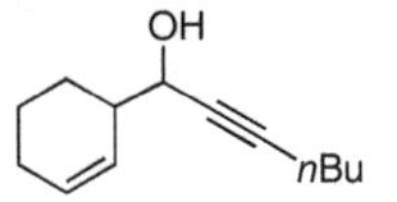	53
COOMe	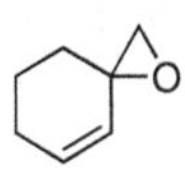	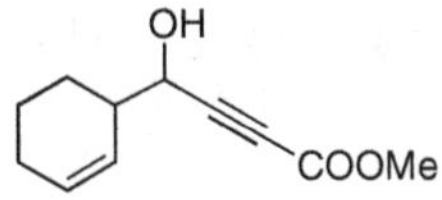	54
n-Bu	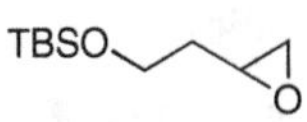	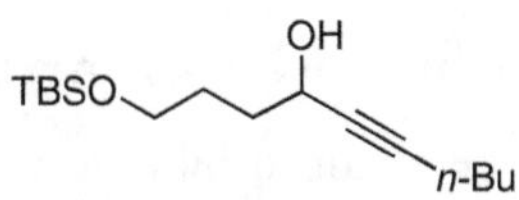	52
COOMe	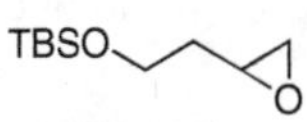	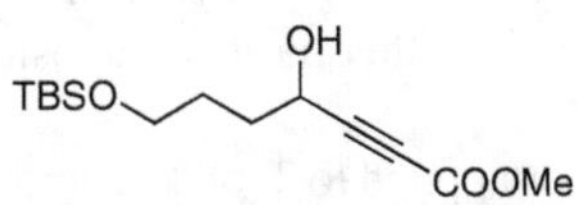	46

Interestingly, 1,2-disubstituted epoxides led to α,α-disubstituted propargyl alcohols, indicating that 1,2 shifts and therefore epoxide-to-aldehyde rearrangements occurred during the reactions (Scheme 10.12). Mechanistic studies by NMR validated this rearrangement and also supported the transmetallation–abstraction mechanism as already proposed (Scheme 10.13).[8]

10.5 sp^2–sp^2 COUPLING REACTIONS PROMOTED BY SILVER SALTS

Such couplings are the most prominent C–C bond-forming reactions in which silver salts have been involved. In such reactions, silver could play a large variety of roles,

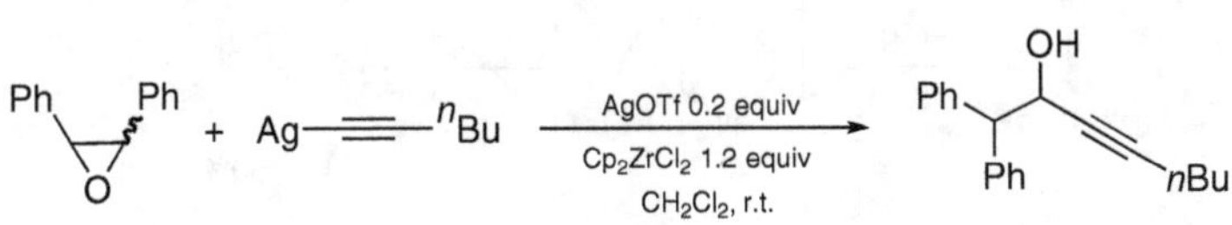

Scheme 10.12

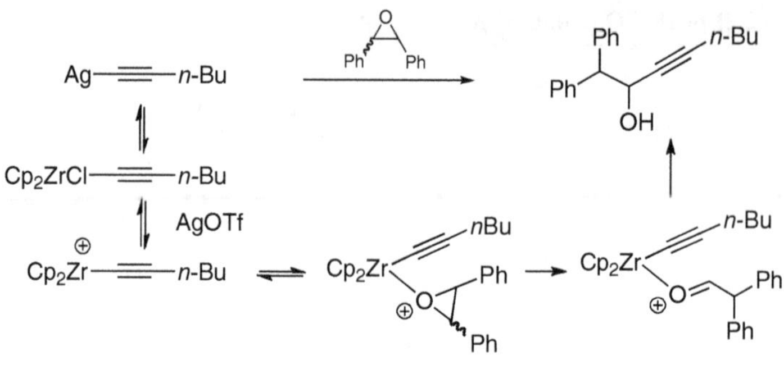

Scheme 10.13

inducing homocoupling or acting as a Lewis acid or as a halide scavenger. Moreover, preformed organosilver species, depending on their stability, can be directly used as reagent for the formation of C–C bonds.

10.5.1 Homocoupling of Vinyl- or Arylsilver Species

Vinylsilvers could be obtained by transmetallation from the corresponding vinyllithiums.[12] More stable than alkylsilvers (see Section 10.2 and Ref. 28), they nevertheless underwent thermal decomposition leading to homocoupling products (Scheme 10.14).

This method was applied to 2- or 4-bromopyridine derivatives and used to produce the corresponding bipyridines, useful as intermediates for herbicide manufacture. Because arylsilvers are more stable, heating was required in this case (Scheme 10.15).[29]

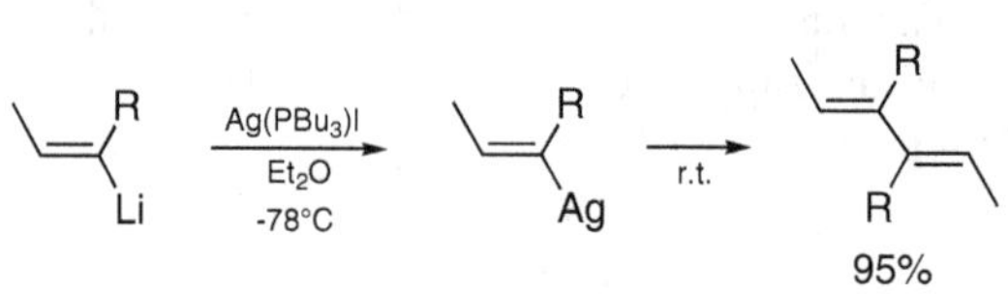

Scheme 10.14

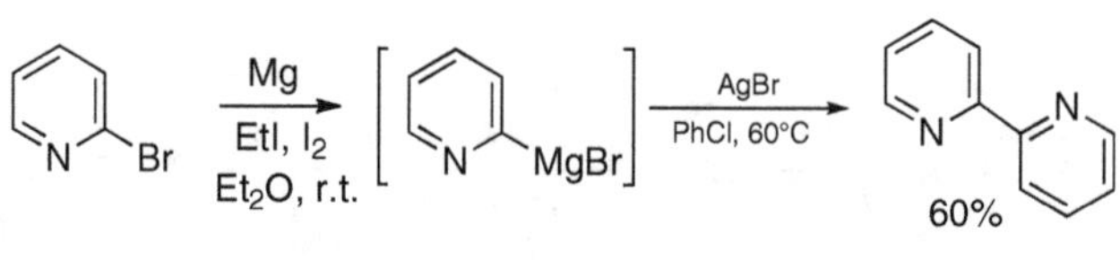

Scheme 10.15

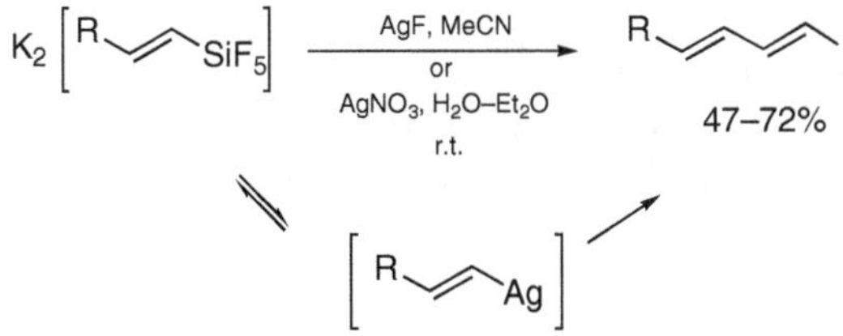

Scheme 10.16

Interestingly, (*E*)-alkenylpentafluorosilicates, readily available from alkynes by hydrosilylation, could also be dimerized to the corresponding symmetric (*E,E*)-1,3-dienes in good yields on treatment with silver fluoride in acetonitrile or with silver nitrate in water/ether (Scheme 10.16).[30] This reaction suggested that pentafluorosilicates could be transmetallated to silver, leading to a vinylsilver that dimerized.

Boranes are also a source of organosilvers through transmetallation (see Section 10.2) and thus, have been coupled in the presence of silver oxide, but the reaction proved more efficient when chromous chloride was added as catalyst.[19] Arylboranes led to biaryls, and vinylboranes gave the corresponding dienes, but noticeably, with neither isomerization of the double bond nor cyclization (Scheme 10.17). Interestingly, these results confirmed other reports (see Section 10.2) suggesting the absence of free radicals as intermediates in the homocoupling of organosilver compounds.

10.5.2 Organosilver Species as Nucleophilic Reagents

Arylargentates are stable species. Their chemistry[31] is so far not well developed, but they have been used as nucleophilic reagents, mainly in transmetallation processes.[32,33] With (η^3-allyl)palladium complexes, new complexes were formed, and, on addition of benzoquinone, reductive elimination occurred. In this reaction, the aryl group was transferred to the allyl system, providing allylbenzene derivatives (Scheme 10.18).

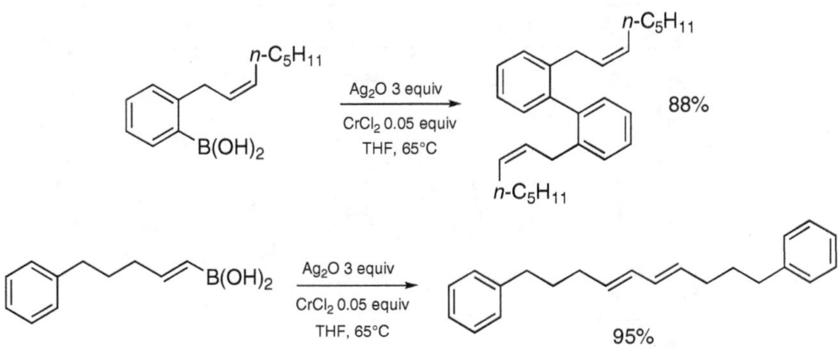

Scheme 10.17

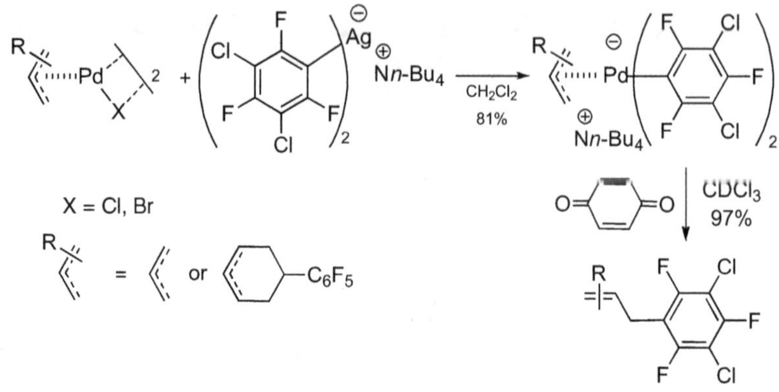

Scheme 10.18

10.5.3 Silver as a Lewis Acid Reagent

With its d^{10} electronic configuration and its f orbitals available, silver(I) ion can act as a Lewis acid,[34] and relativistic effects reinforce such behavior.[35] Moreover, silver salts are able to act as either or both a σ-Lewis acid or a π-Lewis acid, with a slight preference for σ coordination over π coordination, as revealed by calculations. This preference is higher for nitrogen than for oxygen Lewis bases (Scheme 10.19).[36]

On this basis, silver-catalyzed aldolization and allylation reactions have been developed, expanding the technology for creating a C–C bond using silver (see Chapter 9). Asymmetric versions have also been developed.[7d,37]

In related reactions, Snapper and Hoveyda et al. reported an asymmetric version of Mannich-type reactions catalyzed by silver acetate.[38,39] Propargyl imines proved to be susceptible to nucleophilic addition of silylenol ethers derived from acetate esters in the

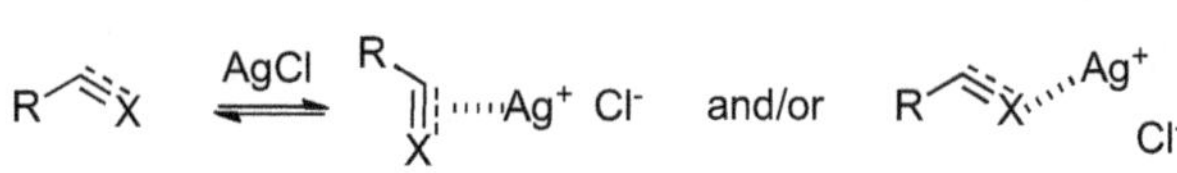

	ΔH_f (kcal/mol)		ΔH_f (kcal/mol)
Ph—≡	-22.6	Cy—≡	-25.2
Ph—CH=CH$_2$	-24.4	Cy—CH=CH$_2$	-34.7
Ph—CH=O	-26.4	Cy—CH=O	-26.0
Ph—CH=NH	-39.6	Cy—CH=NH	-40.4

Scheme 10.19 Computed heats of formation (B3LYP/SDD, kcal/mol) of various substrates with AgCl (from Ref. 36).

TABLE 10.3. Asymmetric Mannich Reactions of Enol Ethers with Alkynyl Imines

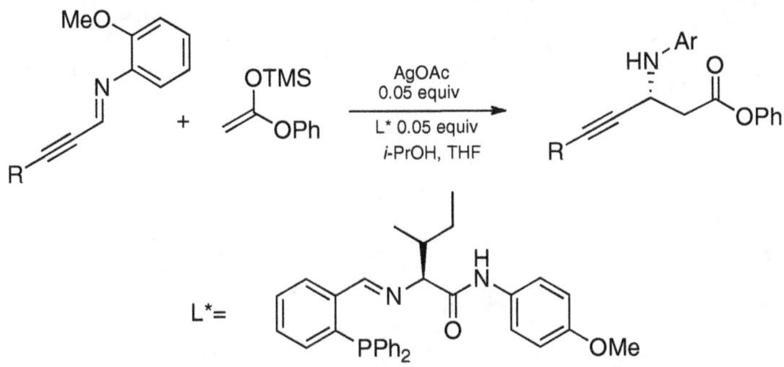

R	Yield (%)	ee (%)
Ph	91	94
$SiMe_3$	84	92
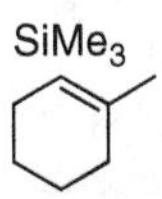	89	90
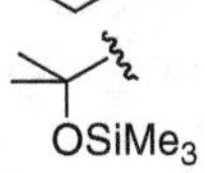	67	84

presence of silver acetate and a ligand derived from an amino acid (Table 10.3). In this reaction, chiral β-aminoesters were obtained in high yield and good enantioselectivity. It is worth noting that propargyl imines are ideal substrates for both σ and π coordination with silver ion, the dual Lewis acidities of silver acting in a cooperative way.

This methodology was applied to the asymmetric synthesis of (−)-sedamine (Scheme 10.20). An imine produced in situ from *o*-anisidine and methyl-5-oxopentanoate was alkylated with an enol silane in the presence of silver acetate as catalyst,

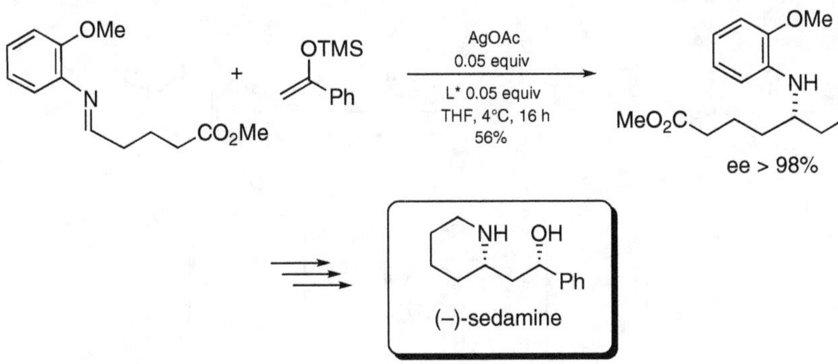

Scheme 10.20

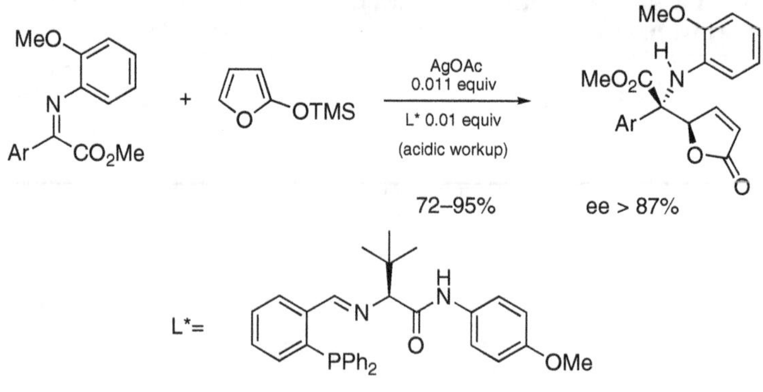

Scheme 10.21

together with a chiral phosphine ligand. This transformation gave a β-aminoketone in 56% yield with enantiomeric excess higher than 98%.[39]

Under similar conditions, the same authors were able to control two stereogenic centers in an asymmetric vinylogous Mannich reaction. Indeed, treatment of imines derived from aryl α-ketoesters with siloxyfuran under related conditions gave functionalized γ-butenolides with high diastereo- and enantioselectivities (Scheme 10.21).[40]

Li et al. reported that β-diketones reacted with styrenes on treatment with silver triflate. The product structure suggested that such reactions proceeded through nucleophilic addition of the enol derived from β-diketones to the silver π complex of styrene (Scheme 10.22).[41]

Very recently, an intramolecular version with alkynyl β-diketones or β-diketoesters has been reported (Scheme 10.23).[42] With such substrates, the reaction required a mixture of silver tetrafluoroborate and copper triflate as catalysts. Here, also, nucleophilic addition of a ketoenol to a π complex seemed responsible for an *exo-dig* cyclization, but depending on substituents, exocyclic or endocyclic alkenes were obtained.

Similarly, β-diketones have been shown to regioselectively add to furans in the presence of silver triflate (Scheme 10.24).[43]

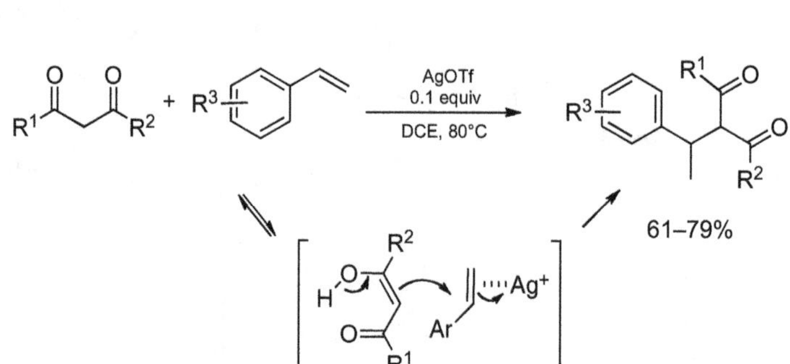

Scheme 10.22

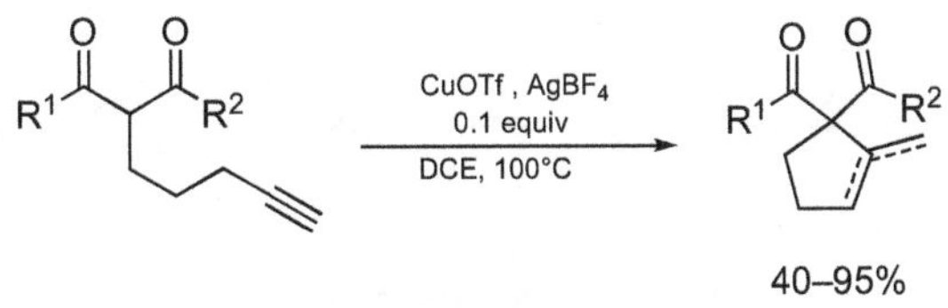

Scheme 10.23

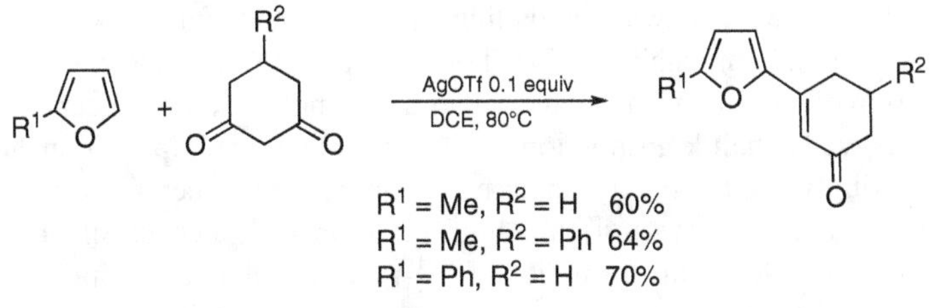

Scheme 10.24

10.5.4 Silver as a Halogen Scavenger

One of the most prominent roles of silver in the formation of carbon–carbon bonds is due to another typical property of silver: its halophilicity. Silver ions exhibit strong Lewis acid–base interaction with halides, and silver halides are often insoluble materials, especially in organic solvents (Table 10.4).

Silver salts have thus been used to abstract halogen(s) in organic or organometallic molecules and in metal complexes. In metal-catalyzed coupling reactions, the main use of silver is to generate a metal intermediate that has an unsaturated coordination sphere by halide abstraction. This resulting metal complex is thus able to coordinate other species, including unsaturated organic compounds. Depending on the nature of the ligand, various reactions can then occur (Scheme 10.25). A vast number of examples have been now described, and it is not possible to mention all of them here. Only a few are reported, emphasizing the probable role of silver.

TABLE 10.4. Molar Solubility Product (K_{sp}) of Common Silver Salts in Water at 25°C

Silver Salts	K_{sp}
AgCl	1.77×10^{-10}
AgBr	5.35×10^{-13}
AgI	8.52×10^{-17}
AgCN	5.97×10^{-17}
AgOAc	1.94×10^{-3}
Ag_3PO_4	8.89×10^{-17}
Ag_2CO_3	8.46×10^{-12}

Source: Reference 44.

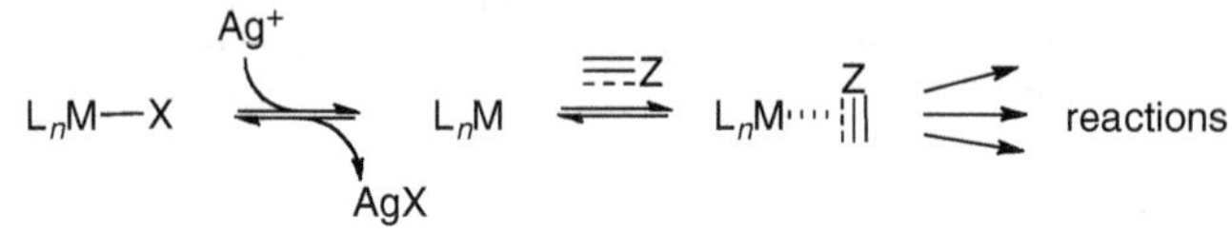

Scheme 10.25

10.5.4.1 Silver in Pd-Catalyzed Couplings

Such processes are well known in palladium chemistry and quite important, because of the role of palladium-catalyzed C–C bond formation in organic chemistry.[3]

Abstraction of chloride from a vinylpalladium complex by silver acetate has been reported,[45] as well as halide abstraction by silver triflate from arylpalladium halides.[46] More interestingly in the context of creating C–C bonds, silver perchlorate was able to promote the reaction of (η^1-aryl)palladium bromide with dienes. Silver-promoted bromide abstraction led to the formation of (η^1-η^2-enyl)palladium complexes, which evolved further through the regioselective formation of a C–C bond between the aryl group and the former diene. Reactions with nonconjugated dienes suggested that the reaction proceeds via carbometallation of the less crowded double bond. Isomerization and β elimination led to a (η^3-allyl)palladium complex (Scheme 10.26).[47]

In a similar but more complex reaction, dienes carrying a β-diketo side chain have been alkylated with various vinyl bromides and cyclized to the corresponding alkylated cyclopropyl alkenes in the presence of silver carbonate (Scheme 10.27). With a relative *cis* stereochemistry for the newly created C–C bonds, the mechanism is

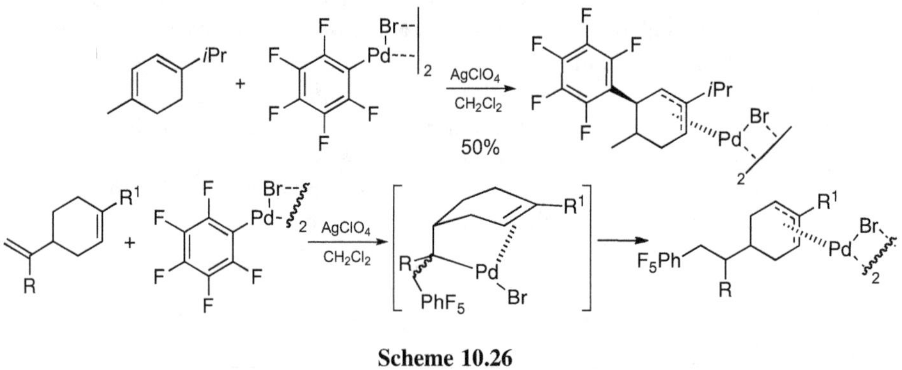

Scheme 10.26

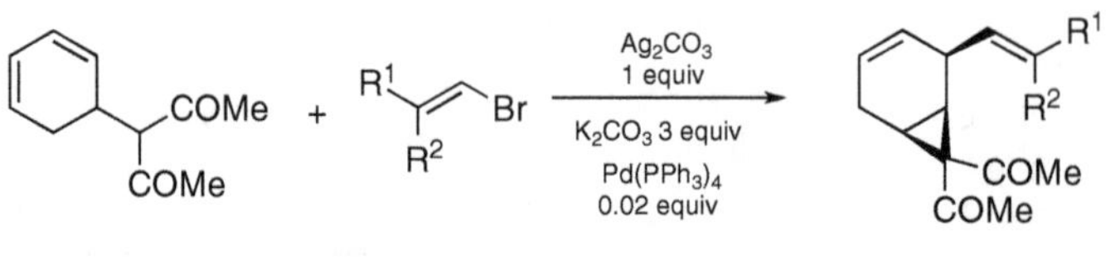

Scheme 10.27

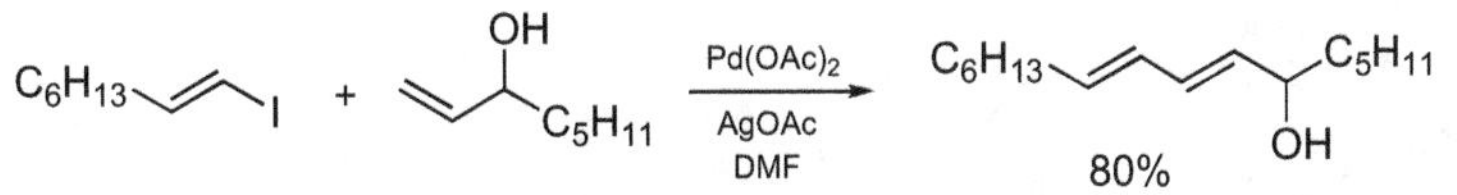

Scheme 10.28

not obvious and probably different from those of the related reactions mentioned above; a palladium enolate has been proposed for this reaction.[48]

In the latter reactions, at least one step corresponded to a carbometallation reaction. Such carbometallations also arose in Heck reactions, and indeed, several conditions set up for such Heck reactions relied on silver salts.[49] In these reactions, silver salts only seemed to abstract halide from intermediate palladium species; their role could, however, be more complex.

The first example was described in the mid-1980s independently by Jeffery, Hallberg, and others in their work on Heck coupling. The former demonstrated the beneficial role of silver salts in the formation of dienols from allylic alcohols and vinyl iodides, as well as the role of counterion or phase transfer conditions (Scheme 10.28).[50] The latter found that silver salts accelerated the reaction between aryl iodide and allyl or vinyl silanes, reinforcing regioselectivity and avoiding desilylation (Scheme 10.29).[51]

Such methodology was successfully applied to the synthesis of isopanepoxydone, a secondary metabolite isolated from the basidiomycete *Panus conchatus* (Scheme 10.30).[52]

The natural polyenic lactone (−)-callystatin A, isolated in 1997 from the marine sponge *Callyspongia truncata* and exhibiting remarkable growth inhibition of cancer cells ($IC_{50} \sim$ 10–20 pg/mL), was also prepared using this Heck-like reaction under Jeffery conditions (Scheme 10.31).[53]

As part of studies pertaining to the total synthesis of the complex hexacyclic alkaloid gelsemine, Overman and Abelman applied the Heck reaction to the cyclization of *N*-(2-halophenyl)enamides.[54] They found that the addition of silver salt suppressed alkene isomerization, a frequent problem in Heck reactions, and dramatically accelerated the reaction (Scheme 10.32).

These results led other investigators to propose that the Heck reaction proceeds via an ionic pathway or a neutral variant, depending on the starting materials and on the

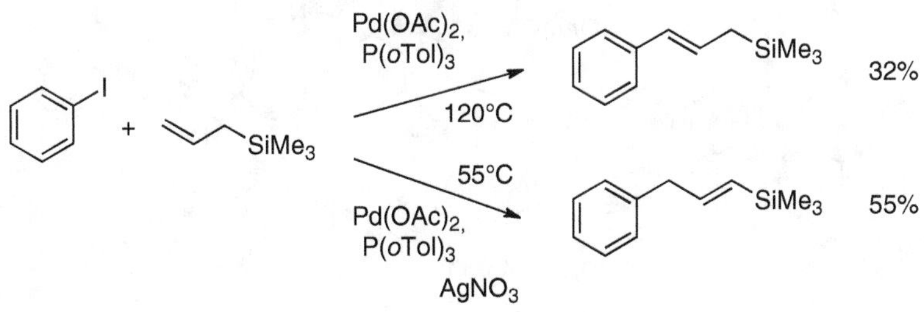

Scheme 10.29

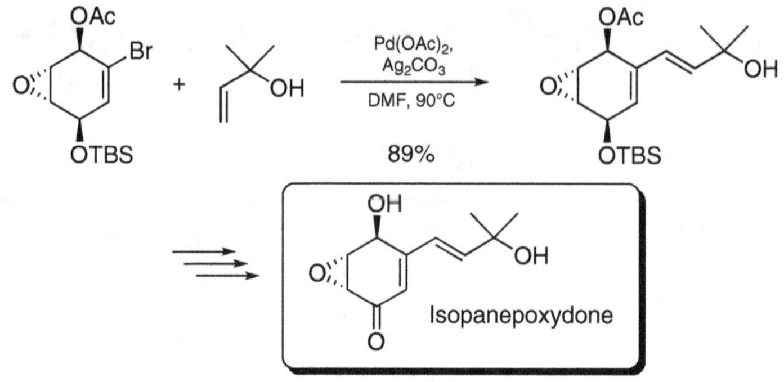

Scheme 10.30

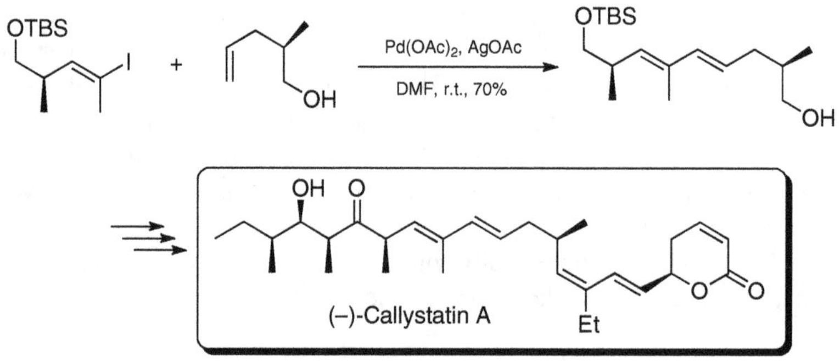

Scheme 10.31

conditions,[49,55] although an alternative pathway that involves a Pd^{II}/Pd^{IV} redox system has been proposed.[49,56] The ionic pathway would be rendered favorable when silver salts are used, or when starting from vinyl or aryl triflates.[57]

Controlling the coordination of palladium intermediates with silver salts was the key to developing highly diastereo- and enantioselective Heck reactions.[58] For example, the presence of silver phosphate in the key cyclization step during the synthesis of various alkaloids reversed the stereochemistry at the newly created spiro center (Scheme 10.33).[59]

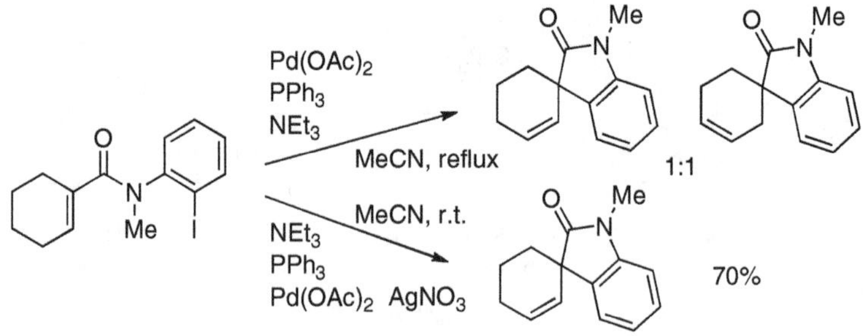

Scheme 10.32

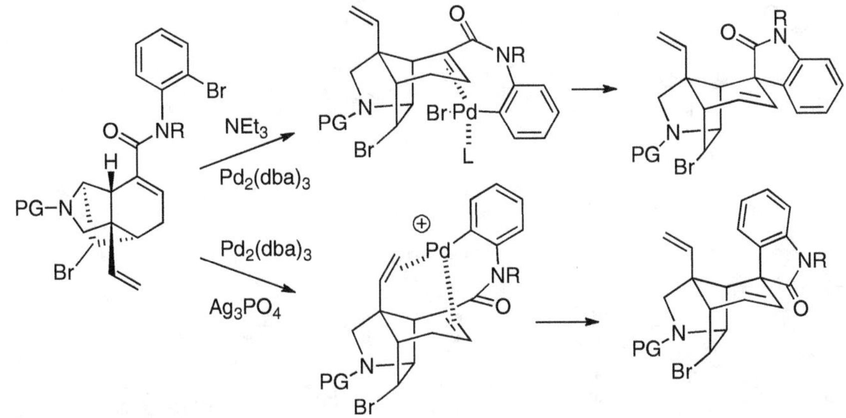

Scheme 10.33

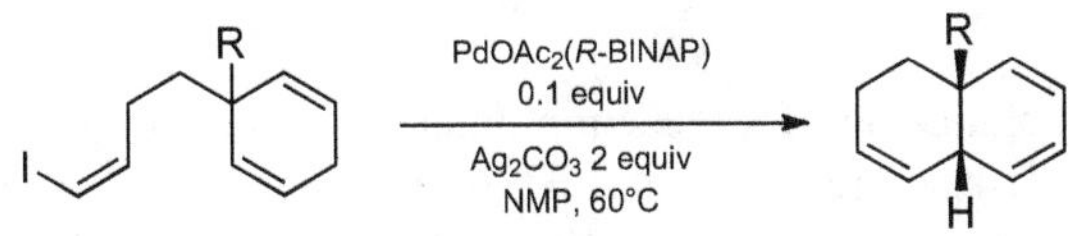

R = COOMe, CH_2OTBS, CH_2OAc

55–74%
33–46% ee

Scheme 10.34

The Shibasaki cyclization of *meso*-cyclohexa-1,4-dienes in the presence of a chiral palladium complex and silver carbonate in 1-methyl-2-pyrrolidinone is probably the first example of an enantioselective Heck reaction (Scheme 10.34).[60] The enantiomeric excess could be slightly improved by replacing the chiral phosphine ligand BINAP with the corresponding less coordinating arsine ligand, as well as replacing silver carbonate with silver phosphate.[61,62]

With such possibilities for regio-and stereocontrol, Pd–Ag-catalyzed Heck reactions have been and still are widely applied to total synthesis. However, just a few more examples are provided here. Brown et al. described a very convergent enantioselective synthesis of (−)-galanthamine (Scheme 10.35), an Amaryllidaceae

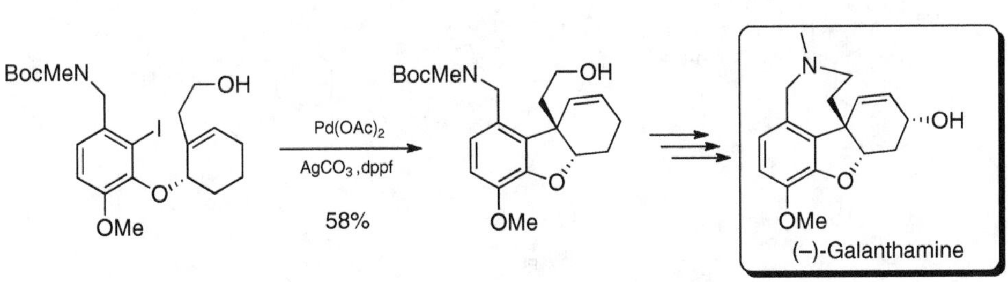

Scheme 10.35

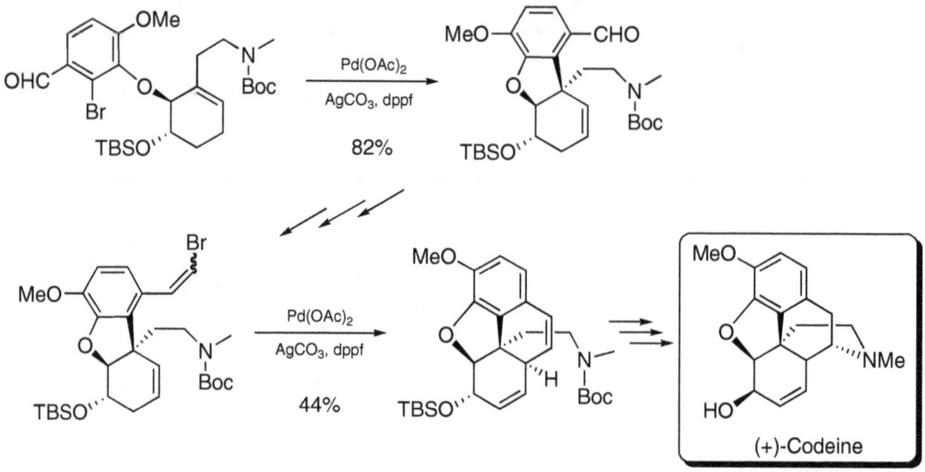

Scheme 10.36

alkaloid that is in clinical use for the symptomatic treatment of Alzheimer's disease. Its preparation has been realized starting from isovanillin.[63] The Heck reaction offered a convenient and ingenious way for the formation of three of the four rings of galanthamine.

Hudlicky et al. described an elegant strategy based on two intramolecular Heck coupling reactions for the enantioselective synthesis of codeine from 2-phenylbromoethane (Scheme 10.36).[64]

Palladium/silver-catalyzed Heck reactions have usually involved vinyl or aryl halides and alkenes, but these reaction conditions were also extended to allenes. Indeed, Zenner and Larock[65] showed that simple alkyl allenes readily reacted with aryl and vinyl iodide derivatives in the presence of palladium acetate or chloride and silver phosphate. Moreover, the reaction could be rendered asymmetric using chiral ligands; the best one was a bisoxazolidine derivative (Scheme 10.37).

More recently, substituted allenyl sulfones were also engaged in Pd/Ag-catalyzed Heck-type reactions. Treated with palladium tetrakis(triphenylphosphine) and a mixture of silver and potassium carbonate in DMF at 80°C, such allenyl sulfones

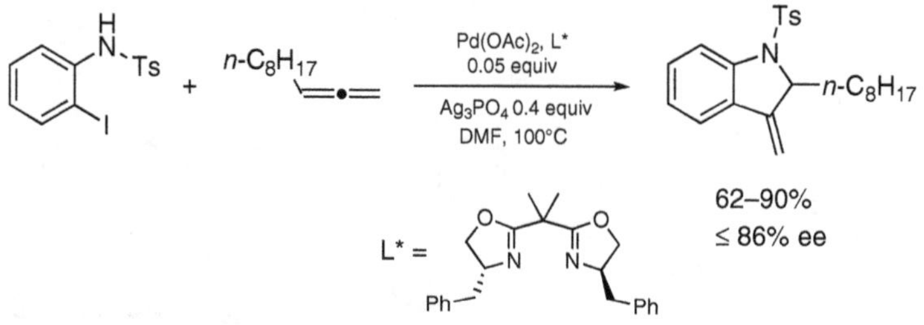

Scheme 10.37

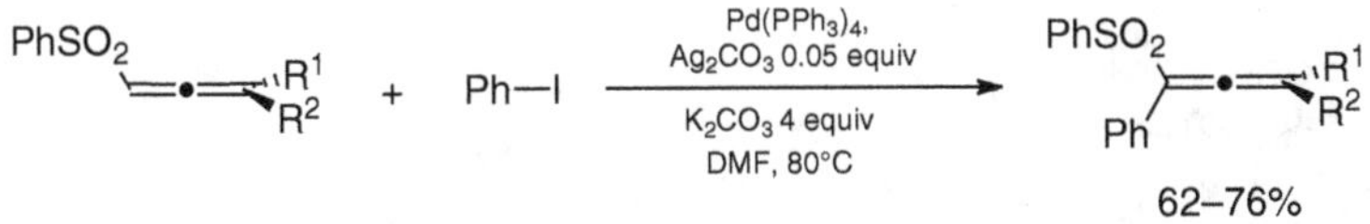

Scheme 10.38

gave more substituted allenyl sulfones (Scheme 10.38).[66] The regioselectivity is very peculiar compared to similar processes already described in the literature; the roles of the sulfonyl group and the silver salt remain unclear.

With such benefits offered by silver salts in these Heck-type reactions, chemists looked for similar improvements in other palladium-catalyzed couplings and related reactions.

Stille reactions have been improved by the addition of silver salts. For example, 5-iodouracil could be coupled with stannylated pyridine, forming 5-(2-pyridinyl) uracil in good yield (Scheme 10.39).[67]

Under similar conditions, addition of silver oxide in dimethylacetamide allowed the production of hindered 1,2-substituted binaphthyl derivatives in high yields (Scheme 10.40).[68] These compounds were useful precursors for the preparation of asymmetric reagents exhibiting axial chirality.

Silver salts are also common activators in numerous Suzuki coupling reactions. The earliest example could probably be found in the Kishi's palytoxin synthesis. Silver oxide as well as thallium hydroxide provided dramatic rate enhancements in the couplings of vinylboronic acids (Scheme 10.41).[69] Both thallium and silver ions are most probably abstracting halide in palladium intermediates, but silver is clearly the most efficient. Moreover, with the right counterion, the silver salt also acts as a mild base and activator.

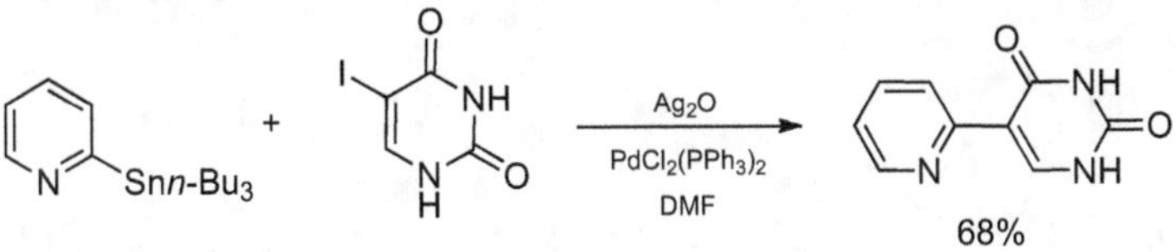

Scheme 10.39

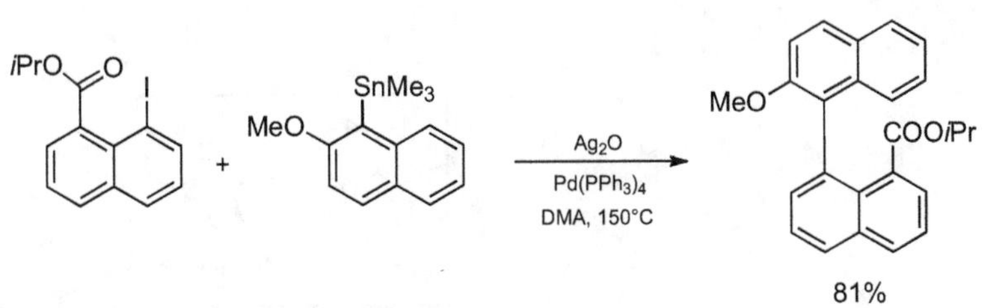

Scheme 10.40

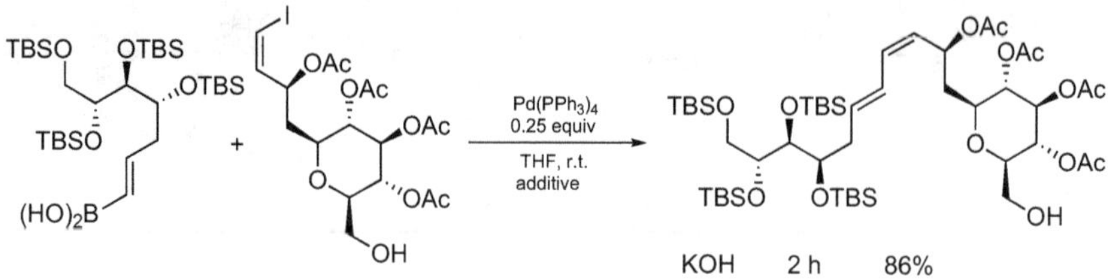

Additive	Time	Yield
KOH	2 h	86%
TlOH	30 min	94%
Ag_2O	5 min	92%

Scheme 10.41

In both Stille and Suzuki couplings, the role of silver salts may be not only as halide scavengers but also as promoters in the transmetallation step. In both coupling types, rate enhancements due to silver could be observed (see, e.g., Scheme 10.41). Indeed, the role that silver could play in Stille reactions could be similar to the one observed with copper.[70] In the Suzuki reaction, the boronic acid species must be activated in order to transfer its organic moiety in the transmetallation step, and an oxygen atom intervenes in this process.[71,72] The silver ion, and especially its basic counterion, could act in this capacity. For example, silver oxide was used to good effect in the synthesis of (+)-fostriecin to promote the Suzuki cross-coupling, finalizing the assembly of the carbon backbone (Scheme 10.42).[73]

Interestingly, Hiyama couplings, similar to Stille and Suzuki couplings but involving silanes instead of stannanes or borates, required stoichiometric amounts of silver salts.[74] However, the silanes must be activated to obtain effective transformations, and usually, vinyl- or arylsilanols are used in such couplings. Among the silver salts, the oxide proved again to be the most active promoter (Scheme 10.43).[75]

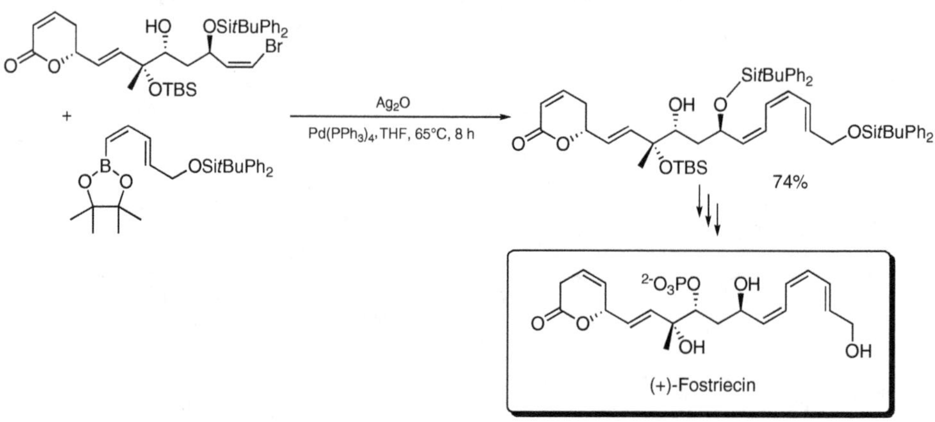

Scheme 10.42

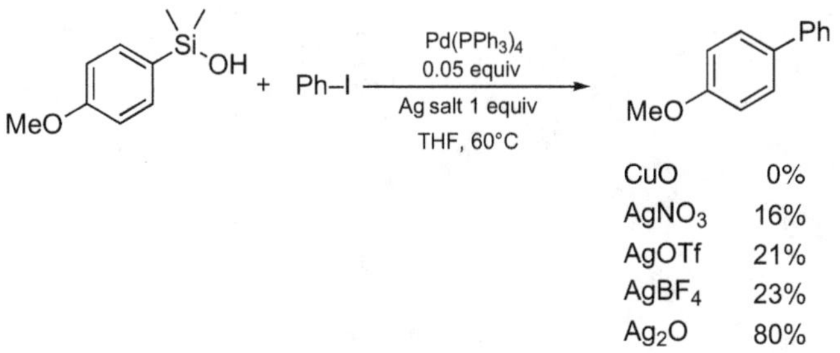

Scheme 10.43

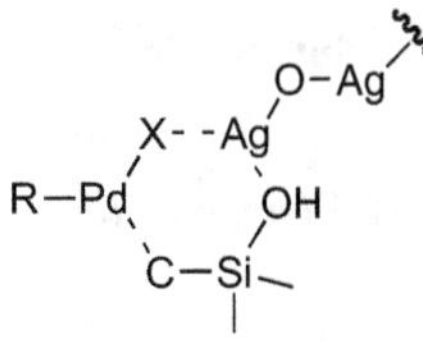

Scheme 10.44

The free hydroxyl group at the silicon center played a key role in this reaction, since aryltrimethylsilane and aryldimethylsilyloxysilane did not react while silanediols or triols[76] and polyalkenylsiloxanes did.[77]

As before, a silver ion and its counterion could act together in the transmetallation step, such as depicted in Scheme 10.44. Such a hypothesis seemed supported by the discovery in 2006 that 2-pyridinyl, but not 3-pyridinyl, allylsilanes reacted under similar conditions and transferred the pyridine moiety to various aryl iodides (Scheme 10.45).[78] In this coupling, the nitrogen atom of the pyridinyl moiety probably played a role similar to that of the hydroxyl group in the silanol-based Hiyama couplings.

In an analogous way, 2-trimethylsilylpyridine has been engaged in coupling with aryl iodides in the presence of a stoichiometric amount of silver oxide and a catalytic amount of tetrabutylammonium fluoride and palladium tetrakistriphenylphosphine.[79] Only the iodides substituted by electron-withdrawing groups gave good 2-arylpyridines yields (Scheme 10.46). Interestingly, the concomitant formation of 2,2′-dipyridine suggested the intermediate formation of 2-pyridinyl silver (see Scheme 10.15, Section 10.5.1).

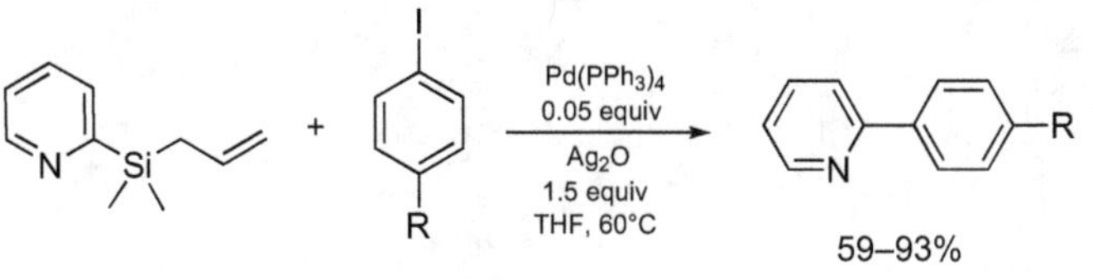

Scheme 10.45

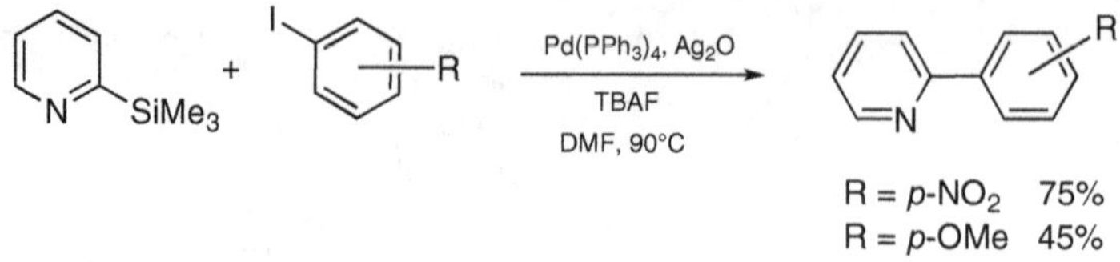

Scheme 10.46

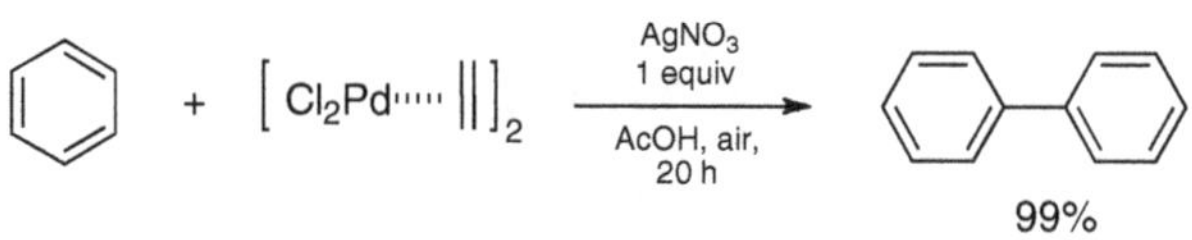

Scheme 10.47

10.5.4.2 Silver in Pd^{II}-Promoted Electrophilic Substitution of Arenes (C-H Activation)

During their work on the arylation of aromatic compounds by substitution, Fujiwara, et al. observed biaryl formation when aromatic compounds were placed in the presence of olefin–palladium complexes and silver nitrate.[80] Developing this reaction as a method for biphenyl synthesis, these authors showed that the more stable the olefin–palladium complex was, the lower the yield. Ethylene dichloropalladium proved to be the best choice, when used with silver nitrate. However, the reaction required stoichiometric amounts of both "catalysts" (Scheme 10.47). Benzene derivatives substituted by electron-donating or -withdrawing groups reacted as well, but a mixture of regioisomers was produced, except for nitrobenzene, which only gave *m,m'*-dinitrobiphenyl.

Similarly, the vinylation of benzene was achieved on treatment with silver acetate and vinylpalladium(II) complexes (Scheme 10.48).[45]

In these reactions, silver ion abstracted chloride from the starting palladium species, rendering the coordination of benzene possible. As a Lewis acid due to its oxidation state, Pd^{II} then initiated electrophilic addition to the aromatic system, leading to C–C bond formation after reductive elimination (Scheme 10.49).

Surprisingly not used in organic synthesis for decades, this methodology is now emerging with the synthesis of biaryls and of functionalized arenes, heterocyclic or not, as the goal.[81]

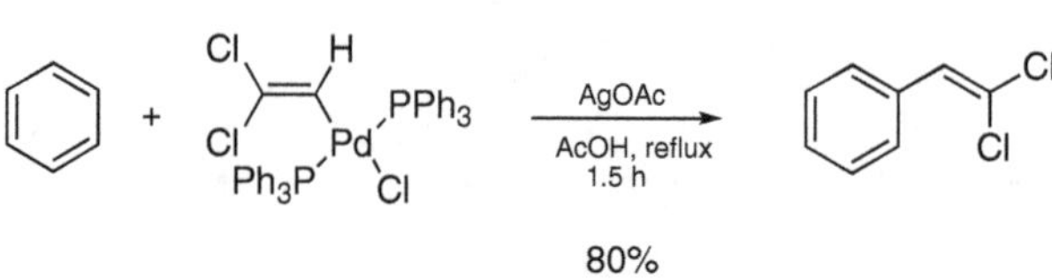

Scheme 10.48

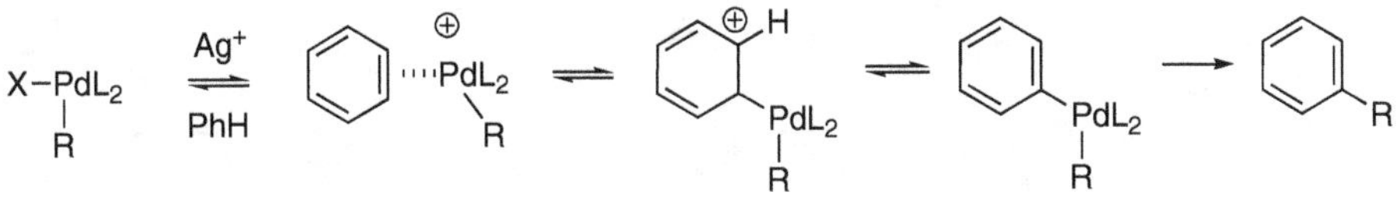

Scheme 10.49

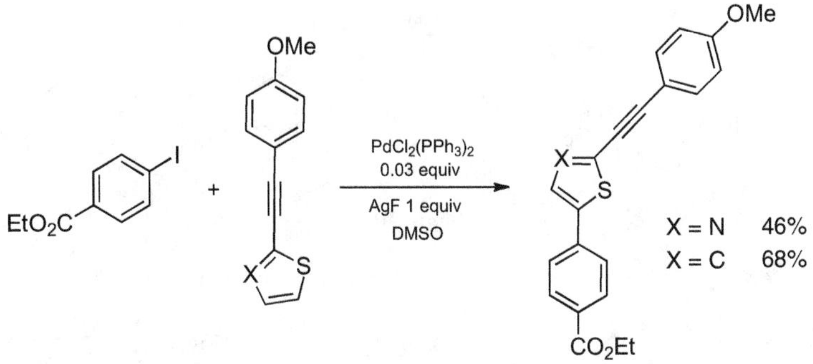

Scheme 10.50

In 2005 and 2006, Mori et al. reported the arylation of heterocycles with aryl iodides using palladium dichloride complexes as catalysts and silver nitrate or fluoride as activators (Scheme 10.50). The best coupling conditions were obtained with a combination of silver nitrate and potassium fluoride.[82]

With organic thin-film transistors, light-emitting devices, and dye-sensitized organic solar cells as the ultimate targets; the same group more recently reported a synthesis of poly(arylated)thiophenes through a series of coupling reactions and a Pd–Ag-promoted arylation (Scheme 10.51).[83]

Applications of this reaction are not limited to advanced materials, but can be applied to natural product synthesis. Indeed, indoles have quite recently (in 2008) been arylated in the presence of palladium acetate and silver oxide (Scheme 10.52).[84]

In the preceding reactions, the arylation was regioselective with an outcome similar to electrophilic aromatic substitution. However, with simple benzene derivatives, mixtures of biaryl derivatives have been obtained (Scheme 10.53).[85] The role of silver trifluoroacetate in these arylations was crucial and, as proposed by the authors, this silver salt could enhance the reactivity and reoxidize Pd species.

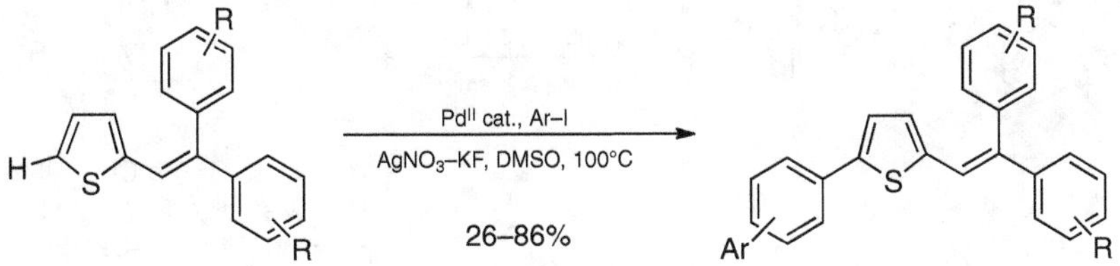

Scheme 10.51

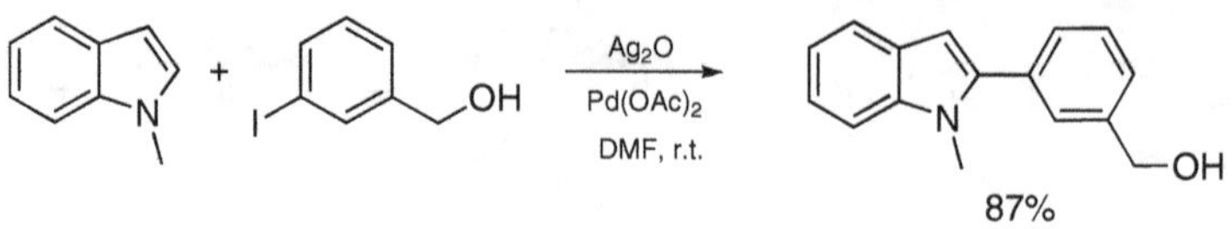

Scheme 10.52

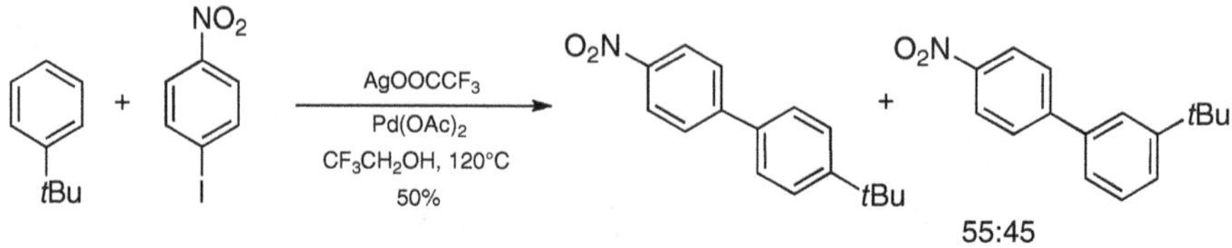

Scheme 10.53

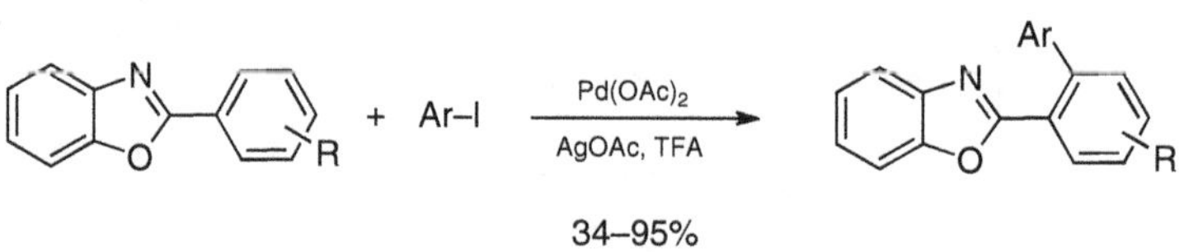

Scheme 10.54

Although discussed in terms of a Pd^0/Pd^{II} versus Pd^{II}/Pd^{IV} mechanism, the lack of regioselectivity suggests an electrophilic aromatic substitution. Other examples, as well as competitive experiments, also supported such a mechanism (see text below).

Such an arylation reaction can nevertheless be rendered regioselective if coordinating atoms are present in the starting molecule at the appropriate position, directing the C–H activation at this position. Thus, 2-arylbenzoxazoles have been regioselectively arylated in the presence of palladium acetate and silver acetate (Scheme 10.54).[86]

Various directing groups could be used in such directed arylations; even carboxylates seemed effective at controlling C–H activation of benzoic acid derivatives (Scheme 10.55).[87]

Interestingly, direct coupling of aromatic systems has been achieved starting from nonactivated arenes. Such a process is based on direct C–H activation of both partners. Although mixture of compounds could be obtained, good regioselectivity was

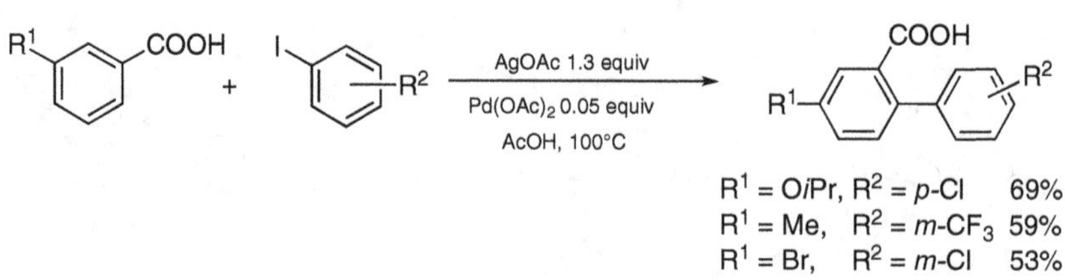

Scheme 10.55

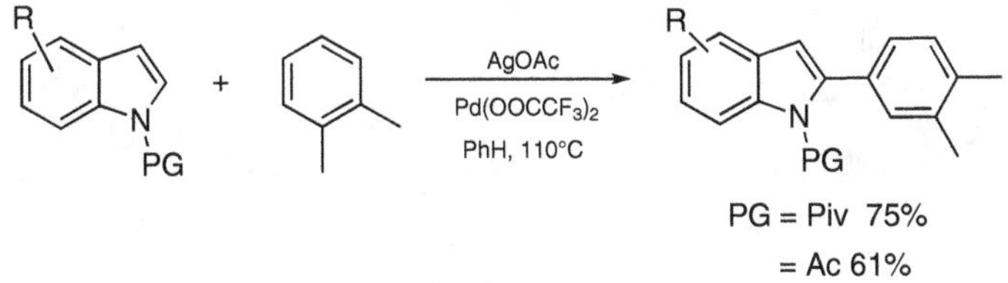

Scheme 10.56

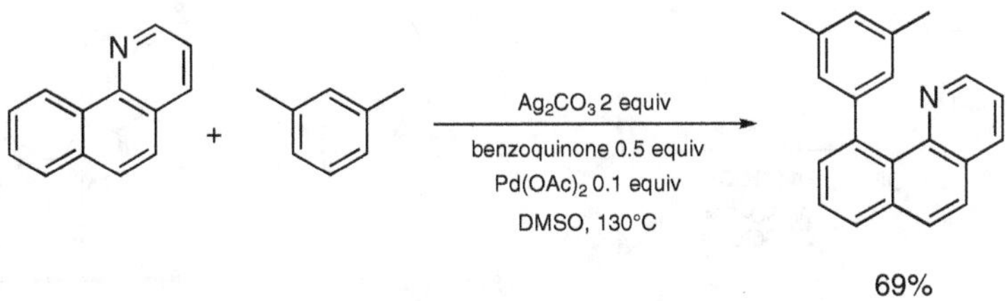

Scheme 10.57

observed by using silver acetate; 2-arylindoles were favored in this case (Scheme 10.56).[88] Interestingly, the reversed regioselectivity, that is, formation of 3-arylindoles, was achieved in the presence of copper acetate as cocatalyst.

As in the preceding arylation, such direct reactions of nonactivated arenes could be completely regioselective with a coordinating substituent. Benzoquinoline has been used for such purposes (Scheme 10.57).[89]

Interestingly, such directed arylations have just been rendered asymmetric by using chiral amino acids as ligands on prochiral diarylpyridinylmethane derivatives (see Scheme 10.10, Section 10.3).[24] Screening revealed that rigid ligands such as 1,1-cyclopropylaminocarboxylates or leucine derivatives *N*-substituted with bulky carbamates were the most effective in inducing high enantiodiscrimination.

10.5.4.3 Silver as Reagent for Decarboxylative Coupling

In the presence of palladium(II) and silver(I) salts, arene carboxylates could be converted to aryl palladium species, which were engaged in Heck coupling reactions.[90] Since the more electrophilic palladium trifluoroacetate proved to be the best catalyst, decarboxylation probably occurred by aromatic electrophilic substitution

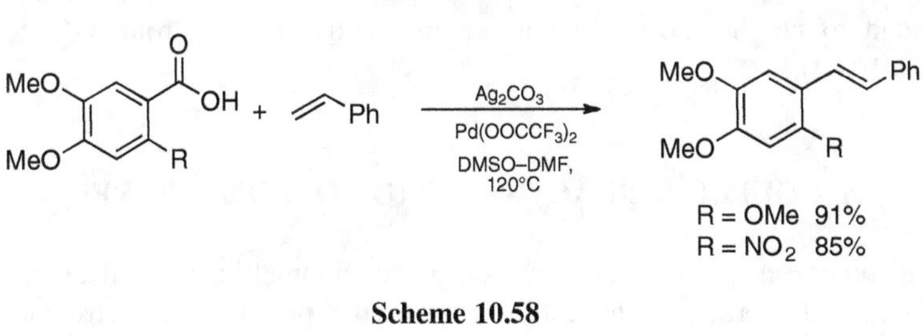

Scheme 10.58

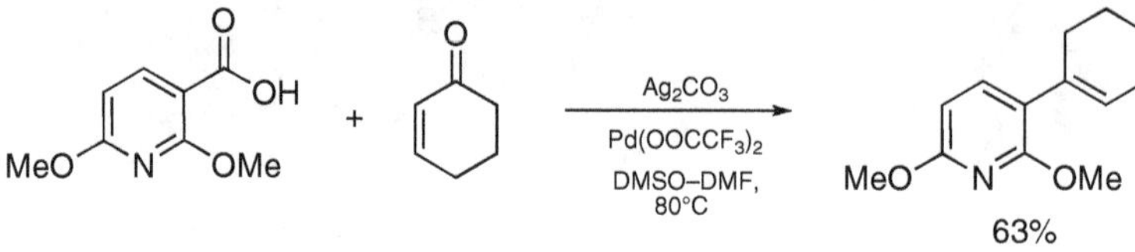

Scheme 10.59

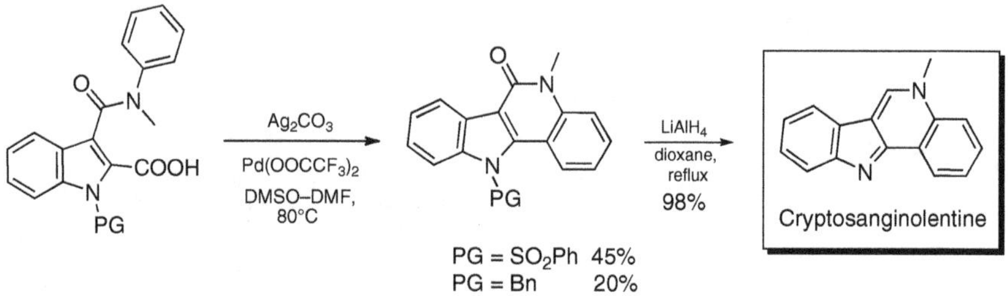

Scheme 10.60

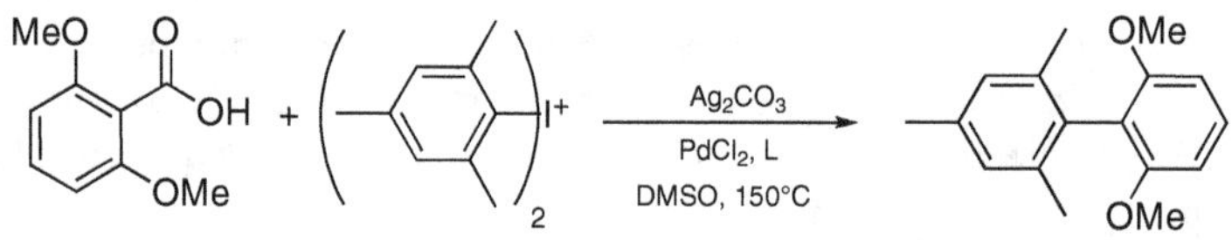

Scheme 10.61

of palladium(II) at the *ipso* position (Scheme 10.58). Silver carbonate dramatically improved this decarboxylation reaction, and DMSO was also critical for the efficiency of the process.

This reaction has been extended to other heteroarene carboxylates and to cycloalkenones, leading to conjugated aryl or heteroaryl enones in good yields (Scheme 10.59).[91]

This reaction has also been applied to the synthesis of indoloquinoline alkaloids (Scheme 10.60).[92]

Similarly, biaryls, including sterically hindered ones, have more recently been obtained by decarboxylative coupling of arenecarboxylic acids and diaryliodonium triflates.[93] The $PdCl_2$/DPEphos system in the presence of silver carbonate in DMSO was found to be the most efficient, giving yields ranging from 37% to 85% (Scheme 10.61).

10.6 *sp*^{2}–*sp* COUPLING REACTIONS PROMOTED BY SILVER SALTS

The title processes are well documented as stoichiometric, as well as catalytic, procedures. In the former, silver acetylides are used or produced in situ from terminal

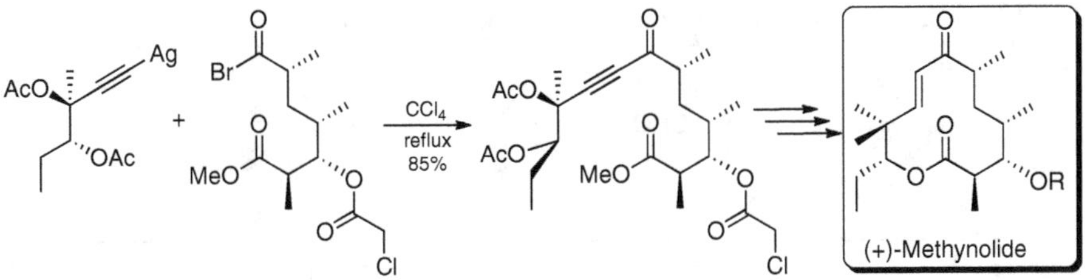

Scheme 10.62

alkynes.[8] In the latter, the formation of a C_{sp^2}–C_{sp} bond is usually performed in association with palladium catalysis.[3]

10.6.1 Organosilver Species as Nucleophilic Reagents

Silver acetylides, easily produced either from the corresponding alkynes[94] or silylated alkynes,[95] were used as nucleophilic reagents toward activated carbonyl groups or iminiums. Silver acetylides exhibit an extreme mildness and a very low basicity, so that they are compatible with a wide range of functional groups and can be produced and used in protic solvents and even in water, rendering them green reagents in green processes.

Since such reactivity has recently been reviewed,[8] only a few examples are reported here. Acyl halides or epoxides can be alkynylated with preformed silver acetylides. Such processes have been applied to the total synthesis of the antibiotic macrolide (+)-methynolide,[96] and in the total synthesis of the antitumor agent FR901464 isolated from a *Pseudomonas* species (Scheme 10.62).[97]

In the latter, the alkynylation proved to be effective only in the presence of dichlorozirconocene and catalytic amounts of silver triflate, suggesting a complex mechanism in which Ag–Zr transmetallation and chloride abstraction by the added silver triflate occurred (Scheme 10.63).[8] Nevertheless, this reaction proved quite general.[98]

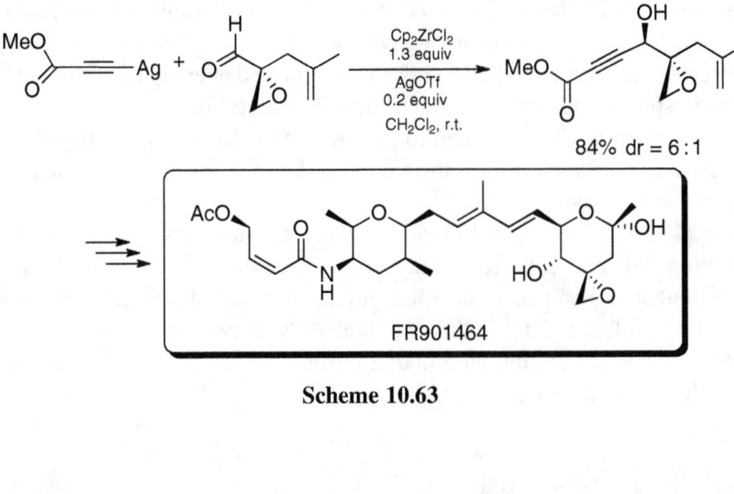

Scheme 10.63

TABLE 10.5. Aldehyde Alkynylation Catalyzed by Ag(I) in Water

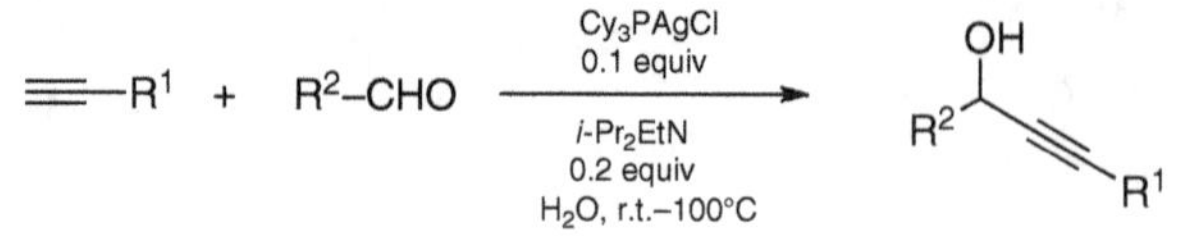

Alkyne	Aldehyde	Conditions	Yield (%)
≡—Ph	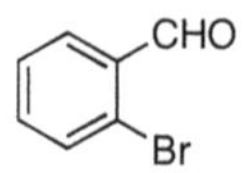	60°C, 12 h	90
≡—Ph	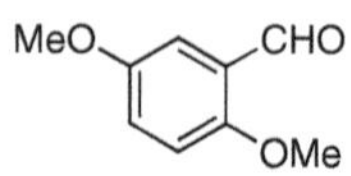	90°C, 12 h	93
≡—Ph	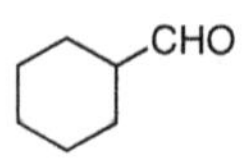	70°C, 2 days	81
≡—Ph	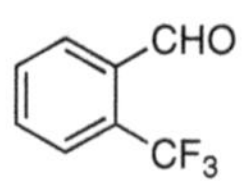	80°C, 6 h	93
≡—*n*-Hex	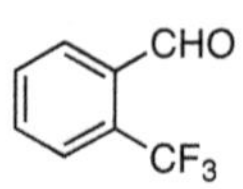	90°C, 1.5 days	77

Interestingly, the first catalytic version of such nucleophilic additions of alkynyl silvers to aldehydes has been described.[99] Indeed, silver chloride in the presence of tricyclohexylphosphine and mild bases such as ethyldiisopropylamine catalyzed the addition of terminal alkynes to aldehydes in good to high yields (Table 10.5). The reaction proved to be almost insensitive to electronic effects; however, alkylacetylenes were less reactive than arylacetylenes. The solvent had a dramatic effect on the reaction course. Control experiments with preformed phenylethynylsilver showed that both phosphine and water activated the silver acetylide.

Similarly, terminal alkynes added to α-iminoesters derived from ethylglyoxylate in the presence of silver salts.[100] In this case, the reaction worked best in apolar solvents (Scheme 10.64).

In a closely related reaction, Li et al.[101] devised a silver-catalyzed three-component coupling between aldehydes, alkynes, and amines. This transformation was really efficient in water and in an ionic liquid, although both required overnight heating at 100°C (Table 10.6). In this reaction, a silver acetylide that was sufficiently nucleophilic to add to the iminium ion that formed by condensation of the amine on the aldehyde was certainly produced.[8]

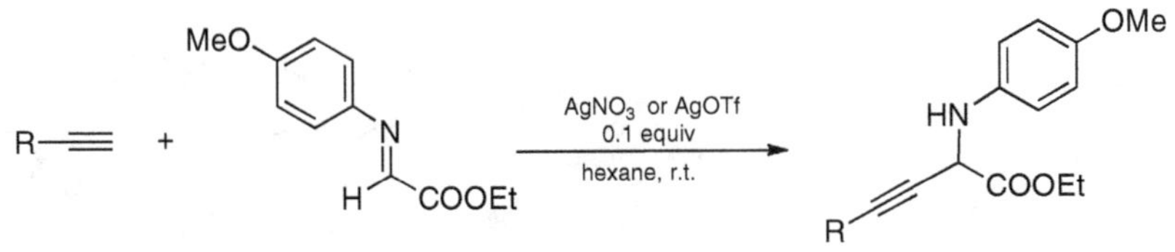

Scheme 10.64

Supported versions of such silver-catalyzed, three-component couplings have been recently reported. Silver oxide on multiwall carbon nanotubes or on alumina,[102] as well as silver-doped zeolites,[103] proved to be efficient and reusable catalysts for such coupling reactions. In the former, water was used as solvent, while in the latter, no solvent was required, making it a truly green process.

Within the same green context, and in an effort to utilize carbon dioxide in organic chemistry, Anastas et al. devised a similar reaction in which the silver acetylide intermediate was trapped by carbon dioxide and the resulting carboxylate was trapped again with benzyl bromide derivatives. The thus-formed ester reacted

TABLE 10.6. Propargylic Amine Formation from Aldehydes and Alkynes

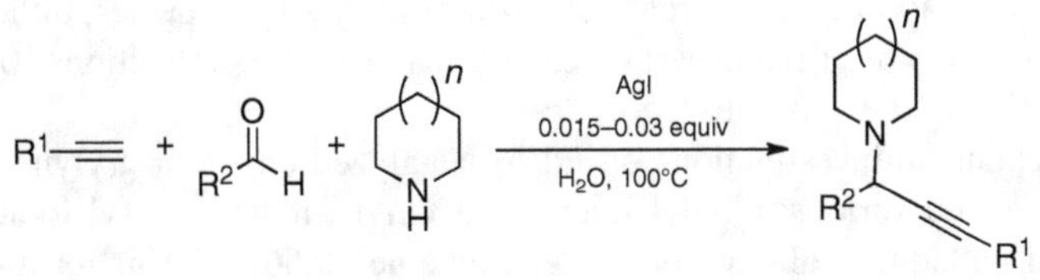

Alkyne	Aldehyde	Amine	Yields (%)
≡—Ph	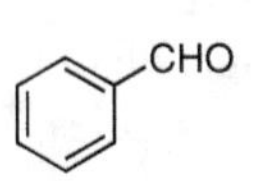	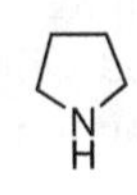	95
≡—Ph	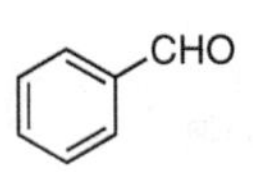	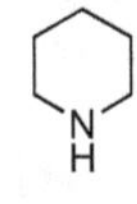	70
≡—Ph	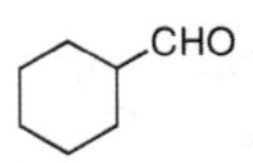	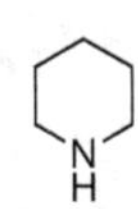	96
≡—Ph	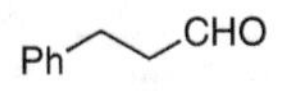	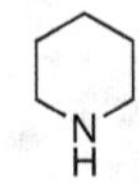	85

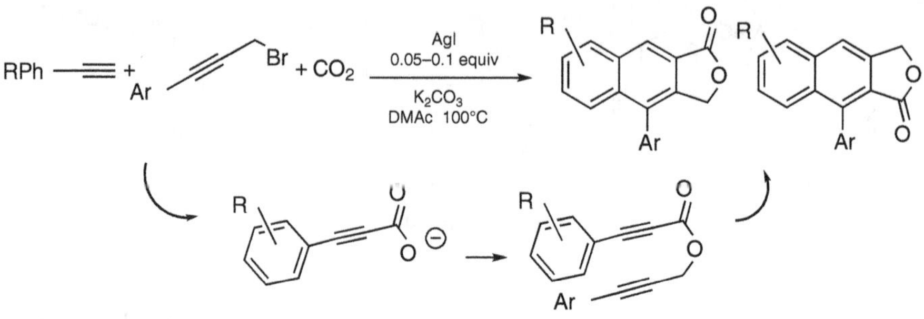

Scheme 10.65

intramolecularly in a Diels–Alder-type reaction, possibly also catalyzed by silver iodide. This reaction offered a new and concise synthesis of arylnaphthalene lactones, analogs of natural lignans (Scheme 10.65).[104]

10.6.2 Organosilver Species in Transmetallations

Combining the mild nucleophilicity of alkynylsilvers with the halogenophilicity of silver ions suggests their use in transmetallation reactions.

This idea led to only a few reactions with alkynylsilver species, although such processes certainly occurred in various catalyzed reactions (see Sections 10.4, 10.5.1, 10.5.2, 10.5.4.1, 10.6.4, and Ref. 8).

For mechanistic investigations on Pd/Ag-catalyzed enyne or arylyne synthesis, Pale et al. mixed various alkynylsilvers with vinyl triflates or aryl iodides in the presence of palladium salts or complexes (Scheme 10.66).[105] This reaction clearly demonstrated that organosilvers can be transmetalated to organopalladium species, even if no halide was present, as demonstrated when starting from vinyl triflates.

In the nucleophilic opening of oxiranes by silver acetylides (see Section 10.4), Koide et al. provided good evidence for transmetallation of silver acetylides to dichlorozirconocene leading to alkynyl zirconium species.[98]

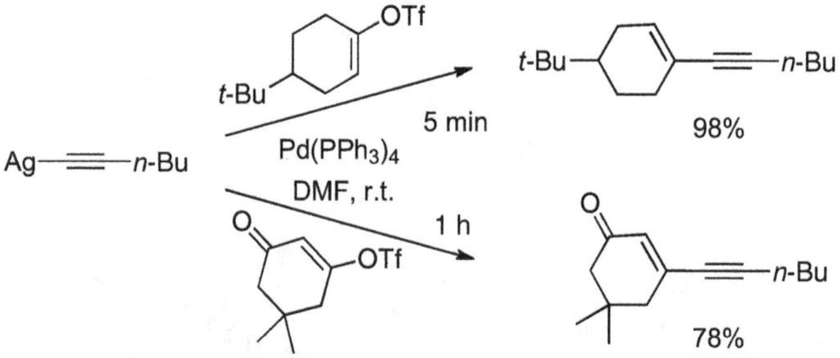

Scheme 10.66

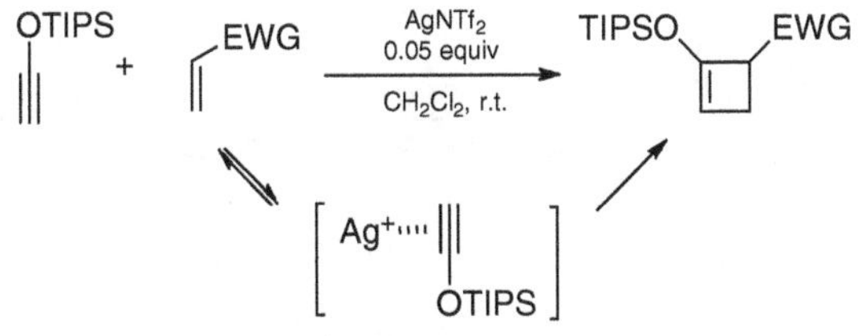

Scheme 10.67

10.6.3 Silver as a Lewis Acid Reagent

Several cycloadditions are known to be promoted by silver salts (see Chapter 2), driven by the Lewis acidity of silver ions. Probably more formal than true cycloadditions, such reactions forming C–C bonds are, indeed, often initiated by the Lewis acidity of silver ions, as revealed by the NMR observation of silver π coordination to siloxyalkynes (Scheme 10.67).[106]

Similarly, γ- or δ-alkynylaldehydes proved to be more π-coordinated than σ-*O*-coordinated in the presence of silver hexafluoroantimonate in solution, as shown again by NMR.[107] It is worth noting that such experimental observations disagree with theoretical calculations (see Section 10.5.3 and Ref. 36). Interestingly, ω-alkynyl–aldehydes reacted intramolecularly, yielding a product formally resulting from a [2 + 2] cycloaddition and a retro[2 + 2] cycloaddition (Scheme 10.68). This reaction could also be applied to intermolecular coupling of alkynes to aldehydes. The fact that tetrafluoroboric acid, as well as trifluoroborane–etherate, gave the same products, suggested that the active species is nevertheless the σ-coordinated species.

The intramolecular carbostannylation reported in 2007 by Echavarren et al. and related reactions are also initiated by the π-Lewis acidity of silver ions.[108] Silver salts catalyzed the cyclization of alkynylated allylstannanes to the corresponding 1-stannylmethylene-2-vinyl five-membered rings (Scheme 10.69). Interestingly, the more electrophilic the salt is, the more rapid the reaction is; the hexafluroroantimonate

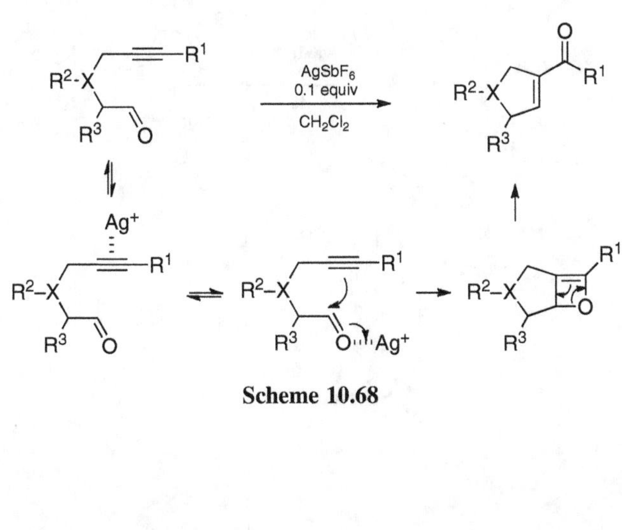

Scheme 10.68

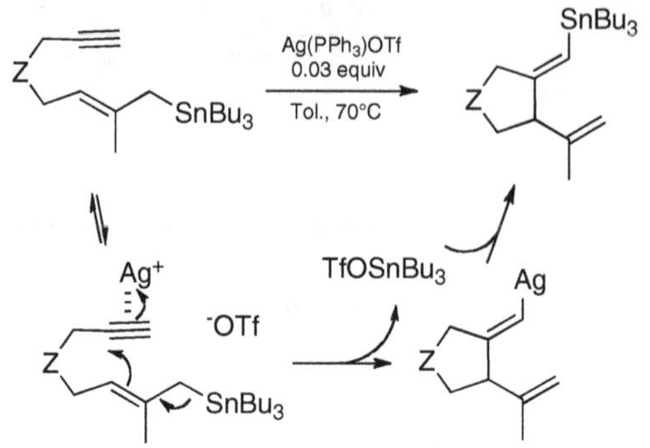

Scheme 10.69

affords the product in one minute. However, more consistent results were achieved with the triphenylphosphine complex of silver triflate.

A similar process is probably responsible for the cyclization of *N*-sulfonyl and *N*-Boc alkynylated enamines. For such reactions, silver triflate was also the best catalyst, leading in high yields to azahydrindane systems, although as a mixture of isomers (Scheme 10.70).[109]

Although not described as such, this reactivity of silver as a Lewis acid in C–C bond formation via enamines was already known and actually described in the synthesis of complex indole alkaloids. A *N*-sulfonyldienamine embedded within a polycyclic indole ring system added to the trimethylsilylated propargyl arm of this system, leading in high yield to a new six-membered ring (Scheme 10.71).[110]

10.6.4 Organosilver Species as Intermediates in Catalyzed Enyne or Arylyne Synthesis

Alkynylsilvers are often intermediates in various coupling reactions catalyzed by both silver and palladium salts, leading to enynes or arylynes.

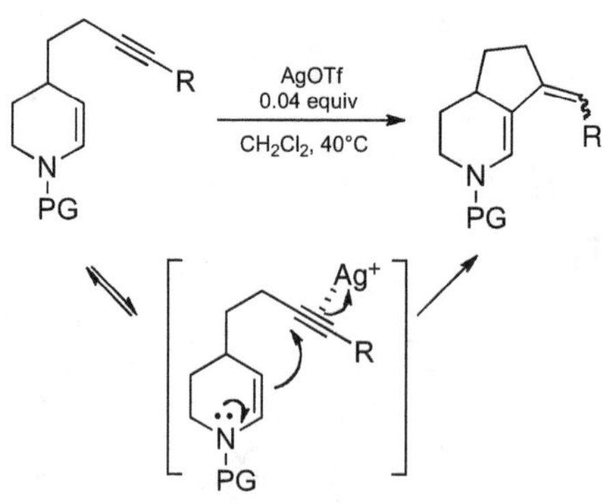

Scheme 10.70

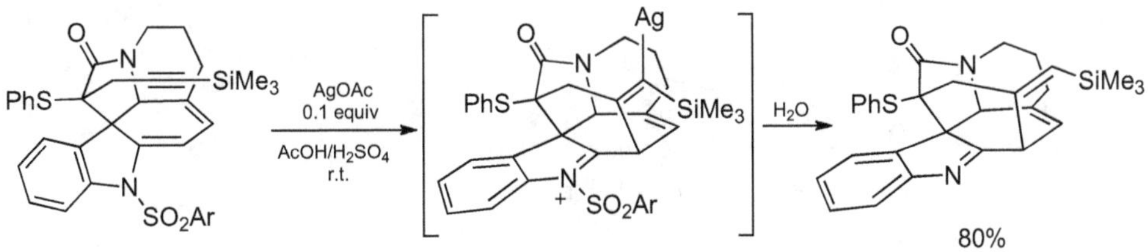

Scheme 10.71

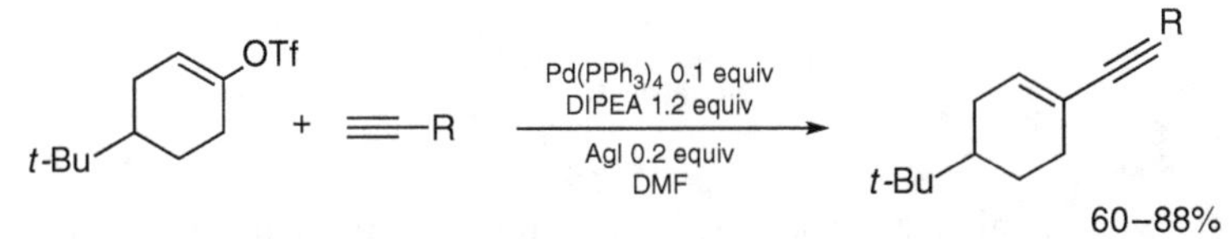

Scheme 10.72

In designing a convergent and versatile synthesis toward antitumoral dienediynes, Pale et al. had to develop alternatives to the classical Sonogashira coupling reaction.[111–113] Most silver salts proved effective cocatalysts in such couplings, but the halides were the most efficient, giving coupling products very cleanly and in good yields, even at room temperature and with the most sensitive compounds (Scheme 10.72). Moreover, under these conditions, the formation of homocoupling products, common side products resulting from Glaser-type reactions when using Cu^{I} catalysts,[114–116] was totally suppressed.

Further investigations including ^{109}Ag NMR confirmed the in situ formation of alkynyl silver through π complexation and proton abstraction.[117] They also revealed that the best palladium catalyst, namely, palladium tetrakis(triphenylphosphine), played a key role, in that a phosphine liberated by decoordination ended up on the alkynyl silver, stabilizing it and rendering it more soluble.[118] The resulting alkynyl silver then entered the palladium catalytic cycle at the transmetallation step (See section 10.6.2).[105]

To avoid a deprotection step, as well as to solve synthetic problems, Pale et al. modified their Pd–Ag catalytic conditions to directly couple 1-trialkylsilyl-1-alkynes, first with AgF as cocatalyst, expecting that silylated alkynes would be directly converted into silver acetylides with this catalyst (Scheme 10.73).[119]

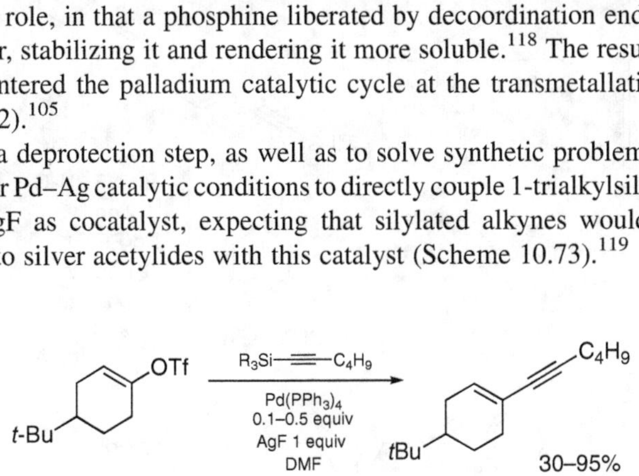

Scheme 10.73

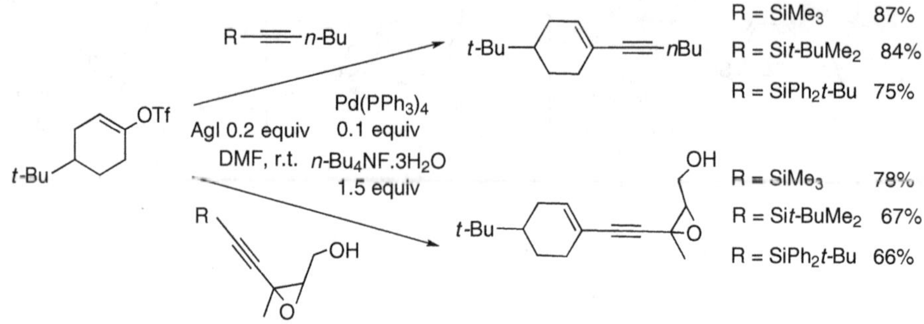

Scheme 10.74

Tetrabutylammonium fluoride trihydrate eventually proved to be the best activator for this goal, and various 1-trialkylsilyl-1-alkynes could be coupled under such conditions (Scheme 10.74).[120,121]

Moreover, alkynes protected with various silyl groups could be distinguished and selectively coupled depending on conditions. 1-Trimethylsilyl-1-alkynes were selectively coupled in the presence of other silylated acetylenic moieties using Pd $(PPh_3)_4$–AgCl as the catalyst in the presence of a slight excess of methanol and potassium carbonate in DMF at room temperature.[122] Successive coupling reactions could thus be achieved, using differentially protected polyynes,[123] even in a one-pot process (Scheme 10.75).[124]

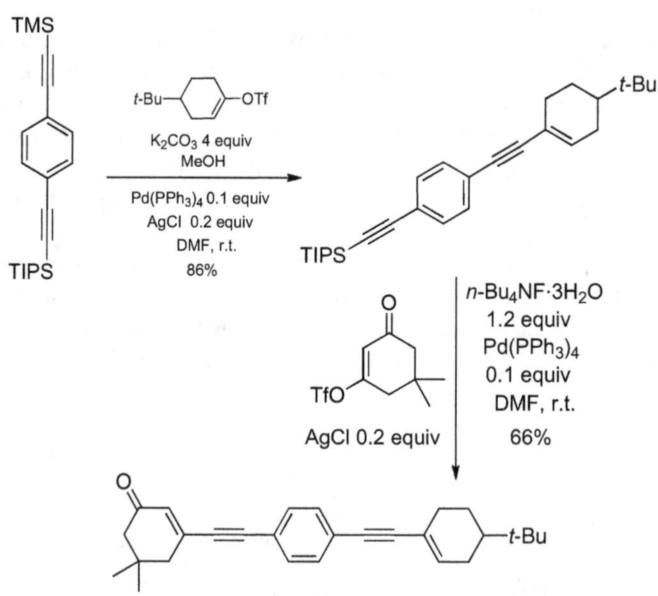

Scheme 10.75

TABLE 10.7. Pd/Ag Coupling Reaction with Triisopropylsilylethynyl Phosphonate

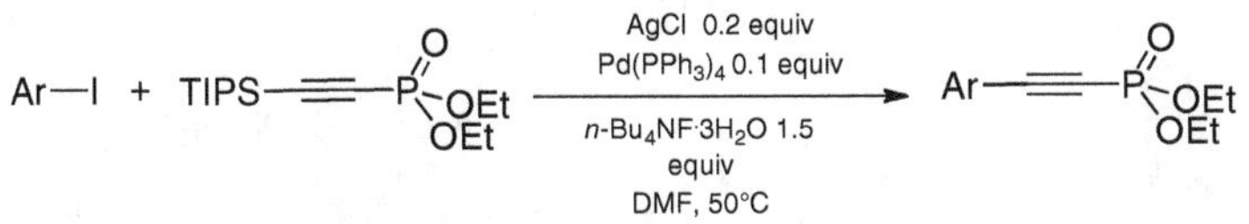

Aryl Halide	Product	Yield (%)
PhI	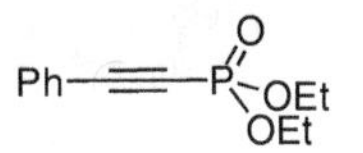	95
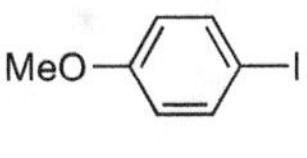	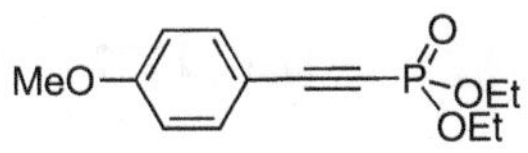	80
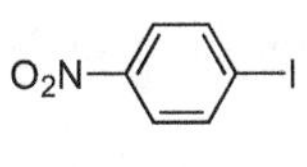	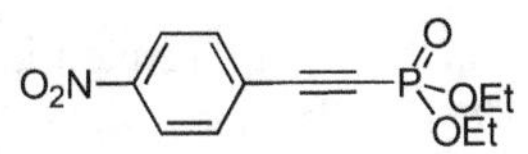	60
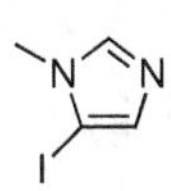	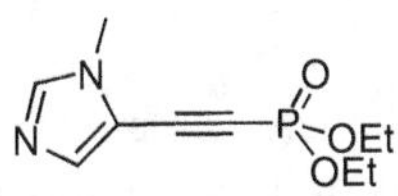	66

Such direct couplings of silylated alkynes have been applied to the synthesis of a variety of aryl- and heteroarylethynyl phosphonates, available only in poor yields with classical routes (Table 10.7).[125]

The same method allowed the preparation of aryl- and heteroaryl propolates directly from trimethylsilylpropiolates (Scheme 10.76).[125]

These methods have also been applied to the development of a novel route to conjugated diynes, directed toward natural polyacetylenic products. Substituted

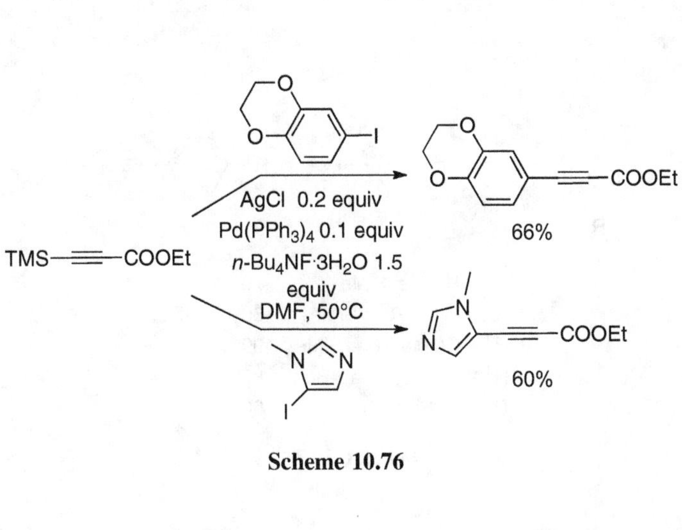

Scheme 10.76

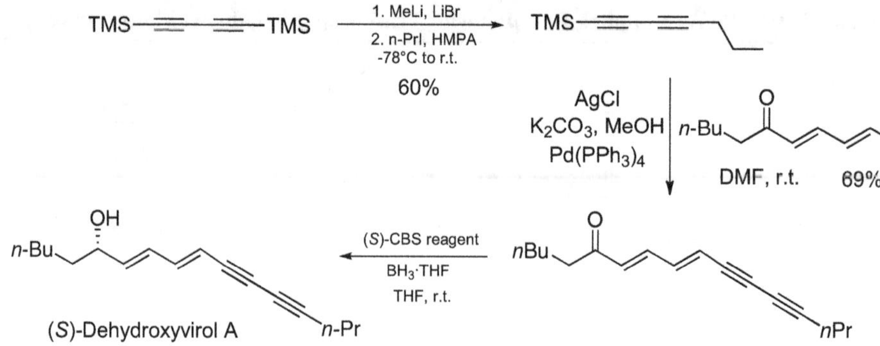

Scheme 10.77

conjugated diynes were obtained through sequential coupling of 1,4-bis(trimethylsilyl)butadiyne (Scheme 10.77).[126–128]

Silver salts having a basic counterion also proved successful in various enyne or arylyne syntheses by coupling.

Conditions based on stoichiometric amounts of silver oxide have been developed by Mori et al. for the synthesis of arylated alkynes from terminal alkynes and aryl iodides. Under such conditions, neither silylated alkynes nor aryl bromides or triflates did not undergo coupling (Scheme 10.78).[129,130]

With silver carbonate as cocatalyst and with a palladium(II) complex as catalyst, enynyl phosphonates have been successfully obtained (Scheme 10.79).[131]

In 2007, Wu et al. showed that arylboronic acids and the corresponding boronates reacted with terminal alkynes in the presence of a palladacycle as catalyst and silver oxide (Scheme 10.80).[132] As in the Mori reaction using basic silver salts, stoichiometric amounts of silver ion were required for high efficiency.

Under similar conditions, but with the addition of chlorides, Nagasaka et al. were able to directly engage 1-trimethylsilyl-1-alkynes in coupling reactions with aryl

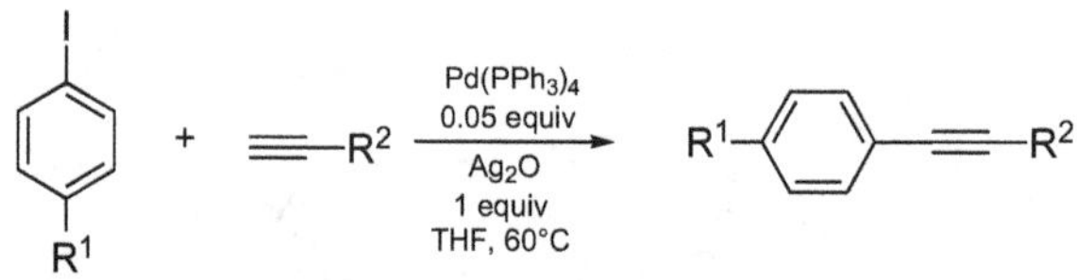

R^1 = H, R^2 = $SiMe_3$	79%
R^1 = OMe, R^2 = $SiMe_3$	60%
R^1 = COMe, R^2 = $SiMe_3$	99%
R^1 = COMe, R^2 = *t*-Bu	87%
R^1 = H, R^2 = Ph	60%

Scheme 10.78

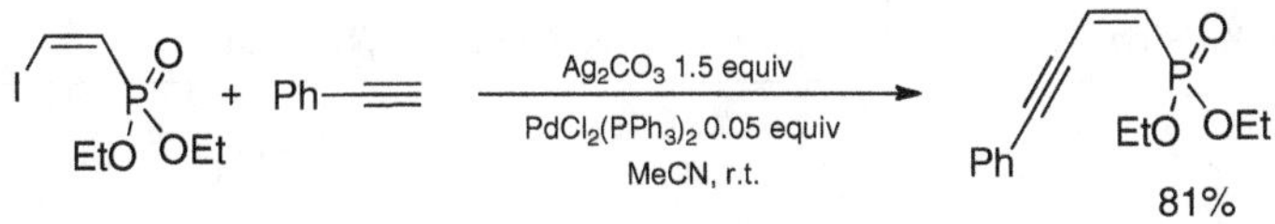

Scheme 10.79

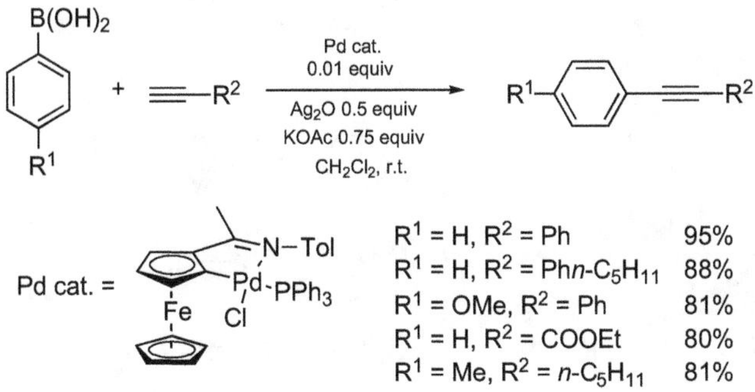

Scheme 10.80

iodides in the presence of silver carbonate. As for the Mori reaction, only iodides (not bromides or triflates) were converted to the expected arylynes. Stoichiometric amounts of silver ion were also required (Scheme 10.81).[133]

Under essentially the same conditions, Jeong et al. were able to prepare disubstituted trifluoromethylated enediynes, either directly through a dicoupling reaction from bis(trimethylsilyl)enediynes or through successive couplings from mono(trimethyl silyl)enediynes (Scheme 10.82).[134]

The first direct coupling of terminal alkynes with aryl iodides or bromides *without palladium* was reported by Wang and Li in 2006.[135] Silver iodide and triphenylphosphane in polar solvents proved to be the best catalyst combination, while potassium carbonate proved to be the better base, giving diarylacetylenes in high yields (Table 10.8).

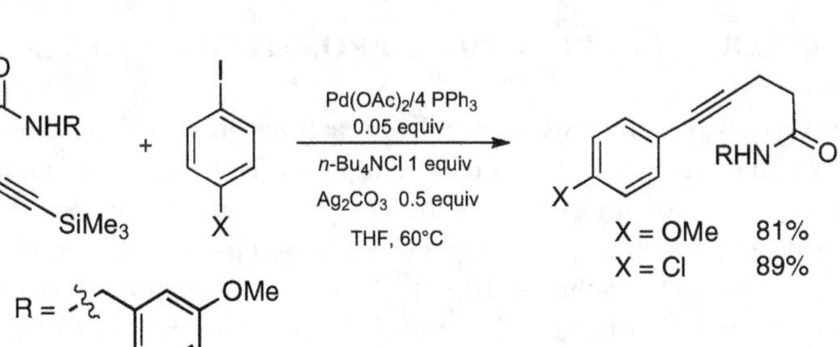

Scheme 10.81

TABLE 10.8. Silver-Catalyzed Palladium-Free Coupling Reactions

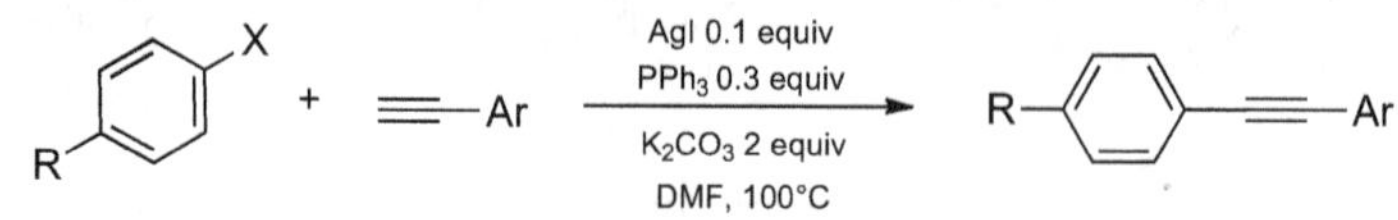

Arylhalide	Ar	Product	Yields (%)
PhI	*p*-MePh	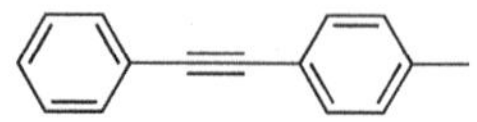	96
PhI	*p*-FPh	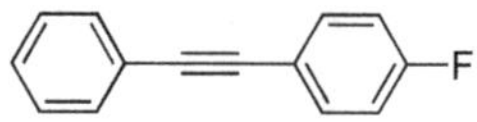	98
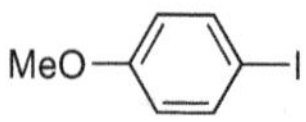	Ph	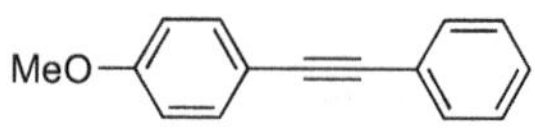	92
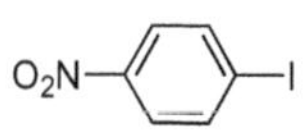	Ph	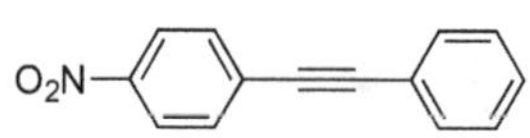	99
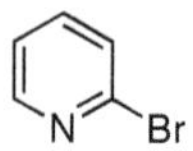	Ph		99
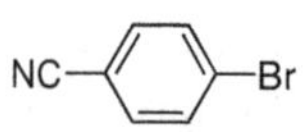	Ph	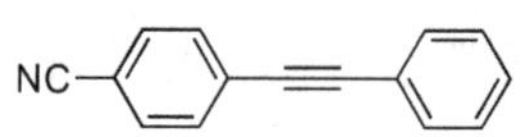	92

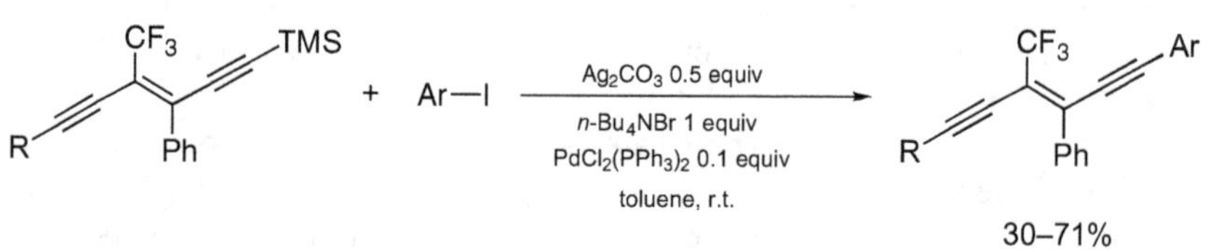

Scheme 10.82

10.7 *sp–sp* COUPLING REACTIONS PROMOTED BY SILVER SALTS

There are very few reports of *sp–sp* coupling reactions promoted by silver salts in the literature, and it seems that only a single example has been reported so far. This work described the homocoupling of aliphatic acetylenes on a solid support by using a catalytic system combining silver tosylate, cupric chloride, and tetramethylethylenediamine (TMEDA) (Scheme 10.83).[136] All other methods for this coupling reaction proved to be unsuccessful. Control experiments showed that the coupling did not occur with copper or silver salts alone or when cupric tosylate was associated with silver tosylate. These results suggested that copper(II) chloride was required in

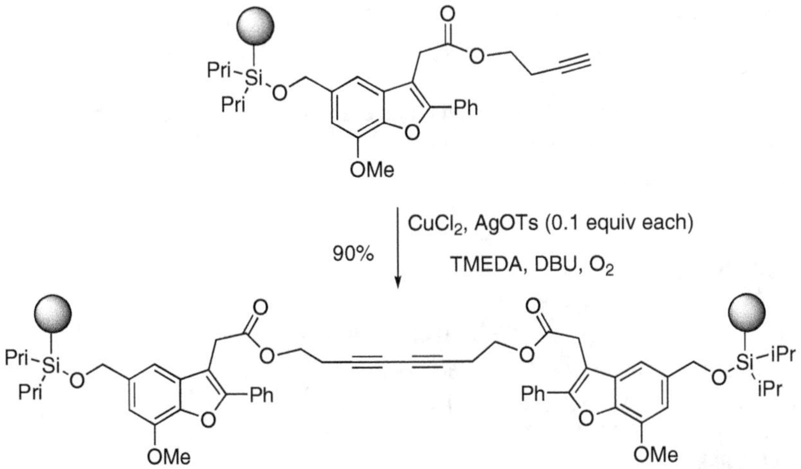

Scheme 10.83

addition to the complex formed from silver tosylate and TMEDA. The authors proposed that silver ion activated the alkyne by complexation, facilitating the formation of a dinuclear copper(II) acetylide complex, which directly collapsed to the homocoupling product, releasing copper(I).

10.8 CONCLUSION

Although neglected in organic chemistry for a century, organosilver species and silver salts have been the subject of increasing interest since the end of the twentieth century and clearly even more so in the past decade. Silver salts have been implied in various C–C bond-forming reactions, and, as shown in this chapter, almost all kinds of coupling have been documented, in either stoichiometric or catalytic processes. These Ag-promoted coupling reactions are based on very different methods and mechanisms, depending on the way that silver is involved in the reaction.

On one hand, organosilver species can now be produced from a wide variety of other reagents through transmetallation (organolithium and magnesium, organoboranes and borates, organosilicates) or directly from alkynes for alkynylsilvers. Organosilver species can be produced in situ and used as nucleophilic reagents or in homocoupling reactions. They can also be transmetallated to other metals, such as zirconium and palladium, and thus can be involved in other reactions, such as cross-couplings and alkylations.

On the other hand, silver salts can act as a mild Lewis acid, promoting various reactions such as allylations, aldolizations, cycloadditions, and cyclizations. Silver salts can also be used as halide scavengers, acting as cocatalysts in cross-coupling reactions catalyzed by other metals, especially palladium. In the latter context, the exact role of silver salts is far from clear and may be more complex than just halide abstraction, possibly with a role in transmetallation. In reactions with alkynes, it has been established that silver salts catalyze the in situ formation of alkynylsilvers as intermediates.

Organosilver species and silver salts thus exhibit a wide variety of behaviors, and further work will probably unravel new roles in organic chemistry, especially for the formation of C–C bonds.

REFERENCES

1. Frankland, E., *Liebigs Ann. Chem.* **1848**, *71*, 171–213.
2. Grignard, V., *Compt. Rend. Acad. Sci.* **1900**, *130*, 1322–1325.
3. (a) De Meijere, A.; Diederich, F., *Metal–Catalyzed Cross–Coupling Reactions*, Wiley-VCH, Weinheim, **2004**; (b) Tsuji, J., *Palladium Reagents and Catalysts*, Wiley, Hoboken, NJ, **2004**.
4. Lipshutz, B. H.; Yamamoto, Y. Coinage metals in organic synthesis, *Chem. Rev.* **2008**, *108*, 2793–3442.
5. 3-Ethoxyprop-1-ynyl silver and phenylethynyl silver were reported as early as 1865 and 1870; see (a) Berend, M.; Liebermann, C., *Liebigs Ann. Chem.* **1865**, *135*, 259–290; (b) Glaser, C., *Liebigs Ann. Chem.* **1870**, *154*, 157–171; (c) Chevastelon, R., *Compt. Rend. Acad. Sci.* **1897**, *124*, 1364.
6. Patai, S.; Rappoport, Z., *Organic Derivatives of Gold and Silver*, Wiley-VCH, Weinheim, **2003**.
7. (a) Weibel, J.-M.; Blanc, A.; Pale, P., *Chem. Rev.* **2008**, *108*, 3149–3173; (b) Álvarez-Corral, M.; Muñoz-Dorado, M.; Rodríguez-García, I., *Chem. Rev.* **2008**, *108*, 3174–3198; (c) Yamamoto, Y., *Chem. Rev.* **2008**. *108*, 3199–3222; (d) Naodovic, M.; Hisashi Yamamoto, H., *Chem. Rev.* **2008**, *108*, 3132–3148.
8. Halbes-Letinois, U.; Weibel, J.-M.; Pale, P., *Chem. Soc. Rev.* **2007**, *36*, 759–769.
9. (a) Gardner, J. H.; Bergstrom, P., *J. Am. Chem. Soc.* **1929**, *51*, 3375–3377; (b) Gardner, J. H.; Joseph, L., *J. Am. Chem. Soc.* **1939**, *61*, 2551–2552; (c) Biclkey, E. A.; Gardner, J. H., *J. Org. Chem.* **1940**, *5*, 126–132.
10. Kharash, M. S.; Reinmuth, O., *Grignard Reagents of Nonmetallic Substances*, Prentice-Hall, New-York, **1954**.
11. (a) Tamura, M.; Kochi, J., *J. Am. Chem. Soc.* **1971**, *93*, 1483–1485; (b) Kochi, J., *J. Organomet. Chem.* **2002**, *653*, 11–19.
12. Whitesides, G. M.; Casey, C. P., *J. Am. Chem. Soc.* **1966**, *88*, 4541–4543.
13. Nagano, T.; Hayashi, T., *Chem. Lett.* **2005**, *34*, 1152–1153.
14. Someya, H.; Ohmiya, H.; Yorimitsu, H.; Oshima, K., *Org. Lett.* **2008**, *10*, 969–971.
15. Snyder, H. R.; Kuck, J. A.; Johnson, J. R., *J. Am. Chem. Soc.* **1938**, *60*, 105–111.
16. Brown, H. C.; Verbrugge, C.; Snyder, C. H., *J. Am. Chem. Soc.* **1961**, *83*, 1001.
17. Brown, H. C.; Hebert, N.; Snyder, C. H., *J. Am. Chem. Soc.* **1961**, *83*, 1001–1002.
18. Brown, H. C.; Snyder, C. H., *J. Am. Chem. Soc.* **1961**, *83*, 1002–1003.
19. Falck, J. R.; Mohapatra, S.; Bondela, M.; Venkataraman, S. K., *Tetrahedron Lett.* **2002**, *43*, 8149–8151.

20. Chen, H.; Deng, M.-Z., *J. Org. Chem.* **2000**, *65*, 4444–4446.
21. Arentsen, K.; Caddick, S.; Georey, F.; Cloke, N.; Herring, A. P.; Hitchcock, P. B., *Tetrahedron Lett.* **2004**, *45*, 3511–3515.
22. Yao, M.-L.; Deng, M.-Z., *J. Org. Chem.* **2000**, *65*, 5034–5036.
23. Chen, X.; Goodhue, C. E.; Yu, J.-Q., *J. Am. Chem. Soc.* **2006**, *128*, 12634–12635.
24. Shi, B.-F.; Maugel, N.; Zhang, Y.-H.; Yu, J.-Q., *Angew. Chem. Int. Ed.* **2008**, *47*, 4882–4886.
25. Isabelle, M. E.; Leitch, L. C., *Can. J. Chem.* **1958**, *36*, 440–448.
26. (a) Pouwer, R. H.; Williams, C. M.; Raine, A. L.; Harper, J. B., *Org. Lett.* **2005**, *7*, 1323–1325; (b) Pouwer, R. H.; Harper, J. B.; Vyakaranam, K.; Michl, J.; Williams, C. M.; Jessen, C. H.; Bernhardt, P. V., *Eur. J. Org. Chem.* **2007**, 241–248.
27. Albert, B. J.; Koide, K., *J. Org. Chem.* **2008**, *73*, 1093–1098.
28. (a) Noltes, J. G.; Van Koten, G.,in Comprehensive Organometallic Chemistry I., Wilkinson, G.; Abel, E. W.; Stone, F. G. A.,eds., Pergamon Press, Oxford, **1982**, Vol. 2, pp. 709–763; (b) Van Koten, G.; James, S. L.; Jastrzebski, J. T. B. H., in Comprehensive Organometallic Chemistry II, Abel, E. W.; Stone, F. G. A.; Wilkinson, G. eds., Pergamon Press, Oxford, **1996**, Vol. 3, pp. 57–133.
29. Lang, G. H.,US Patent 3,962,268. (**1976**).
30. Tamao, K.; Matsumoto, H.; Kakui, T.; Kumada, M., *Tetrahedron Lett.* **1979**, *13*, 1137–1140.
31. (a) Usón, R.; Laguna, A.; Abad, J. A., *J. Organomet. Chem.* **1983**, *246* 341–345; (b) Kronenburg, C. M. P.; Jastrzebski, J. T. B. H.; Boersma, J.; Lutz, M.; Spek, A. L.; van Koten, G., *J. Am. Chem. Soc.* **2002**, *124*, 11675–11683.
32. Albéniz, A. C.; Espinet, P.; López-Cimas, O.; Martín-Ruiz, B., *Chem. Eur. J.* **2004**, *11*, 242–252.
33. Albéniz, A. C.; Espinet, P.; Martín-Ruiz, B., *Chem. Eur. J.* **2001**, *7*, 2481–2489.
34. Bagus, P. S.; Lee, Y. S.; Pitzer, K. S., *Chem. Phys. Lett.* **1975**, *33*, 408–411.
35. (a) Pyykkö, P., *Angew. Chem. Int. Ed.* **2004**, *43*, 4412–4456; (b) Gorin, D. J.; Toste, F. D., *Nature* **2007**, *446*, 395–403.
36. Yamamoto, Y., *J. Org. Chem.* **2007**, *72*, 7817–7831.
37. Yanasigawa, A.; Arai, T., *Chem. Commun.* **2008**, 1165–1172.
38. Josephsohn, N. S.; Carswell, E. L.; Snapper, M. L.; Hoveyda, A. H., *Org. Lett.* **2005**, *7*, 2711–2713.
39. Josephsohn, N. S.; Snapper, M. L.; Hoveyda, A. H., *J. Am. Chem. Soc* **2004**, *126*, 3734–3735.
40. Wieland, L. C.; Vieira, E. M.; Snapper, M. L.; Hoveyda, A. H., *J. Am. Chem. Soc* **2008**, *131*, 570–576.
41. Yao, X.; Li, C.-J., *J. Org. Chem.* **2005**, *70*, 5752–5755.
42. Deng, C.-L.; Zou, T.; Wang, Z.-Q.; Song, R.-J.; Li, J.-H., *J. Org. Chem.* **2009**, *74*, 412–414.
43. Arcadi, A.; Alfonsi, M.; Marinelli, F., *J. Organomet. Chem.* **2007**, *692*, 5322–5326.
44. Lide, D. R., ed., *CRC Handbook of Chemistry and Physics*, Chemical Rubber Co., **1999**.
45. Moritani, I.; Fujiwara, Y.; Danno, S., *J. Organomet. Chem.* **1971**, *27*, 279–282.

46. (a) Grove, D. M.; van Koten, G.; Louwen, J. N.; Noltes, J. G.; Spek, A. L., *J. Am. Chem. Soc.* **1982**, *104* 6609–6616; (b) Rimml, H.; Venanzi, L. M., *J. Organomet. Chem.* **1984**, *260*, C52–C54.
47. Albeniz, A. C.; Espinet, P.; Lin, Y.-S., *Organometallics* **1995**, *14*, 2977–2986.
48. Yeh, M.-C. P.; Tsao, W.-C.; Wang, Y.-J.; Pai, H.-F., *Organometallics* **2007**, *26*, 4271–4277.
49. (a) Larhed, M.; Hallberg, A., Intermolecular Heck reaction: Scope, mechanism, and other fundamental aspects of the intermolecular Heck reaction, in *Handbook of Organopalladium Chemistry for Organic Synthesis*, Negishi, E. I. ed., Wiley, **2003**; (b) Whitcombe, N. J.; Hii, K. K.; Gibson, S. E., *Tetrahedron* **2001**, *57*, 7449–7476.
50. (a) Jeffery, T., *J. Chem. Soc. Chem. Commun.*, **1984**, 1287–1289; (b) Jeffery, T., *Tetrahedron Lett.* **1985**, *26*, 2667–2670; (c) Jeffery, T., *Tetrahedron Lett.* **1990**, *31*, 6641–6644; (d) Jeffery, T., *J. Chem. Soc. Chem. Commun.* **1991**, 324–325.
51. (a) Karabelas, K.; Westerlund, C.; Hallberg, A., *J. Org. Chem.* **1985**, *50*, 3896–3900; (b) Karabelas, K.; Hallberg, A., *J. Org. Chem.* **1986**, *51*, 5286–5290; (c) Karabelas, K.; Hallberg, A., *J. Org. Chem.* **1988**, *53*, 4909–4914.
52. Shotwell, J. B.; Hu, S.; Medina, E.; Abe, M.; Cole, R.; Crews, C. M.; Wood, J. L., *Tetrahedron Lett.* **2000**, *41*, 9639–9643.
53. Kalesse, M.; Chary, K. P.; Quitschalle, M.; Burzlaff, A.; Kasper, C.; Scheper, T., *Chem. Eur. J.* **2003**, *9*, 1129–1136.
54. Abelman, M. M.; Oh, T.; Overman, L. E., *J. Org. Chem.* **1987**, *52*, 4130–4133.
55. Cabri, W.; Candiani, I., *Acc. Chem. Res.* **1995**, 2, 2–7.
56. Sundermann, A.; Uzan, O.; Martin, J. M. L., *Chem. Eur. J.* **2001**, *7*, 1703–1711.
57. (a) Jutand, A., *Appl. Organomet. Chem.* **2004**, *18*, 574–582; (b) Jutand, A., *Pure Appl. Chem.* **2004**, *76*, 565–576.
58. Ashimori, A.; Overman, L. E., *J. Org. Chem.* **1992**, *57*, 4571–4572.
59. Madin, A.; Overman, L. E., *Tetrahedron Lett.* **1992**, *33*, 4859–4862.
60. Sato, Y.; Sodeoka, M.; Shibasaki, M., *J. Org. Chem.* **1989**, *54*, 4738–4739.
61. Sato, Y.; Honda, T.; Shibasaki, M., *Tetrahedron Lett.* **1992**, *33*, 2593–2596.
62. Sato, Y.; Watanabe, S.; Shibasaki, M., *Tetrahedron Lett.* **1992**, *33*, 2589–2592.
63. Satcharoen, V.; McLean, N. J.; Kemp, S. C.; Camp, N. P.; Brown, R. C. D., *Org. Lett.* **2007**, *9*, 1867–1869.
64. Omori, A. T.; Finn, K. J.; Leisch, H.; Carroll, R. J.; Hudlicky, T., *Synlett* **2007**, 2859–2862.
65. Zenner, J. M.; Larock, R. C., *J. Org. Chem.* **1999**, *64*, 7312–7322.
66. Fu, C.; Ma, S., *Org. Lett.* **2005**, *7*, 1605–1607.
67. Malm, J.; Björk, P.; Gronowitz, S.; Hörnfeldt, A.-B., *Tetrahedron Lett.* **1992**, *33*, 2199–2202.
68. Fukui, H.; Fukushi, Y.; Tahara, S., *Tetrahedron Lett.* **2003**, *44*, 4063–4065.
69. Uenishi, Y.; Beau, J.-M.; Armstrong, R. W.; Kishi, Y., *J. Am. Chem. Soc.* **1987**, *109*, 4756–4758.
70. Espinet, P.; Echavarren, A. M., *Angew. Chem. Int. Ed.* **2004**, *43*, 4704–4734.
71. Ridgway, B. H.; Woerpel, K. A., *J. Org. Chem.* **1998**, *63*, 458–460.
72. Matos, K.; Soderquist, J. A., *J. Org. Chem.* **1998**, *63*, 461–470.
73. Reddy, Y. K.; Falck, J. R., *Org. Lett.* **2002**, *4*, 969–971.

74. For a review of related coupling without silver activators, see Handy, C. J., Manoso, A. S., McElroy, W. T., Seganish, W. M., DeShong, P., *Tetrahedron* **2005**, *61*, 12201–12225.
75. Hirabayashi, K.; Kawashima, J.; Nishibara, Y.; Mori, A.; Hiyama, T., *Org. Lett.* **1999**, *1*, 299–301.
76. Hirabayashi, K.; Mori, A.; Kawashima, J.; Suguro, M.; Nishibara, Y.; Hiyama, T., *J. Org. Chem.* **2000**, *65*, 5342–5349.
77. Suguro, M.; Mori, A., *Synlett* **2001**, 845–847.
78. Nokami, T.; Tomida, Y.; Kamei, T.; Itami, K.; Yoshida, J.-I., *Org. Lett.* **2006**, *8*, 729–731.
79. Napier, S.; Marcuccio, S. M.; Tye, H.; Whittaker, M., *Tetrahedron Lett.* **2008**, *49*, 6314–6315.
80. Fujiwara, Y.; Moritani, I.; Ikegami, K.; Tanaka, R.; Teranishi, S., *Bull. Chem. Soc. Jpn.* **1970**, *43*, 863–867.
81. For a review on biaryl synthesis, see Bringmann, G.; Walter, R.; Weirich, R.. *Angew. Chem. Int. Ed.* **1990**, *29*, 977–991.
82. (a) Kobayashi, K.; Sugie, A.; Masabumi, T.; Masui, K.; Mori, A., *Org. Lett.* **2005**, *7*, 5083–5085; (b) Kobayashi, K.; Ahmed, M. S. M.; Mori, A., *Tetrahedron* **2006**, *62*, 9548–9553.
83. Arai, N.; Miyaoku, T.; Teruya, S.; Mori, A., *Tetrahedron Lett.* **2008**, *49*, 1000–1003.
84. Lebrasseur, N.; Larrosa, I., *J. Am. Chem. Soc.* **2008**, *130*, 2926–2927.
85. Lu, W.; Qin, C., *J. Org. Chem.* **2008**, *73*, 7424–7427.
86. Yang, F.; Wu, Y.; Zhu, Z.; Zhang, J.; Li, Y., *Tetrahedron* **2008**, *64*, 6782–6787.
87. Chiong, H. A.; Pham, Q.-N.; Daugulis, O., *J. Am. Chem. Soc.* **2007**, *129*, 9879–9884.
88. Stuart, D. R.; Villemure, E.; Fagnou, K., *J. Am. Chem. Soc.* **2007**, *129*, 12072–12073.
89. Hull, K. L.; Sanford, M. S., *J. Am. Chem. Soc.* **2007**, *129*, 11904–11905.
90. Myers, A. G.; Tanaka, D.; Mannion, M. R., *J. Am. Chem. Soc.* **2002**, *124*, 11250–11251.
91. Tanaka, D.; Myers, A. G., *Org. Lett.* **2004**, *6*, 433–436.
92. Miki, Y.; Kuromatsu, M.; Miyatake, H.; Hamamoto, H., *Tetrahedron Lett.* **2007**, *48*, 9093–9095.
93. Becht, J.-M.; Le Drian, C., *Org. Lett.* **2008**, *10*, 3161–3164.
94. Davies, R. B.; Scheiber, D. H., *J. Am. Chem. Soc.* **1956**, *78*, 1675–1678.
95. Viterisi, A.; Orsini, A.; Weibel, J.-M.; Pale, P., *Tetrahedron Lett.* **2006**, *47*, 2779–2781.
96. Inanaga, J.; Katsuki, T.; Takimoto, S.; Ouchida, S.; Inoue, K.; Nakano, A.; Okukado, N.; Yamaguchi, M., *Chem. Lett.* **1979**, 1021–1024.
97. Albert, B. J.; Sivaramakrishnan, A.; Naka, T.; Czaicki, N. L.; Koide, K., *J. Am. Chem. Soc.* **2007**, *129*, 2648–2659.
98. (a) Naka, T.; Koide, K., *Tetrahedron Lett.* **2003**, *44*, 443–445; (b) Shahi, S. P.; Koide, K., *Angew. Chem. Int. Ed.* **2004**, *43*, 2525–2527.
99. Yao, X.; Li, C.-J., *Org. Lett.* **2005**, *7*, 4395–4398.
100. Ji, J.-X.; Au-Yeung, T. L. T.; Wu, J.; Chiu, W. Y.; Chan, A. S. C., *Adv. Synth. Catal.* **2004**, *346*, 42–44.
101. (a) Wei, C.; Li, Z.; Li, C.-J., *Org. Lett.* **2003**, *5*, 4473–4475; (b) Li, Z.; Wei, C.; Chen, L.; Varma, R. S.; Li, C.-J., *Tetrahedron Lett.* **2004**, *45*, 2443–2447.
102. Zhou, Y.; He, T.; Wang, Z., *Arkivoc* **2008**, *xiii*, 80–90.

103. Maggi, R.; Bello, A.; Oro, C.; Sartori, G.; Soldi, L., *Tetrahedron* **2008**, *64*, 1435–1439.
104. Eghbali, N.; Eddy, J.; Anastas, P., *J. Org. Chem.* **2008**, *73*, 6932–6935.
105. Bertus, P.; Dillinger, S.; Pale, P., *Org. Lett.* **2001**, *3*, 1661–1664.
106. Sweis, R. F.; Schramm, M. P.; Kozmin, S. A., *J. Am. Chem. Soc.* **2004**, *126*, 7442–7443.
107. Rhee, J. U.; Krische, M., *Org. Lett.* **2005**, *7*, 2493–2495.
108. Porcel, S.; Echavarren, A. M., *Angew. Chem. Int. Ed.* **2007**, *46*, 2672–2676.
109. Harrison, T. J.; Dake, G. R., *Org. Lett.* **2004**, *6*, 5023–5026.
110. Magnus, P.; Schultz, J., *Tetrahedron Lett.* **1986**, *27*, 655–658.
111. Bertus, P.; Pale, P., *Tetrahedron Lett.* **1996**, *37*, 2019–2022.
112. Bertus, P.; Pale, P., *Tetrahedron Lett.* **1997**, *38*, 8193–8196.
113. Bertus, P.; Pale, P., *J. Organomet. Chem.* **1998**, *567*, 173–180.
114. Glaser, C., *Ber. Dtsch. Chem. Ges.* **1869**, *2*, 422.
115. Hay, A., *J. Org. Chem.* **1960**, *25*, 1275–1276.
116. Siemsen, P.; Livingston, R. C.; Diederich, F., *Angew. Chem. Int. Ed.* **2000**, *39*, 2632–2657.
117. Halbes-Létinois, U.; Pale, P.; Berger, S., *J. Org. Chem.* **2005**, *70*, 9185–9190.
118. Halbes-Létinois, U.; Pale, P.; Berger, S., *Magn. Reson. Chem.* **2004**, *42*, 831–834.
119. Halbes, U.Ph.D. thesis, Univ. Louis Pasteur, **2003**.
120. Bertus, P.; Halbes, U.; Pale, P., *Eur. J. Org. Chem.* **2001**, 4391–4393.
121. Halbes, U.; Bertus, P.; Pale, P., *Tetrahedron Lett.* **2001**, *42*, 8641–8644.
122. Halbes, U.; Pale, P., *Tetrahedron Lett.* **2002**, *43*, 2039–2042.
123. Halbes-Létinois, U.; Vasiliev, A.; Pale, P., *Eur. J. Org. Chem.* **2005**, 2828–2834.
124. Halbes, U.; Pale, P., *J. Organomet. Chem.* **2003**, *687*, 420–424.
125. Lecerclé, D.; Mothes, C.; Taran, F., *Synth. Commun.* **2007**, *37*, 1301–1311.
126. Fiandanese, V.; Bottalico, D.; Marchese, G.; Punzi, A., *Tetrahedron Lett.* **2003**, *44*, 9087–9090.
127. Fiandanese, V.; Bottalico, D.; Marchese, G.; Punzi, A., *Tetrahedron* **2006**, *62*, 5126–5132.
128. Fiandanese, V.; Bottalico, D.; Cardellicchio, C.; Marchese, G.; Punzi, A., *Tetrahedron* **2005**, *61*, 4551–4556.
129. Mori, A.; Kawashima, J.; Shimada, T.; Suguro, M.; Hirabayashi, K.; Nishibara, Y., *Org. Lett.* **2000**, *2*, 2935–2937.
130. Ahmed, M. S. M.; Sekiguchi, A.; Shimada, T.; Kawashima, J.; Mori, A., *Bull. Chem. Soc. Jpn.* **2005**, *78*, 327–330.
131. (a) Huang, X.; Zhang, C.; Lu, X., *Synthesis* **1995**, 769–771; (b) Peng, A.-Y.; Ding, Y.-X., *Org. Lett.* **2005**, *7*, 3299–3301.
132. Yang, F.; Wu, Y.-J., *Eur. J. Org. Chem.* **2007**, 3476–3479.
133. Koseki, Y.; Omino, K.; Anzai, S.; Nagasaka, T., *Tetrahedron Lett.* **2000**, *41*, 2377–2380.
134. Jeon, H. H.; Son, J. B.; Choi, J. H.; Jeong, I. H., *Tetrahedron Lett.* **2007**, *48*, 627–631.
135. Li, P.-H.; Wang, L., *Synlett* **2006**, 2261–2265.
136. Liao, Y.; Fathi, R.; Yang, Z., *Org. Lett.* **2003**, *5*, 909–912.

11

SUPRAMOLECULAR CHEMISTRY OF SILVER

WEI-YIN SUN AND ZHENG-SHUAI BAI

Coordination Chemistry Institute, State Key Laboratory of Coordination Chemistry, School of Chemistry and Chemical Engineering, Nanjing National Laboratory of Microstructures, Nanjing University, China

JIN-QUAN YU

Department of Chemistry, The Scripps Research Institute, La Jolla, California

11.1 INTRODUCTION

Since the late 1980s, there has been tremendous progress and great achievements in supramolecular chemistry; especially after Lehn, Pedersen, and Cram won the Nobel Prize in Chemistry in 1987, which can be considered a milestone in the development of the field.[1] The results reported to date demonstrate that the simultaneous assembly of multiple components (e.g., organic ligands and metal salts) is a very powerful method for the construction of compounds with specific structures and properties.[2] It is evident that there are two important aspects of so-called metal-directed or metal-mediated assembly: (1) owing to their diverse properties (structure, flexibility, coordination ability, etc.), organic ligands can significantly influence the formation

Silver in Organic Chemistry Edited by Michael Harmata

and structure of the resulting complex; and (2) different metal centers can have different coordination numbers, geometries, and intrinsic properties, so the choice of metal salt is critical. Accordingly, metal complexes with a variety of structures and properties can be achieved through the assembly of rationally designed organic ligands with particular metal salts. As a part of this larger theme, this chapter focuses on the supramolecular chemistry of silver(I).

In the area of silver(I) coordination chemistry, it is known that several different coordination numbers and geometries exist.[3] In addition to well-known two- (linear), three- (triangular or T-shaped), and four- (tetrahedral or square planar) coordinate silver(I) complexes, there are five-, six-, or even seven- and eight-coordinate structures (Fig. 11.1).[4–14] Furthermore, in complexes containing multiple metal centers, there can be argentophilic silver(I)–silver(I) interactions with strength on the order of hydrogen bonds when the metal– metal distance is shorter than the Ag–Ag van der Waals contact distance of 3.44 Å.[15,16] In combining the variety in coordination environment with the possibility for strong metal–metal interactions, a diverse range of silver(I) complexes with specific topologies and structures can be obtained.[17]

This chapter describes the structures and properties of supramolecular silver(I) complexes with specific topologies such as cages, tubes, catenanes and polycatenanes, rotaxanes and polyrotaxanes, and multidimensional frameworks, as synthesized by reactions of various silver(I) salts with predesigned organic ligands.

11.2 CAGE-LIKE COMPLEXES

In contrast to "planar" molecules, such as M_2L_2 macrocycles, M_3L_3 triangles, M_4L_4 squares (Fig. 11.2), isolated individual cages (also referred as *cones*, *capsules*, *boxes*,

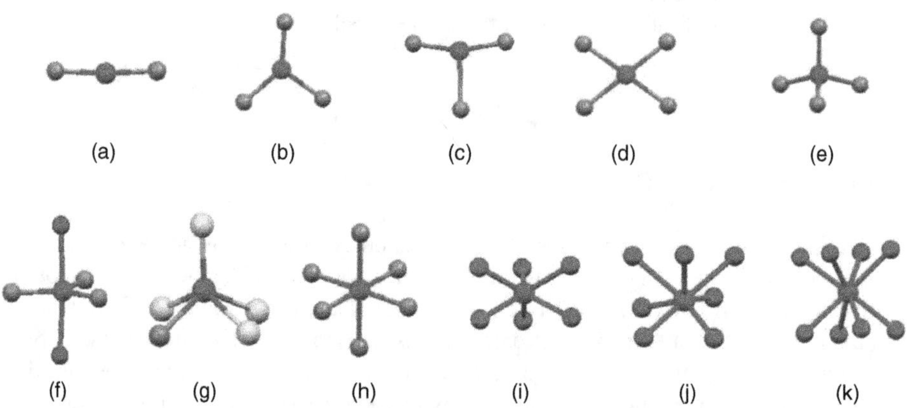

Figure 11.1. Schematic representation of typical silver(I) (central atom) coordination environments with different coordination numbers and geometries: (a) linear two-coordinate; (b) triangular three-coordinate; (c) T-shaped three-coordinate; (d) square planar four-coordinate; (e) tetrahedral four-coordinate; (f) trigonal bipyramidal five-coordinate; (g) tetragonal pyramidal five-coordinate; (h) octahedral six-coordinate; (i) trigonal prism six-coordinate; (j) seven-coordinate; (k) tetragonal prism eight-coordinate.

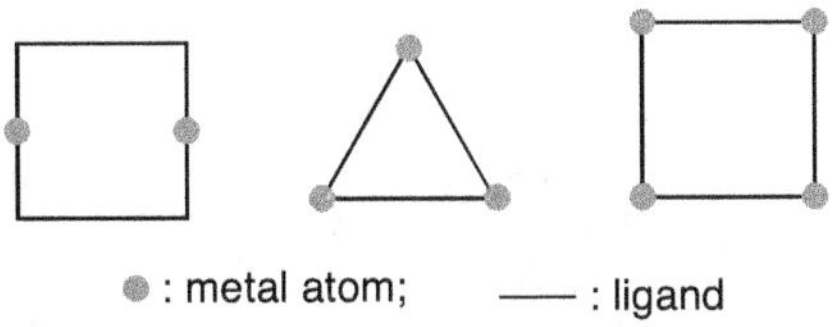

Figure 11.2. Schematic representation of "planar" M_2L_2 macrocycles, M_3L_3 triangles and M_4L_4 squares, formed through metal–ligand coordination.

prisms, *cylinders*, or *containers* depending on shape) have discrete three-dimensional (3D) hollow structures and normally have inner cavities with defined shapes and sizes. Accordingly, these cavities can perform a variety of functions, including hosting ions or uncharged molecules, stabilizing unstable species and regulating or promoting specific reactions, such as the [2 + 2] intermolecular photodimerization of olefins.[18] With this breadth of utility, such cage-like compounds have attracted great interest among chemists.

Studies have shown that these metal-containing cage-like compounds can be constructed efficiently through the assembly of structurally defined building blocks.[19] However, it should be emphasized that cage-like complexes are isolated molecular compounds, containing metal–ligand coordination interactions that are enclosed within a defined space. This space contains a finite number of metal centers and ligands, in contrast to the repeat units of coordination polymers, which form networks that extended infinitely. To achieve this structural motif in the case of a metal atom that has more than two coordination positions, partial coordination of the metal by chelating ligands such as ethylenediamine (en), 2,2′-bipyridine (bipy), or diethylenetriamine may be required to prevent the formation of infinite polymeric compounds. For example, multicomponent assembly of *cis*-protected, square planar metal complexes of the type $[M(en)](NO_3)_2$ or $[M(bipy)](NO_3)_2$ [M = Pd(II) or Pt(II)] with multidentate organic ligands leads to the formation of cage-like complexes with nanometer-scale inner cavities.[20] In contrast to palladium(II) and platinum(II), as mentioned above, silver(I) can have a linear geometry with two coordination sites, and as a result usually no protection via additional chelating ligands is needed for the construction of silver(I) cage-like compounds.

The organic ligands are also important in the construction of silver(I) cage-like complexes. The most popular ligands are compounds with nitrogen donor groups (e. g., pyridyl, imidazolyl, pyrazolyl). Such ligands can be divided into two types according to their relative conformational variability.[21] One type is the "flexible" ligand, which can adopt different conformations and coordination modes according to the geometric requirement of the given metal ion. Such flexible ligands can form metal complexes with a variety of structures, making it difficult to predict the structure of the resulting complexes. The other class consists of "rigid" ligands, which have a lower degree of conformational variation when they coordinate with metal ions. Both flexible and rigid ligands have been used to build silver(I) cages.

Flexible imidazole-containing tripodal ligands [e.g., 1,3,5-tris(imidazol-1-ylmethyl)benzene and 1,3,5-tris(imidazol-1-ylmethyl)-2,4,6-trimethylbenzene (**1**)]

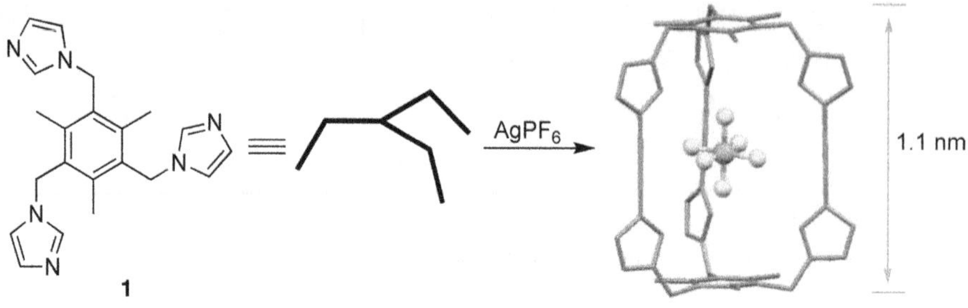

Figure 11.3. An assembly reaction and the resulting crystal structure of the $[Ag_3(\mathbf{1})_2]^{3+}$ cylindrical cage with the inclusion of PF_6^-. Hydrogen atoms and anions outside the cage have been omitted for clarity.

(Fig. 11.3) have been shown to be versatile components in the construction of metal complexes with specific topologies and structures when reacted with different metal salts.[21–24] Reactions of ligand **1** with various silver(I) salts have been systematically investigated, and cage-like complexes have been successfully isolated.[12,25–28] For example, when ligand **1** was reacted with silver(I) hexafluorophosphate [$AgPF_6$] in a solution of methanol and acetonitrile at room temperature, a cage complex $[Ag_3(\mathbf{1})_2]$-$(PF_6)_3$ resembling a cylinder was obtained containing one hexafluorophosphate anion in its inner cavity (Fig. 11.3).[25] Analysis of the resulting crystal structure revealed that the ligand **1** adopts a *cis,cis,cis* conformation, with two such ligands orienting themselves in a face-to-face fashion. The ligands, in turn, are linked together by 3 two-coordinate Ag(I) atoms through the Ag–N coordination. The height of the cage, namely, the distance between the two central benzene ring planes, is approximately 1.1 nm and the Ag(I)–Ag(I) separation is about 7 Å, implying that the cage has an inner cavity of sufficient size for hexafluorophosphate anion inclusion.

It is worth noting that this anion inclusion by the $[Ag_3(\mathbf{1})_2]^{3+}$ cylindrical cage is also an anion recognition process. As illustrated in Figure 11.4, the same $[Ag_3(\mathbf{1})_2]^{3+}$

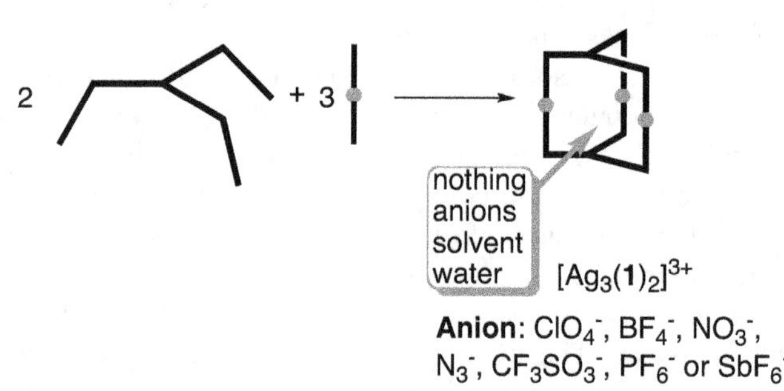

Figure 11.4. Schematic drawing for the assembly reactions of flexible ligand **1** and silver(I) salts with different counteranions to generate $[Ag_3(\mathbf{1})_2]^{3+}$ cylindrical cages with different species included in the inner cavity of the cage.

cage-like framework was obtained by reactions of ligand **1** and silver(I) salts with different counteranions; however, the specific species included in the inner cavity of the cage depends on the corresponding anions.[12,25–28] The cage can be empty (i.e., with nothing included), or it can contain anions, neutral solvent molecules, or water molecules as exhibited in Figure 11.5. Only anions with suitable size and shape can be included inside the cage. In addition to simple anions such as ClO_4^-, BF_4^-, PF_6^-, and SbF_6^-, complex anions such as $[Ag(NO_3)_3]^{2-}$ can also be included (Fig. 11.5c).

Rigid ligands have also been used to build silver(I) cage-like complexes.[29,30] The reaction between the disk-shaped rigid tripodal ligand 1,3,5-tribenzimidazolyl-2,4,6-trimethylbenzene (**2**) (Fig. 11.6) and silver(I) methanesulfonate (AgO_3SCH_3) or silver(I) trifluoromethanesulfonate (AgOTf) was studied by ^1H and ^{19}F NMR and electrospray ionization–time-of-flight (ESI-TOF) mass spectroscopy.[29] The results revealed that two types of cage-like complexes were formed depending on the metal-to-ligand ratio: $[Ag_3(\mathbf{2})_2]^{3+}$ with a triangular shape and $[Ag_4(\mathbf{2})_4]^{4+}$ with a tetrahedral shape (Fig. 11.6). The cages have different cavities of unique shape and size. Only the cavity of the $[Ag_4(\mathbf{2})_4]^{4+}$ cage has the ability to include a guest methanesulfonate or triflate anion, as confirmed by NMR spectroscopy as well as analysis of the X-ray crystal structure (Fig. 11.7). Furthermore, there exists a dynamic

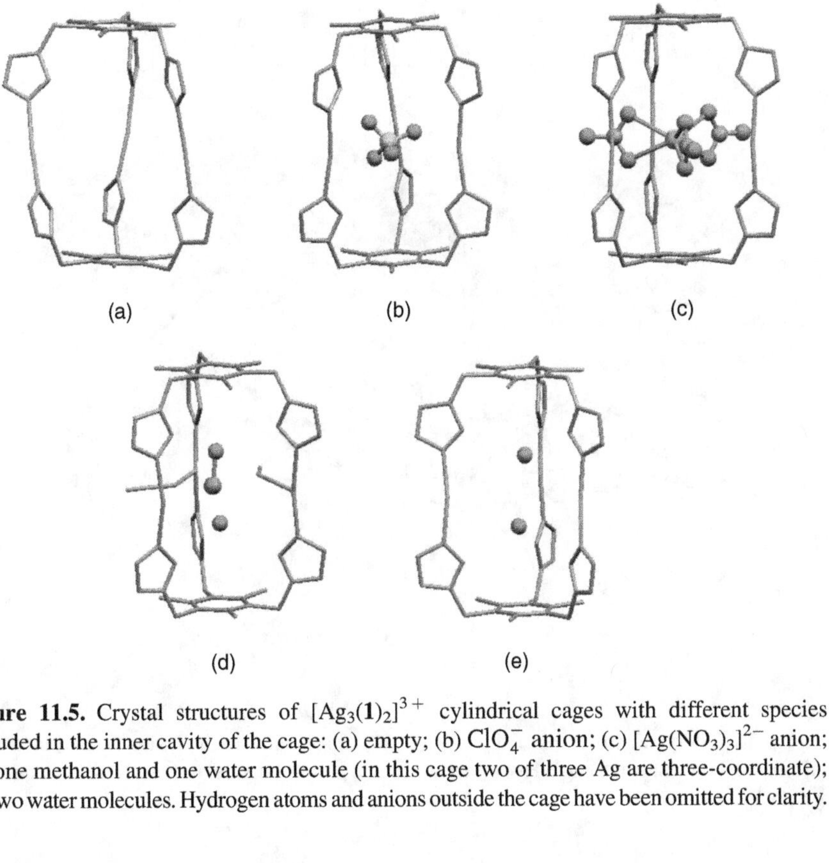

Figure 11.5. Crystal structures of $[Ag_3(\mathbf{1})_2]^{3+}$ cylindrical cages with different species included in the inner cavity of the cage: (a) empty; (b) ClO_4^- anion; (c) $[Ag(NO_3)_3]^{2-}$ anion; (d) one methanol and one water molecule (in this cage two of three Ag are three-coordinate); (e) two water molecules. Hydrogen atoms and anions outside the cage have been omitted for clarity.

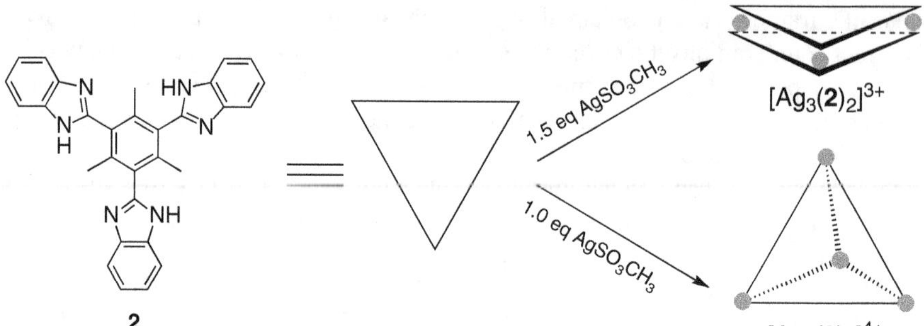

Figure 11.6. Schematic representation of rigid tripodal ligand **2** and the corresponding assembly reaction with silver(I) methanesulfonate to give $[Ag_3(\mathbf{2})_2]^{3+}$ and $[Ag_4(\mathbf{2})_4]^{4+}$ cages.

equilibrium between the $[Ag_3(\mathbf{2})_2]^{3+}$ and $[Ag_4(\mathbf{2})_4]^{4+}$ cages, which can be controlled by the inclusion and release of the guest anion.

When the rigid tridentate ligand 1,3,5-tri-3-pyridyl-2,4,6-tri-*p*-tolyl-benzene (**3**) was reacted with silver(I) triflate or silver(I) hexafluorophosphate, cage-like

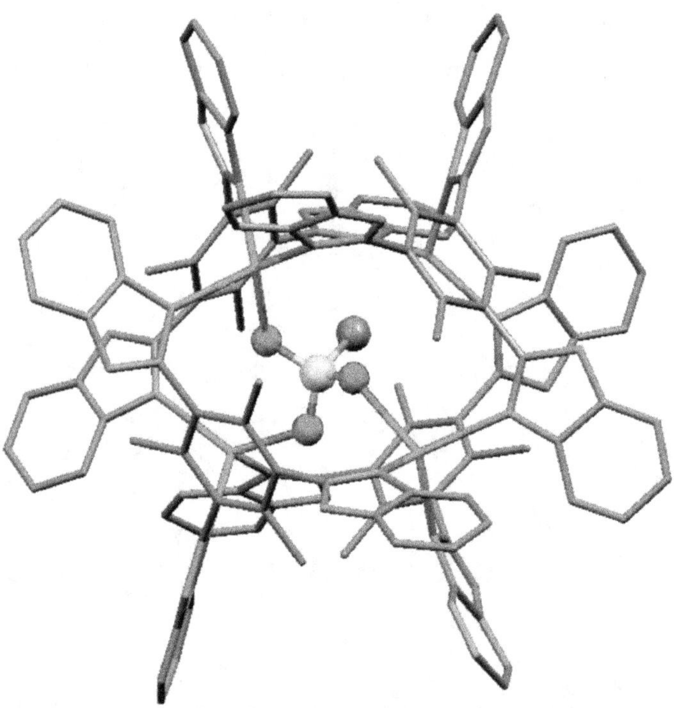

Figure 11.7. Crystal structure of the $[Ag_4(\mathbf{2})_4]^{4+}$ cage with the inclusion of one methanesulfonate anion inside the cage. Hydrogen atoms and anions outside the cage have been omitted for clarity.

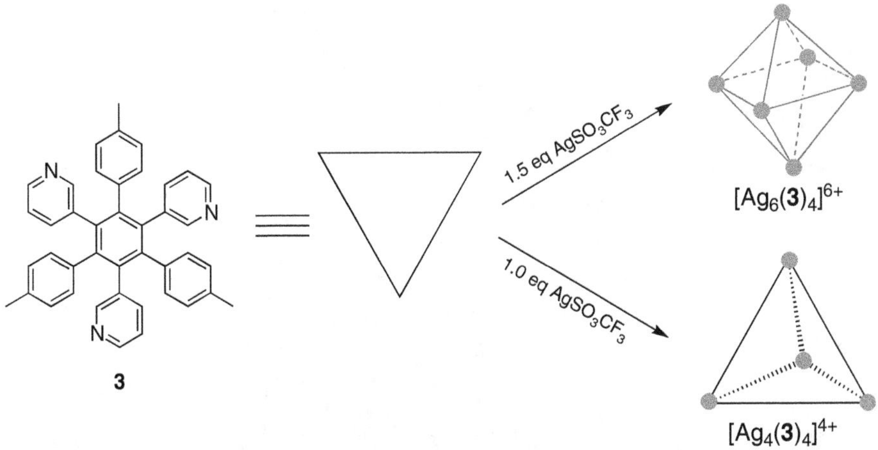

Figure 11.8. Schematic representation of rigid tripodal ligand **3** and the assembly reaction with silver(I) triflate to give $[Ag_6(\mathbf{3})_4]^{6+}$ and $[Ag_4(\mathbf{3})_4]^{4+}$ cages.

complexes $[Ag_6(\mathbf{3})_4]^{6+}$ and $[Ag_4(\mathbf{3})_4]^{4+}$ with octahedral and tetrahedral shapes, respectively, were formed as illustrated in Figure 11.8.[30]

The two cages have different shapes and volumes and possess unique internal cavities and thus have different guest inclusion properties.[30] For example, the $[Ag_4(\mathbf{3})_4]^{4+}$ cage has a high affinity toward a guest adamantane molecule, as evidenced by its X-ray crystal structure (Fig. 11.9), while the $[Ag_6(\mathbf{3})_4]^{6+}$ cage has no interaction with adamantane. Interestingly, interconversion between these two cages can be achieved by changing the metal-to-ligand ratio, thereby effecting the guest molecule's inclusion or release. For instance, the encapsulated guest molecule of adamantane in the $[Ag_4(\mathbf{3})_4]^{4+}$ cage was released on the addition of 2 equiv of silver salt, promoting an accompanying structural conversion of $[Ag_4(\mathbf{3})_4]^{4+}$ to $[Ag_6(\mathbf{3})_4]^{6+}$. Similarly, the $[Ag_4(\mathbf{3})_4]^{4+}$ cage can be recovered from $[Ag_6(\mathbf{3})_4]^{6+}$ by addition of 2 equiv of [2,2,2]-cryptand to trap silver(I). The interconversion between $[Ag_4(\mathbf{3})_4]^{4+}$ and $[Ag_6(\mathbf{3})_4]^{6+}$ is quantitative and fast.[30]

In a 2008 study of silver cages using 1,3,5-tris(2-oxazolinyl)benzene (**4**) as the connecting ligand, assembly reactions of **4** with various silver(I) salts were carried out and silver(I) complexes with different structures were obtained.[31] When ligand **4** was reacted with silver(I) perchlorate in a solution of acetonitrile and chloroform at ambient temperature, a cage-like complex $[Ag_3(\mathbf{4})_2(CH_3CN)_3](ClO_4)_3$ was formed, as shown in Figure 11.10. This structure constitutes a small cage, as indicated by the centroid-to-centroid distance of 3.64 Å between two central benzene ring planes and the Ag–Ag distance of 7.04 Å. Therefore, there is no true cavity inside the cage, and no inclusion properties exist. This particular complex crystallizes in an acentric space group *R*3c, and is second-harmonic generation (SHG) active, with a response of about 0.5 times of that of urea.[31]

Several other silver(I) cage-like complexes have been reported.[32–35] 3,8,13-Tris (4-pyridylmethylamino)-2,7,12-trimethoxy-10,15-dihydro-5*H*-tribenzo[*a*,*d*,*g*]

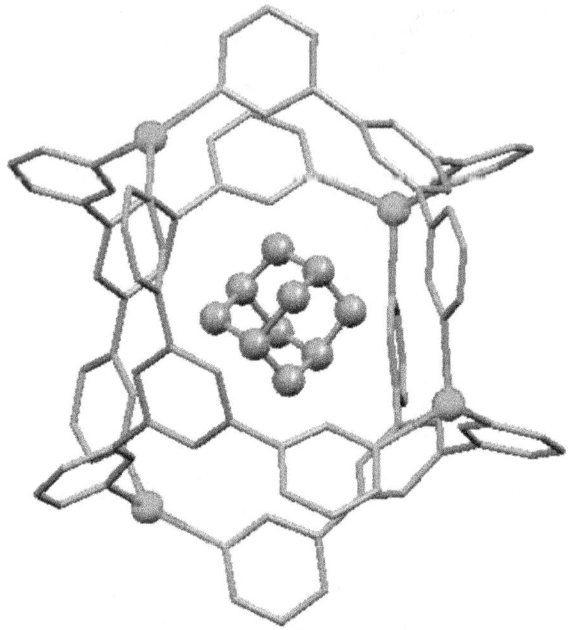

Figure 11.9. Crystal structure of the $[Ag_4(\mathbf{3})_4]^{4+}$ cage with the inclusion of one guest adamantane molecule inside the cage. 2,4,6-Tri-*p*-tolyl groups, hydrogen atoms, and anions have been omitted for clarity.

cyclononene (**5**, Fig. 11.11), a derivative of cyclotriveratrylene (CTV), was reacted with silver(I) tetrafluoroborate ($AgBF_4$) in acetonitrile solution leading to the formation of a stellated tetrahedral cage-like complex $[Ag_4(\mathbf{5})_4(CH_3CN)_4](BF_4)_4{\cdot}7CH_3CN{\cdot}2.8H_2O$ with encapsulation of acetonitrile molecules in the inner cavity of the cage.[32] In addition to the *N*-donor ligands **1–5** mentioned above, assembly reactions of thiacalix[4]arene-*p*-sulfonate (TCAS) with silver(I) and terbium(III) salts were studied by electrospray ionization mass spectrometry under

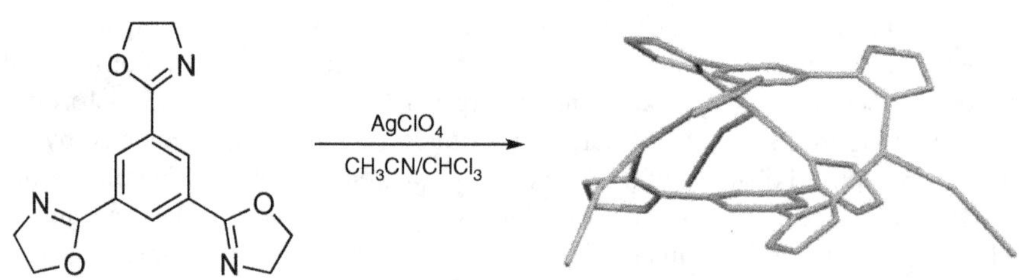

4

$[Ag_3(\mathbf{4})_2(CH_3CN)_3]^{3+}$

Figure 11.10. The assembly reaction and resulting crystal structure of $[Ag_3(\mathbf{4})_2(CH_3CN)_3]^{3+}$. Hydrogen atoms and perchlorate anions have been omitted for clarity.

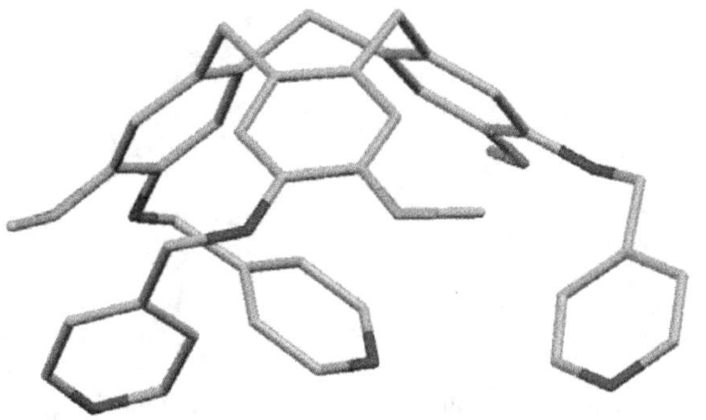

5

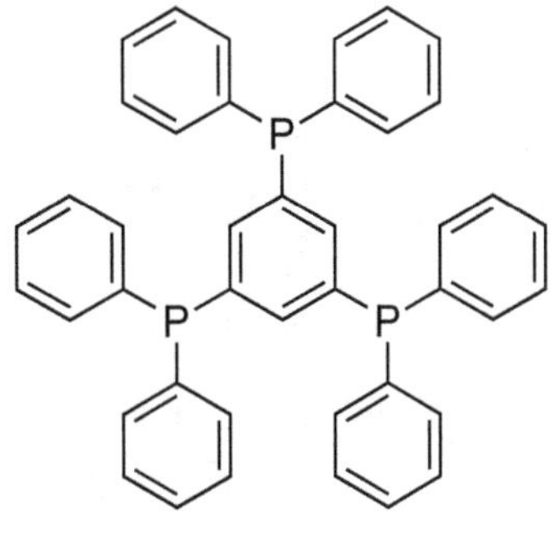

6

Figure 11.11. Ligands **5** and **6** used in the construction of silver(I) cage-like complexes.

varying pH, resulting in cage-like complexes with different compositions and photoluminescent properties.[33] Triphosphine **6** (Fig. 11.11) has also been used to react with silver(I) salts, and the resulting cage-like complexes have been reported.[34]

11.3 TUBE-LIKE COMPOUNDS

Carbon-based nanotubes (CNTs) have attracted a great deal of attention in the fields of chemistry, physics, and materials science and have been extensively studied since their initial discovery in 1991.[36] Much of the excitement in this area of research stems from their unique structures, fundamental electronic and physical properties, and potential applications.[37] Inspired by these carbon nanotube studies, coordination chemists have undertaken significant efforts toward the construction of tube-like complexes through metal–ligand coordination.[38–41] Examples of silver(I) tubular complexes obtained via assembly reactions of predesigned organic ligands with silver(I) salts are described here.[42–44]

A pyridyl- and sulfanyl-containing flexible ligand 2,4,6-tris[(4-pyridyl)methyl-sulfanyl]-1,3,5-triazine (**7**, Fig. 11.12) with multiple coordination sites has been reported,[45] and a nanometer-scale, tube-like silver(I) complex, $[Ag_7(\mathbf{7})_4](ClO_4)_2(NO_3)_5(DMF)_2$, was isolated following an assembly reaction of ligand **7** with silver nitrate ($AgNO_3$) and silver perchlorate ($AgClO_4$) salts in *N,N*-dimethylformamide (DMF) solution.[42] As revealed by X-ray crystallographic analysis, shown in Figure 11.13a, two molecules of **7** with a *cis,cis,cis* conformation are linked together by 3 two-coordinate silver(I) atoms to form a $[Ag_3(\mathbf{7})_2]^{3+}$ unit. In turn, two of these units are further linked by two Ag(I) atoms through Ag–N (the nitrogen of triazine, not of pyridine in **7**) and Ag–S coordination to generate one $[Ag_7(\mathbf{7})_4]^{7+}$ tube with dimensions of $1.34 \times 0.96 \times 0.89$ nm (Figs. 11.13b and 11.13c). There are two DMF

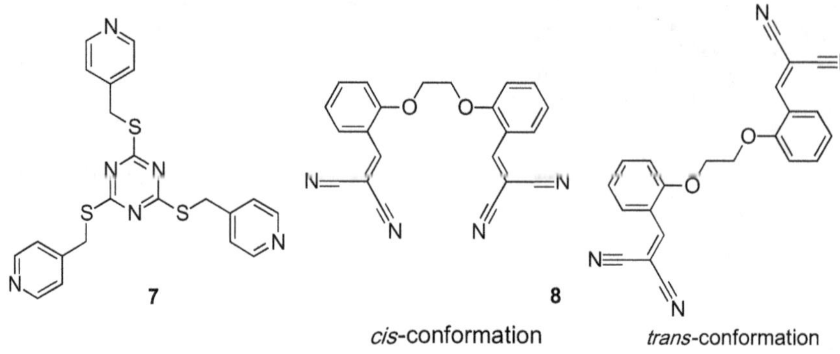

Figure 11.12. Multidentate organic ligands **7** and **8** (with *cis* and *trans* conformations) used in the construction of silver(I) tube-like complexes.

molecules and two perchlorate anions in the large cavity of the tube (Fig. 11.13b). Furthermore, an infinite one-dimensional (1D) chain was formed by $[Ag_7(\mathbf{7})_4]^{7+}$ tubes sharing the Ag(I) atoms, as exhibited in Figure 11.13d. The Ag(I) is four-coordinate with a square planar geometry and an N_2S_2 donor set. This is a unique 1D chain structure with numerous nanosized tubes linked by silver(I) atoms.[42]

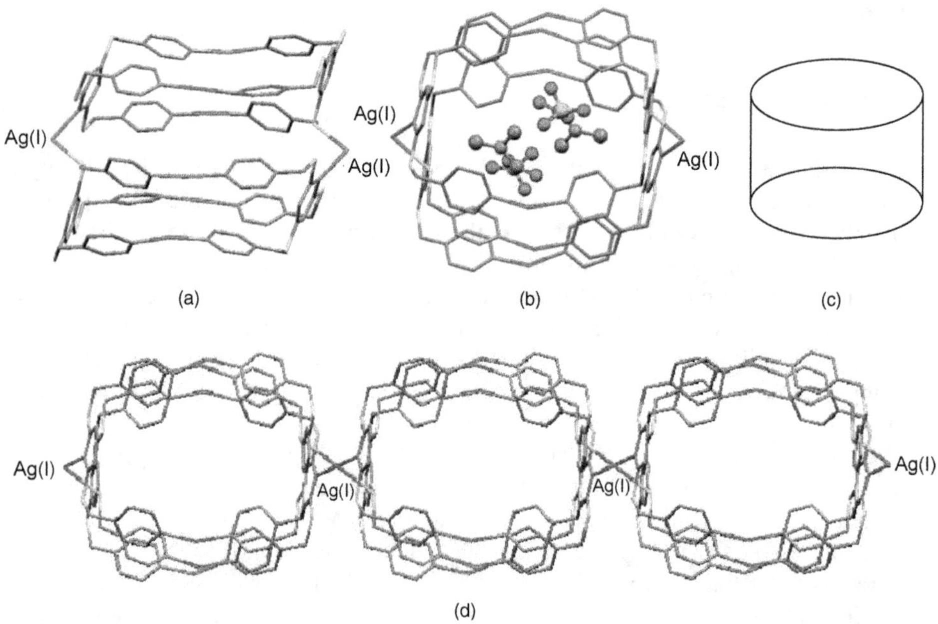

Figure 11.13. (a) Crystal structure of two $[Ag_3(\mathbf{7})_2]^{3+}$ units linked by two Ag(I) atoms through Ag–N and Ag–S coordination to give a $[Ag_7(\mathbf{7})_4]^{7+}$ unit; (b) $[Ag_7(\mathbf{7})_4]^{7+}$ tube with inclusion of two DMF molecules and two perchlorate anions; (c) schematic drawing of the tube; (d) $[Ag_7(\mathbf{7})_4]^{7+}$ tubes linked by four-coordinate silver(I) atoms to form an infinite 1D chain. The hydrogen atoms, solvent molecules, and anions have been omitted for clarity.

Another silver(I) tube-like complex, $[Ag(\mathbf{8})]BF_4{\cdot}0.5(C_6H_6)$, was obtained through the combination of ethyleneglycol ether–bridging tetradentate ligand **8** (Fig. 11.12) with silver(I) tetrafluoroborate.[43] The flexible ligand **8** can have both *cis* and *trans* conformations (Fig. 11.12) through the rotation of the central C–C single bond. Interestingly, these different conformations can be fine-tuned by adjusting the reaction temperature.[43]

When ligand **8** was reacted with $AgBF_4$ in a solution of dichloromethane and benzene at 30°C, complex $[Ag(\mathbf{8})]BF_4{\cdot}0.5C_6H_6$ with a 1D tube structure was obtained in which the ligand **8** adopted a *cis* conformation (Figs. 11.14a and 11.14b). However, when the same reaction was carried out at 0°C, the ligand **8** was found to adopt a *trans* conformation, and complex $[Ag_2(\mathbf{8})(H_2O)](BF_4)_2$ with a two-dimensional (2D) porous structure was formed (Fig. 11.14c).[43] It is immediately obvious that the 1D tube formed from ligand **8** (Figs. 11.14a and 11.14b) is different from the complex obtained from ligand **7** (Fig. 11.13d), as the $[Ag_7(\mathbf{7})_4]^{7+}$ tube has a defined length while the $[Ag(\mathbf{8})]^+$ structure is an infinite 1D chain tube with internal dimensions of approximately 1.2×0.82 nm.

11.4 POLYCATENANES WITH SILVER(I)

Catenanes and polycatenanes are an interesting class of compounds with interlocking units, as schematically represented in Figure 11.15. These structures have attracted a great deal of attention from chemists due to their topological novelty, which offers the potential for possible application in molecular switches, machines, and devices.[46] A number of catenanes and polycatenanes have been prepared by assembly reactions of metal salts with rationally predesigned organic ligands;[47–53] and two kinds of silver(I) polycatenanes with bis(2-methylimidazol-1-yl)methane (**9**) and 1,5-(3-oxapentane) diyl-bis(isonicotinate) (**10**) (Fig. 11.16) have been reported.[51–53]

When ligand **9** was reacted with $AgNO_3$ and $AgBF_4$, complexes with the formulas $[Ag_4(\mathbf{9})_4](NO_3)_4{\cdot}3H_2O$ and $[Ag_4(\mathbf{9})_4](BF_4)_4{\cdot}H_2O$ were obtained, respectively. The results of structural analysis showed that they have the same cationic framework structure. As exhibited in Figure 11.17a, it is clear that four two-coordinate silver(I) atoms link four molecules of **9** through Ag–N coordination to form M_4L_4 macrocycles, which interlock with each other to generate an infinite 1D polycatenane.[51,52] This is an example of transformation of individual macrocycles to an infinite 1D chain through catenation. It is worth noting that the formation of such interlocking polycatenanes can be influenced by subtle variation of the ligand and counteranion. For example, a complex with a simple 1D zigzag chain structure was isolated by reaction of the $AgNO_3$ with bis(imidazol-1-yl)methane.[51] On the other hand, a discrete complex with a macrometallocyclic structure of $[Ag_6(\mathbf{9})_6](SO_3CF_3)_6$ was produced by reaction of ligand **9** with AgOTf as shown in Figure 11.17b.[52]

Another example is the catenation leading to the formation of a 2D network from 1D chains (Fig. 11.18).[53] One molecule of ligand **10** links to two silver(I) atoms, and one metal atom, in turn, coordinates with four molecules of **10** to give an infinite 1D

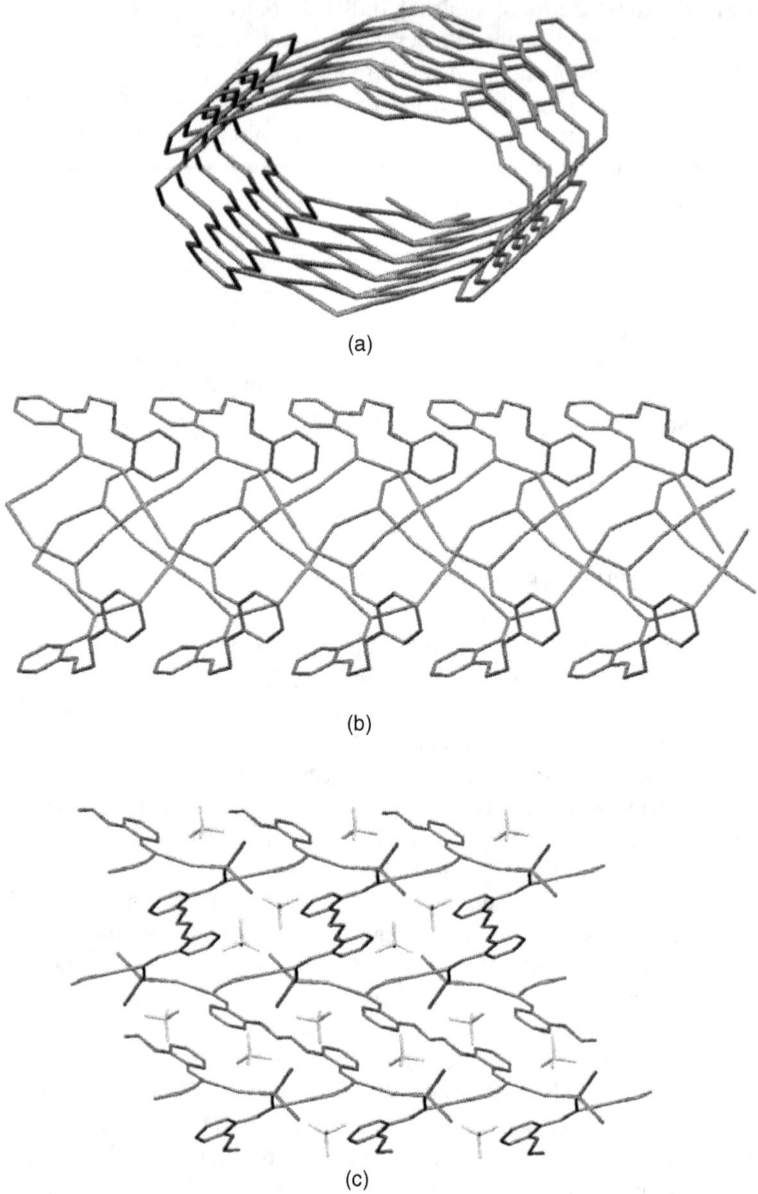
(a)

(b)

(c)

Figure 11.14. Top (a) and side (b) views of an infinite 1D [Ag(**8**)]$BF_4 \cdot 0.5C_6H_6$ tube obtained at 30°C (hydrogen atoms, anions, and solvent molecules have been omitted for clarity); (c) 2D network structure of [Ag_2(**8**)(H_2O)]$(BF_4)_2$ isolated at 0°C (hydrogen atoms have been omitted for clarity).

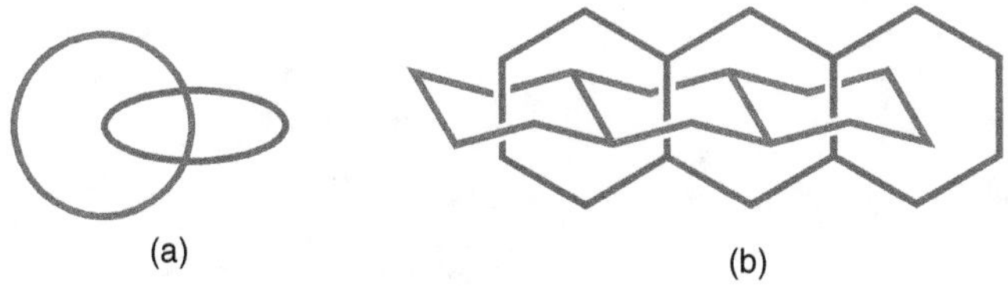

Figure 11.15. Schematic drawing of [2]-catenane (a) and infinite 1D [n]-polycatenane (b).

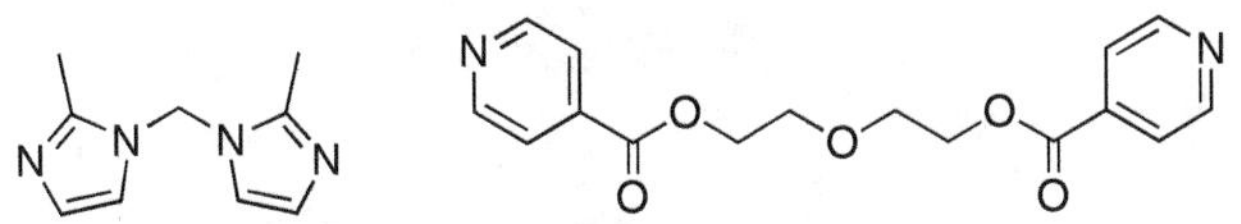

9 10

Figure 11.16. Structures of organic ligands **9** and **10** used for construction of polycatenaned silver(I) complexes.

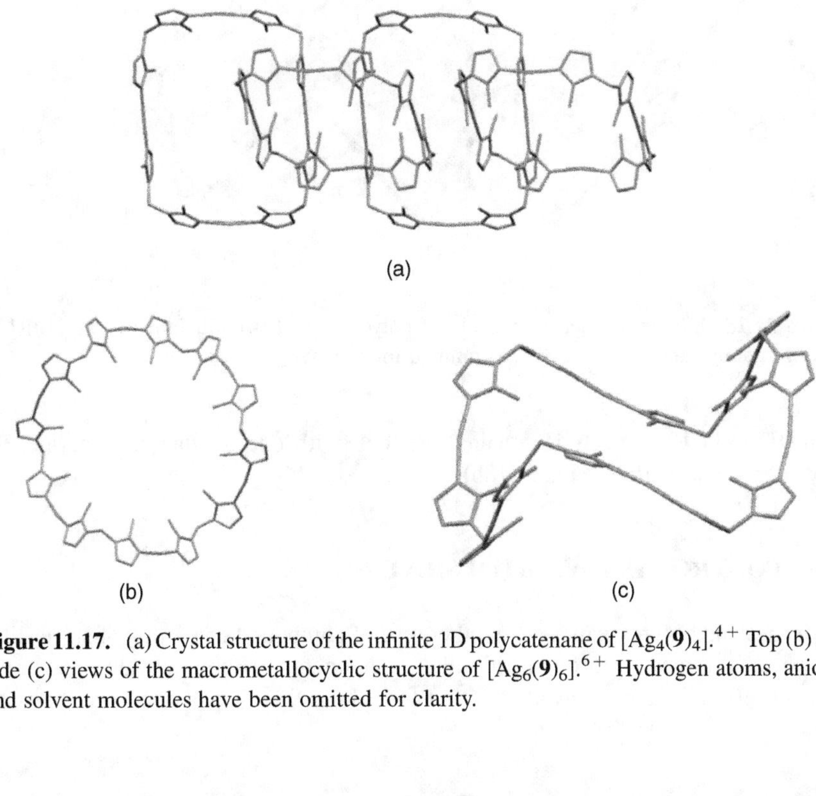

Figure 11.17. (a) Crystal structure of the infinite 1D polycatenane of $[Ag_4(\mathbf{9})_4]^{4+}$. Top (b) and side (c) views of the macrometallocyclic structure of $[Ag_6(\mathbf{9})_6]^{6+}$. Hydrogen atoms, anions, and solvent molecules have been omitted for clarity.

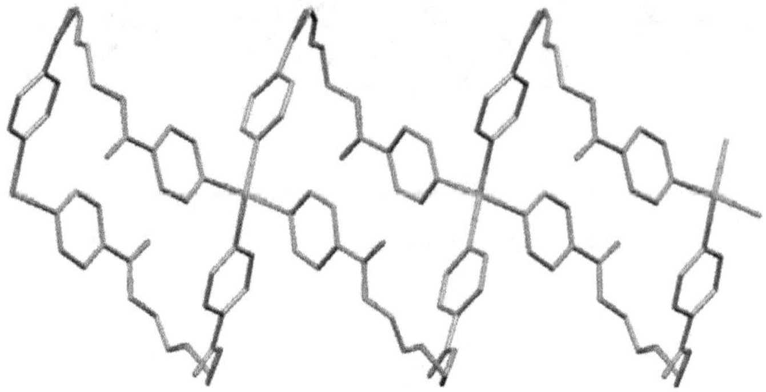

(a)

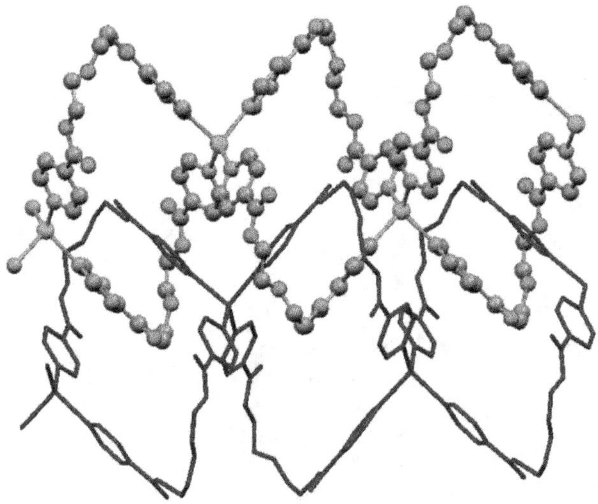

(b)

Figure 11.18. Infinite 1D chain (a) and 2D polycatenaned (b) structures of $[Ag(\mathbf{10})_2]^+$. Hydrogen atoms and anions have been omitted for clarity.

chain (Fig. 11.18a). Such 1D chains interlock with one another to generate 2D polycatenane structures (Fig. 11.18b).

11.5 POLYROTAXANES WITH SILVER(I)

Rotaxanes and polyrotaxanes are another type of compound consisting of string-like components threaded through ring-like components without covalent bonding (Fig. 11.19) and have received attention for a long time.[46] Reported ring parts of

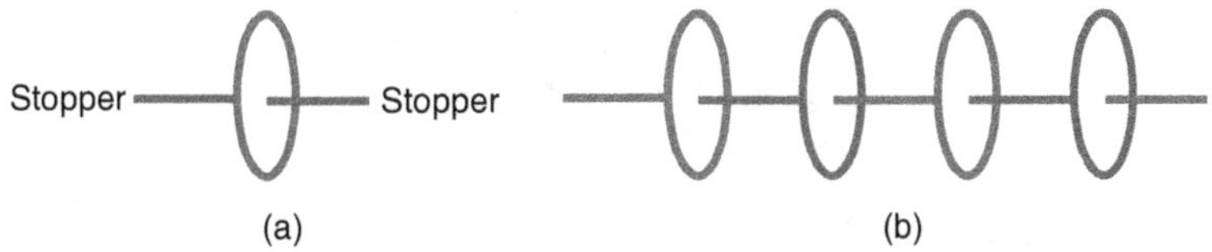

Figure 11.19. Schematic drawing of a rotaxane (a) and an infinite 1D polyrotaxane (b).

rotaxanes and polyrotaxanes include crown ethers, cyclophanes, cyclodextrins, calixarenes, and cucurbiturils as well as the metal complexes themselves.[46,54–57]

Cucurbit[6]uril (CB[6]), with its six glycoluril units linked by 12 methylene bridges (Fig. 11.20a), has been found to be an efficient building block in the preparation of rotaxanes and polyrotaxanes.[54] Organic compounds with diamine units are well suited to serve as the string part of rotaxanes, since the protonated diamine moiety can incorporate CB[6] via N–H–O hydrogen-bonding interactions. For example, *N,N′*-bis(3-pyridylmethyl)-1,6-hexanediamine (**11**) and *N,N′*-bis(3-pyridylmethyl)-1,4-benzenedimethylamine (**12**) (Fig. 11.20b) were reacted with CB[6] followed by $AgNO_3$ under acidic conditions to yield polyrotaxanes; the effects of the flexibility and length of the ligands on the formation and structure of the resulting polyrotaxanes have also been explored.[56] The protonated ligand (H_2**12**) was included in the inside cavity of CB[6] to give pseudorotaxanes, which are further connected by silver(I) through Ag–N coordination, and an infinite 1D polyrotaxane was ultimately formed (Fig. 11.21).[56]

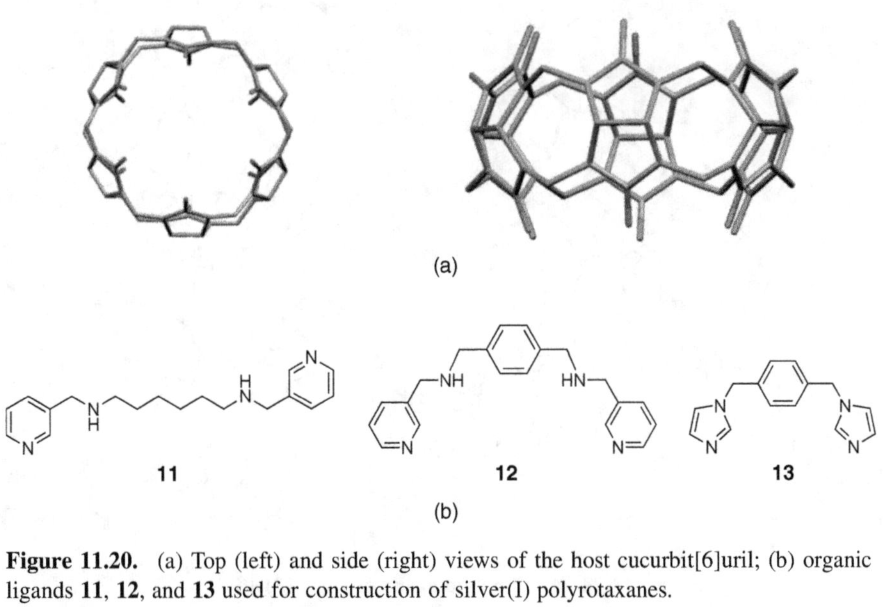

Figure 11.20. (a) Top (left) and side (right) views of the host cucurbit[6]uril; (b) organic ligands **11**, **12**, and **13** used for construction of silver(I) polyrotaxanes.

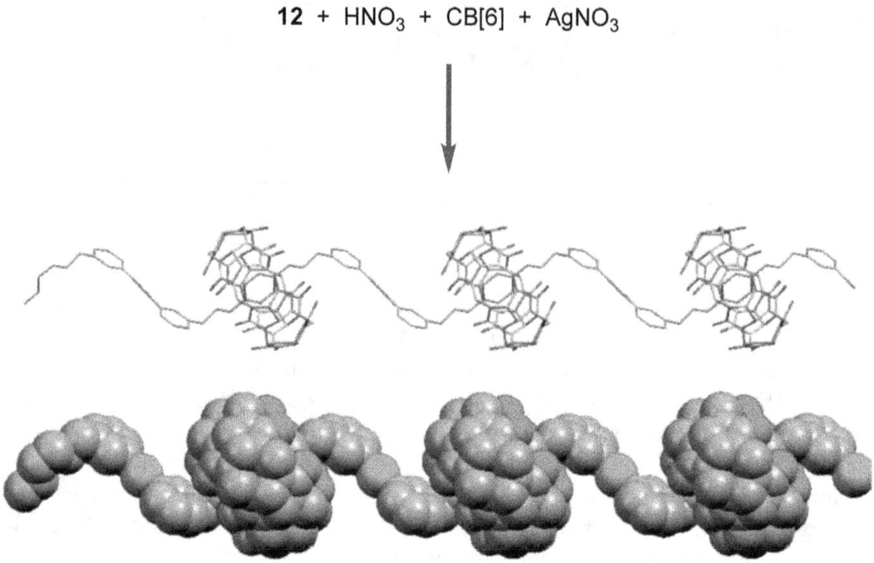

Figure 11.21. Formation of the infinite 1D polyrotaxane of $[Ag(H_2\mathbf{12})(CB[6])](NO_3)_3 \cdot 10H_2O$ and the corresponding crystal structure.

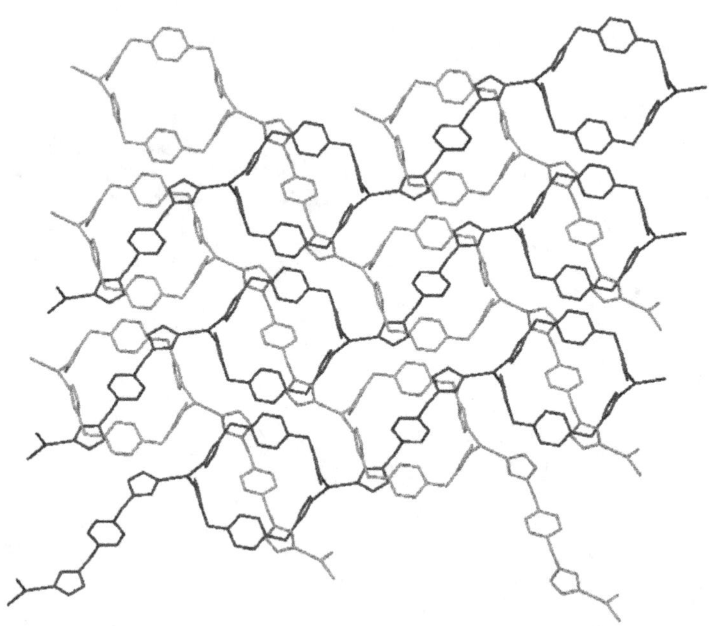

Figure 11.22. A 2D polyrotaxane of $[Ag_2(\mathbf{13})_3](NO_3)_2$.

Polyrotaxanes can also be formed through interlocking coordination polymers, a typical example of which has been reported by Robson and coworkers.[57] 1,4-Bis (imidazol-1-yl-methyl)benzene (**13**) (Fig. 11.20b) coordinates with silver(I) to give 1D polymeric chains that are further knitted together to provide a 2D polyrotaxane (Fig. 11.22).

11.6 SILVER(I) COORDINATION POLYMERS WITH SPECIFIC TOPOLOGY

In addition to the infinite tubes, polycatenanes, and polyrotaxanes, all of which are polymeric complexes, a great number of silver(I) coordination polymers with 1D, 2D and 3D framework structures have been reported more recently. Selected examples and their specific topologies of some of these are listed in Table 11.1. Schematic structures of the corresponding organic ligands are shown in Figure 11.23.[9,58–75]

An infinite 1D complex $\{[Ag(\mathbf{14})(\mathbf{15})]\}_n$ with a single-stranded helical chain structure was obtained through the reaction of 6-amino-1-naphthalenesulfonic acid (H-**14**) and 2-methylpyrazine (**15**) with Ag_2CO_3.[58] Furthermore, when tetra- and hexaethyleneglycol-spaced ligands, 1,11-(3,6,9-trioxaundecane)diyl-bis(nicotinate) (**16**) and 1,17-(3,6,9,12,15-pentaoxaheptadecane)diyl-bis(nicotinate) (**17**), were reacted with silver(I) salts, double-stranded helical complexes (as exhibited in Fig. 11.24) were formed.[59]

Various 2D networks with common (4,4) and (6,3) topologies have been reported. The silver(I) complex $\{[Ag(\mathbf{18})(NO_3)]\}_n$ was formed from 1,3-bis(2-pyridylamino-methyl)benzene (**18**) and $AgNO_3$, and similarly, $\{[Ag(\mathbf{19})_{2/3}]_3[Ag(\mathbf{19})]_2(CF_3SO_3)_5\}_n$ was isolated from the reaction of 1,3,5-tris(4-((2-ethyl-benzimidazol-1-yl)methyl) phenyl)benzene (**19**) with AgO_3SCF_3. These two complexes have (4,4) and (6,3) topologies, respectively.[60,61] When the imidazole-containing tripodal ligand 1,3-bis-(1-imidazolyl)-5-(imidazol-1-ylmethyl)benzene (**20**) was reacted with $[Ag(NH_3)_2]$ ClO_4 in an aqueous methanol solution, the twofold interpenetrated complex $\{[Ag(\mathbf{20})]ClO_4\}_n$ with 4.8^2 topology was generated (Fig. 11.25).[62]

To date, a number of different topologies have been reported for silver(I) complexes with 3D structures, as listed in Table 11.1.[9,63–75] The reaction of a silver–ammonia solution with 3,3′,5,5′-tetramethyl-4,4′-bipyrazole (H$_2$-**21**) provides a polymeric complex $\{[Ag_2(\mathbf{21})]\}_n$ with a fourfold interpenetrated (10,3)-*a* framework structure.[63] The complex $\{[Ag_4(\mathbf{22})_2](NO_3)_4{\cdot}MeCN{\cdot}2H_2O\}_n$ with interpenetrating (10,3)-*b* nets was prepared from the reaction of *N*′,*N*′-bis[1-(pyrazine-2-yl)ethylidene]benzildihydrazone (**22**) with $AgNO_3$.[64] Another type of (10,3) topology is (10,3)-*d*, which has been observed in the complex $\{[Co(\mathbf{23})_3AgBF_4]\}_n$ [**23** = 5-(4-pyridyl)-4,6-dipyrrinato].[9] In this case, the $[Co(\mathbf{23})_3]$ acts as building block and is further connected by Ag(I) to give a 3D framework with twofold interpenetration, as illustrated in Figure 11.26.

The reaction of 1H-1,2,4-triazole-1-acetic acid (H-**24**) with $AgNO_3$ in the presence of sodium hydroxide under hydrothermal conditions gave $\{[Ag(\mathbf{24})]\}_n$. This complex

TABLE 11.1. Selected Silver(I) Coordination Polymers with Specific Topology

Complex	Coordination Number and Geometry of Ag(I)	Topology	Reference
$\{[Ag(\mathbf{14})(\mathbf{15})]\}_n$	3, trigonal	1D single-stranded helix	58
$\{[Ag(\mathbf{16})](BF_4)\}_n/\{[Ag(\mathbf{17})](PF_6)\}_n$	2, linear	1D double-stranded helix	59
$\{[Ag(\mathbf{18})(NO_3)]\}_n$	4, tetrahedral	2D (4,4) network	60
$\{[Ag(\mathbf{19})_{2/3}]_3[Ag(\mathbf{19})]_2(CF_3SO_3)_5\}_n$	2, linear; 3, trigonal	2D (6,3) network	61
$\{[Ag(\mathbf{20})]ClO_4\}_n$	3, trigonal	2D (4.8^2) network	62
$\{[Ag_2(\mathbf{21})]\}_n$	2, linear	3D (10,3)-*a* or $SrSi_2$	63
$\{[Ag_4(\mathbf{22})_2](NO_3)_4{\cdot}MeCN{\cdot}2H_2O\}_n$	4, tetrahedral	3D (10,3)-*b* or $ThSi_2$	64
$\{[Co(\mathbf{23})_3AgBF_4]\}_n$	4, trigonal pyramidal	3D (10,3)-*d*	9
$\{[Ag(\mathbf{24})]\}_n$	4, tetrahedral	3D diamond	65
$\{[Ag_6(\mathbf{25})_4Cl](OH){\cdot}6H_2O\}_n$	2, linear	3D 4.14^2 (dia-*f*)	66
$\{[Ag(\mathbf{26})]\ CH_3OH\}_n$	5, tetragonal pyramidal	3D $SrAl_2$ (sra)	67
$\{[Ag(\mathbf{27})_3](NO_3)\}_n$	6, octahedral	3D α-Po (pcu)	68
$\{[Ag(\mathbf{28})](NO_3)(CH_3OH)\}_n$	4, tetrahedral	3D $(8,^3_4)$-*b*	69
$\{[Ag_2(\mathbf{29})](ClO_4)_2\}_n$	4, tetrahedral	3D pyrite (pyr)	70
$\{[Ag_2(\mathbf{30})](ClO_4)_2\}_n$	4, tetrahedral	3D rutile (rtl)	71
$\{[Ag(\mathbf{31})](ClO_4){\cdot}4.5H_2O\}_n$	6, trigonal prism	3D (8,3)-*b*	71
$\{[Ag(\mathbf{32})_2](SbF_6)\}_n$	4, tetrahedral	3D coesite (coe)	72
$\{Fe(\mathbf{33})_2[Ag(CN)_2]_2{\cdot}2/3H_2O\}_n$	2, linear	3D NbO	73
$\{Fe(\mathbf{34})[Ag(CN)_2][Ag_2(CN)_3]\}_n$	2, linear	3D (6,6)	74
$\{[Ag(\mathbf{35})](BF_4)_{1/2}(OH)_{1/2}{\cdot}(CH_3CN)_{1/4}\}_n$	4, square planar	3D lvt	75
$\{[Ag(\mathbf{35})](BF_4){\cdot}1.5H_2O\}_n$	4, tetrahedral	3D PtS	75

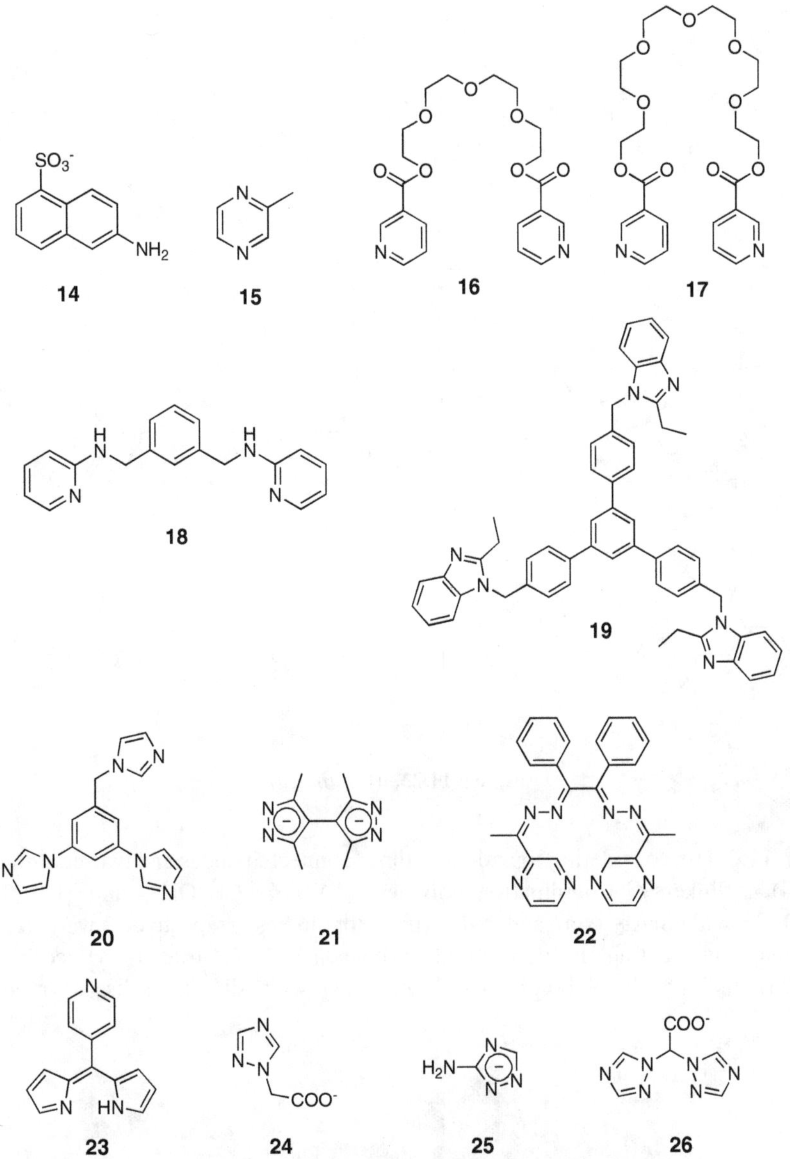

Figure 11.23. Structures of organic ligands **14–35** used for construction of the silver(I) coordination polymers listed in Table 11.1.

has a 3D structure with a typical diamond topology in which the silver(I) has a tetrahedral coordination geometry and serves as a four-connector.[65] The polymeric complex $\{[Ag_6(\mathbf{25})_4Cl](OH){\cdot}6H_2O\}_n$ was obtained by reaction of an ammonia solution of AgCl and 3-amino-1,2,4-triazole (H-**25**). The 3D framework structure was confirmed by X-ray crystallographic analysis exhibiting 4.14^2 (dia-*f*) topology

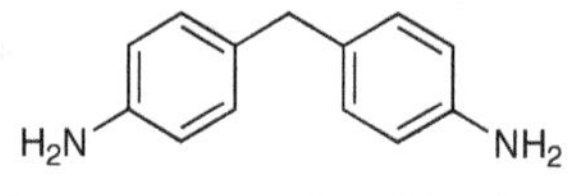

27

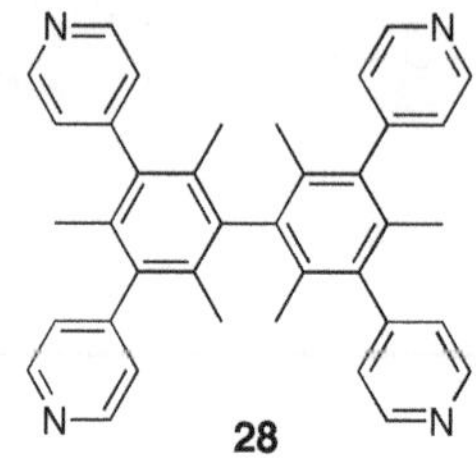

28

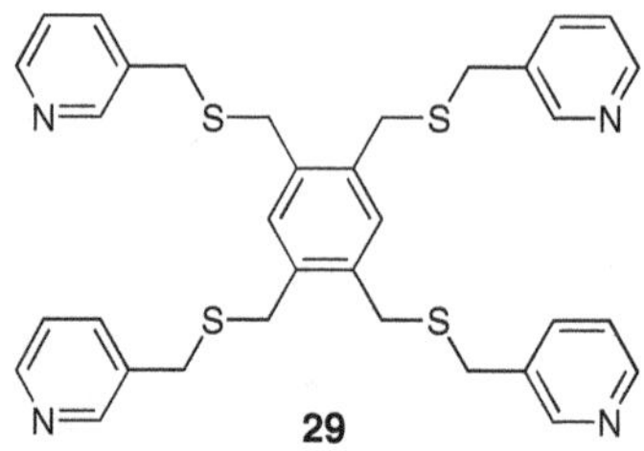

29

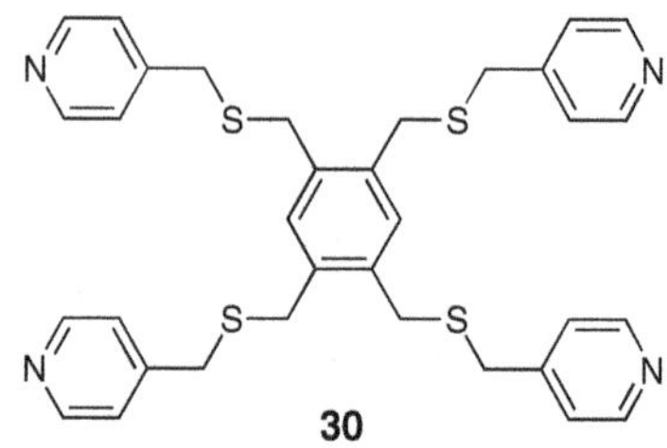

30

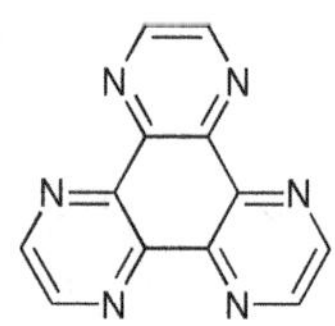

31

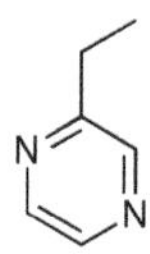

32

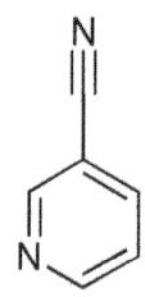

33

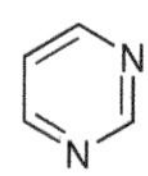

34

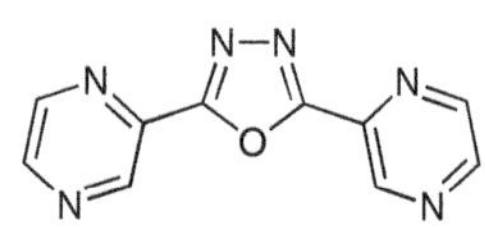

35

Figure 11.23. (*Continued*).

(Fig. 11.27) by considering ligand **25** as three-connected nodes and two-coordinated Ag(I) as linkers.[66] Coordination polymers $\{[Ag(\mathbf{26})]\cdot CH_3OH\}_n$ and $\{[Ag(\mathbf{27})_3](NO_3)\}_n$ with $SrAl_2$ (sra) and α-Po (pcu) topologies, respectively, were readily obtained by reacting the corresponding ligands bis(1,2,4-triazol-1-yl)acetic acid (H-**26**) and 4,4′-diaminodiphenylmethane (**27**) with silver(I) salts, as reported

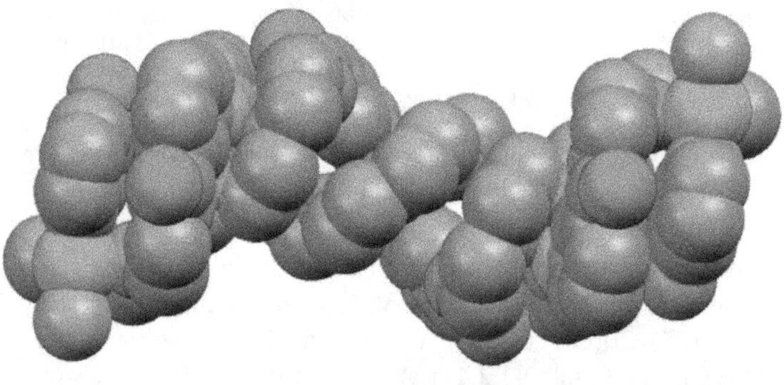

Figure 11.24. Space-filling model of the double-stranded helix $\{[Ag(\mathbf{16})](BF_4)\}_n$.

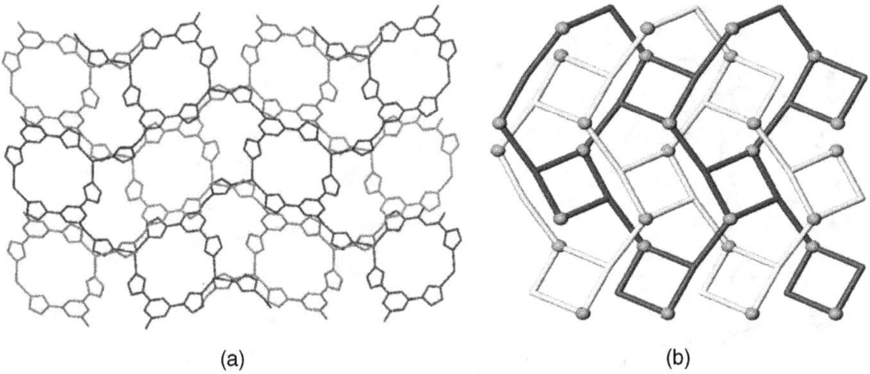

Figure 11.25. Parallel interpenetration of two 4.8^2 nets of $\{[Ag(\mathbf{20})]ClO_4\}_n$, crystal structure (a) and schematic representation (b).

by Du and Ciani.[67,68] The topology of $(8,^3_4)$-b was predicted by Wells in the late 1970s, and a silver(I) complex $\{[Ag(\mathbf{28})](NO_3)(CH_3OH)\}_n$ with such a topology was synthesized by using 3,3′,5,5′-tetrakis(4-pyridyl)bimesityl (**28**) as a ligand.[69]

Hanton and coworkers reported silver(I) complexes with different topologies induced by the ligands 1,2,4,5-tetrakis(3-pyridylmethylsulfanylmethyl)benzene

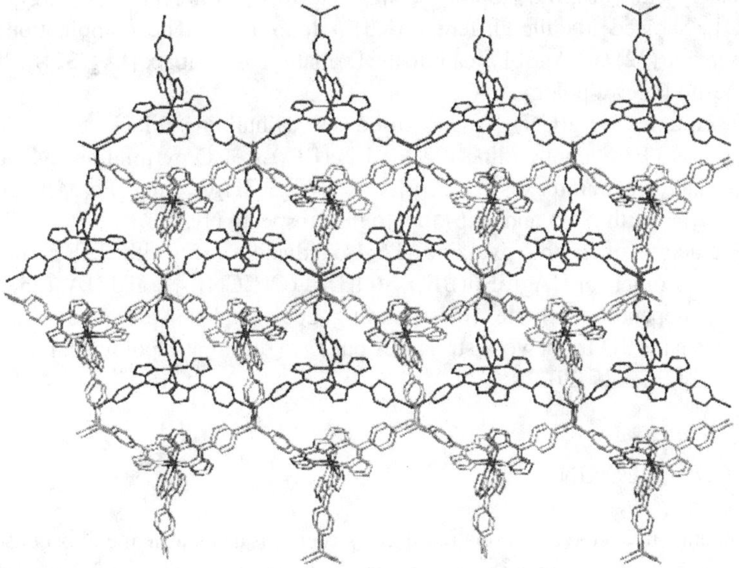

Figure 11.26. Crystal structure of $\{[Co(\mathbf{23})_3AgBF_4]\}_n$ with (10,3)-*d* topology and twofold interpenetration.

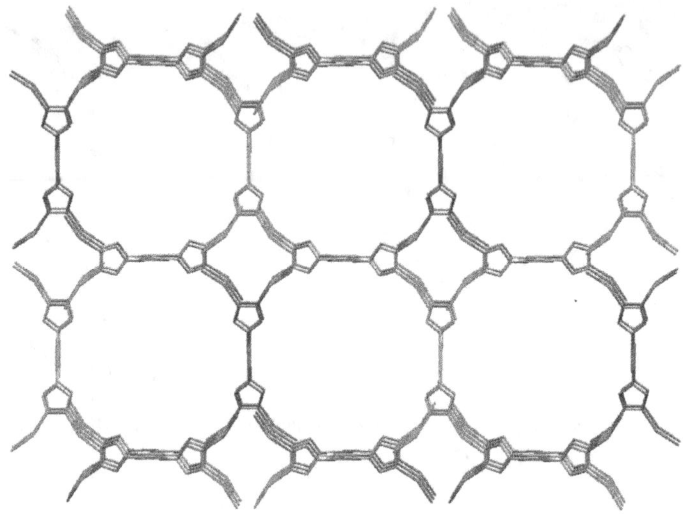

Figure 11.27. Three-dimensional 4.14^2 (dia-*f*) net observed in the crystal structure of $\{[Ag_6(\mathbf{25})_4Cl](OH)\cdot 6H_2O\}_n$.

(**29**) and 1,2,4,5-tetrakis(4-pyridylmethylsulfanylmethyl)benzene (**30**).[70] The complexes $\{[Ag_2(\mathbf{29})](ClO_4)_2\}_n$ and $\{[Ag_2(\mathbf{30})](ClO_4)_2\}_n$ have 3D structures with pyrite (pyr) and rutile (rtl) topologies, as shown in Figure 11.28. The compound 1,4,5,8,9,12-hexaazatriphenylene (**31**) was reacted with different silver(I) salts, and complexes with (10,3)-*a* topology were formed; however, $\{[Ag(\mathbf{31})](ClO_4)\cdot 4.5H_2O\}_n$ showed a quite different (8,3)-*b* topology.[71] The combination of 2-ethylpyrazine (**32**) and $AgSbF_6$ leads to the formation of complex $\{[Ag(\mathbf{32})_2](SbF_6)\}_n$ with coesite (coe) topology.[72]

The complex $[Ag(CN)_2]^-$ was used as a building block to react with $Fe(BF_4)_2\cdot 6H_2O$/3-cyanopyridine (**33**) and $Fe(BF_4)_2\cdot 6H_2O$/pyrimidine (**34**) leading to the formation of complexes $\{Fe(\mathbf{33})_2[Ag(CN)_2]_2\cdot 2/3H_2O\}_n$ and $\{Fe(\mathbf{34})[Ag(CN)_2][Ag_2(CN)_3]\}_n$ with NbO and (6,6) topologies, respectively.[73,74]

The reaction of 2,5-bis(pyrazine)-1,3,4-oxadiazole (**35**) with $AgBF_4$ provided topological isomers of $\{[Ag(\mathbf{35})](BF_4)_{1/2}(OH)_{1/2}\cdot(CH_3CN)_{1/4}\}_n$ and $\{[Ag(\mathbf{35})](BF_4)\cdot 1.5H_2O\}_n$ depending on the corresponding guest and solvent medium.[75] Both complexes have 3D framework structures but have different topologies: lvt and PtS, respectively (Fig. 11.29).

11.7 CONCLUSION

More recent studies have demonstrated that silver(I) is a versatile metal because of its variable coordination number and geometry. As a consequence of these inherent properties, a wide variety of structures, ranging from individual cage-like molecular complexes to infinite 1D, 2D, and 3D architectures with specific topologies, have been

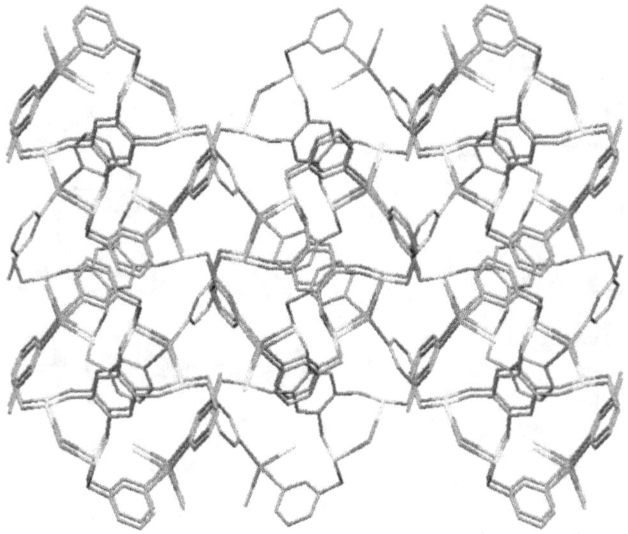

(a)

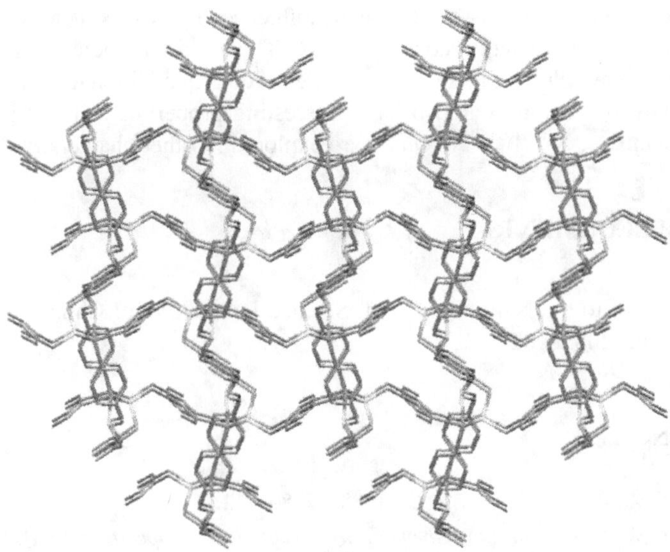

(b)

Figure 11.28. $\{[Ag_2(\mathbf{29})](ClO_4)_2\}_n$ (a) and $\{[Ag_2(\mathbf{30})](ClO_4)_2\}_n$ (b) with pyrite (pyr) and rutile (rtl) topologies, respectively.

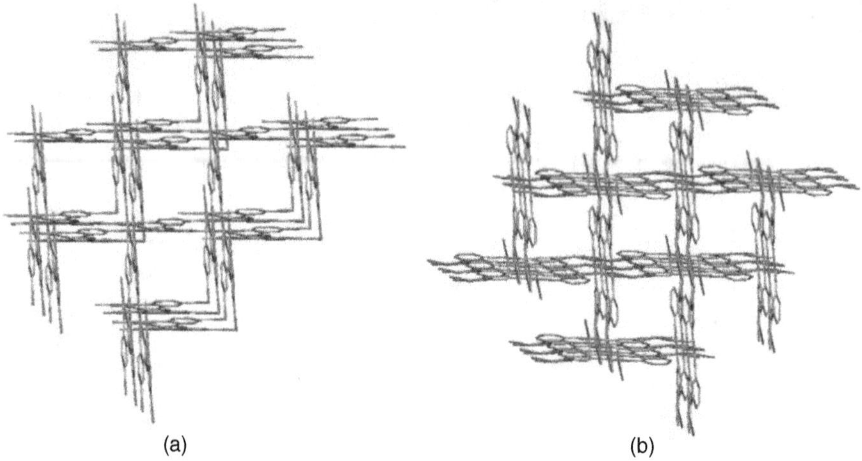

Figure 11.29. Topological isomers of $\{[Ag(\mathbf{35})](BF_4)_{1/2}(OH)_{1/2}\cdot(CH_3CN)_{1/4}\}_n$ (a) and $\{[Ag(\mathbf{35})](BF_4)\cdot 1.5H_2O\}_n$ (b) with lvt and PtS topologies, respectively.

obtained, as described in this chapter. Despite this progress, the challenge of predicting the exact structure of the assembled products remains because of the complicated assembly processes that can be influenced by factors such as the chosen counteranion and the reaction conditions.[76–78] Thus, further studies that systematically explore the influence of such factors are needed. In addition to their structural diversity, silver(I) complexes also have interesting properties and applications in organic chemistry,[26,30,31,63,79,80] which are explored in other chapters in this book.

ACKNOWLEDGMENTS

We are grateful to the National Natural Science Foundation of China for financial support (Grant 20828002).

REFERENCES

1. Lehn, J. M., *Angew. Chem. Int. Ed.* **1988**, *27*, 89–112.
2. Lehn, J. M., *Supramolecular Chemistry: Concepts and Perspectives*, VCH, Weinheim, **1995**.
3. Young, A. G.; Hanton, L. R., *Coord. Chem. Rev.* **2008**, *252*, 1346–1386.
4. For an example of two-coordinated linear silver(I), see Fitchett, C.M.; Steel, P.J., *Dalton Trans.* **2006**, 4886–4888.
5. For an example of three-coordinated triangular silver(I), see Chiang, L. M.; Yeh, C. W.; Chan, Z. K.; Wang, K. M.; Chou, Y. C.; Chen, J. D.; Wang, J. C.; Lai, J. Y., *Cryst. Growth Des.* **2008**, *8*, 470–477.

6. For an example of three-coordinated T-shaped silver(I), see Ganguly, S.; Chattopadhyay, S.; Sinha, C.; Chakravorty, A., *Inorg. Chem.* **2000**, *39*, 2954–2956.
7. For an example of four-coordinated square planar silver(I), see Reger, D. L.; Gardinier, J. R.; Smith, M. D., *Inorg. Chem.* **2004**, *43*, 3825–3832.
8. For an example of four-coordinated tetrahedral silver(I), see Argent, S. P.; Adams, H.; Riis-Johannessen, T.; Jeffery, J. C.; Harding, L. P.; Clegg, W.; Harringtond, R. W.; Ward, M. D., *Dalton Trans.* **2006**, 4996–5013.
9. For an example of five-coordinated trigonal bipyramidal silver(I), see Halper, S. R.; Do, L.; Stork, J. R.; Cohen, S. M., *J. Am. Chem. Soc.* **2006**, *128*, 15255–15268.
10. For an example of five-coordinated tetragonal pyramidal silver(I), see Habata, Y.; Noto, K.; Osaka, F., *Inorg. Chem.* **2007**, *46*, 6529–6534.
11. For an example of six-coordinated octahedral silver(I), see Niu, C. Y.; Wu, B. L.; Zheng, X. F.; Zhang, H. Y.; Li, Z. J.; Hou, H. W., *Dalton Trans.* **2007**, 5710–5713.
12. For an example of six-coordinated trigonal prism silver(I), see Fan, J.; Zhu, H. F.; Okamura, T. A.; Sun, W. Y.; Tang, W. X.; Ueyama, N., *Chem. Eur. J.* **2003**, *9*, 4724–4731.
13. For an example of seven-coordinated silver(I), see Prince, P. D.; Steed, J. W., *Supramol. Chem.* **1998**, *10*, 155–158.
14. For an example of eight-coordinated tetragonal prism silver(I), see Jones, P. G.; Gries, T.; Grutzmacher, H.; Roesky, H. W.; Schimkowiak, J.; Sheldrick, G. M., *Angew. Chem. Int. Ed.* **1984**, *23*, 376–376.
15. Katz, M. J.; Sakai, K.; Leznoff, D. B., *Chem. Soc. Rev.* **2008**, *37*, 1884–1895.
16. Xu, J.; Yuan, Q.; Bai, Z. S.; Su, Z.; Sun, W. Y., *Inorg. Chem. Commun.* **2009**, *12*, 58–61.
17. Steel, P. J.; Fitchett, C. M., *Coord. Chem. Rev.* **2008**, *252*, 990–1006.
18. Sun, W. Y.; Yoshizawa, M.; Kusukawa, T.; Fujita, M., *Curr. Opin. Chem. Biol.* **2002**, *6*, 757–764.
19. Fujita, M.; Umemoto, K.; Yoshizawa, M.; Fujita, N.; Kusukawa, T.; Biradha, K., *Chem. Commun.* **2001**, 509–518.
20. Fujita, M.; Tominaga, M.; Hori, A.; Therrien, B., *Acc. Chem. Res.* **2005**, *38*, 371–380.
21. Sun, W. Y. in *New Development in Organometallic Chemistry Research*, Cato, M. A., ed., Nova Science Publishers, New York, **2006**, Chapter 2, pp. 37–61.
22. Xu, G. C.; Ding, Y. J.; Okamura, T. A.; Huang, Y. Q.; Liu, G. X.; Sun, W. Y.; Ueyama, N., *Cryst. Eng. Comm.* **2008**, *10*, 1052–1062.
23. Xu, G. C.; Ding, Y. J.; Huang, Y. Q.; Liu, G. X.; Sun, W. Y., *Micropor. Mesopor. Mater.* **2008**, *113*, 511–522.
24. Wang, X. F.; Lv, Y.; Okamura, T. A.; Kawaguchi, H.; Wu, G.; Sun, W. Y.; Ueyama, N., *Cryst. Growth Des.* **2007**, *7*, 1125–1133.
25. Liu, H. K.; Huang, X.; Lu, T.; Wang, X.; Sun, W. Y.; Kang, B. S., *Dalton Trans.* **2008**, 3178–3188.
26. Sun, W. Y.; Fan, J.; Okamura, T. A.; Xie, J.; Yu, K. B.; Ueyama, N., *Chem. Eur. J.* **2001**, *7*, 2557–2562.
27. Sun, W. Y.; Fan, J.; Yu, K. B., *Acta Chim. Sinca.* **2001**, *59*, 1102–1105.
28. Fan, J.; Sun, W. Y.; Okamura, T. A.; Xie, J.; Tang, W. X.; Ueyama, N., *New J. Chem.* **2002**, *26*, 199–201.
29. Hiraoka, S.; Yi, T.; Shiro, M.; Shionoya, M., *J. Am. Chem. Soc.* **2002**, *124*, 14510–14511.

30. Hiraoka, S.; Harano, K.; Shiro, M.; Shionoya, M., *Angew. Chem. Int. Ed.* **2005**, *44*, 2727–2731.

31. Huang, Y. Q.; Shen, Z. L.; Okamura, T. A.; Wang, Y.; Wang, X. F.; Sun, W. Y.; Yu, J. Q.; Ueyama, N., *Dalton Trans.* **2008**, 204–213.

32. Sumby, C. J.; Hardie, M. J., *Angew. Chem. Int. Ed.* **2005**, *44*, 6395–6399.

33. Iki, N.; Ohta, M.; Horiuchi, T.; Hoshino, H., *Chem. Asian J.* **2008**, *3*, 849–853.

34. Zhang, J.; Xu, X.; James, S. L., *Chem. Commun.* **2006**, 4218–4220.

35. Willans, C. E.; Anderson, K. M.; Junk, P. C.; Barbour, L. J.; Steed, J. W., *Chem. Commun.* **2007**, 3634–3636.

36. Iijima, S., *Nature* **1991**, *354*, 56–58.

37. Special issue on carbon nanotubes, *Acc. Chem. Res.* **2002**, *35*, 997–1113.

38. Fan, J.; Zhu, H. F.; Okamura, T. A.; Sun, W. Y.; Tang, W. X.; Ueyama, N., *Inorg. Chem.* **2003**, *42*, 158–162.

39. Harada, R.; Matsuda, Y.; Okawa, H.; Kojima, T., *Angew. Chem. Int. Ed.* **2004**, *43*, 1825–1828.

40. Tashiro, S.; Tominaga, M.; Kusukawa, T.; Kawano, M.; Sakamoto, S.; Yamaguchi, K.; Fujita, M., *Angew. Chem. Int. Ed.* **2003**, *42*, 3267–3270.

41. Yamaguchi, T.; Tashiro, S.; Tominaga, M.; Kawano, M.; Ozeki, T.; Fujita, M., *Am. Chem. Soc.* **2004**, *126*, 10818–10819.

42. Hong, M. C.; Zhao, Y. J.; Su, W. P.; Cao, R.; Fujita, M.; Zhou, Z. Y.; Chan, A. S. C., *Angew. Chem. Int. Ed.* **2000**, *39*, 2468–2470.

43. Dong, Y. B.; Jiang, Y. Y.; Li, J.; Ma, J. P.; Liu, F. L.; Tang, B.; Huang, R. Q.; Batten, S. R., *J. Am. Chem. Soc.* **2007**, *129*, 4520–4521.

44. Kleina, C.; Graf, E.; Hosseini, M. W.; Cianb, A. D.; Fischer, J., *Chem. Commun.* **2000**, 239–240.

45. Hong, M. C.; Zhao, Y. J.; Su, W. P.; Cao, R.; Fujita, M.; Zhou, Z. Y.; Chan, A. S. C., *J. Am. Chem. Soc.* **2000**, *122*, 4819–4820.

46. Sauvage, J. P.; Dietrich-Buchecker C., eds., *Molecular Catenanes, Rotaxanes, and Knots*, Wiley-VCH, Weinheim, **1999**.

47. Vignon, S. A.; Wong, J.; Tseng, H. R.; Stoddart, J. F., *Org. Lett.* **2004**, *6*, 1095–1098.

48. Hori, A.; Sawada, T.; Yamashita, K.; Fujita, M., *Angew. Chem. Int. Ed.* **2005**, *44*, 4896–4899.

49. Zhu, H. F.; Fan, J.; Okamura, T. A.; Sun, W. Y.; Ueyama, N., *Cryst. Growth Des.* **2005**, *5*, 289–294.

50. Fan, J.; Sun, W. Y.; Okamura, T. A.; Tang, W. X.; Ueyama, N., *Inorg. Chim. Acta* **2004**, *357*, 2385–2389.

51. Jin, C. M.; Lu, H.; Wu, L. Y.; Huang, J., *Chem. Commun.* **2006**, 5039–5041.

52. Jin, C. M.; Wu, L. Y.; Lu, H.; Xu, Y., *Cryst. Growth Des.* **2008**, *8*, 215–218.

53. Sagué, J. L.; Fromm, K. M., *Cryst. Growth Des.* **2006**, *6*, 1566–1568.

54. Kim, K., *Chem. Soc. Rev.* **2002**, *31*, 96–107.

55. Fan, J.; Sun, W. Y.; Okamura, T. A.; Zheng, Y. Q.; Sui, B.; Tang, W. X.; Ueyama, N., *Cryst. Growth Des.* **2004**, *4*, 579–584.

56. Wang, Z. B.; Zhu, H. F.; Zhao, M.; Li, Y. Z.; Okamura, T. A.; Sun, W. Y.; Chen, H. L.; Ueyama, N., *Cryst. Growth Des.* **2006**, *6*, 1420–1427.

57. Hoskins, B. F.; Robson, R.; Slizys, D. A., *J. Am. Chem. Soc.* **1997**, *119*, 2952–2953.
58. Liu, H. Y.; Wu, H.; Ma, J. F.; Song, S. Y.; Yang, J.; Liu, Y. Y.; Su, Z. M., *Inorg. Chem.* **2007**, *46*, 7299–7311.
59. Jouaiti, A.; Hosseini, M. W.; Kyritsakas, N., *Chem. Commun.* **2003**, 472–473.
60. Deng, Z. P.; Zhu, L. N.; Gao, S.; Huo, L. H.; Ng, S. W., *Cryst. Growth Des.* **2008**, *8*, 3277–3284.
61. Zhang, X. L.; Guo, C. P.; Yang, Q. Y.; Lu, T. B.; Tong, Y. X.; Su, C. Y., *Chem. Mater.* **2007**, *19*, 4630–4632.
62. Fan, J.; Sun, W. Y.; Okamura, T. A.; Tang, W. X.; Ueyama, N., *Inorg. Chem.* **2003**, *42*, 3168–3175.
63. Zhang, J. P.; Horike, S.; Kitagawa, S., *Angew. Chem. Int. Ed.* **2007**, *46*, 889–892.
64. Bai, Y.; Duan, C. Y.; Cai, P.; Dang, D. B.; Meng, Q. J., *Dalton. Trans.* **2005**, 2678–2680.
65. Hu, T. L.; Du, W. P.; Hu, B. W.; Li, J. R.; Bu, X. H.; Cao, R., *Cryst. Eng. Comm.* **2008**, *10*, 1037–1043.
66. Zhang, J. P.; Lin, Y. Y.; Zhang, W. X.; Chen, X. M., *J. Am. Chem. Soc.* **2005**, *127*, 14162–14163.
67. Du, M.; Zang, Z. H.; Tang, L. F.; Wang, X. G.; Zhao, X. J.; Batten, S. R., *Chem. Eur. J.* **2007**, *13*, 2578–2586.
68. Carlucci, L.; Ciani, G.; Proserpioa, D. M.; Portab, F., *Cryst. Eng. Comm.* **2006**, *8*, 696–706.
69. Natarajan, R.; Savitha, G.; Dominiak, P.; Wozniak, K.; Moorthy, J. N., *Angew. Chem. Int. Ed.* **2005**, *44*, 2115–2119.
70. Cordes, D. B.; Hanton, L. R., *Inorg. Chem.* **2007**, *46*, 1634–1644.
71. Abrahams, B. F.; Jackson, P. A.; Robson, R., *Angew. Chem. Int. Ed.* **1998**, *37*, 2656.
72. Carlucci, L.; Ciani, G.; Proserpio, D. M.; Rizzato, S., *J. Chem. Soc. Dalton Trans.* **2000**, 3821–3827.
73. Galet, A.; Niel, V.; Munoz, M. C.; Real, J. A., *J. Am. Chem. Soc.* **2003**, *125*, 14224–14225.
74. Niel, V.; Thompson, A. L.; Goeta, A. E.; Enachescu, C.; Hauser, A.; Galet, A.; Munoz, M. C.; Real, J. A., *Chem. Eur. J.* **2005**, *11*, 2047–2060.
75. Du, M.; Zhao, X. J.; Guoa, J. H.; Batten, S. R., *Chem. Commun.* **2005**, 4836–4838.
76. Domasevitch, K. V.; Solntsev, P. V.; Gural'skiy, I. A.; Krautscheid, H.; Rusanov, E. B.; Chernegac, A. N.; Howard, J. A. K., *Dalton Trans.* **2007**, 3893–3905.
77. Northrop, B. H.; Yang, H. B.; Stang, P. J., *Chem. Commun.* **2008**, 5896–5908.
78. Gimeno, N.; Vilar, R., *Coord. Chem. Rev.* **2006**, *250*, 3161–3189.
79. Yan, Z. Q.; McCracken, T.; Xia, S. J.; Maslak, V.; Gallucci, J.; Hadad, C. M.; Badji, J. D., *J. Org. Chem.* **2008**, *73*, 355–363.
80. Halbes-Letinois, U.; Weibel, J. M.; Pale, P., *Chem. Soc. Rev.* **2007**, *36*, 759–769.

12

A CRITICAL COMPARISON: COPPER, SILVER, AND GOLD

A. Stephen K. Hashmi

Institute of Organic Chemistry, Ruprecht-Karls-University, Heidelberg, Germany

Silver in Organic Chemistry Edited by Michael Harmata

12.1 INTRODUCTION

In organic synthesis, catalysis by silver is an important methodology. The first time that chemists hear about silver catalysts during their studies is in an introductory organic chemistry course—a technical process, in which the oxidation of ethene to oxirane is catalyzed by silver.[1] This process is not possible with propene, as the allylic position will preferentially be oxidized, but the research since the late 1990s has demonstrated that certain gold catalysts are able to give good selectivities for this top target in industrial chemistry.[2,3]

This basic example demonstrates what will be discussed in this chapter, similarities and differences between the elements of the copper triad, but with a focus on homogeneous catalysis.

All three metals have been used extensively in homogeneous catalysis of organic reactions. Early work focused on copper; thus the catalysis-related literature for this element is abundant.[4] Silver had sustained continuous interest,[5] but never to the extent that copper experienced. Gold is the "youngest" member in the field of catalysis, but is currently (as of 2009) catching up at an incredible rate.[6]

This chapter discusses the similarities and differences that one can find in the publications covering homogeneous catalysis by these metals.

12.2 REACTIONS CATALYZED BY COPPER, SILVER, OR GOLD

12.2.1 Aldehyde–Alkyne–Amine Coupling

The coupling of the three components -*a*ldehyde, *a*lkyne and *a*mine- has been called "A^3 coupling."[7] It delivers propargyl amines, which are important synthetic intermediates and, in contrast to previously used routes, avoids the use of strong-base reagents such as BuLi, ethylmagnesium bromide or LDA, and the need for the kinds of solvents necessary for these reagents.

An initial finding was the catalytic activity of CuI under solvent-free conditions in which an acceleration by microwave irradiation was reported (Scheme 12.1).[8] The reaction probably proceeds through an iminium or imine species; thus there is a close relationship to the historic work of Reppe for the formation of the corresponding propargyl alcohols.

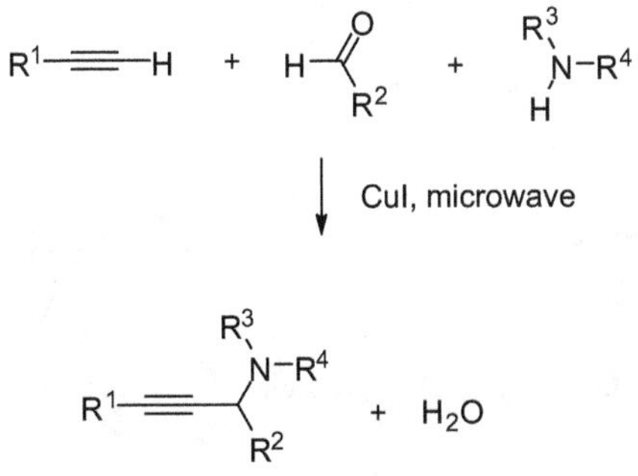

Scheme 12.1. Copper-catalyzed reaction of aldehydes with alkynes and secondary amines.

Subsequent research revealed that with different soluble silver salts only low conversions between 25% and 45% could be obtained, which was due to the in situ reduction of the more noble silver. However, with AgCl, AgBr, and AgI, good catalytic activity was observed.[9] For example, with as little as 0.2 mol% AgI in water at 100°C, excellent yields could be obtained.

Investigation of gold catalysts showed that with $AuBr_3$ a very clean reaction can be observed in water.[10] Similar results were obtained with AuCl and AuI; it was speculated that gold(III) is reduced in situ to gold(I). The choice of $AuBr_3$ for most of the discussion in this chapter ultimately resulted from the fact that at the time of the investigation, on normalization of gold for catalysis, this was the least expensive gold halide commercially available. The use of an ionic liquid, [Bmim][BF_4], as the solvent for these conversions has been reported; in this case copper and gold catalysts tended to deliver trimers of the aliphatic aldehydes as side products.[11]

Only the copper-catalyzed version of this reaction was extended to an asymmetric version (known as "AA^3 coupling"). Both CuOTf and the pybox ligand gave excellent enantioselectivity values of up to 99% ee for anilines as amine components. By anomalous diffraction on a derivative bearing a bromo atom, the absolute configuration of the product could be determined to be (*R*) for the ligand configuration shown in Scheme 12.2.[12] There is no evidence in the literature that chiral gold or silver complexes have also been tested; in the context of the success of chiral gold and silver complexes in the catalytic asymmetric aldol reaction,[6b] this would have been an interesting option.

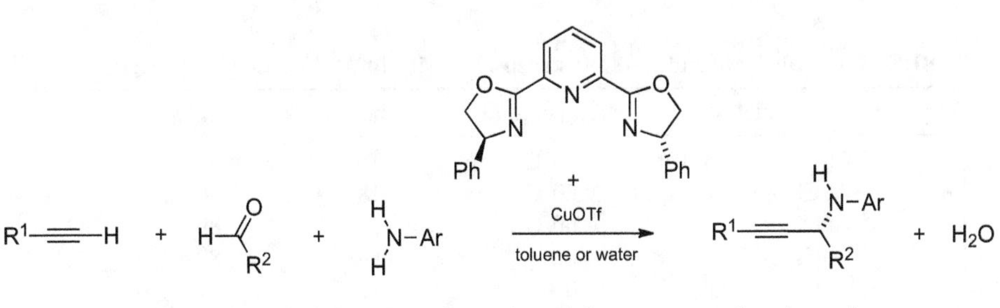

Scheme 12.2. Catalytic asymmetric version of the copper-catalyzed reaction.

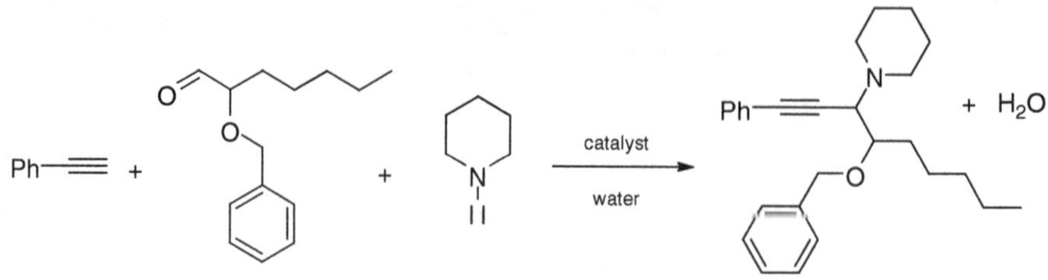

Scheme 12.3. Diastereoselective A^3 coupling.

This kind of comparison has been made only for the diastereoselective version of this reaction.[13] Water was still used as solvent for the conversion of phenylacetylene, 2-benzyloxyheptanal, and piperidine (Scheme 12.3), but the reaction temperature was lowered to 60°C, room temperature, or even 0°C. As shown in Table 12.1, neither with copper(I) halides (entry 1) nor with silver(I) halides (entry 2) could any satisfactory results be obtained. Only gold(I) halides were still reactive enough at these temperatures. With AuCl at 60°C, the diastereomeric ratio (dr) was still low (entry 3); lowering the temperature to room temperature led to a slight increase of the dr (entry 4); and going to 0°C further increased the dr, but at the same time diminished the yield. With AuI at room temperature (entry 5) and 0°C (entry 6), only marginally better results were obtained.

12.2.2 Carbene Insertion Reactions

Cyclohexane is a saturated hydrocarbon in which no regioselectivity problems of C–H insertion can occur. The reaction of cyclohexane with ethyl diazoacetate was investigated in a thorough study by Perez et al.[14] The *N*-heterocyclic carbene ligand IPr (IPr = 1,3-bis(diisopropylphenyl)imidazol-2-yliden) was used, all three compounds IPrMCl (M = Cu, Ag, Au) were inactive as catalysts in cyclohexane (Table 12.2, entry 1). Addition of the sodium BARF salt ($NaB[3,5\text{-}(CF_3)_2C_6H_3]_4$) gave ethyl cyclohexyl acetate as the C–H insertion product of the carbene (Scheme 12.4).

In this case, the best yields were obtained for the gold(I) catalyst (Table 12.2, entry 4), the copper(I) catalyst was second in this ranking (entry 2), with silver(I) affording the lowest yield (74%, entry 3).

TABLE 12.1. Substrate-Induced Diastereoselectivity in A^3 Coupling in Water

Entry	Catalyst	Temperature	Time (h)	Yield (%)	dr
1	CuCl, CuBr, or CuI	60°C	18	—	—
2	AgCl, AgBr, or AgI	60°C	18	—	—
3	AuCl	60°C	18	89	64 : 36
4	AuCl	r.t.	18	89	72 : 28
5	AuCl	0°C	12	57	75 : 25
6	AuI	r.t.	18	100	73 : 27
7	AuI	0°C	12	58	76 : 24

TABLE 12.2. Yields of the Metal-Catalyzed Carbene Insertion

Entry	Catalyst	Yield (%)
1	IPrCuCl, IPrAgCl, or IPrAuCl	—
2	IPrCuCl + $NaBAr'_4$	80
3	IPrAgCl + $NaBAr'_4$	74
4	IPrAuCl + $NaBAr'_4$	84

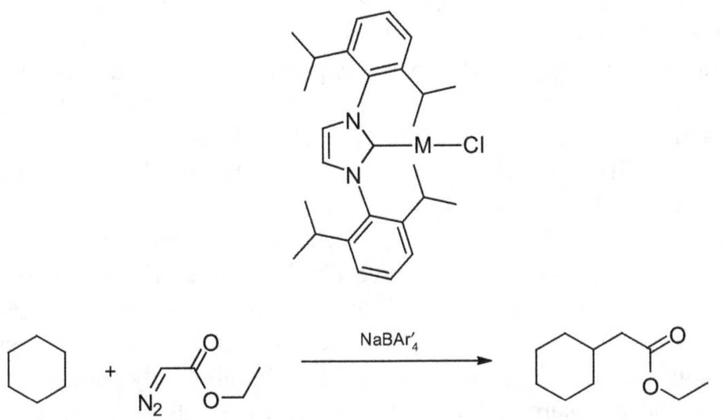

Scheme 12.4. Carbene insertions catalyzed by copper(I), silver(I), and gold(I).

12.2.3 In Silico Comparison of Organocopper(I), Organosilver(I), and Organogold(I)-Ate Compounds

Organocuprate(I) reagents like R_2CuLi readily react with organic electrophiles (E^+), for example, alkyl halides or α,β-unsaturated carbonyl compounds (Scheme 12.5).[15] On the other hand, there are only a few reports where one could assume a participation of organoargentate[16] or organoaurate[17] intermediates in organic reactions.

In a computational study, Nakamura et al.[17] examined these reactions. They had a close look at both steps, the association with the electrophile to form the intermediate R_2CuE, and the subsequent elimination of the product R–E. The higher reactivity of the organocuprate(I) compounds is based on the high-lying 3*d* orbitals; the 4*d* orbitals of the corresponding silver(I) species and the 5*d* orbitals of the gold(I) species are low-lying. Owing to the relativistic effect, the 5*d* orbitals and the 4*d* orbitals are close in energy. In addition, the next step, the elimination of R–E from the intermediate, is easier for copper than for silver or gold. This results from the higher stability of the

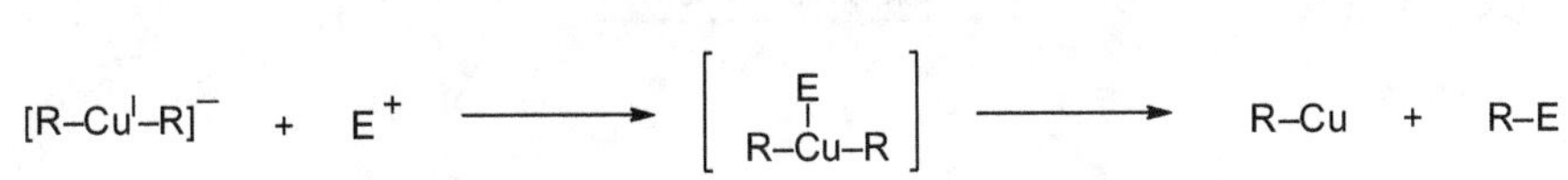

Scheme 12.5. Typical reactions of organocuprates.

carbon–gold bond, which again is a relativistic effect[18] and follows the general rule that the transition metal–carbon bond strength increases from the first-row transition metals to the heavier analogs.

12.2.4 Cyclization of *ortho*-Alkynylbenzaldehydes

This reaction type was intensively investigated by Yoshinori Yamamoto et al., who focused mainly on copper and gold.[19] In the context of an application in the synthesis of azaphilones, the Porco group thoroughly compared the reactivity of the three metals in the conversion of a hydroxylated *ortho*-alkynylbenzaldehyde to the corresponding pyrylium salt (Scheme 12.6, Table 12.3).[20]

The catalyst screening involved control experiments without a catalyst [excluding catalytic activity of the trifluoroacetic acid (TFA), Table 12.3, entry 1]. Copper salts (entries 2 and 3) were effective; after 20 min copper(I) had converted more substrate than had copper(II). Silver(I) (entry 4) was even better. Four different gold(III) catalysts were tested; the acetate (entry 7) was the most active catalyst, followed by the chloride (entry 6) and the bromide (entry 5). In general, this type of screening had been helpful for the authors, as it delivered a highly active catalyst, but in the overall context of this chapter it is not very helpful, as systematic changes are missing. For example, keeping the counterion constant and comparing the different metals in the same oxidation state would have been helpful. Still, this nicely points out one of the difficulties of the field, namely, that in skipping from one entry to the next in the tables, two or more changes may be made simultaneously, without allowing for a proper analysis of the data at the end.

12.2.5 Allenyl Ketones: The Cycloisomerization to Furans

The cycloisomerization of allenyl ketones was initially described as being catalyzed by rhodium(I) or silver(I) by Marshall et al.[21] The activity of copper, silver, and gold for this reaction was first compared in two papers published later (Scheme 12.7).[22] In the case of copper and silver, only a cycloisomerization was observed (Table 12.4, entries 1 and 2); with gold, a dimer is obtained as well (entry 3).

This initial work led to a beautiful synthetic methodology by Gevorgyan et al.[23] They discovered that two different products are formed from bromoallenyl ketones, and again this depends not only on the catalyst but also on the solvent. This time, there are two different monomers; no dimer is observed (Scheme 12.8).

A more detailed view of the selectivities (Table 12.5) reveals a selective formation of the 2,4-disubstituted furan with the copper(I) catalysts (entries 1 and 2), a low

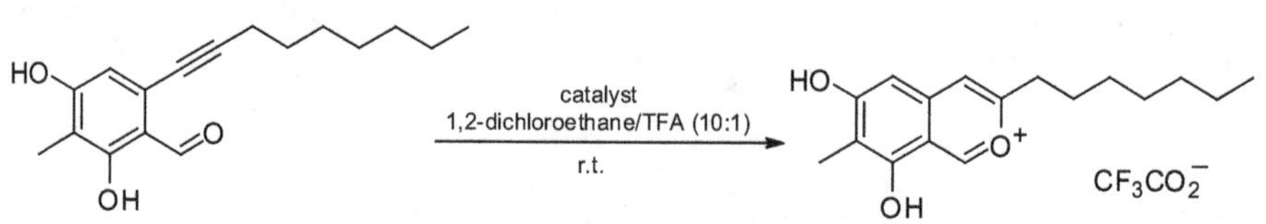

Scheme 12.6. Formation of 2-benzopyrylium salts.

TABLE 12.3. Different Catalysts for the Formation of 2-Benzopyrylium Salts

Entry	Catalyst	Time (min)	Conversion (%)
1	—	20[a]	<1
2	5 mol% $Cu(OTf)_2$	20	41
3	2.5 mol% $[Cu(OTf)]_2$·toluene	20	74
4	5 mol% $AgNO_3$	20	94
5	5 mol% $AuBr_3$	20	48
6	5 mol% $AuCl_3$	20	75
7	5 mol% $Au(OAc)_3$	1	100

[a] Reaction performed at 40°C.

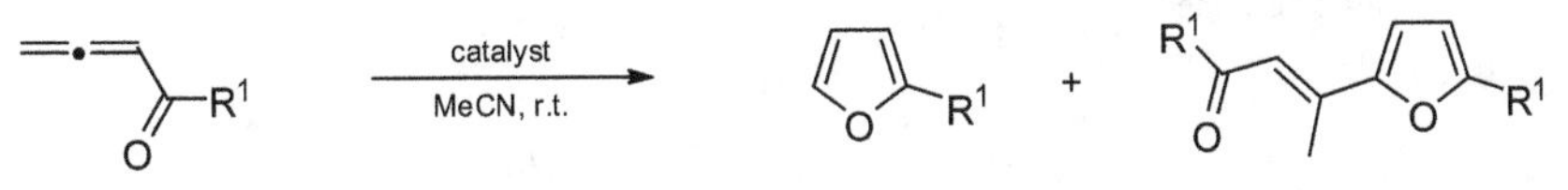

Scheme 12.7. Depending on the metal, different products can be observed.

reactivity with the silver(I) catalysts (entry 3), and a high selectivity for the 2,5-disubstitution with $AuCl_3$ in toluene (entry 4). However in THF, the selectivity is completely inverted (entry 5). With Ph_3AuCl in toluene the selectivity is lower than with $AuCl_3$ in this solvent (entry 6), but with Et_3PAuCl, the selectivity switches again (entry 7).

Similar effects have been observed with the diphenylallenyl ketones (Scheme 12.9).[23] In this case, only one product is formed; one of the phenyl groups has to migrate.

The individual results are listed in Table 12.6. Copper(I) halides are not active for this conversion (entries 1–3); only the copper triflates deliver the product, with copper (II) affording a better yield (entries 4 and 5). All silver salts are active, and again the triflates are superior and in dichloromethane even operate at room temperature

TABLE 12.4. Cycloisomerization or Cycloisomerization/Dimerization of Allenyl Ketones ($R^1 = Me$)

Entry	Catalyst	Conversion (%)	Ratio Monomer : Dimer
1	1 mol% CuCl	100	1 : 0
2	1 mol% $AgNO_3$	100	1 : 0
3	1 mol% $AuCl_3$	94	1 : 1

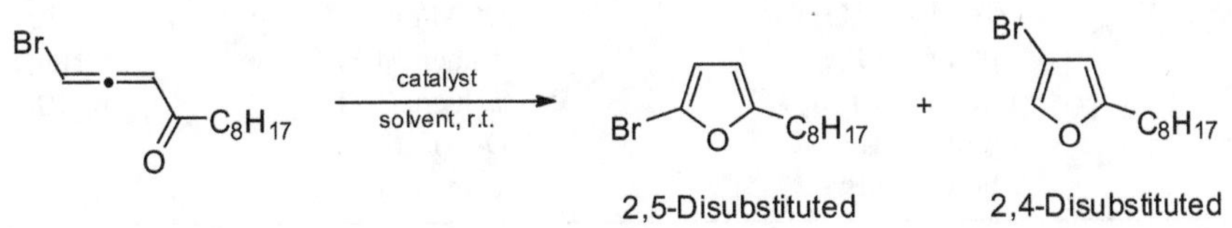

Scheme 12.8. Two different cycloisomers from bromoallenyl ketones.

TABLE 12.5. Product Distribution of the Bromoallenyl Ketone Cyclization

Entry	Catalyst	Solvent/Time	Conversion (%)	Ratio 2,5-Disubstitution : 2,4-Disubstitution
1	10 mol% CuCl	Toluene/1 day	29	0 : 1
2	10 mol% CuI	Toluene/1 day	21	0 : 1
3	5 mol% $AgBF_4$	CH_2Cl_2/1 day	Traces	—
4	1 mol% $AuCl_3$	Toluene/5 min	86	95 : 5
5	1 mol% $AuCl_3$	THF/5 min	78	5 : 95
6	1 mol% Ph_3AuCl	Toluene/9 h	[a]	16 : 84
7	1 mol% Et_3PAuCl	Toluene/9 h	[a]	<1 : 99

[a] Conversion not determined.

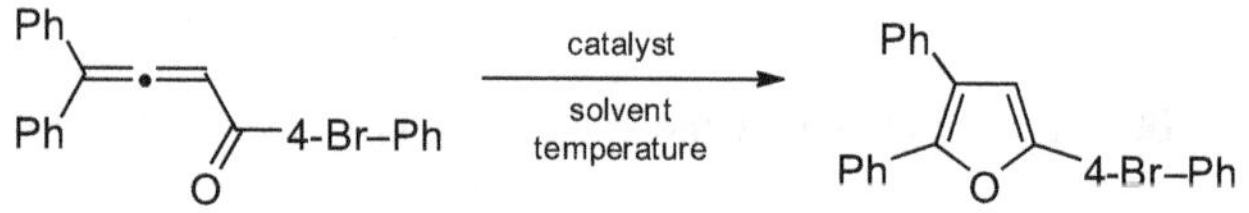

Scheme 12.9. Only one isomer by a 1,2-phenyl shift.

(entries 6–8). Switching to gold(I) shows that AuI is not active (entry 9), while $Ph_3PAuOTf$ is active again, even in DCM at room temperature (entries 10 and 11). On the other hand, $AuBr_3$ leads to a low yield only (entry 12).

12.2.6 A Thiol in the Substrate: The Cyclization of α-Thioallenes

In a spectacular work, Morita and Krause[24] studied the cyclization of allenylthiocarbinols to dihydrothiophenes by copper, silver, and gold catalysts. The reaction was the first example of a gold-catalyzed C–S bond formation and was highly stereo-

TABLE 12.6. Conditions and Yields for the 1,2-Phenyl Shift in Diphenylallenyl Ketones

Entry	Catalyst	Solvent/Temperature	Yield (%)
1	5 mol% CuCl	Toluene/100°C	—
2	5 mol% CuBr	Toluene/100°C	—
3	5 mol% CuI	Toluene/100°C	—
4	5 mol% $[CuOTf]_2$·toluene	Toluene/100°C	42
5	5 mol% $Cu(OTf)_2$	Toluene/100°C	95
6	5 mol% $AgPF_6$	Toluene/100°C	47
7	5 mol% AgOTf	Toluene/100°C	80
8	20 mol% AgOTf	DCM/r.t.	70
9	5 mol% AuI	Toluene/100°C	Traces
10	1 mol% $Ph_3PAuOTf$	Toluene/100°C	100
11	1 mol% $Ph_3PAuOTf$	DCM/r.t.	99
12	5 mol% $AuBr_3$	Toluene/100°C	23

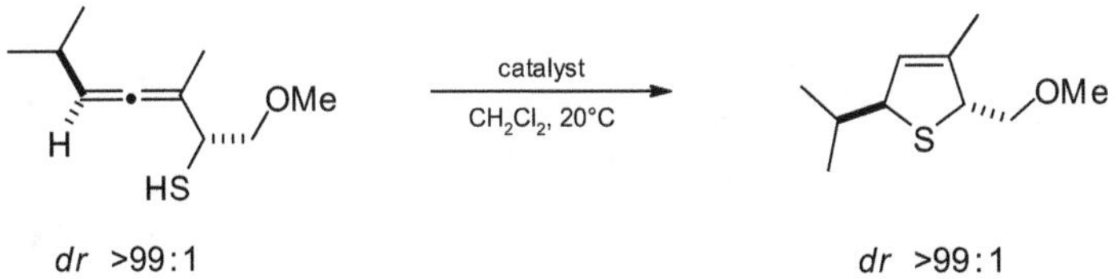

Scheme 12.10. Transition-metal-catalyzed cyclization of an allenylthiocarbinol.

selective, the diastereomerically pure starting material delivering only one diastereomer of the product. In the case of gold(III) chloride, the formation of the disulfide derived from the starting material provided some evidence for an in situ reduction of gold(III) to gold(I). Similar evidence for a possible in situ reduction of gold(III) was obtained in related work on allenyl carbinols (C–C bond formation in the latter case).[25]

The test substrate for the exceptional systematic screening is shown in Scheme 12.10. The results of the study (in Table 12.7) clearly show that the copper(I) halides tested are not active (entries 1 and 2). The same is true for silver chloride and silver tetrafluoroborate (entries 3 and 4). Only gold catalysts show activity, with gold(I) as the chloride and iodide (entries 5–7) as well as the cationic (phosphane)gold(I) species (entry 9), but not as the coordinatively saturated phosphanegold(I) chloro complex. Both gold(III) chloride and bromide are active, but the formation of the disulfide mentioned above, indicated that redox processes changes the nature of the gold complex.

The authors also screened different solvents. Dichloromethane was by far superior to THF, toluene, or hexane.

12.2.7 Furans by Propargyl Claisen Reaction

The Kirsch group studied the propargyl Claisen reaction, which consists of an initial [3,3]-sigmatropic reaction and a subsequent ring closure (Scheme 12.11).[26]

TABLE 12.7. Systematic Comparison of Catalysts

Entry	Catalyst	Time	Yield (%)
1	20 mol% CuCl	1 h	Traces
2	20 mol% CuI	2 days	—
3	15 mol% AgCl	2 days	—
4	15 mol% $AgBF_4$	2 days	Traces
5	5 mol% AuCl	1.5 h	56
6	5 mol% AuI	5 min	88
7	1 mol% AuI	1.5 h	64
8	5 mol% Ph_3PAuCl	7 days	Traces
9	5 mol% Ph_3PAuCl/10 mol% $AgBF_4$	4 h	52
10	5 mol% $AuCl_3$	3 h	58
11	5 mol% $AuBr_3$	20 min	56

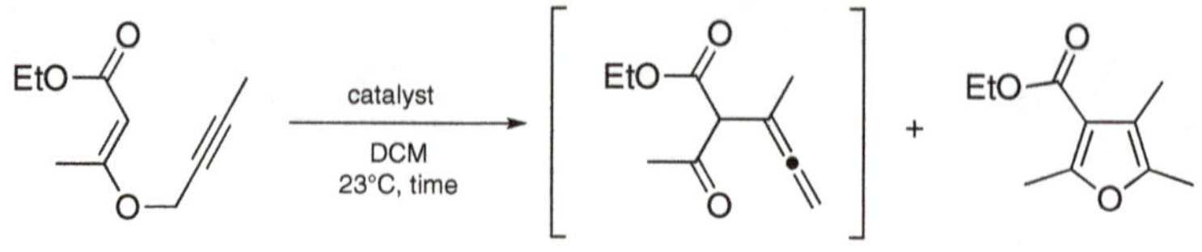

Scheme 12.11. Highly substituted furan heterocycles by coinage metal catalysis.

Again, the authors tested some variations of the metal catalyst; the entries with the copper triad metals are listed in Table 12.8. Copper(I) iodide was inactive (entry 1), silver tetrafluoroborate and $AuCl_3$ both delivered the allene without cycloisomerization within 24 h (entries 2 and 4), and potassium tetrachloroaurate was inactive (entry 3). Without activation, phosphanegold(I) chlorides were inactive (entry 5), but after removal of the chloro ligand by silver salts, excellent yields were obtained in short reaction times of 40 min (entries 6–8).

12.2.8 Tandem Cyclization/Pinacol Rearrangement

Here the substrates are 3-silyloxy-1,5-enynes; the concept was to combine a metal-induced 6-*endo-trig* cyclization with a ring contraction by a sigmatropic rearrangement (Scheme 12.12).[27] From the cationic intermediate, elimination to the biaryl product can be observed in addition to the aldehyde.

The different catalysts tested are shown in Table 12.9. The obligatory CuI (entry 1) delivers a mixture of starting material and the two products, and $AgSbF_6$ does not convert the substrate at all (entry 2). Among the (phosphane)gold(I) complexes, only those with SbF_6 counterions are active (entries 3–7), affording good yields of the aldehyde in selective reactions.

12.2.9 Furanones by Domino Heterocyclization/1,2 Shift

The easily accessible α-hydroxypropargyl ketones deliver furanones (Scheme 12.13).[28]

TABLE 12.8. Cationic Gold(I) Complexes are Crucial for the Rearrangement

Entry	Catalyst	Time	Yield (%)
1	5 mol% CuI	24 h	—
2	10 mol% $AgBF_4$	24 h	100 (only allene)
3	5 mol% $K[AuCl_4]$	24 h	—
4	2 mol% $AuCl_3$	24 h	7 + 47 allene
5	2 mol% Ph_3PAuCl	24 h	—
6	2 mol% Ph_3PAuCl/2 mol% $AgBF_4$	40 min	97
7	2 mol% Ph_3PAuCl/2 mol% $AgSbF_6$	40 min	83
8	2 mol% Ph_3PAuCl/2 mol% AgOTf	40 min	60

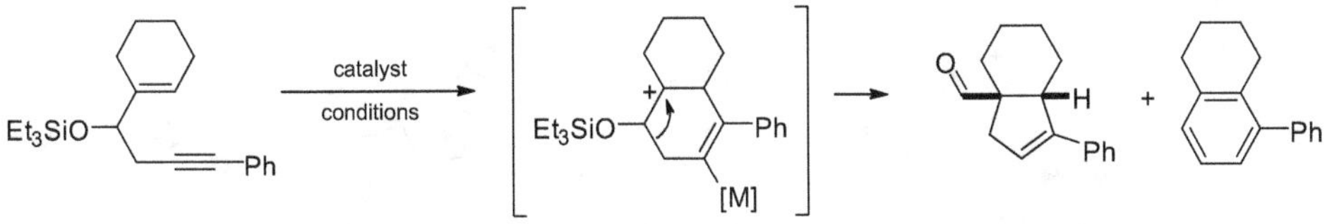

Scheme 12.12. Kirsch's tandem reaction.

TABLE 12.9. Silver Delivers Only Poor Results in the Tandem Sequence

Entry	Catalyst	Conditions	Yield (%) Aldehyde	Yield (%) Biaryl
1	10 mol% CuI	*i*-PrOH, toluene, 100°C, 24 h	21	15
2	5 mol% $AgSbF_6$	*i*-PrOH, DCM, 23°C, 1 h	—	—
3	5 mol% $[(Ph_3PAu)_3O]BF_4$	*i*-PrOH, DCM, 23°C, 24 h	—	—
4	10 mol% Ph_3PAuCl/5 mol% $AgBF_4$	*i*-PrOH, DCM, 23°C, 24 h	—	—
5	10 mol% Ph_3PAuCl/5 mol% $AgSbF_6$	H_2O, DCM, 23°C, 1 h	39	—
6	2 mol% Ph_3PAuCl/1 mol% $AgSbF_6$	*i*-PrOH, DCM, 23°C, 2.5 h	81	—
7	10 mol% Ph_3PAuCl/5 mol% $AgSbF_6$	*i*-PrOH, DCM, 23°C, 10 min	93	—

Here, while the classical CuI (Table 12.10, entry 1) and $AgSbF_6$ (entry 2) deliver some product, the gold(I) complex that was tested does not deliver any (entry 3). Thus, for the oxidation state + 1, silver is the optimum, but it cannot compete with $AuCl_3$, which delivered superior yields (entries 4 and 5), with temperatures slightly above room temperature preferred.

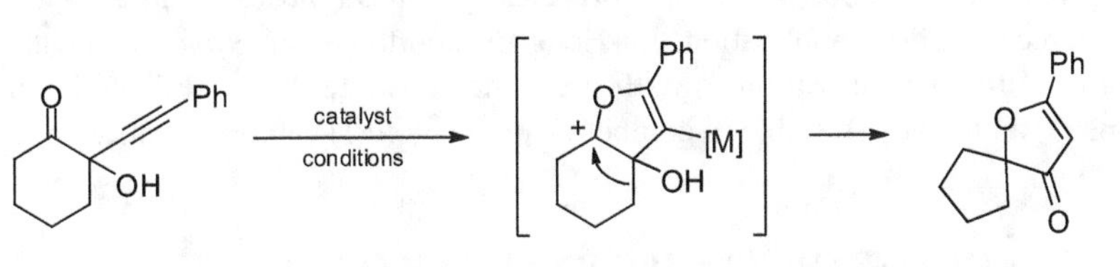

Scheme 12.13. An interesting pathway to furanones.

TABLE 12.10. Gold Outperforms the Other Catalysts, but it has to Be Gold(III)

Entry	Catalyst	Conditions	Yield (%)
1	5 mol% CuI	80°C, DMF, 2 h	22
2	5 mol% $AgSbF_6$	23°C, DCM, 1.5 h	50
3	5 mol% Ph_3PAuBF_4	23°C, DCM, 1.5 h	—
4	5 mol% $AuCl_3$	23°C, DCM, 1.5 h	83
5	5 mol% $AuCl_3$	38°C, DCM, 1.5 h	95

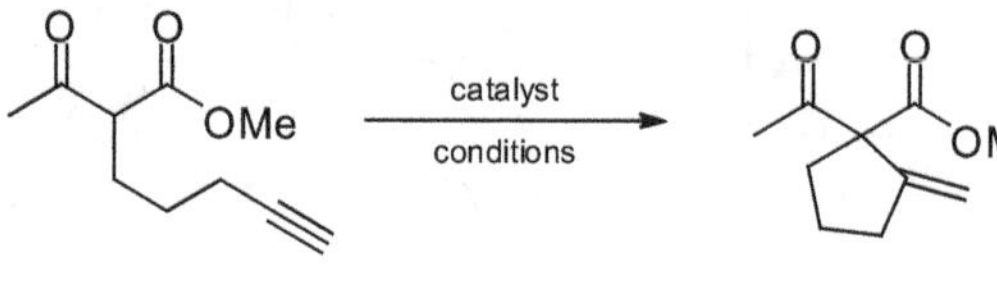

Scheme 12.14. Alkylidenecyclopentanes from 1,3-dicarbonyl compounds.

TABLE 12.11. A 5-*Exo-Dig* Cyclization Delivers Five-Membered Rings

Entry	Catalyst	Conditions	Yield (%)
1	10 mol% AgOTf	DCE, r.t., 18 h	50
2	10 mol% $AuCl_3$	DCE, r.t., 30 min	30
3	10 mol% Ph_3PAuCl	DCE, 60°C, 6 h	—
4	10 mol% $Ph_3PAuOTf$	DCE, r.t., <15 min	>95
5	1 mol%$[(Ph_3PAu)_3O]BF_4$	HOTf, DCE, r.t., <15 min	>95
6	10 mol% $(CuOTf)_2{\cdot}C_6H_6$/10 mol% Ag_2O	DCE, 100°C, 22 h	55
7	10 mol% $(CuOTf)_2{\cdot}C_6H_6$/10 mol% $AgBF_4$	DCE, 100°C, 23 h	100

12.2.10 Conia-ene Reaction

The Conia-ene reaction was brought to the field of gold catalysis by Toste et al.[29] It provided very mild conditions for this cyclization to a functionalized five-membered ring (Scheme 12.14).

In this reaction, silver(I) is successful (Table 12.11, entry 1); gold(III) leads to an unselective conversion (entry 2), but gold(I) is more active and selective. As usual, coordinatively saturated gold(I) complexes fail (entry 3), but cationic gold(I) complexes deliver very good yields in short reaction times (entries 4 and 5).

In a recent (2009) publication, Jin-Heng Li reported a copper/silver-cocatalyzed version of this isomerization.[30] With the same substrate, here the combination of copper(I) and silver(I) is able to compete with the gold(I) catalysts (entries 6 and 7).

12.3 REACTIONS CATALYZED BY SILVER OR GOLD

12.3.1 Cyclization of *N*-Propargylcarboxamides

A mild conversion of a simple class of substrates, the *N*-propargylcarboxamides, is possible with gold catalysts. With gold(I), the alkylidene oxazoles are formed; with gold(III), the aromatic oxazoles are formed.[31] The reactions readily proceed at room temperature, and there are numerous examples of this reaction (Scheme 12.15).

A 2008 publication by Harmata and Huang[32] revealed that for a number of propargyl amide derivatives, 5 mol% of $AgSbF_6$ at room temperature can achieve the same conversion to alkylidene oxazoles in 89–98% yield in minutes! In the case of disubstitution at the propargylic position, the reactions proceeds in DCM; without the

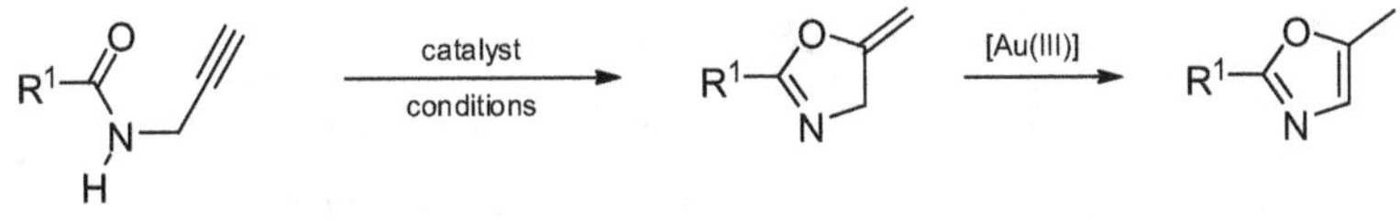

Scheme 12.15. A simple route to the oxazole ring system.

propargylic substitution, $MeNO_2$ was used as solvent. Still, gold catalysts can reach several thousand turnovers in these reactions, while for silver there are 20 turnovers.

12.3.2 Dake's Pyrrole Synthesis

The reaction of homopropargylic ketones with primary amines is an excellent method for the synthesis of highly substituted pyrroles. As shown for one example in Scheme 12.16, the yields often are better with the silver(I) catalyst.[33]

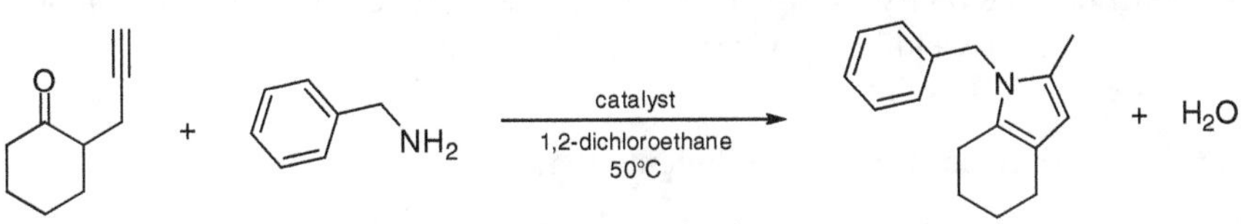

5 mol% AgOTf, 3.5 h 78%

5 mol% AuCl, AgOTf, Ph_3P, 3.5 h 74%

Scheme 12.16. An easy pathway to anellated pyrroles.

12.3.3 Combination with Organocatalysis

The transformation of an electrophilic carbonyl group to a nucleophilic enamine group with the help of a secondary amine was also applied in coinage metal catalysis.[34] The reaction often, in much the same way as do other enamine catalyses, requires quite high catalyst loadings of the secondary amine. A test substrate for catalyst optimization was the malonate shown in Scheme 12.17.

In a complex system like this, several control experiments are necessary. A characteristic set is listed in Table 12.12. Without any of the catalysts (entry 1), or with only one of the catalysts (entries 2 and 3), no product is formed. Only the combination of the gold and the organocatalyst affords the desired product (entry 4).

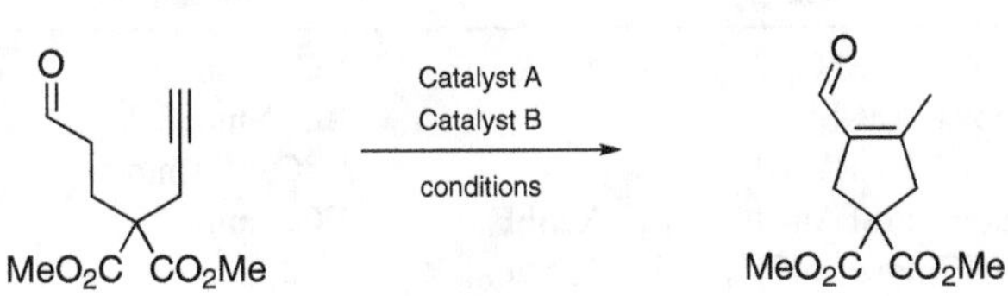

Scheme 12.17. Enamine formation and activation of the triple bond lead to ring closure.

TABLE 12.12. Selected Experiments for Optimization

Entry	Catalyst A	Catalyst B	Conditions	Yield (%)
1	—	—	120°C, toluene, 24 h	RSM[a]
2	10 mol% $[(Ph_3PAu)_3O]BF_4$	—	70°C, $CDCl_3$, 6 h	RSM
3	—	$HN(i\text{-}Pr)_2^b$	70°C, $CDCl_3$, 6 h	RSM
4	10 mol% $[(Ph_3PAu)_3O]BF_4$	$HN(i\text{-}Pr)_2^b$	70°C, $CDCl_3$, 3 h	86
5	10 mol% $AgSbF_6$	$HN(i\text{-}Pr)_2^b$	70°C, $CDCl_3$, 24 h	16
6	10 mol% AgOTf	$HN(i\text{-}Pr)_2^b$	70°C, $CDCl_3$, 2 h	59

[a] Recovered starting material.
[b] Catalyst loading: 20 mol%

The use of $AgSbF_6$ instead of the gold(I) catalyst shows only a little conversion after a long reaction time (entry 5). Again, AgOTf is the more reactive silver(I) complex, affording a higher yield in significantly less time (entry 6).

12.3.4 Vinylallenes Deliver Cyclopentadienes

The electrophilic cyclization of vinylallenes delivers cyclopentadienes with interesting substitution patterns.[35] The anellated cyclopentadienes shown in Scheme 12.18 can be prepared in this way. Silver and gold catalysts have been compared using this substrate.

Some of the catalysts tested are listed in Table 12.13. With only a silver(I) catalyst, the substrate decomposed (entry 1), while $AuCl_3$ at least afforded 30% of the product (entry 2). As expected, the coordinatively saturated gold(I) complex gave

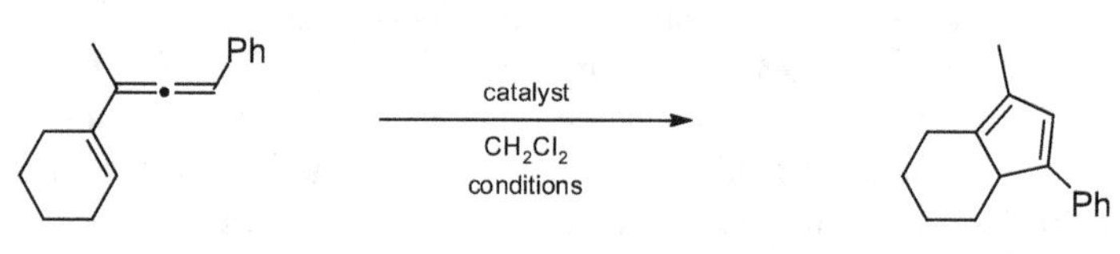

Scheme 12.18. Tetrasubstituted cyclopentadienes from silver and gold catalysis.

TABLE 12.13. Different Efforts to Initiate the Cyclization

Entry	Catalyst	Conditions	Yield (%)
1	5 mol% $AgSbF_6$	0°C, 5 min	Decomposition
2	5 mol% $AuCl_3$	0°C, 5 min	30
3	5 mol% Ph_3PAuCl	23°C, 180 min	—
4	2 mol% Ph_3PAuCl/2 mol% $AgSbF_6$	0°C, 1 min	97
5	2 mol% Ph_3PAuCl/2 mol% $AgSbF_6$	−20°C, 5 min	93
6	1 mol% Ph_3PAuCl/1 mol% $AgSbF_6$	0°C, 1 min	96

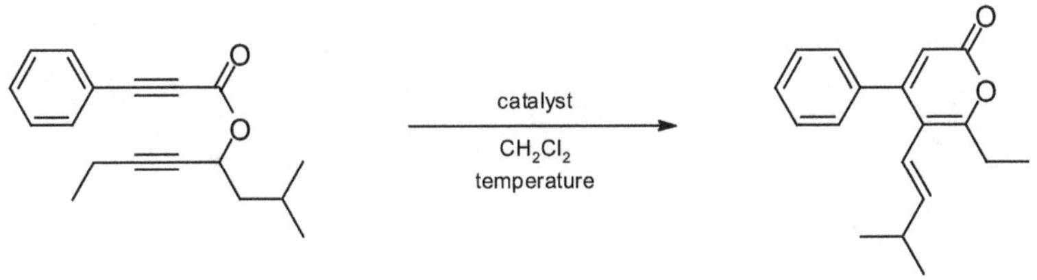

Scheme 12.19. Difficult substitution patterns at α-pyrones made possible.

no conversion even after long reaction times at room temperature (entry 3), while the cationic gold(I) complex generated in situ by chloride abstraction was very successful under a range of conditions, including lower catalyst loadings (entries 4–6).

12.3.5 α-Pyrones by a Cascade Reaction

From propiolic acid propargyl esters an elegant route to α-pyrones has been developed by Schreiber and coworkers.[36] The conversion is shown in Scheme 12.19.

Out of a long table with reaction conditions and catalysts tested, the entries listed in Table 12.14 are the most important ones in this context. The silver(I) salt used delivered small amounts of the desired product only at reflux temperature (entries 1 and 2). Only the gold(I) catalyst gave a high yield of the α-pyrone; a higher temperature was preferable (entries 3 and 4).

12.3.6 Dihydrofurans from Propargyl Esters

The synthesis of dihydrofurans from propargyl esters is another reaction in which both gold and silver have been employed. In early work, Shigemasa et al.[37] reported the efficient chirality transfer in the cycloisomerization of monoesters of butynediols. They used 8–15 mol% of $AgBF_4$ as catalyst in benzene at 80°C in the dark.

The corresponding gold(I)-catalyzed variation was reported by Gagosz and coworkers.[38] A typical reaction is shown in Scheme 12.20. The diastereomeric ratio shifts only slightly in the direction of the thermodynamically more stable *trans* product; the enantioselectivity remains unchanged, and overall only 2 mol% of catalyst at room temperature is needed.

TABLE 12.14. Cationic Gold(I) Complexes Dominate This Reaction

Entry	Catalyst	Temperature	Yield (%)
1	5 mol% $AgSbF_6$	r.t.	Trace
2	5 mol% $AgSbF_6$	40°C	11[a]
3	5 mol% Ph_3PAuCl/5 mol% $AgSbF_6$	r.t.	61
4	5 mol% Ph_3PAuCl/5 mol% $AgSbF_6$	40°C	81

[a] Decomposition of most of the starting material.

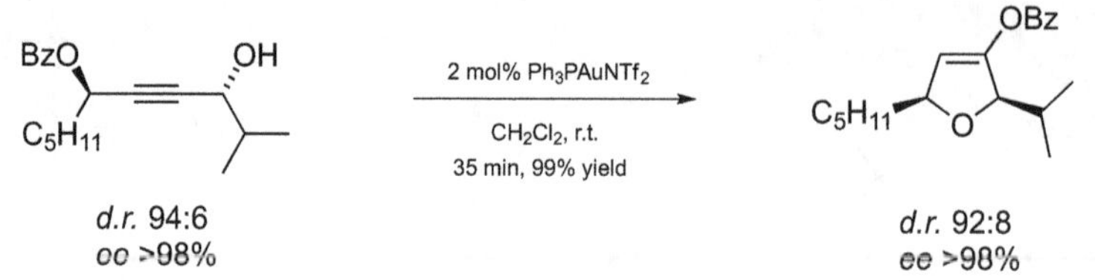

Scheme 12.20. Stereoselective synthesis of 2,5-dihydrofurans.

12.3.7 Methylene Butyrolactones by Addition of Carboxylates to Alkynes

The silver-catalyzed synthesis of methylene butyrolactones has been known for a long time;[39] the field was revived by Genêt more recently.[40] Scheme 12.21 shows that the silver(I) catalysis required 10 mol% of catalyst and 110°C, but then gave a very good yield of product. Scheme 12.22 displays the corresponding investigation of Genet and coworkers; the different conditions and catalysts tested are shown in Table 12.15.

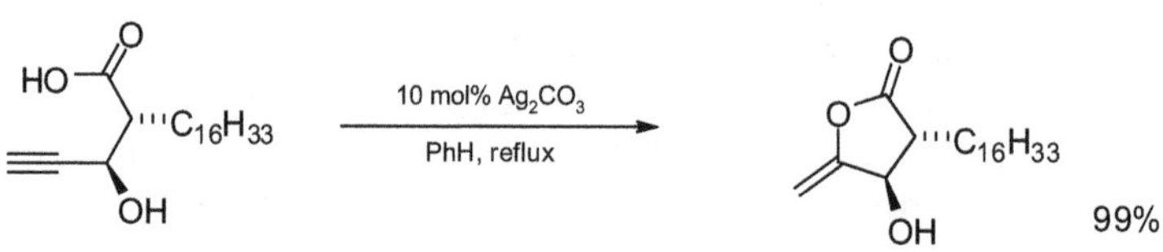

Scheme 12.21. Pale's silver-catalyzed cyclization.

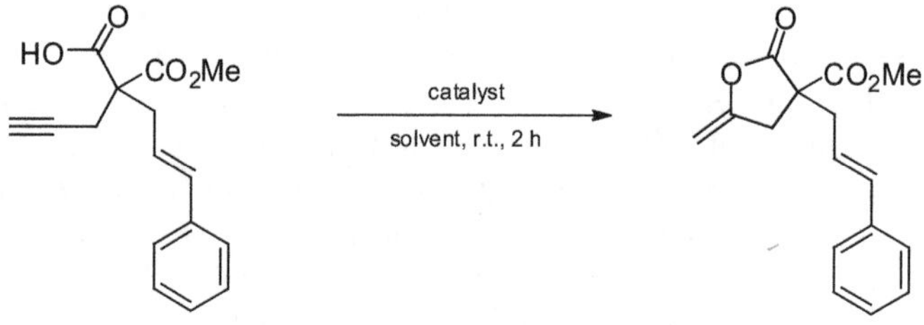

Scheme 12.22. Genêt's gold-catalyzed cyclization.

TABLE 12.15. Lactone Formation

Entry	Catalyst	Solvent	Yield (%)
1	5 mol% AgOTf	MeCN	10
2	5 mol% $AuCl_3$	MeCN	84
3	5 mol% AuCl	MeCN	90
4	5 mol% AuCl	Toluene	60
5	5 mol% AuCl	DCM	63

Silver(I) (entry 1) is significantly less active than gold(III) (entry 2) or gold(I) (entry 3), but again the table lists two changes at a time, the metal and the counterion. Tests with two more solvents (entries 4 and 5) did not give superior results.

12.3.8 Hydroarylation of Allenes

In the context of the total synthesis of (−)-rhazinilam, an intramolecular hydroarylation of an allene (Scheme 12.23) was investigated.[41] The authors report that the Ag(I) complexes were completely ineffective, while $AuCl_3$ delivered 27% of the product and a mixture of 5 mol% $AuCl_3$/20 mol% AgOTf gave 82%. The optimum was then reached with 5 mol% $Ph_3PAuOTf$ (92% yield). In all cases, the diastereoselectivity was better than 92 : 8.

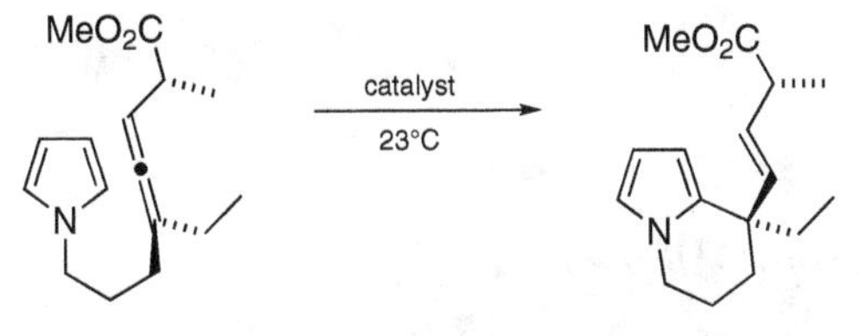

Scheme 12.23. Anellated pyrroles by hydroarylation of allenes.

12.3.9 Different Products by Silver and Gold Catalysts

12.3.9.1 The Epoxide–Alkyne Reaction

For the reaction of the epoxide shown in Scheme 12.24 with a gold catalyst, the formation of a 1,2-dicarbonyl compound was reported; while with the silver salt, the acylindene was obtained.[42]

It is difficult to put this observation into a wider context, as this substrate type usually gives only the acylindenes with the gold catalysts as well.[43]

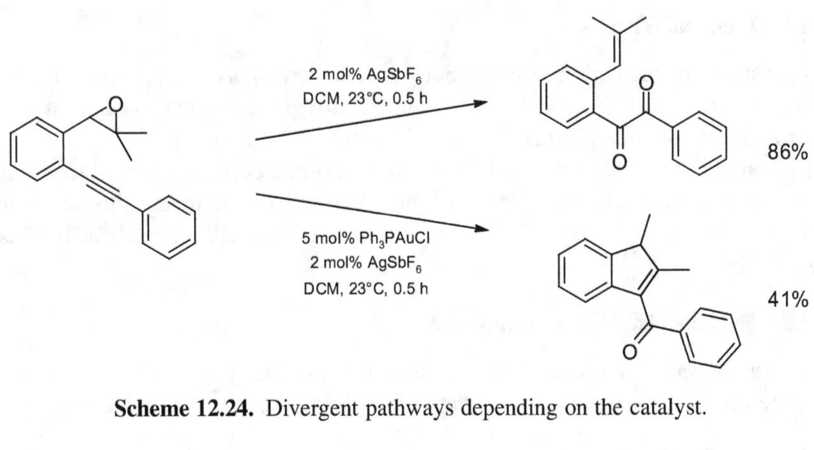

Scheme 12.24. Divergent pathways depending on the catalyst.

TABLE 12.16. A Remarkable Change in Selectivity

		Yield (%)	
Entry	Catalyst	Five-Membered Ring	Six-Membered Ring
1	5 mol% $AuCl_3$	—	—
2	5 mol% Ph_3PAuCl	—	25
3	5 mol% Ph_3PAuCl/5 mol% $AgSbF_6$	—	>95
4	5 mol% $AgSbF_6$	—	>95
5	5 mol% AgOCN	40	60
6	5 mol% Ag_2O	100	0%

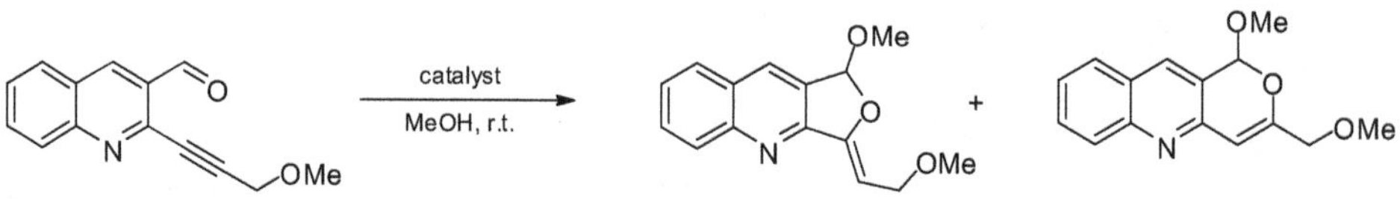

Scheme 12.25

12.3.9.2 The Carbonyl–Alkyne Reaction

With a quite specific class of compounds, Belmont and coworkers reported a selectivity switch when changing the catalyst.[44] The reaction is shown in Scheme 12.25.

With gold(III), no conversion was observed (Table 12.16, entry 1); the coordinatively saturated gold(I) complex gave a low yield of the 6-*endo-dig* cyclization product (entry 2). The cationic gold(I) complex with a free coordination site gives an excellent yield with the same selectivity (entry 3); the same is true for $AgSbF_6$ alone (entry 4). But this is not true for all silver(I) catalysts; with a systematic increase in the pK_a value of the conjugate acid of the silver counterion, increased portions of the 5-*exo-dig* product were produced (entries 5 and 6).

12.4 REACTIONS CATALYZED BY COPPER OR SILVER

12.4.1 General Trends

A comparison of the carbonyl frequencies of a tris(pyrazolyl borate) complexes of copper(I) [$Tp^{Br3}Cu(CO)$, $\nu(CO) = 2105\ cm^{-1}$] provides good evidence for a lower electron density at silver(I) [$Tp^{Br3}Cu(CO)$, $\nu(CO) = 2157\ cm^{-1}$].[14a]

By comparison of the vibrational spectra of cationic copper(I) and silver(I) tetra (acetonitrile) complexes, the different bond strengths of the metal nitrogen bonds have been determined and the influence of this bond strength on catalysis has been shown to be strong.[45]

12.4.2 Pyrroles by Hydroamination

An investigation comparing different hydroamination catalysts for the conversion of enamine alkynes to pyrroles is illustrated in Scheme 12.26.[46] In this test, different

TABLE 12.17. Some Catalysts Tested for the Reaction in Scheme 12.26

Entry	Catalyst	Yield (%)
1	4 mol% CuO	13
2	4 mol% $CuAc_2$	33
3	4 mol% $Cu(NO_3)_2$	53
4	4 mol% $CuCl_2$	65
5	4 mol% Ag_2O	6
6	4 mol% $AgOAc$	7
7	4 mol% $AgNO_3$	14
8	4 mol% $AgCl$	9
9	4 mol% $Zn(NO_3)_2$	99

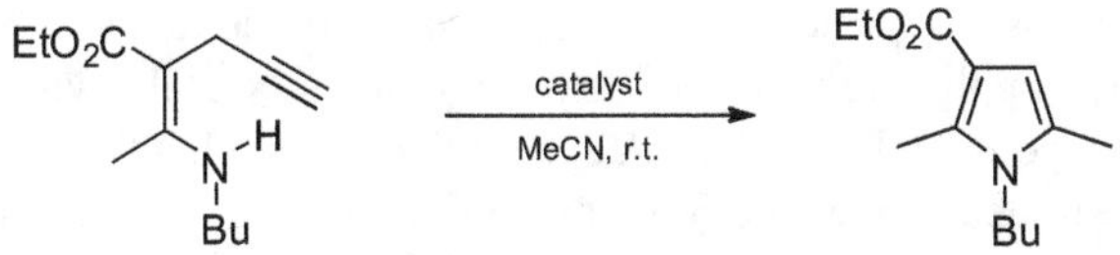

Scheme 12.26. Trisubstituted pyrroles by hydroamination.

copper(II) salts were compared to different silver(I) salts. It remains unclear why neither copper(I) nor gold catalysts were tested (the paper appeared in 2008). Table 12.17 shows these results. In the end, copper(II) was more reactive than silver(I), but the most effective systems were based on zinc(II) (entry 9, Table 2.17).

In another more recent publication, the effect of copper(I) or silver(I) on the intramolecular addition of a sulfonamide to an alkyne was explored. In the case of silver, the reaction temperature was lower and the reaction times were shorter, and the best results were finally obtained with ethanol as solvent, but this was then never tested for copper.[47]

12.4.3 Copper/Silver Cocatalysis

Another version of the Conia-ene reaction is the focus of this research. Here the reaction of γ-alkynic β-ketoesters was tested.[48] The screening resulted in a copper(I)/silver(I) cocatalysis system. It remains unclear why the authors did not at least run some reactions with gold(I) catalysts, too, especially since they even cite the work on the gold-catalyzed Conia-ene reaction in the introduction (Section 12.1).

12.4.4 Carbonylations

A paper dating from 1976 investigated the reaction of olefins or alcohols with cationic copper(I) and silver(I) complexes of carbon monoxide.[49] The remarkable outcome of these kinetic measurements is that there is no difference between the catalytic behavior of $Cu(CO)_n^+$ and $Ag(CO)_2^+$.

12.5 CONCLUSION

Most of the investigations in the area of coinage metal catalysis only provide data that do not allow a sound comparison of the catalytic activity or selectivity of the different metal complexes. In an often random manner, not only the metal but also the counterions and/or ligands and/or solvents are changed, and perhaps the reaction temperature as well. One gets the impression that the complexes that were at hand were the ones to be tested, not the ones that scientifically make sense at that point of the investigation.

Still, as a general pattern, it seems that gold catalysts often have higher activity and operate at lower temperatures and lower catalyst loadings.

A vast number of reactions still proceed with one catalyst, but apparently not with the other; there are too many reactions to list here, but often some of the crucial control experiments are missing. Much more work needs to be done here.

For industrial applications where several coinage metals work as catalysts, certainly the cheaper metal is preferred, even if the catalyst loading would be higher, but for sensitive molecules, where functional group tolerance is an issue, the heavier analogs might be more interesting and useful.

REFERENCES

1. Boxhoorn, G. (Shell Internationale Research Maatschappij B.V., Netherlands), Patent EP 255975, **1988**, p. 8.
2. Sinha, A. K.; Seelan, S.; Okumura, M.; Akita, T.; Tsubota, S.; Haruta, M., *J. Phys. Chem. B* **2005**, *109*, 3956–3965.
3. For a detailed discussion, see the only review that covers both heterogeneous and homogeneous catalysis by gold: Hashmi, A. S. K., Hutchings, G., *Angew. Chem.* **2006**, *118*, 8064–8105; *Angew. Chem. Int. Ed.* **2006**, *45*, 7896–7936.
4. (a) Alexakis, A.; Bäckvall, J. E.; Krause, N.; Pmie, O.; Diguez, M., *Chem. Rev.* **2008**, *108*, 2793–2795; (b) Harutyunyan, S. R.; den Hartog, T.; Geurts, K.; Minnaard, A. J.; Feringa, B. L., *Chem. Rev.* **2008**, *108*, 2824–2852; (c) Shibasaki, M.; Kanai, M., *Chem. Rev.* **2008**, *108*, 2853–2873; (d) Yamada, K.-I.; Tomioka, K., *Chem. Rev.* **2008**, *108*, 2874–2886; (e) Stanley, L. M.; Sibi, M. P., *Chem. Rev.* **2008**, *108*, 2887–2902; (f) Poulsen, T. B.; Jørgensen, K. A., *Chem. Rev.* **2008**, *108*, 2903–2915; (g) Deutsch, C.; Krause, N.; Lipshutz, B. H., *Chem. Rev.* **2008**, *108*, 2916–2927; (h) Meldal, M.; Tornøe, C. W., *Chem. Rev.* **2008**, *108*, 2952–3015; (i) Evano, G.; Blanchard, N.; Toumi, M., *Chem. Rev.* **2008**, *108*, 3054–3131.
5. (a) Naodovic, M.; Yamamoto, H., *Chem. Rev.* **2008**, *108*, 3132–3148; (b) Weibel, J.-M.; Blanc, A.; Pale, P., *Chem. Rev.* **2008**, *108*, 3149–3173; (c) Alvarez-Corral, M.; Muñoz-Dorado, M.; Rodrgíuez-Garca, I., *Chem. Rev.* **2008**, *108*, 3174–3198; (d) Yamamoto, Y., *Chem. Rev.* **2008**, *108*, 3199–3222.
6. (a) Sawamura, M.; Ito, Y., *Chem. Rev.* **1992**, *92*, 857–871; (b) Sawamura, M.; Ito, Y., in *Catalytic Asymmetric Synthesis*, 2nd ed., Ojima, I., ed., Wiley-VCH, Weinheim, **2000**, pp. 493–512; (c) Dyker, G., in *Organic Synthesis Highlights V.;* Schmalz, H.-G.; Wirth, T., eds., Wiley-VCH, Weinheim, **2004**, pp. 48–55; (d) Hashmi, A. S. K., *Gold Bull.* **2003**, *36*, 3–9; (e) Hashmi, A. S. K., *Gold Bull.* **2004**, *37*, 51–65; (f) Arcadi, A.; Di Giuseppe, S.,

Curr. Org. Chem. **2004**, *8*, 795–812; (g) Hoffmann-Röder, A.; Krause, N., *Org. Biomol. Chem.* **2005**, *3*, 387–391; (h) Widenhoefer, R.; Han, X., *Eur. J. Org. Chem.*, **2006**, 4555–4563; (i) Hashmi, A. S. K.; Hutchings, G., *Angew. Chem.* **2006**, *118*, 8064–8105; *Angew. Chem. Int. Ed.* **2006**, *45*, 7896–7936; (j) Hashmi, A. S. K., *Chem. Rev.* **2007**, *107*, 3180–3211; (k) Fürstner, A.; Davies, P. W., *Angew. Chem.* **2007**, *119*, 3478–3519; *Angew. Chem. Int. Ed.* **2007**, *46*, 3410–3449; (l) Gorin, D. J.; Toste, F. D., *Nature* **2007**, *446*, 395–403; (m) Shen, C. H., *Tetrahedron* **2008**, *64*, 3885–3903; (n) Skouta, R.; Li, C.-J., *Tetrahedron* **2008**, *64*, 4917–4938; (o) Muzart, J., *Tetrahedron* **2008**, *64*, 5815–5849; (p) Li, Z.; Brouwer, C.; He, C., *Chem. Rev.* **2008**, *108*, 3239–3265; (q) Arcadi, A., *Chem. Rev.* **2008**, *108*, 3266–3325; (r) Jiménez-Núnez, E.; Echavarren, A. M., *Chem. Rev.* **2008**, *108*, 3326–3350; (s) Gorin, D. J.; Sherry, B. D.; Toste, F. D., *Chem. Rev.* **2008**, *108*, 3351–3378; (t) Hashmi, A. S. K.; Rudolph, M., *Chem. Soc. Rev.* **2008**, *37*, 1766–1775.

7. Wei, C.; Li, Z.; Li, C.-J., *Synlett* **2004**, 1472–1483.
8. (a) Kabalka, G. W.; Wang, L.; Pagni, R. M., *Synlett* **2001**, 676–678; (b) Sharifi, A.; Farhangian, H.; Mohsenzadeh, F.; Naimi-Jamal, M. R., *Monatsh. Chem.* **2002**, *133*, 199–204; (c) Wang, L.; Li, P.-H., *Chin. J. Chem.* **2003**, *21*, 710–713.
9. Wei, C. M.; Li, Z. G.; Li, C. J., *Org. Lett.* **2003**, *5*, 4473–4475.
10. Wei, C. M.; Li, C. J., *J. Am. Chem. Soc.* **2003**, *125*, 9584–9585.
11. Li, Z.; Wei, C.; Chen, L.; Varma, R. S.; Li, C.-J., *Tetrahedron Lett.* **2004**, *45*, 2443–2446.
12. (a) Wei, C. M.; Li, C. J., *J. Am. Chem. Soc.* **2002**, *124*, 5638–5639; (b) Wei, C. M.; Mague, J. T.; Li, C. J., *Proc. Natl. Acad. Sci. USA* **2004**, *101*, 5749–5754.
13. Huang, B.; Yao, X.; Li, C.-J., *Adv. Synth. Catal.* **2006**, *348*, 1528–1532.
14. (a) Mar Díaz-Requejo, M.; Pérez, P. J., *J. Organomet. Chem.* **2005**, *690*, 5441–5450; (b) Fructos, M. R.; de Frémont, P.; Scott, N. M.; Nolan, S. P.; Diaz-Requejo, M. M.; Pérez, P. J., *Angew. Chem.* **2005**, *117*, 5418–5422; *Angew., Chem., Int., Ed.* **2005**, *44*, 5284–5288.
15. (a) Posner, G. H., *Org. React.* **1972**, *19*, 1–113; (b) Marshall, J. A., *Chem. Rev.* **1989**, *89*, 1503–1511.
16. (a) James, P. F.; O'Hair, R. A. J., *Org. Lett.* **2004**, *6*, 2761–2764; (b) Pale, P.; Chuche, J., *Tetrahedron Lett.* **1987**, *28*, 6447–6448; (c) Dalla, V.; Pale, P., *Tetrahedron Lett.* **1994**, *35*, 3525–3528; (d) Dalla, V.; Pale, P., *New J. Chem.* **1999**, *23*, 803–805; (e) Dillinger, S.; Bertus, P.; Pale, P., *Org. Lett.* **2001**, *3*, 1661–1664; (f) Paul, A. M.; Bent, B. E., *J. Catal.* **1994**, *147*, 264–271.
17. Nakanishi, W.; Yamanaka, M.; Nakamura, E., *J. Am. Chem. Soc.* **2005**, *127*, 1446–1453.
18. (a) Tai, H. C.; Krossing, I.; Seth, M.; Deubel, D. V., *Organometallics* **2004**, *23*, 2343–2349; (b) Pyykko, P., *J. Am. Chem. Soc.* **1995**, *117*, 2067–2070.
19. (a) Asao, N., *Synlett* **2006**, 1645–1656; (b) Patil, N. T.; Yamamoto, Y., *Arkivoc* **2007**, 6–19; for related work, see (c) Dyker G.; Hildebrandt, D.; Liu, J.; Merz, K., *Angew. Chem.* **2003**, *115*, 4536–4538; *Angew. Chem. Int. Ed.* **2003**, *42*, 4399–4402.
20. Zhu, J.; Germain, A. R.; Porco, J. A., *Angew. Chem.* **2004**, *116*, 1259–1263; *Angew., Chem., Int., Ed.* **2004**, *43*, 1239–1243.
21. Marshall, J. A.; Robinson, E. D., *J. Org. Chem.* **1990**, *55*, 3450–3451.
22. (a) Hashmi, A. S. K., *Angew. Chem.* **1995**, *107*, 1749–1751; *Angew. Chem. Int. Ed.* **1995**, *34*, 1581–1583; (b) Hashmi, A. S. K.; Schwarz, L.; Choi, J.-H.; Frost, T. M., *Angew. Chem.* **2000**, *112*, 2382–2385; *Angew., Chem., Int., Ed.* **2000**, *39*, 2285–2288.

23. Dudnik, A. S.; Sromek, A. W.; Rubina, M.; Kim, J. T.; Kel'in, A. V.; Gevorgyan, V., *J. Am. Chem. Soc.* **2008**, *130*, 1440–1452.

24. Morita, N.; Krause, N., *Angew. Chem.* **2006**, *118*, 1930–1933; *Angew., Chem., Int., Ed.***2006**, *45*, 1897–1899.

25. Hashmi, A. S. K.; Blanco, M. C.; Fischer, D.; Bats, J. W., *Eur. J. Org. Chem.* **2006**, 1387–1389.

26. Suhre, M. H.; Reif, M.; Kirsch, S. F., *Org. Lett.* **2005**, *7*, 3925–3927.

27. Kirsch, S. F.; Binder, J. T.; Crone, B.; Duschek, A.; Haug, T. T.; Liébert, C.; Menz, H., *Angew. Chem.* **2007**, *119*, 2360–2363; *Angew. Chem. Int. Ed.* **2007**, *46*, 2310–2313.

28. Kirsch, S. F.; Binder, J. T.; Liébert, C.; Menz, H., *Angew. Chem.* **2006**, *118*, 6010–6013; *Angew. Chem. Int. Ed.* **2006**, *45*, 5878–5880.

29. (a) Kennedy-Smith, J. J.; Staben, S. T.; Toste, F. D., *J. Am. Chem. Soc.* **2004**, *126*, 4526–4527; (b) Staben, S. T.; Kennedy-Smith, J. J.; Toste, F. D., *Angew. Chem.* **2004**, *116*, 5464–5466; *Angew. Chem. Int. Ed.* **2004**, *43*, 5350–5352.

30. Deng, C.-L.; Zou, T.; Wang, Z.-Q.; Song, R.-J.; Li, J.-H., *J. Org. Chem.* **2009**, *74*, 412–414.

31. (a) Hashmi, A. S. K.; Weyrauch, J. P.; Frey, W.; Bats, J. W., *Org. Lett.* **2004**, *6*, 4391–4394; (b) Milton, M. D.; Inada, Y.; Nishibayashi, Y.; Uemura, S., *Chem. Commun.* **2004**, 2712–2713; (c) Dorsch, D.; Burgdorf, L. T.; Gericke, R.; Beier, N.; Mederski, W., PCT International Patent Application WO 2005123688 A2, CAN 144:69837, **2005**; (d) Hashmi, A. S. K.; Rudolph, M.; Schymura, S.; Visus, J.; Frey, W., *Eur. J. Org. Chem.* **2006**, 4905–4909.

32. Harmata, M.; Huang, C., *Synlett* **2008**, 1399–1401.

33. Harrison, T. J.; Kozak, J. A.; Corbella-Pané, M.; Dake, G. R., *J. Org. Chem.* **2006**, *71*, 4525–4529.

34. Binder, J. T.; Crone, B.; Haug, T. T.; Menz, H.; Kirsch, S. F., *Org. Lett.* **2008**, *10*, 1025–1028.

35. Lee, J. H.; Toste, F. D., *Angew. Chem.* **2007**, *119*, 930–932; *Angew. Chem. Int. Ed.* **2007**, *46*, 912–914.

36. Luo, T.; Schreiber, S. L., *Angew. Chem.* **2007**, *119*, 8398–88401; *Angew. Chem. Int. Ed.* **2007**, *46*, 8250–8253.

37. Shigemasa, Y.; Yasui, M.; Ohrai, S.-I.; Sasaki, M.; Sashiwa, H.; Saimoto, H., *J. Org. Chem.* **1991**, *56*, 910–912.

38. Buzas, A.; Istrate, F.; Gagosz, F., *Org. Lett.* **2006**, *8*, 1957–1959.

39. (a) Dalla V.; Pale P., *Tetrahedron Lett.* **1994**, *35*, 3525–3528; for related work with allenes, see Pale, P.; Chuche, J., *Tetrahedron Lett.* **1987**, *28*, 6447–6448.

40. Genin, E.; Tullec, P. Y.; Antoniotti, S.; Brancour, C.; Genêt, J.-P.; Michelet, V., *J. Am. Chem. Soc.* **2006**, *128*, 3112–3113.

41. Liu, Z.; Wasmuth, A. S.; Nelson, S. G., *J. Am. Chem. Soc.* **2006**, *128*, 10352–10353.

42. Lin, G.-Y.; Li, C.-W.; Hung, S.-H.; Liu, R.-S., *Org. Lett.* **2008**, *10*, 5059–5062.

43. Hashmi, A. S. K.; Bührle, M.; Salathé, R.; Bats, J. W., *Adv. Synth. Catal.* **2008**, *350*, 2059–2064.

44. Godet, Th.; Vaxelaire, C.; Michel, C.; Milet, A.; Belmont, P., *Chem. Eur. J.* **2007**, *13*, 5632–5641.

45. Zhang, Y.; Sun, W.; Freund, C.; Santos, A. M.; Herdtweck, E.; Mink, J.; Kühn, F. E., *Inorg. Chim. Acta* **2006**, *359*, 4723–4729.
46. Prior, A. M.; Robinson, R. S., *Tetrahedron Lett.* **2008**, *49*, 411–414.
47. Barange, D. K.; Nishad, T. C.; Swamy, N. K.; Bandameedi, V.; Kumar, D.; Sreekanth, B. R.; Vyas, K.; Pal, M., *J. Org. Chem.* **2007**, *72*, 8547–8550.
48. Deng, C.-L.; Song, R.-J.; Guo, S.-M.; Wang, Z.-Q.; Li, J.-H., *Org. Lett.* **2007**, *9*, 5111–5114.
49. Souma, Y.; Sano, H., *Bull. Chem. Soc. Jpn.* **1976**, *49*, 3296–3299.

INDEX

www.ingramcontent.com/pod-product-compliance
Lightning Source LLC
LaVergne TN
LVHW011904200926
848484LV00002B/527